ERGODIC CONTROL OF DIFFUSION PROCESSES

This comprehensive volume on ergodic control of diffusions highlights intuition alongside technical arguments. A concise account of Markov process theory is followed by a complete development of the fundamental issues and formalisms in control of diffusions. This then leads to a comprehensive treatment of ergodic control, a problem that straddles stochastic control and the ergodic theory of Markov processes.

The interplay between the probabilistic and ergodic-theoretic aspects of the problem, notably the asymptotics of empirical measures on one hand, and the analytic aspects leading to a characterization of optimality via the associated Hamilton–Jacobi–Bellman equation on the other, is clearly revealed. The more abstract controlled martingale problem is also presented, in addition to many other related issues and models.

Assuming only graduate-level probability and analysis, the authors develop the theory in a manner that makes it accessible to users in applied mathematics, engineering, finance and operations research.

Encyclopedia of Mathematics and Its Applications

This series is devoted to significant topics or themes that have wide application in mathematics or mathematical science and for which a detailed development of the abstract theory is less important than a thorough and concrete exploration of the implications and applications.

Books in the **Encyclopedia of Mathematics and Its Applications** cover their subjects comprehensively. Less important results may be summarized as exercises at the ends of chapters. For technicalities, readers can be referred to the bibliography, which is expected to be comprehensive. As a result, volumes are encyclopedic references or manageable guides to major subjects.

All the titles listed below can be obtained from good booksellers or from Cambridge University Press. For a complete series listing visit

http://www.cambridge.org/mathematics

88 T. Mora *Solving Polynomial Equation Systems I*
89 K. Bichteler *Stochastic Integration with Jumps*
90 M. Lothaire *Algebraic Combinatorics on Words*
91 A. A. Ivanov and S. V. Shpectorov *Geometry of Sporadic Groups II*
92 P. McMullen and E. Schulte *Abstract Regular Polytopes*
93 G. Gierz *et al. Continuous Lattices and Domains*
94 S. R. Finch *Mathematical Constants*
95 Y. Jabri *The Mountain Pass Theorem*
96 G. Gasper and M. Rahman *Basic Hypergeometric Series, 2nd edn*
97 M. C. Pedicchio and W. Tholen (eds.) *Categorical Foundations*
98 M. E. H. Ismail *Classical and Quantum Orthogonal Polynomials in One Variable*
99 T. Mora *Solving Polynomial Equation Systems II*
100 E. Olivieri and M. Eulália Vares *Large Deviations and Metastability*
101 A. Kushner, V. Lychagin and V. Rubtsov *Contact Geometry and Nonlinear Differential Equations*
102 L. W. Beineke and R. J. Wilson (eds.) with P. J. Cameron *Topics in Algebraic Graph Theory*
103 O. J. Staffans *Well-Posed Linear Systems*
104 J. M. Lewis, S. Lakshmivarahan and S. K. Dhall *Dynamic Data Assimilation*
105 M. Lothaire *Applied Combinatorics on Words*
106 A. Markoe *Analytic Tomography*
107 P. A. Martin *Multiple Scattering*
108 R. A. Brualdi *Combinatorial Matrix Classes*
109 J. M. Borwein and J. D. Vanderwerff *Convex Functions*
110 M.-J. Lai and L. L. Schumaker *Spline Functions on Triangulations*
111 R. T. Curtis *Symmetric Generation of Groups*
112 H. Salzmann, T. Grundhöfer, H. Hähl and R. Löwen *The Classical Fields*
113 S. Peszat and J. Zabczyk *Stochastic Partial Differential Equations with Lévy Noise*
114 J. Beck *Combinatorial Games*
115 L. Barreira and Y. Pesin *Nonuniform Hyperbolicity*
116 D. Z. Arov and H. Dym *J-Contractive Matrix Valued Functions and Related Topics*
117 R. Glowinski, J.-L. Lions and J. He *Exact and Approximate Controllability for Distributed Parameter Systems*
118 A. A. Borovkov and K. A. Borovkov *Asymptotic Analysis of Random Walks*
119 M. Deza and M. Dutour Sikirić *Geometry of Chemical Graphs*
120 T. Nishiura *Absolute Measurable Spaces*
121 M. Prest *Purity, Spectra and Localisation*
122 S. Khrushchev *Orthogonal Polynomials and Continued Fractions*
123 H. Nagamochi and T. Ibaraki *Algorithmic Aspects of Graph Connectivity*
124 F. W. King *Hilbert Transforms I*
125 F. W. King *Hilbert Transforms II*
126 O. Calin and D.-C. Chang *Sub-Riemannian Geometry*
127 M. Grabisch *et al. Aggregation Functions*
128 L. W. Beineke and R. J. Wilson (eds.) with J. L. Gross and T. W. Tucker *Topics in Topological Graph Theory*
129 J. Berstel, D. Perrin and C. Reutenauer *Codes and Automata*
130 T. G. Faticoni *Modules over Endomorphism Rings*
131 H. Morimoto *Stochastic Control and Mathematical Modeling*
132 G. Schmidt *Relational Mathematics*
133 P. Kornerup and D. W. Matula *Finite Precision Number Systems and Arithmetic*
134 Y. Crama and P. L. Hammer (eds.) *Boolean Functions*
135 V. Berthé and M. Rigo (eds.) *Combinatorics, Automata and Number Theory*
136 A. Kristály, V. D. Rădulescu and C. Varga *Variational Principles in Mathematical Physics, Geometry and Economics*
137 J. Berstel and C. Reutenauer *Noncommutative Rational Series with Applications*
138 B. Courcelle *Graph Structure and Monadic Second-Order Logic*
139 M. Fiedler *Matrices and Graphs in Geometry*
140 N. Vakil *Real Analysis through Modern Infinitesimals*
141 R. B. Paris *Hadamard Expansions and Hyperasymptotic Evaluation*
142 Y. Crama and P. L. Hammer *Boolean Functions*

ENCYCLOPEDIA OF MATHEMATICS AND ITS APPLICATIONS

Ergodic Control of Diffusion Processes

ARI ARAPOSTATHIS
University of Texas, Austin

VIVEK S. BORKAR
Tata Institute of Fundamental Research, Mumbai

MRINAL K. GHOSH
Indian Institute of Science, Bangalore

CAMBRIDGE
UNIVERSITY PRESS

CAMBRIDGE
UNIVERSITY PRESS

Shaftesbury Road, Cambridge CB2 8EA, United Kingdom

One Liberty Plaza, 20th Floor, New York, NY 10006, USA

477 Williamstown Road, Port Melbourne, VIC 3207, Australia

314–321, 3rd Floor, Plot 3, Splendor Forum, Jasola District Centre, New Delhi – 110025, India

103 Penang Road, #05–06/07, Visioncrest Commercial, Singapore 238467

Cambridge University Press is part of Cambridge University Press & Assessment, a department of the University of Cambridge.

We share the University's mission to contribute to society through the pursuit of education, learning and research at the highest international levels of excellence.

www.cambridge.org
Information on this title: www.cambridge.org/9780521768405

First published 2012

A catalogue record for this publication is available from the British Library

Library of Congress Cataloging-in-Publication data
Arapostathis, Ari, 1954–
Ergodic control of diffusion processes / Ari Arapostathis, Vivek S. Borkar, Mrinal K. Ghosh.
p. cm. – (Encyclopedia of mathematics and its applications ; 143)
Includes bibliographical references and indexes.
ISBN 978-0-521-76840-5
1. Diffusion processes. 2. Ergodic theory. I. Borkar, Vivek S. II. Ghosh, Mrinal K., 1956–
III. Title. IV. Series.
QA274.75.A73 2011
519.2´33 – dc23 2011031547

ISBN 978-0-521-76840-5 Hardback

Dedicated to our parents:

Theodore and Helen,

Shripad and Sarita,

Sudhir Kumar and Uma.

Contents

Preface

Ergodic is a term appropriated from physics that derives from the Greek words $\acute{\epsilon}\rho\gamma o\nu$ and $o\delta\acute{o}\varsigma$, meaning "work" and "path." In the context of controlled Markov processes it refers to the problem of minimizing a time averaged penalty, or cost, over an infinite time horizon. It is of interest in situations when transients are fast and therefore relatively unimportant, and one is essentially comparing various possible equilibrium behaviors. One typical situation is in communication networks, where continuous time and space models arise as scaled limits of the underlying discrete state and/or time phenomena.

Ergodic cost differs from the simpler "integral" costs such as finite horizon or infinite horizon discounted costs in several crucial ways. Most importantly, one is looking at a cost averaged over infinite time, whence any finite initial segment is irrelevant as it does not affect the cost. This counterintuitive situation is also the reason for the fundamental difficulty in handling this problem analytically – one cannot use for this problem the naive dynamic programming heuristic because it is perforce based on splitting the time horizon into an initial segment and the rest. One is thus obliged to devise altogether different techniques to handle the ergodic cost. One of them, the more familiar one, is to treat it as a limiting case of the infinite horizon discounted cost control problem as the discount factor tends to zero. This "vanishing discount" approach leads to the correct dynamic programming, or "Hamilton–Jacobi–Bellman" (HJB) equation for the problem, allowing one to characterize optimal control policies at least in the "nicer" situations when convenient technical hypotheses hold. It also forms the basis of the approach one takes in order to do what one can in cases when these hypotheses do not hold. Dynamic programming, though the most popular "classical" approach to control problems, is not the only one. An alternative approach that is gaining currency, particularly as it allows one to handle some nonclassical variants and because of the numerical schemes it facilitates, is that based on casting the control as an infinite dimensional convex (in fact, *linear*) program. Our treatment of ergodic control straddles both lines of thought, often combining them to advantage.

Historically, the theory of ergodic control was first developed for discrete time and discrete state space Markov chains (see Arapostathis *et al.* [6] for an extensive survey and historical account). The results therein, as always, are suggestive of what one might expect in the continuous time continuous state space situation. That in itself, however, is of little help, as the technical difficulties in carrying out the program are highly nontrivial. Not surprisingly, the theory in the latter case has been slow to follow, some of it being of a fairly recent vintage. This book gives a comprehensive, integrated treatment of these developments. It begins with the better understood "nondegenerate, complete observations" case, and leads to the more difficult issues that arise when either the nondegeneracy assumption or the assumption of complete observations fail.

Our focus is primarily on controlled *diffusion processes*, a special class of Markov processes in continuous time and space. We build the basic theory of such processes in Chapter 2, following a quick review of the relevant aspects of ergodic theory of Markov processes in Chapter 1. This itself will serve as a comprehensive account of the probabilistic theory of controlled diffusions. It is an update on an earlier monograph [28], and appears in one place at this level of generality and extent for the first time. It forms the backdrop for the developments in the rest of this monograph and will also serve as useful source material for stochastic control researchers working on other problems. Chapter 3 gives a complete account of the relatively better understood case of controlled diffusions when the state is observed and the diffusion matrix is nondegenerate. The latter intuitively means that the noise enters all components of the state space evenly. The smoothing properties of the noise then yield sufficient regularity of various averages of interest. This permits us to use to our advantage the theory of nondegenerate second order elliptic partial differential equations. Our pursuits here are typical of all control theory: existence of optimal controls and necessary and sufficient conditions for optimality. We employ the infinite dimensional convex (in fact, linear) programming perspective for the former and the vanishing discount paradigm for the latter. The theory here is rich enough that it allows us to be a bit more ambitious and handle some nonclassical classes of problems as well with some additional work. One of these, that of switching diffusions, is treated in Chapter 5. This incorporates "regime-switching" phenomena and involves some nontrivial extensions of the theory of elliptic PDEs used in Chapter 3, to systems of elliptic PDEs. Chapter 4, in turn, studies several other spin-offs: the first is constrained and multi-objective problems and is followed by singularly perturbed ergodic control problems involving two time-scales, the aim being to justify a lower dimensional approximation obtained by averaging the slow dynamics over the equilibrium behavior of the fast components. It also studies diffusions in a bounded domain, and works out an example in detail.

With Chapter 6, we enter the vastly more difficult terrain of degeneracy. This chapter in particular is devoted to developing the abstract framework of ergodic control of martingale problems that will form the backdrop of subsequent chapters. A key development in this chapter is an important generalization of Echeverria's celebrated criterion for stationary distributions to controlled martingale problems on a Polish space. The rest of the chapter is devoted to characterizing extremal solutions, the abstract linear programming formulation, and the existence theorems for optimal solutions in various classes of interest. These results are specialized in Chapter 7 to controlled diffusions with degeneracy, where they lead to existence theorems for optimal controls with little extra effort. For two special cases, viz., the so-called "asymptotically flat" diffusions and "partially degenerate" diffusions, we also derive dynamic programming principles. This takes the form of an HJB equation, albeit interpreted in the viscosity sense. Chapter 8 considers nondegenerate diffusions under partial observations. The standard approach, summarized in this chapter, is to reduce the problem to a completely observed case by moving over to the "sufficient statistics" such as the regular conditional law of the state given observed quantities till the time instant in question. The evolution of this law is given by the stochastic PDE describing the associated measure-valued nonlinear filter, thus making it a completely observed but infinite dimensional and in a certain sense, degenerate problem. This formulation, however, does allow us to apply the theory of Chapter 6 and develop the existence theory for optimal controls, as well as to derive a martingale dynamic programming principle under suitable hypotheses.

The epilogue in Chapter 9, an important component of this work, sketches several open issues. This being a difficult problem area has its advantages as well – there is still much work left to be done, some of it crying out for novel approaches and proof techniques. We hope that this book will spur other researchers to take up some of these issues and see successful resolution of at least some of them in the near future.

We extend our heartfelt thanks to our numerous friends, colleagues and students, far too numerous to list, who have helped in this project in a variety of ways. Special thanks are due to Professor S.R.S. Varadhan, who, in addition to his pervasive influence on the field as a whole, also made specific technical recommendations in the early stages of our work in this area, which had a critical impact on the way it developed.

This project would not have been possible without the support of the Office of Naval Research under the Electric Ship Research and Development Consortium, and the kind hospitality of the Tata Institute of Fundamental Research in Mumbai. Vivek Borkar also thanks the Department of Science and Technology, Govt. of India, for their support through a J. C. Bose Fellowship during the course of this work.

It has been both a challenge and a pleasure for us to write this book. We particularly cherish the time we spent together in Mumbai working on it together, the hectic schedule punctuated by the culinary delights of south Mumbai restaurants and cafes.

Frequently Used Notation

The sets of real numbers, integers and natural numbers are denoted by $\mathbb{R}$, $\mathbb{Z}$ and $\mathbb{N}$ respectively. Also, $\mathbb{R}_+$ ($\mathbb{Z}_+$) denotes the set of nonnegative real numbers (nonnegative integers). We use ":=" to mean "is defined as," and "≡" to mean "identically equal to." If f and g are real-valued functions (or real numbers), we define

$$
\begin{aligned}
f \wedge g &:= \min\{f, g\}, \qquad & f \vee g &:= \max\{f, g\},\\
f^{+} &:= f \vee 0, \qquad & f^{-} &:= (-f) \vee 0.
\end{aligned}
$$

For a subset A of a topological space, ∂A denotes its boundary, A^c its complement and $\bar{A}$ its closure. The indicator function of A is denoted by $\mathbb{I}_A$. In the interest of readability for a set that is explicitly defined by an expression $\{\cdot\}$ we denote its indicator function as $\mathbb{I}\{\cdot\}$.

The Borel σ-field of a topological space E is denoted by $\mathscr{B}(E)$. Metric spaces are in general viewed as equipped with their Borel σ-field, and therefore the notation $\mathscr{P}(E)$ for the set of probability measures on $\mathscr{B}(E)$ of a metric space E is unambiguous. The set of bounded real-valued measurable functions on a metric space E is denoted by $\mathcal{B}(E)$. The symbol $\mathbb{E}$ always denotes the expectation operator, and $\mathbb{P}$ the probability.

The standard Euclidean norm is denoted as $|\cdot|$. The term *domain* in $\mathbb{R}^d$ refers to a non-empty open connected subset of the Euclidean space $\mathbb{R}^d$. If D and D' are domains in $\mathbb{R}^d$, we use the notation $D \Subset D'$ to indicate that $\bar{D} \subset D'$. Also $|D|$ stands for the Lebesgue measure of a bounded domain D. We introduce the following notation for spaces of real-valued functions on a domain $D \subset \mathbb{R}^d$. The space $L^p(D)$, for $p \in [1, \infty)$, stands for the Banach space of (equivalence classes) of measurable functions f satisfying $\int_D |f(x)|^p \, \mathrm{d}x < \infty$, and $L^\infty(D)$ is the Banach space of functions that are essentially bounded in D. The space $C^k(D)$ ($C^\infty(D)$) refers to the class of all functions whose partial derivatives up to order k (of any order) exist and are continuous, $C_c^k(D)$ is the space of functions in $C^k(D)$ with compact support, and $C_0^k(\mathbb{R}^d)$ ($C_b^k(\mathbb{R}^d)$) is the subspace of $C^k(\mathbb{R}^d)$ consisting of those functions whose derivatives from order 0 to k vanish at infinity (are bounded). Also, the space $C^{k,r}(D)$ is the class of all functions whose partial derivatives up to order k are Hölder continuous of order r. Therefore $C^{0,1}(D)$ is precisely the space of Lipschitz continuous functions on D.

We adopt the notation $\partial_i := \frac{\partial}{\partial x_i}$, $\partial_{ij} := \frac{\partial^2}{\partial x_i \partial x_j}$, and $\partial_t := \frac{\partial}{\partial t}$. Also we adopt the standard summation rule for repeated indices, i.e., repeated subscripts and superscripts are summed from 1 through d. For example,

$$a^{ij}\partial_{ij}\varphi + b^i\partial_i\varphi := \sum_{i,j=1}^{d} a^{ij}\frac{\partial^2\varphi}{\partial x_i \partial x_j} + \sum_{i=1}^{d} b^i \frac{\partial\varphi}{\partial x_i}.$$

The standard Sobolev space of functions on D whose generalized derivatives up to order k are in $L^p(D)$, equipped with its natural norm, is denoted by $\mathscr{W}^{k,p}(D)$, $k \geq 0$, $p \geq 1$. The closure of $C_c^\infty(D)$ in $\mathscr{W}^{k,p}(D)$ is denoted by $\mathscr{W}_0^{k,p}(D)$. If B is an open ball, then $\mathscr{W}_0^{k,p}(B)$ consists of all functions in $\mathscr{W}^{k,p}(B)$ which, when extended by zero outside B, belong to $\mathscr{W}^{k,p}(\mathbb{R}^d)$.

In general if $\mathcal{X}$ is a space of real-valued functions on D, $\mathcal{X}_{\mathrm{loc}}$ consists of all functions f such that $\varphi f \in \mathcal{X}$ for every $\varphi \in C_c^\infty(D)$. In this manner we obtain the spaces $L^p_{\mathrm{loc}}(D)$ and $\mathscr{W}^{2,p}_{\mathrm{loc}}(D)$.

1

Markov Processes and Ergodic Properties

1.1 Introduction

This book, as the title indicates, is about ergodic control of diffusion processes. The operative words here are *ergodic*, *control* and *diffusion processes*. We introduce two of these, diffusion processes and ergodic theory, in this chapter. It is a bird's eye view, sparse on detail and touching only the highlights that are relevant to this work. Further details can be found in many excellent texts and monographs available, some of which are listed in the bibliographical note at the end. The next level issue of control is broached in the next chapter.

We begin with diffusion processes, which is a special and important subclass of Markov processes. But before we introduce Markov processes, it is convenient to recall some of the framework of the general theory of stochastic processes which provides the backdrop for it.

Let $(\Omega, \mathfrak{F}, \mathbb{P})$ be a complete probability space, or in other words, Ω is a set called the sample space, $\mathfrak{F}$ is a σ-field of subsets of Ω (whence $(\Omega, \mathfrak{F})$ is a *measurable* space), and $\mathbb{P}$ is a probability measure on $(\Omega, \mathfrak{F})$. *Completeness* is a technicality that requires that any subset of a set having zero probability be included in $\mathfrak{F}$. A random process $\{X_t\}$ defined on $(\Omega, \mathfrak{F}, \mathbb{P})$ is a family of random variables indexed by a time index t which can be discrete (e.g., $t = 0, 1, 2, \dots$) or continuous (e.g., $t \geq 0$). We shall mostly be interested in the continuous time case. For notational ease, we denote by X the entire process. For each $t \in \mathbb{R}_+$, X_t takes values in a Polish space S, i.e., a separable Hausdorff space whose topology is metrizable with a complete metric. This is a convenient level of generality to operate with, because an amazingly large body of basic results in probability carry over to Polish spaces and most of the spaces one encounters in the study of random processes are indeed Polish. Let $d(\cdot\,,\cdot)$ be a complete metric on S. For any Polish space S, $\mathscr{P}(S)$ denotes the Polish space of probability measures on S under the Prohorov topology [32, chapter 2]. Recall that if S is a metric space, then a collection $M \subset \mathscr{P}(S)$ is called *tight* if for every $\varepsilon > 0$, there exists a compact set $K_\varepsilon \subset S$, such that $\mu(K_\varepsilon) \geq 1 - \varepsilon$ for all $\mu \in M$. The celebrated criterion for compactness in $\mathscr{P}(S)$, known as Prohorov's theorem, states that for a metric space S if $M \subset \mathscr{P}(S)$ is tight, then it is relatively compact and, provided S is complete and separable, the converse also holds [24, theorems 6.1 and 6.2, p. 37], [55, p. 104].

We shall usually have in the background a *filtration*, i.e., an increasing family $\{\mathfrak{F}_t\}$ of sub-σ-fields of $\mathfrak{F}$ indexed by t again. Intuitively, $\mathfrak{F}_t$ corresponds to *information available at time t*. Again, for technical reasons we assume that it is complete (i.e., contains all sets of zero probability and their subsets) and right-continuous (i.e., $\mathfrak{F}_t = \cap_{s>t}\mathfrak{F}_s$). We say that X is *adapted* to the filtration $\{\mathfrak{F}_t\}$ if for each t, X_t is $\mathfrak{F}_t$-measurable. A special case of a filtration is the so-called natural filtration of X, denoted by $\mathfrak{F}_t^X$ and defined as the completion of $\cap_{t'>t}\sigma(X_s : s \le t')$. Clearly X is adapted to its own natural filtration. An important notion related to a filtration is that of a *stopping time*. A $[0, \infty]$-valued random variable τ is said to be a stopping time w.r.t. the filtration $\{\mathfrak{F}_t\}$ (the filtration is usually implicit) if for all $t \ge 0$, $\{\tau \le t\} \subset \mathfrak{F}_t$. Intuitively, what this says is that at time t, one knows whether τ has occurred already or not. For example, the first time the process hits a prescribed closed set is a stopping time with respect to its natural filtration, but the last time it does so need not be. We associate with τ the σ-field $\mathfrak{F}_\tau$ defined by

$$\mathfrak{F}_\tau = \{A \in \mathfrak{F} : A \cap \{\tau \le t\} \in \mathfrak{F}_t \text{ for all } t \in [0, \infty)\} .$$

Intuitively, $\mathfrak{F}_\tau$ are the events prior to τ.

So far we have viewed X only as a collection of random variables indexed by t. But for a fixed sample point in Ω, it is also a function of t. The least we shall assume is that it is a measurable function. A stronger notion is progressive measurability, which requites that for each $T > 0$, the function $(t, \omega) \to X_t(\omega)$, $(t, \omega) \in [0, T] \times \Omega$, be measurable with respect to $\mathscr{B}_T \times \mathfrak{F}_T$, where $\mathscr{B}_T$ denotes the Borel σ-field on $[0, T]$. The sub-σ-field of $[0, \infty) \times \Omega$ generated by the progressively measurable processes is known as the progressively measurable σ-field. If a process is adapted to $\mathfrak{F}_t$ and has right or left-continuous paths, then it is progressively measurable [47, p. 89].

There is one serious technicality which has been glossed over here. Two random processes X, X' are said to be versions or modifications of each other if $X_t = X'_t$ a.s. for all t. This defines an equivalence relation and it is convenient to work with such equivalence classes. That is, when one says that X has measurable sample paths, it is implied that it has a version which is so. The stronger equivalence notion of $\mathbb{P}(X_t = X'_t \text{ for all } t \ge 0) = 1$ is not as useful. We shall not dwell on these technicalities too much. See Borkar [32, chapter 6], for details.

We briefly mention the issue of the actual construction of a random process. In practice, a random process is typically described by its finite dimensional marginals, i.e., the laws of $(X_{t_1}, \ldots, X_{t_n})$ for all finite collections of time instants $t_1 < \cdots < t_n$, $n \ge 1$. In particular, all versions of a process have the same finite dimensional distributions. These are perforce consistent, i.e., if $B \subset A$ are two such collections, then the law for B is the induced law from the law for A under the appropriate projection. The celebrated Kolmogorov extension theorem gives us the converse statement: Given a consistent family of such finite dimensional laws, there is a unique probability measure on $S^{[0,\infty)}$ consistent with it. Thus we can let $\Omega = S^{[0,\infty)}$, $\mathfrak{F}$ the product σ-field,

$\mathbb{P}$ denote this unique law, and set $X_t(\omega) = \omega(t)$, where $\omega = \{\omega(t) : t \geq 0\}$ denotes a sample point in Ω. This is called the canonical construction of the random process X. While it appears appealingly simple and elegant, it has its limitations. The most significant limitation is that $\mathfrak{F}$ contains only "countably described sets," i.e., any set in $\mathfrak{F}$ must be describable in terms of countably many time instants. (It is an interesting exercise to prove this.) This eliminates many sets of interest. Thus it becomes essential to "lift" this construction to a more convenient space, such as the space $C([0,\infty); S)$ of continuous functions $[0,\infty) \mapsto S$ or the space $\mathcal{D}([0,\infty); S)$ of functions $[0,\infty) \mapsto S$ that are continuous from the right and have limits from the left at each t (r.c.l.l.). We briefly sketch how to do the former, as that's what we shall need for diffusion processes.

The key result here is that if X is *stochastically continuous*, in other words if

$$\mathbb{P}\left(d(X_s, X_t) > \varepsilon\right) \to 0 \tag{1.1.1}$$

for any $\varepsilon > 0$ as $s \to t$, and for each $T > 0$, the modulus of continuity

$$w_T(X,\delta) := \sup\left\{d(X_s, X_t) : 0 \leq s \leq t \leq T,\ |t-s| < \delta\right\} \to 0 \quad \text{a.s.}, \tag{1.1.2}$$

as $\delta \to 0$, then X has a continuous version. The proof is simple: restrict X to rationals, extend it uniquely to a continuous function on $[0,\infty)$ (which is possible because (1.1.2) guarantees uniform continuity on rationals when restricted to any $[0,T]$, for $T > 0$), and argue using stochastic continuity that this indeed yields a version of X. A convenient test for (1.1.1)–(1.1.2) to hold is the Kolmogorov continuity criterion (see Wong and Hajek [122, pp. 57]), that for each $T > 0$ there exist positive scalars a, b, and c satisfying

$$\mathbb{E}\left[d(X_t, X_s)^a\right] \leq b|t-s|^{1+c} \qquad \forall\, t, s \in [0,T]\,.$$

Note that the above procedure a.s. defines a map that maps an element of Ω to the continuous extension of its restriction to the rationals. Defining the map to be the function that is identically zero on the zero probability subset of Ω that is left out, we have a measurable map $\Omega \mapsto C([0,\infty); S)$. The image μ of $\mathbb{P}$ under this map defines a probability measure on $C([0,\infty); S)$. We may thus realize the continuous version as a *canonically* defined random process X' on the new probability space $(C([0,\infty); S), \mathfrak{G}, \mu)$, where $\mathfrak{G}$ is the Borel σ-field of $C([0,\infty); S)$, as $X'_t(\omega) = \omega(t)$ for $\omega \in C([0,\infty); S)$. An analogous development is possible for the space $\mathcal{D}([0,\infty); S)$ of paths $[0,\infty) \to S$ that are right-continuous and have left limits. This space is Polish: it is separable and metrizable with a complete metric $\mathfrak{d}_S$ (due to Skorohod) defined as follows [55, p. 117]. Let Λ denote the space of strictly increasing Lipschitz continuous surjective maps λ from $\mathbb{R}_+$ to itself such that

$$\gamma(\lambda) := \operatorname*{ess\,sup}_{t\geq 0} |\log \lambda'(t)| < \infty\,.$$

With ρ a complete metric on S, define

$$d_T(x,y,\lambda) := \sup_{t\geq 0} \left[1 \wedge \rho(x(t\wedge T), y(\lambda(t)\wedge T))\right], \quad T>0,$$

$$\mathfrak{d}_s(x,y) := \inf_{\lambda\in\Lambda} \left[\gamma(\lambda) \vee \int_0^\infty \mathrm{e}^{-s} d_s(x,y,\lambda)\,\mathrm{d}s\right].$$

Convergence with respect to $\mathfrak{d}_s$ has the following simple interpretation: $x_n \to x$ in $\mathcal{D}([0,\infty); S)$ if there exists a sequence $\{\lambda_n \in \Lambda : n \in \mathbb{N}\}$ such that $\lambda_n(t) \to t$ uniformly on compacta, and $\sup_{[0,T]} \rho(x_n \circ \lambda_n, x) \to 0$ for all $T > 0$. This topology is known as the Skorohod topology.

A useful criterion due to Chentsov for the existence of a r.c.l.l. version that extends the Kolmogorov continuity criterion is that for any $T > 0$ there exist positive constants a, b, c and C satisfying [66, pp. 159–164]

$$\mathbb{E}\left[|X_t - X_r|^a |X_r - X_s|^b\right] \leq C|t-s|^{1+c}, \quad \forall s<r<t.$$

1.2 Markov processes

Before defining Markov processes, it is instructive to step back and recall what a deterministic dynamical system is. A deterministic dynamical system has as its backdrop a set Σ called the *state space* in which it evolves. Its evolution is given by a *time t map* Φ_t, $t \in \mathbb{R}$, with the interpretation that for $x \in \Sigma$, $x(t) := \Phi_t(x)$ is the position of the system at time t if it starts at x at time 0. The idea is that once at x, the trajectory $\{x(t) : t \geq 0\}$ is completely specified, likewise for $t \leq 0$. This in fact is what qualifies $x(t)$ as the *state at time t* in the sense of physics: $x(t)$ is all you need to know at time t to be able to determine the future trajectory $\{x(s) : s \geq t\}$. Thus $\Phi_0(x) = x$, and $\Phi_t \circ \Phi_s = \Phi_s \circ \Phi_t = \Phi_{s+t}$, i.e., $\{\Phi_t : t \in \mathbb{R}\}$ is a group.

A two-parameter version is possible for time-dependent dynamics, i.e., when the future (or past) trajectory depends on the position x as well as the precise time t_0 at which the trajectory is at x. Thus we need a two-parameter group $\Phi_{s,t}$, satisfying $\Phi_{t,t}(x) = x$, and $\Phi_{s,t} \circ \Phi_{u,s} = \Phi_{u,t}$ for all u, s, and t.

Clearly for stochastic dynamical systems, it does not make sense to demand that the complete future trajectory be determined by the present position. Nevertheless there is a natural generalization of the notions of a state and a dynamical system. We require that the (regular) conditional law of $\{X_s : s \geq t\}$, given $\{X_u : u \leq t\}$, should be the same as its conditional law given X_t alone. In other words, knowing how one arrived at X_t tells us nothing more about the future than what is already known by knowing X_t. From an equivalent definition of conditional independence [32, p. 42], this is equivalent to the statement that $\{X_s : s > t\}$ and $\{X_s : s < t\}$ are conditionally independent given X_t for each t. This definition is symmetric in time. Thus, for example, it is also equivalent to: for each t, the regular conditional law of $\{X_s : s < t\}$, given $\{X_s : s \geq t\}$, is the same as its regular conditional law given X_t. In fact, a more general statement is true: for any $t_1 < t_2$, $\{X_s : s < t_1 \text{ or } s > t_2\}$ and

$\{X(s) : t_1 < s < t_2\}$ are conditionally independent given $\{X_{t_1}, X_{t_2}\}$. This also serves as an equivalent definition. Though not very useful in the present context, this is the definition that extends well to indices more general than time, such as $t \in \mathbb{R}^2$. These are the so-called Markov random fields.

Since finite dimensional marginals completely specify the law of a stochastic process, an economical statement of the Markov property is: For every collection of times $0 \le t \le t_1 < \cdots < t_k < \infty$, and Borel subsets $A_1, \dots, A_k \subset S$, it holds that

$$\mathbb{P}\left(X_{t_i} \in A_i,\ i = 1, \dots, k \mid \mathfrak{F}_t^X\right) = \mathbb{P}\left(X_{t_i} \in A_i,\ i = 1, \dots, k \mid X_t\right).$$

A stronger notion is that of the *strong Markov property*, which requires that

$$\mathbb{P}\left(X_{\tau+t_i} \in A_i,\ i = 1, \dots, k \mid \mathfrak{F}_\tau^X\right) = \mathbb{P}\left(X_{\tau+t_i} \in A_i,\ i = 1, \dots, k \mid X_\tau\right)$$

a.s. on $\{\tau < \infty\}$, for every $(\mathfrak{F}_t^X)$-stopping time τ. If X has the Markov, or strong Markov property, then it said to be a Markov, or a strong Markov process, respectively.

For $t > s$ and $x \in S$, let $P(s, x, t, \mathrm{d}y)$ denote the regular conditional law of X_t given $X_s = x$. This is called the transition probability (kernel) of X. In particular, $P\colon (s, x, t) \mapsto P(s, x, t, S)$ is measurable. By the filtering property of conditional expectations:

$$\mathbb{E}\left[\mathbb{E}\left[f(X_t) \mid \mathfrak{F}_r^X\right] \mid \mathfrak{F}_s^X\right] = \mathbb{E}\left[f(X_t) \mid \mathfrak{F}_s^X\right], \quad s \le r \le t,$$

which, coupled with the Markov property, yields

$$P(s, x, t, \mathrm{d}y) = \int_S P(s, x, r, \mathrm{d}z)\, P(r, z, t, \mathrm{d}y), \quad s \le r \le t.$$

These are called the Chapman–Kolmogorov equations. While the transition probability kernels of Markov processes must satisfy these, the converse is not true [57].

Let $\mathcal{B}(S)$ denote the space of bounded measurable functions $S \mapsto \mathbb{R}$. Define

$$T_{s,t} f(x) := \int P(s, x, t, \mathrm{d}y)\, f(y), \quad t \ge s \ge 0, \quad f \in \mathcal{B}(S).$$

Then by the Chapman–Kolmogorov equations, $\{T_{s,t} : 0 \le s \le t\}$ is a two-parameter semigroup of operators, i.e., it satisfies

$$T_{t,t} = I, \quad T_{r,t} \circ T_{s,r} = T_{s,t}, \quad 0 \le s \le r \le t,$$

where I is the identity operator. This is weaker than the group property for deterministic flows. However, this is inevitable because of the irreversibility of stochastic processes.

Let $C_b(S)$ denote the space of bounded continuous real-valued functions on S. The process X above is said to be *Feller* if $\{T_{s,t} : 0 \le s \le t\}$ maps $C_b(S)$ into $C_b(S)$, and *strong Feller* if it maps $\mathcal{B}(S)$ into $C_b(S)$. The former case is obtained if the transition probability kernel $P(s, x, t, \mathrm{d}y)$ is continuous in the initial state x. The latter requires more – the kernel should have some additional smoothing properties.

For example, when $S = \mathbb{R}$, the Gaussian kernel

$$P(s, x, t, \mathrm{d}y) = \frac{1}{\sqrt{2\pi t}} \mathrm{e}^{-\frac{(x-y)^2}{2(t-s)}} \mathrm{d}y$$

does have this property. More generally, if the transition kernel is of the form

$$P(s, x, t, \mathrm{d}y) = p(s, x, t, y)\lambda(\mathrm{d}y)$$

for a positive measure λ and the density p is continuous in the x variable, then the strong Feller property follows by the dominated convergence theorem.

An important consequence of the Feller property is that it implies the strong Markov property for Markov processes with right-continuous paths. To see this, one first verifies the strong Markov property for stopping times taking values in $\left\{\frac{k}{2^n} : k \geq 0\right\}$ for $n \in \mathbb{N}$ fixed. This follows by a straightforward verification [55, p. 159] which does not require the Feller property. Given a general a.s.-finite stopping time τ, the property then holds for $\tau^{(n)} := \frac{\lfloor 2^n \tau \rfloor + 1}{2^n}$ which is also seen to be a stopping time for all $n \geq 1$. As $n \uparrow \infty$, $\tau^{(n)} \downarrow \tau$. Using the right-continuity of paths along with the Feller property and the a.s. convergence property of reverse martingales, the strong Markov property for τ can be inferred from that for $\tau^{(n)}$.

In the case of Feller processes, we may restrict $\{T_{s,t}\}$ to a semigroup on $C_b(S)$. Of special interest is the case when the transition probability kernel $P(s, x, t, \mathrm{d}y)$ depends only on the difference $r = t - s$. By abuse of terminology, we then write $P(r, x, \mathrm{d}y)$ and also $T_r = T_{s,s+r}$. Then T_t, $t \geq 0$ defines a one-parameter semigroup of operators. A very rich theory of such semigroups is available, with which we deal in the next section. It is worth noting here that this is a special case of the general theory of operator semigroups. These are the so-called Markov semigroups, characterized by the additional properties:

(a) $T_t(\alpha f + \beta g) = \alpha T_t f + \beta T_t g$ for all $\alpha, \beta \in \mathbb{R}$ and $f, g \in C_b(S)$;
(b) $f \geq 0 \Longrightarrow T_t f \geq 0$;
(c) $\|T_t f\| \leq \|f\|$, where $\|f\| := \sup_{s \in S} |f(s)|$;
(d) $T_t \mathbf{1} = \mathbf{1}$, where $\mathbf{1}$ is the constant function $\equiv 1$.

We now give some important examples of Markov processes.

Example 1.2.1 (i) Poisson process: Let $\lambda > 0$. A $\mathbb{Z}_+$-valued stochastic process $N = \{N_t : t \geq 0\}$ is called a Poisson process with parameter λ if

(a) $N_0 = 0$;
(b) for any $0 \leq t_0 < t_1 < \cdots < t_n$,

$$N_{t_1} - N_{t_0}, N_{t_2} - N_{t_1}, \ldots, N_{t_n} - N_{t_{n-1}}$$

are independent, i.e., N has independent increments;
(c) $t \mapsto N_t$ is a.s. right-continuous;

(d) for $t \geq s \geq 0$, $N_t - N_s$ has the Poisson distribution with parameter λ, i.e.,

$$\mathbb{P}(N_t - N_s = n) = \frac{\lambda^n (t-s)^n \mathrm{e}^{-\lambda(t-s)}}{n!}, \qquad n = 0, 1, \ldots$$

It can be verified that N is a Markov process with transition function

$$P(t, m, \{n\}) = \frac{(\lambda t)^{n-m} \mathrm{e}^{-\lambda t}}{(n-m)!}, \qquad n \geq m.$$

(ii) One-dimensional Brownian motion: A real-valued process $\mathbb{B} = \{\mathbb{B}_t : t \geq 0\}$ is called a one-dimensional Brownian motion (or Wiener process) if

(a) $\mathbb{B}_0 = 0$;
(b) $t \mapsto \mathbb{B}_t$ is a.s. continuous;
(c) for $t > s \geq 0$, $\mathbb{B}_t - \mathbb{B}_s$ has the Normal distribution with mean 0 and variance $t - s$;
(d) for any $0 \leq t_0 < t_1 < \cdots < t_n$,

$$\mathbb{B}_{t_1} - \mathbb{B}_{t_0}, \; \mathbb{B}_{t_2} - \mathbb{B}_{t_1}, \ldots, \; \mathbb{B}_{t_n} - \mathbb{B}_{t_{n-1}}$$

are independent.

Then $\mathbb{B}$ is a Markov process with transition function

$$P(t, x, A) = \frac{1}{\sqrt{2\pi t}} \int_A \mathrm{e}^{-\frac{(y-x)^2}{2t}} \mathrm{d}y.$$

(iii) In general, any stochastic process with independent increments is a Markov process.

(iv) d-dimensional Brownian motion: An $\mathbb{R}^d$-valued process $W = \{W_t : t \geq 0\}$, where $W_t = (W_t^1, \ldots, W_t^d)$, is called a d-dimensional Brownian motion if

(a) for each i, $W^i = \{W_t^i : t \geq 0\}$ is a one-dimensional Brownian motion;
(b) for $i \neq j$, the processes W^i and W^j are independent.

Then W is a Markov process with transition function

$$P(t, x, A) = \frac{1}{(2\pi t)^{d/2}} \int_A \mathrm{e}^{-\frac{|y-x|^2}{2t}} \mathrm{d}y.$$

There are several ways of constructing Markov processes. We list the common ones below.

(1) *Via the theorem of Ionescu–Tulcea:* Once an initial law λ at t_0 and the transition probability kernel are prescribed, one can write down the finite dimensional marginals of the process:

$$\mathbb{P}(X_{t_k} \in A_k, \; 0 \leq k \leq n) = \int_{A_0} \lambda(\mathrm{d}y_0) \int_{A_1} P(t_0, y_0, t_1, \mathrm{d}y_1) \cdots \int_{A_n} P(t_{n-1}, y_{n-1}, t_n, \mathrm{d}y_n).$$

These are easily seen to form a consistent family and thus define a unique law for X, by the theorem of Ionescu–Tulcea [93, p. 162]. This may be lifted to a suitable function space such as $C([0,\infty); S)$ by techniques described in Section 1.1.

(2) *Via dynamics driven by an independent increment process:* A typical instance of this is the equation

$$X_t = X_0 + \int_0^t h(X_s)\,\mathrm{d}s + W_t\,, \quad t \geq 0\,,$$

where W is a Brownian motion. Suppose that for a given trajectory of W, this equation has an a.s. unique solution X (this requires suitable hypotheses on h). Then for $t > s$,

$$X_t = X_s + \int_s^t h(X_r)\,\mathrm{d}r + W_t - W_s\,, \quad t \geq 0\,,$$

and therefore X_t is a.s. specified as a functional of X_s and the independent increments $\{W_u - W_s : s \leq u \leq t\}$. The Markov property follows easily from this. Later on we shall see that the a.s. uniqueness property used above holds for a very general class of equations. One can also consider situations where W is replaced by other independent increment processes.

(3) *Via change of measure:* Suppose X is a Markov process constructed canonically on its path space, say $C([0,\infty); S)$. That is, we take $\Omega = C([0,\infty); S)$, $\mathfrak{F}$ its Borel σ-field, and $\mathbb{P}$ the law of X. Then $X_t(\omega) = \omega(t)$ for $t \geq 0$ and $\omega \in \Omega$. Let $X_{[s,t]}$ denote the trajectory segment $\{X_r : s \leq r \leq t\}$. Let $\mathfrak{F}^X_{s,t}$ be the right-continuous completion of $\sigma\{X_r : s \leq r \leq t\}$. A family $\{\Lambda_{s,t} : s < t\}$ of $\mathfrak{F}^X_{s,t}$-measurable random variables is said to be a multiplicative functional if $\Lambda_{r,s}\Lambda_{s,t} = \Lambda_{r,t}$ for all $r < s < t$. If in addition $\{\Lambda_{0,t} : t \geq 0\}$ is a nonnegative martingale with mean equal to one, we can define a new probability measure $\hat{\mathbb{P}}$ on $(\Omega, \mathfrak{F})$ as follows: If $\mathbb{P}_t$ and $\hat{\mathbb{P}}_t$ denote the restrictions of $\mathbb{P}$ and $\hat{\mathbb{P}}$, respectively, to $\mathfrak{F}^X_t$ for $t \geq 0$, then

$$\frac{\mathrm{d}\hat{\mathbb{P}}_t}{\mathrm{d}\mathbb{P}_t} = \Lambda_{0,t}\,, \quad t \geq 0\,. \tag{1.2.1}$$

Since $\mathfrak{F} = \bigvee_{t\geq 0} \mathfrak{F}^X_t$, it follows by the martingale property of $\{\Lambda_{0,t} : t \geq 0\}$ that (1.2.1) consistently defines a probability measure on $(\Omega, \mathfrak{F})$. Let $\mathbb{E}$ and $\hat{\mathbb{E}}$ denote the expectations under $\mathbb{P}$ and $\hat{\mathbb{P}}$, respectively. For any $f \in \mathcal{B}(S)$ and $s < t$, one has the well-known *Bayes formula*

$$\hat{\mathbb{E}}\left[f(X_t) \mid \mathfrak{F}^X_s\right] = \frac{\mathbb{E}\left[f(X_t)\Lambda_{0,t} \mid \mathfrak{F}^X_s\right]}{\mathbb{E}\left[\Lambda_{0,t} \mid \mathfrak{F}^X_s\right]} = \frac{\mathbb{E}\left[f(X_t)\Lambda_{0,t} \mid \mathfrak{F}^X_s\right]}{\Lambda_{0,s}}\,.$$

From the multiplicative property, the right-hand side is simply

$$\mathbb{E}\left[f(X_t)\Lambda_{s,t} \mid \mathfrak{F}^X_s\right] = \mathbb{E}\left[f(X_t)\Lambda_{s,t} \mid X_s\right]\,,$$

in other words, a function of X_s alone. Thus X remains a Markov process under $\hat{\mathbb{P}}$. We shall later see an important instance of this construction when we construct the so-called weak solutions to stochastic differential equations.

(4) *Via approximation:* Many Markov processes are obtained as limits of simpler Markov (or even non-Markov) processes. In the next section, we discuss the semigroup and martingale approaches to Markov processes. These provide the two most common approximation arguments in use. In the semigroup approach, one works with the semigroup $\{T_t\}$ described above. This specifies the transition probability kernel via

$$\int P(t, x, \mathrm{d}y) f(y) = T_t f(x), \quad f \in C_b(S),$$

and therefore also determines the law of X once the initial distribution is given. One often constructs Markov semigroups $\{T_t\}$ as limits (in an appropriate sense) of a sequence of known Markov semigroups $\{T_t^n : n \geq 1\}$ as $n \uparrow \infty$. See Ethier and Kurtz [55] for details and examples. The martingale approach, on the other hand, uses the *martingale characterization*, which characterizes the Markov process X by the property that for a sufficiently rich class of $f \in C_b(S)$ and an operator $\mathcal{L}$ defined on this class, $f(X_t) - \int_0^t \mathcal{L} f(X_s)\,\mathrm{d}s$, $t \geq 0$, is an $(\mathfrak{F}_t^X)$-martingale. As the martingale property is preserved under weak (Prohorov) convergence of probability measures, this allows us to construct Markov processes as limits in law of other Markov or sometimes non-Markov processes. The celebrated *diffusion limit* in queuing theory is a well-known example of this scheme, as are many systems of infinite interacting particles.

We conclude this section by introducing the notion of a *Markov family*. This is a family of probability measures $\{\mathbb{P}_x : x \in S\}$ on $(\Omega, \mathfrak{F})$ along with a stochastic process X defined on it such that the law of X under $\mathbb{P}_x$ for each x is that of a Markov process corresponding to a common transition probability kernel (i.e., with a common functional dependence on x), with initial condition $X_0 = x$. This allows us to study the Markov process under multiple initial conditions at the same time. We denote a Markov family by $(X, (\Omega, \mathfrak{F}), \{\mathbb{P}_x\}_{x \in S})$.

1.3 Semigroups associated with Markov processes

Let E be a Polish space, and $(X, (\Omega, \mathfrak{F}), \{\mathfrak{F}_t\}_{t \in \mathbb{R}_+}, \{\mathbb{P}_x\}_{x \in E})$ a Markov family. We define a one-parameter family of operators $T_t \colon \mathcal{B}(E) \to \mathcal{B}(E)$, $t \in \mathbb{R}_+$, as follows:

$$T_t f(x) := \mathbb{E}_x [f(X_t)] = \int_E P(t, x, \mathrm{d}y) f(y), \quad f \in \mathcal{B}(E). \tag{1.3.1}$$

The following properties are evident

(i) for each t, T_t is a linear operator;
(ii) $\|T_t f\|_\infty \leq \|f\|_\infty$, where $\|\cdot\|_\infty$ is the L^∞-norm;
(iii) T_t is a positive operator, i.e., $T_t f \geq 0$ if $f \geq 0$;
(iv) $T_t \mathbf{1} = \mathbf{1}$, where $\mathbf{1}$ denotes the function identically equal to 1;
(v) $T_0 = I$, where I denotes the identity operator;
(vi) $T_{t+s} = T_t T_s$.

Thus $\{T_t : t \geq 0\}$ is a contractive semigroup of positive operators on $\mathcal{B}(E)$. We next define another semigroup, which is in a sense dual to $\{T_t : t \geq 0\}$.

Let $\mathfrak{M}_s(E)$ denote the space of all finite signed measures on $(E, \mathscr{B}(E))$, under the topology of total variation norm. For $t \in \mathbb{R}_+$, we define $S_t \colon \mathfrak{M}_s(E) \to \mathfrak{M}_s(E)$ by

$$(S_t\mu)(A) := \int_E P(t, x, A)\,\mu(\mathrm{d}x)\,.$$

Then properties (i)–(vi) hold for $\{S_t\}$, under the modification $\|S_t\mu\|_{\mathrm{TV}} \leq \|\mu\|_{\mathrm{TV}}$ for property (ii) and $S_t\mu(E) = \mu(E)$ for property (iv).

Let $f \in \mathcal{B}(E)$. We define an operator $\mathcal{A}$ on $\mathcal{B}(E)$ by

$$\mathcal{A}f = \lim_{t\downarrow 0} \frac{T_t f - f}{t}$$

in L^∞-norm, provided the limit exists. We refer to the set of all such functions f as the domain of $\mathcal{A}$ and denote it by $\mathscr{D}(\mathcal{A})$. The operator $\mathcal{A}$ is called the *infinitesimal generator* of the semigroup $\{T_t\}$. Let

$$\mathcal{B}_0 := \{f \in \mathcal{B}(E) : \|T_t f - f\| \to 0, \text{ as } t \downarrow 0\}\,.$$

It is easy to verify that $\mathcal{B}_0$ is a closed subspace of $\mathcal{B}(E)$ and that $T_t f$ is uniformly continuous in t for each $f \in \mathcal{B}_0$. Also $T_t(\mathcal{B}_0) \subset \mathcal{B}_0$ for all $t \in \mathbb{R}_+$ and $\mathscr{D}(\mathcal{A}) \subset \mathcal{B}_0$. Thus $\{T_t\}$ is a strongly continuous semigroup on $\mathcal{B}_0$. For Feller processes we have $T_t \colon C_b(E) \to C_b(E)$ and the previous discussion applies with $C_b(E)$ replacing $\mathcal{B}(E)$.

The following result is standard in semigroup theory.

Proposition 1.3.1 *For a strongly continuous semigroup $\{T_t\}$ on a Banach space $\mathcal{X}$ with generator $\mathcal{A}$, the following properties hold:*

(i) *if $f \in \mathcal{X}$, then $\int_0^t T_s f\,\mathrm{d}s \in \mathscr{D}(\mathcal{A})$ for all $t \in \mathbb{R}_+$, and*

$$T_t f - f = \mathcal{A}\int_0^t T_s f\,\mathrm{d}s\,;$$

(ii) *if $f \in \mathscr{D}(\mathcal{A})$, then $T_t f \in \mathscr{D}(\mathcal{A})$ for all $t \in \mathbb{R}_+$, and*

$$\frac{\mathrm{d}}{\mathrm{d}t}T_t f = \mathcal{A}T_t f = T_t\mathcal{A}f\,,$$

or equivalently,

$$T_t f - f = \int_0^t \mathcal{A}T_s f\,\mathrm{d}s = \int_0^t T_s\mathcal{A}f\,\mathrm{d}s\,, \quad t \in \mathbb{R}_+\,.$$

We describe briefly the construction of a semigroup from its infinitesimal generator $\mathcal{A}$. For this purpose we introduce the notion of the *resolvent* of a semigroup $\{T_t\}$. This is a family of operators $\{\mathcal{R}_\lambda\}_{\lambda>0}$ on $\mathcal{B}(E)$ defined by

$$\mathcal{R}_\lambda f(x) = \int_0^\infty \mathrm{e}^{-\lambda t}T_t f(x)\,\mathrm{d}t = \mathbb{E}_x\left[\int_0^\infty \mathrm{e}^{-\lambda t}f(X_t)\,\mathrm{d}t\right], \quad x \in E\,.$$

By Laplace inversion, $\{\mathcal{R}_\lambda f : \lambda > 0\}$ determines $T_t f$ for almost every t. Hence, if the map $t \mapsto T_t f$ is continuous, then $T_t f$ is determined uniquely. Thus the resolvent determines the semigroup $\{T_t\}$ on $\mathcal{B}_0$. In addition, if $\mathcal{B}_0$ separates the elements of $\mathfrak{M}_s(E)$, i.e., if, for $\mu \in \mathfrak{M}_s(E)$,

$$\int_E f \, d\mu = 0 \qquad \forall f \in \mathcal{B}_0$$

implies $\mu = 0$, then the resolvent determines the semigroup uniquely, which in turn determines the Markov family via (1.3.1).

The properties of $\{T_t\}$ are reflected in $\{\mathcal{R}_\lambda\}$. In particular, $\|T_t\| \le 1$ corresponds to

$$\|\mathcal{R}_\lambda\| \le \lambda^{-1}, \tag{1.3.2}$$

the positivity of T_t implies that of $\mathcal{R}_\lambda$, and the semigroup property of $\{T_t\}$ corresponds to

$$\mathcal{R}_\mu \mathcal{R}_\lambda f = (\mu - \lambda)^{-1}(\mathcal{R}_\lambda f - \mathcal{R}_\mu f), \quad \mu \neq \lambda. \tag{1.3.3}$$

The property in (1.3.3) is known as the resolvent equation. From (1.3.3) it follows that $\mathcal{R}_\lambda(\mathcal{B}(E)) = \mathcal{R}_\mu(\mathcal{B}(E))$ for all $\lambda, \mu > 0$, and that for $f \in \mathcal{B}(E)$

$$\mathcal{R}_\lambda f = \mathcal{R}_\mu(f + (\mu - \lambda)\mathcal{R}_\lambda f).$$

Also $\mathcal{R}_\lambda(\mathcal{B}(E)) \subset \mathcal{B}_0$, and

$$\|f - \lambda \mathcal{R}_\lambda f\| \xrightarrow[\lambda \to \infty]{} 0 \qquad \forall f \in \mathcal{B}_0. \tag{1.3.4}$$

Equation (1.3.4) corresponds to the property $T_0 = I$.

The restriction of $\mathcal{R}_\lambda$ on $\mathcal{B}_0$ is the inverse of $(\lambda I - \mathcal{A})$ defined on $\mathcal{D}(\mathcal{A})$. Indeed, it holds that:

(i) if $f \in \mathcal{B}_0$, then $\mathcal{R}_\lambda f \in \mathcal{D}(\mathcal{A})$ and

$$(\lambda I - \mathcal{A})\mathcal{R}_\lambda f = f; \tag{1.3.5}$$

(ii) if $f \in \mathcal{D}(\mathcal{A})$, then

$$\mathcal{R}_\lambda(\lambda I - \mathcal{A})f = f. \tag{1.3.6}$$

It follows by properties (1.3.5)–(1.3.6) that the operator $\mathcal{A}$ determines the semigroup uniquely on $\mathcal{B}_0$. Indeed, for $\lambda > 0$, we obtain $\mathcal{R}_\lambda f$ via (1.3.5), which in turn determines $T_t f$ via Laplace inversion. Also since $\mathcal{B}_0$ is closed and $\mathcal{D}(\mathcal{A}) = \mathcal{R}_\lambda(\mathcal{B}_0)$ by (1.3.5)–(1.3.6), it follows from (1.3.4) that $\mathcal{D}(\mathcal{A})$ is dense in $\mathcal{B}_0$. It is straightforward to show that the operator $\mathcal{A}$ is closed.

The celebrated theorem of Hille–Yosida provides the necessary and sufficient conditions for an operator on a Banach space to be the infinitesimal generator of a strongly continuous contraction semigroup.

Theorem 1.3.2 (Hille–Yosida) *Let $\mathcal{X}$ be a Banach space and $\mathcal{L}$ a linear operator on $\mathcal{X}$. Then $\mathcal{L}$ is the infinitesimal generator of a strongly continuous contraction semigroup on $\mathcal{X}$ if and only if*

(i) *$\mathscr{D}(\mathcal{L})$ is dense in $\mathcal{X}$;*
(ii) *$\mathcal{L}$ is* dissipative, *i.e., $\|\lambda f - \mathcal{L}f\| \geq \lambda\|f\|$, for every $f \in \mathscr{D}(\mathcal{L})$;*
(iii) *$\mathscr{R}(\lambda I - \mathcal{L}) = \mathcal{X}$ for some $\lambda > 0$.*

An equivalent statement of the Hille–Yosida theorem in Rogers and Williams [102] uses the resolvent $\{\mathcal{R}_\lambda : \lambda > 0\}$ as the starting point. Specifically, a contraction resolvent $\{\mathcal{R}_\lambda : \lambda > 0\}$ is defined as a family of bounded linear operators on $\mathcal{X}$ satisfying (1.3.2) and (1.3.3). It is said to be strongly continuous if (1.3.4) holds. The alternative version of the Hille–Yosida theorem from [102, p. 237] states that such a family corresponds to a unique strongly continuous contraction semigroup $\{T_t : t \geq 0\}$ such that

$$\mathcal{R}_\lambda f = \int_0^\infty e^{-\lambda t} T_t f \, dt \qquad \forall t > 0,\ f \in \mathcal{B}(E),$$

and

$$T_t f = \lim_{\lambda\uparrow\infty} e^{-\lambda t} \sum_{n=0}^{\infty} \frac{(\lambda t)^n (\lambda \mathcal{R}_\lambda f)^n}{n!}. \tag{1.3.7}$$

The generator $\mathcal{A}$ is in turn given by

$$\mathcal{A}f = \lim_{\lambda\uparrow\infty} \lambda(\lambda\mathcal{R}_\lambda f - f) \tag{1.3.8}$$

whenever the limit exists, with $\mathscr{D}(\mathcal{A})$ being precisely the set of those f for which it does.

The two versions of the Hille–Yosida theorem are related in the spirit of Tauberian theorems. The advantage of the resolvent based viewpoint is that (1.3.7)–(1.3.8) often lead to useful approximation procedures; the developments of Section 6.3 are of this kind. The original statement of the Hille–Yosida theorem given above, however, is more fundamental from a dynamic point of view. In dynamical systems (and Markov processes after all are stochastic dynamical systems), one wants a characterization of the dynamics in terms of an "instantaneous" prescription that leads to the global dynamic behavior through an appropriate integration procedure. The generator provides precisely such a prescription. It renders precise the intuition $T_t = e^{t\mathcal{A}}$, which is straightforward for bounded $\mathcal{A}$, but highly nontrivial otherwise. Unfortunately the latter is the case for all but the simplest Markov processes (see the examples below).

Example 1.3.3 (i) Let $E = \mathbb{Z}_+$ and $N = \{N_t : t \geq 0\}$ be a Poisson process with parameter λ and $N_0 = 0$. Then for $f \in \mathcal{B}(\mathbb{Z}_+)$ the infinitesimal generator $\mathcal{A}$ is given by

$$\mathcal{A}f(x) = \lambda(f(x+1) - f(x)), \quad x \in \mathbb{Z}_+. \tag{1.3.9}$$

Conversely, if $\mathcal{A}$ is defined by (1.3.9), then since $\mathcal{A}$ is a bounded linear operator on $\mathcal{B}(\mathbb{Z}_+)$ the corresponding semigroup $\{T_t\}$ is given by

$$T_t f(x) = e^{t\mathcal{A}} f(x) = \sum_{n=0}^{\infty} \frac{(\lambda t)^n e^{-\lambda t}}{n!} f(x+n), \quad x \in \mathbb{Z}_+,$$

and the transition function takes the form

$$P(t, x, \{y\}) = \frac{(\lambda t)^{y-x} \mathrm{e}^{-\lambda t}}{(y-x)!}, \quad y \geq x.$$

(ii) Let $E = \{1, \dots, N\}$ and $X = \{X_t : t \geq 0\}$ be a time-homogeneous Markov chain with transition probabilities

$$P_{ij}(t) = \mathbb{P}(X_{t+s} = j \mid X_s = i), \quad t \geq 0.$$

We assume that

$$\lim_{t \downarrow 0} P_{ij}(t) = \begin{cases} 1 & \text{if } i = j, \\ 0 & \text{if } i \neq j. \end{cases}$$

If $P(t) = [P_{ij}(t)]$ denotes the transition matrix as a function of time, then

$$P(t+s) = P(t)P(s), \quad t, s \geq 0,$$

and it follows that P is continuous in t. Moreover, P is differentiable, i.e.,

$$\lim_{t \downarrow 0} \frac{P_{ij}(t)}{t} = \lambda_{ij}, \quad i \neq j,$$

$$\lim_{t \downarrow 0} \frac{P_{ii}(t) - 1}{t} = \lambda_{ii}.$$

for some $\lambda_{ij} \in [0, \infty)$, $i \neq j$, and $\lambda_{ii} = -\sum_{j \neq i} \lambda_{ij}$.

An alternate description of the Markov chain can be provided via

$$\mathbb{P}(X_{t+h} = j \mid X_t = i) = \lambda_{ij} h + o(h), \quad i \neq j,$$

$$\mathbb{P}(X_{t+h} = i \mid X_t = i) = 1 + \lambda_{ii} h + o(h),$$

which yields a characterization of the chain in terms of the sojourn times, which are defined as the random intervals that the chain spends in any given state. If the chain starts at $X_0 = i$, then the sojourn time in the state i is exponentially distributed with parameter $-\lambda_{ii}$. The chain then jumps to a state $j \neq i$ with probability $-\frac{\lambda_{ij}}{\lambda_{ii}}$. The sequence of states visited by X, denoted by $\{Y_0, Y_1, \dots\}$, is itself a discrete Markov chain, called the *embedded chain*. Conditioned on the states $\{Y_0, Y_1, \dots\}$, the successive sojourn times $\tau_0, \tau_1, \dots$ are independent and exponentially distributed random variables with parameters $-\lambda_{Y_0 Y_0}, -\lambda_{Y_1 Y_1}, \dots$, respectively.

Define the $N \times N$ matrix $\Lambda = [\lambda_{ij}]$. Then

$$\Lambda = \lim_{t \downarrow 0} \frac{P(t) - I}{t}.$$

Hence, Λ is the infinitesimal generator of the Markov chain X. Conversely, if $\Lambda = [\lambda_{ij}]$ is an $N \times N$ matrix with $\lambda_{ij} \geq 0$ for $i \neq j$, and $\sum_j \lambda_{ij} = 0$, then we can construct a Markov chain $X = \{X_t\}$ taking values in E having transition matrix $P(t) = \mathrm{e}^{t\Lambda}$.

(iii) Let $W = \{W_t : t \geq 0\}$ be a standard d-dimensional Wiener process. Then the domain of the infinitesimal generator $\mathcal{A}$ contains the space $C_c^2(\mathbb{R}^d)$ of twice-continuously differentiable functions with compact support, and for $f \in C_c^2(\mathbb{R}^d)$, $\mathcal{A}$ takes the form

$$\mathcal{A}f(x) = \tfrac{1}{2}\Delta f(x),$$

where Δ denotes the Laplace operator.

1.4 The martingale approach

Consider an E-valued Feller process X and the associated semigroup $\{T_t\}$ on $C_b(E)$. Integrating the semigroup equation

$$\frac{\mathrm{d}}{\mathrm{d}t} T_t f = T_t \mathcal{A} f, \quad f \in \mathcal{D}(\mathcal{A}),$$

we obtain

$$T_t f(x) = f(x) + \int_0^t T_s \mathcal{A} f(x)\,\mathrm{d}s.$$

That is,

$$\mathbb{E}\left[f(X_t) - f(x) - \int_0^t \mathcal{A}f(X_s)\,\mathrm{d}s \;\middle|\; X_0 = x\right] = 0.$$

More generally,

$$\mathbb{E}\left[f(X_t) - f(X_s) - \int_s^t \mathcal{A}f(X_r)\,\mathrm{d}r \;\middle|\; \mathfrak{F}_s^X\right]$$
$$= \mathbb{E}\left[f(X_t) - f(X_s) - \int_s^t \mathcal{A}f(X_r)\,\mathrm{d}r \;\middle|\; X_s\right] = 0,$$

where the first equality follows from the Markov property. In other words

$$f(X_t) - \int_0^t \mathcal{A}f(X_s)\,\mathrm{d}s, \quad t \geq 0, \tag{1.4.1}$$

is an $(\mathfrak{F}_t^X)$-martingale.

The martingale approach to Markov processes pioneered by Stroock and Varadhan uses this property to characterize Markov processes, essentially providing a converse to the above statement. The operator $\mathcal{A}$ is then said to be an *extended generator* of the process and the functions f for which (1.4.1) holds form the domain of the extended generator, denoted by $\tilde{\mathcal{D}}(\mathcal{A})$ [55].

Some advantages of this approach are

(1) Whereas $\mathcal{D}(\mathcal{A})$ in Section 1.3 is sometimes difficult to pin down, $\tilde{\mathcal{D}}(\mathcal{A})$ can often be chosen in a more convenient manner. This is demonstrated in Section 1.6.

(2) It allows us to use the machinery of martingale theory for analyzing Markov processes. For example:

(2a) Continuous time martingales have versions that are r.c.l.l., a standard consequence of the forward and reverse martingale convergence theorems. The martingale approach then allows one to deduce the r.c.l.l. property for the Markov process in question [55, Theorem 3.6, p. 178].

(2b) Often limit theorems such as the central limit theorem for functions of a Markov process can be deduced from the corresponding limit theorem for the associated martingale [55, Chapter 7].

(2c) There are also simple and elegant tests for tightness and convergence in the martingale approach for Markov processes [55, Section 3.9].

Given $\mathcal{A}\colon \tilde{\mathscr{D}}(\mathcal{A}) \to C_b(E)$ and $\nu \in \mathscr{P}(E)$, we say that a process X (or rather its law) solves the martingale problem for $(\mathcal{A}, \nu)$ if $\mathscr{L}(X_0) = \nu$ and for all $f \in \tilde{\mathscr{D}}(\mathcal{A})$, (1.4.1) holds. It is said to be the unique solution thereof if this X is unique in law. In that case X can be shown to be Markov [55, p. 184]. More generally, a family $\{\mathbb{P}_x\}_{x\in E} \subset \mathscr{P}(\mathcal{D}[0,\infty); E)$, solves the martingale problem for $\mathcal{A}$ if for each x, $\mathbb{P}_x$ solves the martingale problem for $(\mathcal{A}, \delta_x)$. The martingale problem is well-posed if for each x, the solution $\mathbb{P}_x$ is unique. In this case $\{\mathbb{P}_x\}$ is a Markov family.

A key result in this development is the following [55, theorem 4.1, p. 182].

Theorem 1.4.1 *Suppose $\mathcal{A}$ is linear, dissipative and for some $\lambda > 0$, the closure of the range of $\lambda I - \mathcal{A}$ separates points of E. Then the martingale problem for $(\mathcal{A}, \nu)$, $\nu \in \mathscr{P}(E)$, has a unique solution, which moreover is a Markov process.*

1.5 Ergodic theory of Markov processes

In this section we first present a self-contained treatment of the ergodic properties of a measure preserving transformation and related issues. Then we study the ergodic behavior of a stationary process, and conclude by addressing the ergodic properties of a Markov process.

1.5.1 Ergodic theory

Ergodic theory originated in the attempts to render rigorous the statistical mechanical notion that "time averages equal ensemble averages." It was put on a firm mathematical footing by Birkhoff, von Neumann and others. Their original formulations were in the framework of *measure preserving transformations*, a natural formulation of time invariant phenomena if we consider time shifts as measure preserving. Specialized to stochastic processes, and even more so, to Markov processes, the theory has a much richer structure that subsumes, among other things, the classical law of large numbers. We outline here the general framework of ergodic theory and specialize later to stationary processes and Markov processes.

We begin with some basic concepts.

Definition 1.5.1 Let $(E, \mathfrak{E}, \mu)$ be a probability space. An $\mathfrak{E}$-measurable transformation $T: E \to E$ is said to be measure preserving if

$$\mu(T^{-1}(A)) = \mu(A) \qquad \forall A \in \mathfrak{E}. \tag{1.5.1}$$

If a transformation T preserves a measure μ in the sense of (1.5.1), then μ is often referred to as an invariant (probability) measure of T. Let

$$\mathcal{M}_T(E) := \{\mu \in \mathscr{P}(E) : \mu \circ T^{-1} = \mu\}, \tag{1.5.2}$$

i.e., $\mathcal{M}_T(E)$ is the set of all invariant probability measures of T. If E is a compact metric space and $T: E \to E$ is continuous, then $\mu \mapsto \mu \circ T^{-1}$ is a continuous affine map from a convex compact subset of $C(E)^*$ to itself. Thus by the Schauder fixed point theorem [51, p. 456], $\mu \mapsto \mu \circ T^{-1}$ has a fixed point, and it follows that $\mathcal{M}_T(E)$ is non-empty. Alternatively, let $\nu \in \mathscr{P}(E)$, and define

$$\mu_n := \frac{1}{n} \sum_{k=0}^{n-1} \nu \circ T^{-k}.$$

Since E is compact, $\{\mu_n\}$ is tight. It is easy to show that any limit point of $\{\mu_n\}$ in $\mathscr{P}(E)$ belongs to $\mathcal{M}_T(E)$ (Krylov–Bogolioubov theorem [119, corollary 6.9.1, p. 152]).

Let T be a measure preserving transformation on a probability space $(E, \mathfrak{E}, \mu)$. A set $A \in \mathfrak{E}$ is called T-invariant if $T^{-1}A = A$. In such a case the study of T can be split into the study of the transformations $T|_A$ and $T|_{A^c}$.

Definition 1.5.2 Let $(E, \mathfrak{E}, \mu)$ be a probability space. A measure preserving transformation $T: E \to E$ is called ergodic if for any T-invariant set $A \in \mathfrak{E}$, $\mu(A) = 0$ or $\mu(A) = 1$.

From the perspective of invariant measures, a T-invariant probability measure μ is called ergodic if for any T-invariant set $A \in \mathfrak{E}$, $\mu(A) = 0$ or $\mu(A) = 1$. Ergodicity can also be expressed in a functional language: a μ-preserving transformation T is ergodic if and only if every T-invariant real-valued function f (i.e., $f \circ T = f$) is constant a.s. with respect to μ.

Theorem 1.5.3 (Birkhoff ergodic theorem) *Let T be a measure preserving transformation on a probability space $(E, \mathfrak{E}, \mu)$ and $f \in L^1(\mu)$. Then*

$$\lim_{n\to\infty} \frac{1}{n} \sum_{k=0}^{n-1} f(T^k(x)) = \mathbb{E}[f \mid \mathfrak{I}] \quad \mu\text{-a.s.},$$

where $\mathfrak{I}$ is the μ-completion of the σ-field $\{A \in \mathfrak{E} : T^{-1}A = A\}$.

The proof can be found in several texts, e.g., [119].

Corollary 1.5.4 *If T defined on $(E, \mathfrak{E}, \mu)$ is ergodic, and $f \in L^1(\mu)$, then*

$$\lim_{n\to\infty} \frac{1}{n} \sum_{k=0}^{n-1} f(T^k(x)) = \int_E f \, \mathrm{d}\mu \qquad \mu\text{-a.s.}$$

Useful properties of the set of invariant probability measures $\mathcal{M}_T(E)$ defined in (1.5.2) are provided by the following theorem.

Theorem 1.5.5 *Let E be compact and $T\colon E \to E$ continuous. Then*

(i) *$\mathcal{M}_T(E)$ is a convex and compact subset of $\mathscr{P}(E)$;*
(ii) *μ is an extreme point of $\mathcal{M}_T(E)$ if and only if μ is ergodic, i.e., if T is an ergodic measure preserving transformation on $(E, \mathfrak{E}, \mu)$;*
(iii) *if μ, $\nu \in \mathcal{M}_T(E)$ are both ergodic and $\mu \neq \nu$, then μ and ν are mutually singular.*

Proof (i) The convexity of $\mathcal{M}_T(E)$ is obvious. Also, it is easy to show that it is closed, and hence compact by Prohorov's theorem.

(ii) If $\mu \in \mathcal{M}_T(E)$ is not ergodic, then there exists $A \in \mathfrak{E}$ such that $T^{-1}A = A$ and $0 < \mu(A) < 1$. Define μ_1 and μ_2 in $\mathscr{P}(E)$ by

$$\mu_1(B) = \frac{\mu(B \cap A)}{\mu(A)}, \qquad \mu_2(B) = \frac{\mu(B \cap A^c)}{\mu(A^c)}, \qquad B \in \mathfrak{E}.$$

It is straightforward to verify that μ_1 and μ_2 belong to $\mathcal{M}_T(E)$, and since

$$\mu = \mu(A)\mu_1 + (1 - \mu(A))\mu_2,$$

μ is not an extreme point of $\mathcal{M}_T(E)$. Conversely, suppose $\mu \in \mathcal{M}_T(E)$ is ergodic and $\mu = a\mu_1 + (1-a)\mu_2$ for some $\mu_1, \mu_2 \in \mathcal{M}_T(E)$, and $a \in (0,1)$. Clearly $\mu_1 \ll \mu$. Let

$$A = \left\{x \in E : \tfrac{d\mu_1}{d\mu}(x) < 1\right\}.$$

Then

$$\begin{aligned}\int_{A\cap T^{-1}A} \frac{d\mu_1}{d\mu}\, d\mu + \int_{A\setminus T^{-1}A} \frac{d\mu_1}{d\mu}\, d\mu &= \mu_1(A)\\ &= \mu_1(T^{-1}A)\\ &= \int_{A\cap T^{-1}A} \frac{d\mu_1}{d\mu}\, d\mu + \int_{T^{-1}A\setminus A} \frac{d\mu_1}{d\mu}\, d\mu,\end{aligned}$$

which implies

$$\int_{A\setminus T^{-1}A} \frac{d\mu_1}{d\mu}\, d\mu = \int_{T^{-1}A\setminus A} \frac{d\mu_1}{d\mu}\, d\mu.$$

Since $\frac{d\mu_1}{d\mu} < 1$ on $A \setminus T^{-1}A$, $\frac{d\mu_1}{d\mu} \geq 1$ on $T^{-1}A \setminus A$, and

$$\begin{aligned}\mu(A \setminus T^{-1}A) &= \mu(A) - \mu(A \cap T^{-1}A)\\ &= \mu(T^{-1}A) - \mu(A \cap T^{-1}A)\\ &= \mu(T^{-1}A \setminus A),\end{aligned}$$

it follows that

$$\mu(A \setminus T^{-1}A) = \mu(T^{-1}A \setminus A) = 0.$$

If $\mu(A) = 1$, then

$$\mu_1(E) = \int_A \tfrac{d\mu_1}{d\mu}\,d\mu < \mu(A) = 1\,,$$

a contradiction. Thus $\mu(A) = 0$. Similarly, if $B = \left\{x \in E : \tfrac{d\mu_1}{d\mu}(x) > 1\right\}$, following the same arguments we show that $\mu(B) = 0$. Therefore $\mu_1 = \mu$, which in turn implies $\mu_1 = \mu_2$, and thus μ is an extreme point.

(iii) Using the Lebesgue decomposition theorem, we can find a unique $a \in [0, 1]$ and a unique pair $\mu_1, \mu_2 \in \mathscr{P}(E)$ such that $\mu = a\mu_1 + (1 - a)\mu_2$, where $\mu_1 \ll \nu$ and μ_2 is singular with respect to ν. However,

$$\mu = \mu \circ T^{-1} = a\mu_1 \circ T^{-1} + (1 - a)\mu_2 \circ T^{-1}\,,$$

and since $\mu_1 \circ T^{-1} \ll \nu \circ T^{-1} = \nu$ and $\mu_2 \circ T^{-1}$ is singular with respect to $\nu \circ T^{-1} = \nu$, the uniqueness of the decomposition implies that $\mu_1, \mu_2 \in \mathcal{M}_T(E)$. Since μ is an extreme point of $\mathcal{M}_T(E)$, we have $a \in \{0, 1\}$. If $a = 0$, then $\mu = \mu_2$ and μ is singular with respect to ν. If $a = 1$, then $\mu \ll \nu$, and we can argue as in (ii) to conclude that $\mu = \nu$, a contradiction. □

Let $\mathcal{M}_T^e(E)$ denote the set of extreme points of $\mathcal{M}_T(E)$.

Definition 1.5.6 Let $\mathcal{X}$ be a Hausdorff, locally convex, topological vector space and $C \subset \mathcal{X}$ a convex, compact, metrizable subset. An element $x \in C$ is called the barycenter of a measure $\mu \in \mathscr{P}(C)$ if

$$f(x) = \int_C f\,d\mu$$

for all continuous affine $f : C \to \mathbb{R}$.

We state without proof a theorem of Choquet [95].

Theorem 1.5.7 (Choquet) *Let C be a convex, compact, metrizable subset of a locally convex topological vector space. Then each $x \in C$ is the barycenter of some $\mu \in \mathscr{P}(C)$ supported on the set of extreme points of C.*

The metrizability of C ensures that the set of its extreme points is G_δ in C and hence measurable, whereas the compactness of C ensures that it is non-empty [44, p. 138].

Lemma 1.5.8 *For each $\mu \in \mathcal{M}_T(E)$, there exists a unique $\xi \in \mathscr{P}(\mathcal{M}_T(E))$, with $\xi(\mathcal{M}_T^e(E)) = 1$, such that*

$$\int_E f(x)\,\mu(dx) = \int_{\mathcal{M}_T^e(E)} \left(\int_E f(x)\,\nu(dx)\right)\xi(d\nu) \qquad \forall f \in C(E)\,.$$

Proof Since $\mathcal{M}_T(E)$ is convex and compact, the existence of such a ξ follows from Choquet's theorem.

To show uniqueness, note that if μ_1 and μ_2 are T-invariant, so is $\mu_1 - \mu_2$, hence also $(\mu_1 - \mu_2)^+$ and $(\mu_1 - \mu_2)^-$. It follows that the measures $|\mu_1 - \mu_2| = (\mu_1 - \mu_2)^+ + (\mu_1 - \mu_2)^-$,

$\mu_1 \vee \mu_2 = \frac{1}{2}(\mu_1 + \mu_2 + |\mu_1 - \mu_2|)$ and $\mu_1 \wedge \mu_2 = \frac{1}{2}(\mu_1 + \mu_2 - |\mu_1 - \mu_2|)$ are also T-invariant. Therefore $\mathcal{M}_T(E)$ is a Choquet simplex [44, p. 160], and by a theorem of Choquet and Meyer [44, p. 163], ξ is unique. □

We write

$$\mu(A) = \int_{\mathcal{M}_T^e(E)} \nu(A)\,\xi(\mathrm{d}\nu)\,, \quad A \in \mathfrak{E}\,,$$

and call this the ergodic decomposition of μ.

We have thus far treated ergodicity for a single measure preserving transformation. Next we extend the study to a semigroup of measure preserving transformations.

Let $(E, \mathfrak{E}, \mu)$ be a probability space and $\{T_t : t \geq 0\}$ a semigroup of measure preserving transformations on E. Note then that

$$\int_E f(T_t(x))\,\mu(\mathrm{d}x) = \int_E f(x)\,\mu(\mathrm{d}x) \qquad \forall f \in L^1(\mu)\,.$$

Assume that $(t, x) \mapsto T_t(x)$ is measurable with respect to $\mathscr{B}(\mathbb{R}_+) \times \mathfrak{E}$. We have the following extension of Birkhoff's ergodic theorem for $\{T_t\}$.

Theorem 1.5.9 *Let $\{T_t : t \geq 0\}$ be a semigroup of measure preserving transformations on a probability space $(E, \mathfrak{E}, \mu)$. Define the operator $\mathcal{S}_t$ by*

$$\mathcal{S}_t f(x) := \frac{1}{t} \int_0^t f(T_s(x))\,\mathrm{d}s\,.$$

Then for every $f \in L^1(\mu)$ there exists $f^* \in L^1(\mu)$ such that

$$\lim_{t\to\infty} \mathcal{S}_t f = f^* \quad \mu\text{-a.s.}, \tag{1.5.3}$$

and $f^* = \mathbb{E}\,[f \mid \mathfrak{I}]$, where $\mathfrak{I}$ is the μ-completion of the σ-field

$$\{A \in \mathfrak{E} : T_t^{-1} A = A\,,\ \forall t \in \mathbb{R}_+\}\,.$$

In particular, $\mathbb{E}\,[f \circ T_t \mid \mathfrak{I}] = \mathbb{E}\,[f \mid \mathfrak{I}]$ for all $t \in \mathbb{R}_+$. Moreover, if $f \in L^p(\mu)$, with $p \in [1, \infty)$, then

$$\left\|\mathcal{S}_t f - f^*\right\|_{L^p(\mu)} \xrightarrow[t\to\infty]{} 0\,. \tag{1.5.4}$$

Proof Without loss of generality assume that $f \geq 0$. Define $f_1(x) := \int_0^1 f(T_s(x))\,\mathrm{d}s$ for $x \in E$. Then

$$\begin{aligned}\int_E f_1(x)\,\mu(\mathrm{d}x) &= \int_E \left[\int_0^1 f(T_s(x))\,\mathrm{d}s\right] \mu(\mathrm{d}x)\\ &= \int_0^1 \left[\int_E f(T_s(x))\,\mu(\mathrm{d}x)\right] \mathrm{d}s\\ &= \int_E f(x)\,\mu(\mathrm{d}x)\,.\end{aligned}$$

Therefore $f_1 \in L^1(\mu)$. Since T_1 is measure preserving, and $T_1^n = T_n$ for any $n \in \mathbb{N}$, we have

$$\sum_{k=0}^{n-1} f_1(T_1^k(x)) = \sum_{k=0}^{n-1} \int_0^1 f(T_{k+s}(x))\,\mathrm{d}s = \int_0^n f_1(T_s(x))\,\mathrm{d}s\,.$$

Hence, by Theorem 1.5.3,

$$f^*(x) := \lim_{n\to\infty} \frac{1}{n} \int_0^n f(T_s(x))\,\mathrm{d}s \quad \text{exists } \mu\text{-a.s.} \tag{1.5.5}$$

For $t \geq 1$, let $\lfloor t \rfloor$ denote the integer part of t, i.e., $\lfloor t \rfloor \leq t < \lfloor t \rfloor + 1$. We have

$$\frac{1}{\lfloor t \rfloor + 1} \int_0^{\lfloor t \rfloor} f(T_s(x))\,\mathrm{d}s \leq \frac{1}{t} \int_0^t f(T_s(x))\,\mathrm{d}s \leq \frac{1}{\lfloor t \rfloor} \int_0^{\lfloor t \rfloor + 1} f(T_s(x))\,\mathrm{d}s\,. \tag{1.5.6}$$

Since $\frac{\lfloor t \rfloor}{\lfloor t \rfloor + 1} \to 1$, it follows from (1.5.5) – (1.5.6) that the limit in (1.5.3) exists μ-a.s. on E. Also, since

$$\int_A f\,\mathrm{d}\mu = \int_A f \circ T_t\,\mathrm{d}\mu \qquad \forall A \in \mathfrak{I}\,,$$

we have, for every bounded f,

$$\begin{aligned}
\int_A f^*\,\mathrm{d}\mu &= \int_A \left[\lim_{t\to\infty} \frac{1}{t} \int_0^t f \circ T_s\,\mathrm{d}s\right] \mathrm{d}\mu \\
&= \lim_{t\to\infty} \frac{1}{t} \int_0^t \left[\int_A f \circ T_s\,\mathrm{d}\mu\right] \mathrm{d}s \\
&= \int_A f\,\mathrm{d}\mu\,.
\end{aligned}$$

Thus $f^* = \mathbb{E}\,[f \mid \mathfrak{I}]$.

Since $\|\mathcal{S}_t f\|_{L^p(\mu)} \leq \|f\|_{L^p(\mu)}$, $\mathcal{S}_t$ is a bounded linear operator on $L^p(\mu)$, $p \in [1, \infty)$. If f is bounded, then (1.5.4) holds by dominated convergence. If $f \in L^p(\mu)$ for some $p \in [1, \infty)$, then for every $\varepsilon > 0$ there exists a bounded function f_ε on E such that $\|f - f_\varepsilon\|_{L^p(\mu)} \leq \frac{\varepsilon}{3}$. Also since $\mathcal{S}_t f_\varepsilon$ is Cauchy in $L^p(\mu)$, there exists $T = T(\varepsilon, f_\varepsilon)$ such that $\|\mathcal{S}_t f_\varepsilon - \mathcal{S}_{t'} f_\varepsilon\|_{L^p(\mu)} \leq \frac{\varepsilon}{3}$, for all $t, t' > T$. Thus, we have

$$\begin{aligned}
\left\|\mathcal{S}_t f - \mathcal{S}_{t'} f\right\|_{L^p(\mu)} &\leq \left\|\mathcal{S}_t(f - f_\varepsilon)\right\|_{L^p(\mu)} + \left\|\mathcal{S}_t f_\varepsilon - \mathcal{S}_{t'} f_\varepsilon\right\|_{L^p(\mu)} + \left\|\mathcal{S}_{t'}(f - f_\varepsilon)\right\|_{L^p(\mu)} \\
&\leq \varepsilon\,, \quad \forall t, t' > T(\varepsilon, f_\varepsilon)\,.
\end{aligned}$$

Therefore $\mathcal{S}_t f$ is Cauchy in $L^p(\mu)$ and hence it converges. It is evident that the limit is f^*. □

Theorem 1.5.9 admits a more general form as follows [108, p. 2].

Theorem 1.5.10 *Let $\{T_t : t \geq 0\}$ be a semigroup of measure preserving transformations on a σ-finite measurable space $(E, \mathfrak{E}, \mu)$, and $f, g \in L^1(\mu)$, $g > 0$. Suppose*

$\int_0^t g(T_s(x))\,\mathrm{d}s \to \infty$ *as* $t \to \infty$, μ*-a.s. Then there exists* $f_g^* \in L^1(\mu)$ *such that*

$$\lim_{t\to\infty} \frac{\int_0^t f(T_s(x))\,\mathrm{d}s}{\int_0^t g(T_s(x))\,\mathrm{d}s} = f_g^*(x) \quad \mu\text{-a.s.}$$

Moreover, $f_g^*(x) = f_g^*(T_t(x))$ μ*-a.s. for all* $t > 0$, *and*

$$\int_E f_g^*(x) g(x)\,\mu(\mathrm{d}x) = \int_E f(x)\,\mu(\mathrm{d}x)\,.$$

Definition 1.5.11 A semigroup $\{T_t : t \geq 0\}$ of measure preserving transformations on a probability space $(E, \mathfrak{E}, \mu)$ is called ergodic if $\mu(A) \in \{0, 1\}$ for all $A \in \mathfrak{I}$.

If $\{T_t\}$ is ergodic, then μ is often referred to as an ergodic invariant probability measure for the semigroup $\{T_t\}$. Also note that if $\{T_t\}$ is ergodic, then any $\mathfrak{I}$-measurable function is constant μ-a.s.

Remark 1.5.12 (i) If $\{T_t\}$ is ergodic and $f \in L^1(\mu)$, then

$$\lim_{t\to\infty} \frac{1}{t} \int_0^t f(T_s(x))\,\mathrm{d}s = \int_E f(x)\,\mu(\mathrm{d}x)\,.$$

(ii) The analogue of Theorem 1.5.5 holds for a semigroup $\{T_t\}$ of measure preserving transformations.

1.5.2 Ergodic theory of random processes

Recall that a random process is called stationary if its law is invariant under time shifts. We now consider the implications of ergodic theory for stationary stochastic r.c.l.l. processes with state space S. The space of sample paths $E = \mathcal{D}(\mathbb{R}; S)$ with the Skorohod topology is itself a Polish space. It is well known that any Polish space E can be homeomorphically embedded into a G_δ subset of $[0, 1]^\infty$ [32, p. 2]. The closure of its image can then be viewed as a compactification $\bar{E}$ of E, i.e., E is densely continuously embedded in the compact Polish space $\bar{E}$. Then $\mathcal{M} := \mathscr{P}(\bar{E})$ is compact by Prohorov's theorem. Let $\mathcal{M}_s^0$ denote the subset of $\mathcal{M}$, corresponding to stationary probability measures on E, and $\mathcal{M}_s$ denote its closure in $\mathcal{M}$. The sets $\mathcal{M}_s$ and $\mathcal{M}_s^0$ are convex and $\mathcal{M}_s$ is compact. Let $\mathcal{M}_e$, $\mathcal{M}_e^0$ denote the set of extreme points of $\mathcal{M}_s$, $\mathcal{M}_s^0$, respectively. That $\mathcal{M}_e$ is non-empty follows from compactness of $\mathcal{M}$. By Choquet's theorem [44, p. 140] (see also Theorem 1.5.7, p. 18), every $\mu \in \mathcal{M}_s$ is the barycenter of a probability measure ξ_μ on $\mathcal{M}_e$, i.e., it satisfies

$$\int_{\bar{E}} f(z)\,\mu(\mathrm{d}z) = \int_{\mathcal{M}_e} \left(\int_{\bar{E}} f(z)\,\nu(\mathrm{d}z) \right) \xi_\mu(\mathrm{d}\nu) \qquad \forall f \in C_b(\bar{E})\,.$$

By Theorem 1.5.5 (iii), the elements of $\mathcal{M}_e$ are mutually singular. Consequently, as in Lemma 1.5.8, ξ_μ is uniquely determined by μ.

Lemma 1.5.13 *If* $\mu \in \mathcal{M}_s^0$, *then* $\xi_\mu\left(\mathcal{M}_e \setminus \mathcal{M}_e^0\right) = 0$.

Proof If $\xi_\mu(\{\nu \in \mathcal{M}_e : \nu(\bar{E} \setminus E) > 0\}) > 0$, then $\mu(\bar{E} \setminus E) > 0$, a contradiction. Thus, $\xi_\mu\left((\mathcal{M}_e^0)^c\right) = 0$. In particular $\mathcal{M}_e^0 \neq \varnothing$. □

We now turn to Markov processes. Let $P(t, x, \mathrm{d}y)$ be a transition kernel on a Polish space S, and X denote a Markov process governed by this kernel.

Definition 1.5.14 $\mu \in \mathscr{P}(S)$ is said to be an *invariant probability measure* (or *stationary distribution*) for the kernel P if

$$\int_S \mu(\mathrm{d}x)P(t, x, \mathrm{d}y) = \mu(\mathrm{d}y).$$

We let $\mathscr{H}$ denote the set of all such μ, and $\mathscr{I}$ denote the set of stationary laws of X corresponding to $\mathscr{L}(X_0) \in \mathscr{H}$.

Assuming P is Feller, the following are some immediate consequences of this definition:

(i) if, for some $t \in \mathbb{R}$, the law of X_t is μ, then it is so for all $t \in \mathbb{R}$, and X is a stationary process;
(ii) the sets $\mathscr{H}$ and $\mathscr{I}$, if non-empty, are closed and convex.

The existence of an invariant probability measure can be characterized as follows.

Theorem 1.5.15 *Let S be a Polish space, and $P\colon \mathbb{R}_+ \times S \to \mathscr{P}(S)$ a transition probability. Then there exists an invariant probability measure for P, only if for some $x_0 \in S$ the following holds: for every $\varepsilon > 0$ there exists a compact set $K_\varepsilon \subset S$ such that*

$$\liminf_{t\to\infty} \frac{1}{t}\int_0^t P(s, x_0, K_\varepsilon^c)\,\mathrm{d}s < \varepsilon. \tag{1.5.7}$$

Provided P has the Feller property, this condition is also sufficient for the existence of an invariant probability measure.

Proof Suppose that $\mu \in \mathscr{P}(S)$ is invariant. Since S is Polish, μ is tight. Let K_n be a sequence of compact sets such that $\mu(K_n^c) \to 0$ as $n \to \infty$. Fix a sequence $t_k \uparrow \infty$, and define

$$\varphi_n(x) := \liminf_{k\to\infty} \frac{1}{t_k}\int_0^{t_k} P(s, x, K_n^c)\,\mathrm{d}s.$$

Then, since

$$\mu(K_n^c) = \int_S \mu(\mathrm{d}x)\left[\frac{1}{t_k}\int_0^{t_k} P(s, x, K_n^c)\,\mathrm{d}s\right], \quad k \geq 0,$$

taking limits we obtain

$$\mu(K_n^c) \geq \int_S \varphi_n(x)\,\mu(\mathrm{d}x),$$

which implies that along some subsequence φ_n converges to 0 μ-a.s. Thus for n large enough (1.5.7) holds. This completes the proof of necessity. To show sufficiency, it follows from (1.5.7) that there exists a sequence $t_k \uparrow \infty$ and an increasing sequence of compact sets K_n such that

$$\sup_{k\geq 0} \frac{1}{t_k}\int_0^{t_k} P(s, x_0, K_n^c)\, \mathrm{d}s \xrightarrow[n\to\infty]{} 0 .$$

Consequently the sequence of measures

$$\mu_k(A) := \frac{1}{t_k}\int_0^{t_k} P(s, x_0, A)\, \mathrm{d}s\,, \quad A \in \mathscr{B}(S)\,,$$

is tight, and hence it converges along some subsequence, also denoted by $\{t_k\}$, to some $\mu \in \mathscr{P}(S)$. By Feller continuity, for $f \in C_b(S)$,

$$\begin{aligned}\int_S \mu(\mathrm{d}x) T_t f(x) &= \lim_{k\to\infty} \frac{1}{t_k}\int_0^{t_k} \mathrm{d}s \int_S P(s, x_0, \mathrm{d}x) T_t f(x)\\ &= \lim_{k\to\infty} \frac{1}{t_k}\int_0^{t_k} T_{s+t} f(x_0)\, \mathrm{d}s\,. \end{aligned} \tag{1.5.8}$$

The right-hand side of (1.5.8) can be decomposed as

$$\begin{aligned}\lim_{k\to\infty} \frac{1}{t_k}\int_0^{t_k} T_{s+t} f(x_0)\, \mathrm{d}s &= \lim_{k\to\infty} \frac{1}{t_k}\left[\int_0^{t_k} T_r f(x_0)\, \mathrm{d}r - \int_0^t T_r f(x_0)\, \mathrm{d}r \right.\\ &\qquad\qquad \left. + \int_{t_k}^{t_k+t} T_r f(x_0)\, \mathrm{d}r\right]\\ &= \int_S f(x)\mu(\mathrm{d}x)\,, \end{aligned} \tag{1.5.9}$$

and it follows by (1.5.8) – (1.5.9) that μ is an invariant probability measure. □

Note that although we are addressing Markov processes on $[0, \infty)$, in what follows we work with the bi-infinite path space $\mathcal{D}(\mathbb{R}; S)$ instead of $\mathcal{D}([0, \infty); S)$. Given a stationary probability distribution on the latter, its image under the shift operator θ_{-t} defines a stationary distribution on $\mathcal{D}([-t, \infty); S)$ for each $t \geq 0$. These distributions are mutually consistent due to stationarity. Thus they define a unique stationary probability measure on $\mathcal{D}(\mathbb{R}; S)$. Conversely, a stationary probability measure on $\mathcal{D}(\mathbb{R}; S)$ uniquely specifies a stationary probability measure on $\mathcal{D}([0, \infty); S)$ through its image under the map that restricts functions in $\mathcal{D}(\mathbb{R}; S)$ to $\mathcal{D}([0, \infty); S)$. Thus there exists a one-to-one correspondence between the two. Also note that extremality, ergodicity and other relevant properties needed here translate accordingly.

Given a transition kernel P on S of a Markov process, we extend it to $\bar{S}$ by defining $P(t, x, \mathrm{d}y) = \delta_x(\mathrm{d}y)$ for $x \in \bar{S} \setminus S$. Then Theorem 1.5.7 yields the following lemma.

Lemma 1.5.16 *The law of a stationary Markov process with transition kernel P is the barycenter of a probability measure on $\mathscr{I}_e$, the set of extreme points of $\mathscr{I}$.*

We refer to $\mathscr{I}_e$ as the set of *extremal stationary laws*.

Theorem 1.5.17 *Every extremal stationary law is ergodic.*

Proof Suppose not. Then for some $\xi \in \mathscr{I}_e$, there exist mutually singular ξ^1 and ξ^2 in $\mathcal{M}_s^0$ and $a \in (0,1)$ such that $\xi = a\xi^1 + (1-a)\xi^2$. Let μ, μ^1 and μ^2 in $\mathscr{P}(S)$ denote the marginals of ξ, ξ^1 and ξ^2, respectively, on S. Then $\mu = a\mu^1 + (1-a)\mu^2$. Let A be a measurable subset of $\mathcal{D}(\mathbb{R};S)$ such that $\xi^1(A) = 1 = \xi^2(A^c)$, which is possible because $\xi^1 \perp \xi^2$. Clearly $\xi^1, \xi^2 \ll \xi$ and

$$\Lambda^1 := \frac{\mathrm{d}\xi^1}{\mathrm{d}\xi} = \frac{\mathbb{I}_A}{a}, \qquad \Lambda^2 := \frac{\mathrm{d}\xi^2}{\mathrm{d}\xi} = \frac{\mathbb{I}_{A^c}}{1-a},$$

where $\mathbb{I}_A$ denotes the indicator function of the set A. Let $\{\mathfrak{F}_t\}$ denote the canonical filtration of Borel subsets of $\mathcal{D}((-\infty,t];S)$, completed with respect to ξ. Let ξ_t, ξ_t^1 and ξ_t^2 denote the restrictions of ξ, ξ^1 and ξ^2, respectively, to $(\mathcal{D}(\mathbb{R};S), \mathfrak{F}_t)$, $t \in \mathbb{R}$. Then

$$\Lambda_t^i := \frac{\mathrm{d}\xi_t^i}{\mathrm{d}\xi_t} = \mathbb{E}[\Lambda^i \mid \mathfrak{F}_t] \quad \xi\text{-a.s.}, \quad i = 1,2,$$

where the expectation is with respect to ξ. By the martingale convergence theorem, $\Lambda_t^i \to \Lambda^i$, a.s., $i = 1,2$. Since Λ_t^i, $t \in \mathbb{R}$, is stationary, we have

$$\Lambda_t^1 = \frac{\mathbb{I}_A}{a}, \qquad \Lambda_t^2 = \frac{\mathbb{I}_{A^c}}{1-a}, \qquad t \in \mathbb{R}.$$

In particular, on A,

$$\frac{\Lambda_{t+s}^1}{\Lambda_t^1} = 1 \quad \xi\text{-a.s.}, \qquad \forall t, s \in \mathbb{R}. \tag{1.5.10}$$

Let $\nu_{t,s} = \nu_{t,s}(\cdot \mid \omega)$ and $\nu_{t,s}^i = \nu_{t,s}^i(\cdot \mid \omega)$ denote the regular conditional law of $\{X_{t+r} : 0 \le r \le s\}$ given $\mathfrak{F}_t$, under ξ and ξ^i, respectively, $i = 1,2$. Write ξ_{t+s} and ξ_{t+s}^i as

$$\xi_{t+s}(\mathrm{d}\omega', \mathrm{d}\omega'') = \xi_t(\mathrm{d}\omega')\,\nu_{t,s}(\mathrm{d}\omega'' \mid \omega'),$$
$$\xi_{t+s}^i(\mathrm{d}\omega', \mathrm{d}\omega'') = \xi_t^i(\mathrm{d}\omega')\,\nu_{t,s}^i(\mathrm{d}\omega'' \mid \omega'), \quad i = 1,2,$$

for $\omega' \in \mathcal{D}((-\infty,t];S)$ and $\omega'' \in \mathcal{D}((t,t+s];S)$. Since $\xi_t^i \ll \xi_t$ for all $t \in \mathbb{R}$, we have $\nu_{t,s}^i \ll \nu_{t,s}$, ξ_t^i-a.s. Therefore

$$\frac{\mathrm{d}\xi_{t+s}^i}{\mathrm{d}\xi_{t+s}}(\omega', \omega'') = \frac{\mathrm{d}\xi_t^i}{\mathrm{d}\xi_t}(\omega')\frac{\mathrm{d}\nu_{t,s}^i}{\mathrm{d}\nu_{t,s}}(\omega', \omega'') \quad \xi_t^i\text{-a.s.}, \quad i = 1,2,$$

i.e.,

$$\Lambda_{t+s}^i(\omega', \omega'') = \Lambda_t^i(\omega')\frac{\mathrm{d}\nu_{t,s}^i}{\mathrm{d}\nu_{t,s}}(\omega', \omega'') \quad \xi_t^i\text{-a.s.}, \quad i = 1,2.$$

By (1.5.10), $\Lambda_{t+s}^i = \Lambda_t^i$, ξ^i-a.s., and hence $\frac{\mathrm{d}\nu_{t,s}^i}{\mathrm{d}\nu_{t,s}} = 1$, ξ_t^i-a.s. for $i = 1,2$. Therefore, $\nu_{t,s}^i = \nu_{t,s}$, ξ_t^i-a.s. for $i = 1,2$, and it follows that the regular conditional law of X_{t+s} given X_t is $P(s, X_t, \mathrm{d}y)$, ξ^i-a.s., $i = 1,2$. Thus $\xi^i \in \mathscr{I}$, $i = 1,2$. Since $\xi^1 \neq \xi^2$ this contradicts the fact that ξ is an extreme point. Thus ξ must be ergodic. □

The proof of Theorem 1.5.17 depends crucially on working with $\mathcal{D}(\mathbb{R}; S)$ instead of $\mathcal{D}([0,\infty); S)$. From now on, however, we revert to $\mathcal{D}([0,\infty); S)$, keeping in mind that the foregoing continues to hold in view of the correspondence between stationary measures on the two spaces, as noted earlier.

Let $\mathfrak{I}$ denote the sub-σ-field of $\mathscr{B}(\mathcal{D}([0,\infty); S))$ consisting of those sets that are invariant under the shift operator θ_t, $t \in [0,\infty)$, completed with respect to $\xi \in \mathscr{I}$, i.e.,

$$\mathfrak{I} := \left\{A \in \mathscr{B}(\mathcal{D}([0,\infty); S)) : \xi(A \,\triangle\, \theta_t^{-1}(A)) = 0\,,\ \forall t \in [0,\infty)\right\}.$$

The following is immediate from Theorem 1.5.9.

Theorem 1.5.18 *For $\xi \in \mathscr{I}$ and $f \in L^1(\xi)$,*

$$\frac{1}{T}\int_0^T f(X \circ \theta_t)\,\mathrm{d}t \xrightarrow[T\to\infty]{a.s.} \mathbb{E}_\xi\left[f(X) \mid \mathfrak{I}\right],$$

and for $\xi \in \mathscr{I}_e$ and $f \in L^1(\xi)$,

$$\frac{1}{T}\int_0^T f(X \circ \theta_t)\,\mathrm{d}t \xrightarrow[T\to\infty]{a.s.} \int_{\mathcal{D}(\mathbb{R};S)} f\,\mathrm{d}\xi\,.$$

Moreover, if $f \in L^p(\xi)$, $p \geq 1$, then convergence in $L^p(\xi)$ also holds.

Remark 1.5.19 We are primarily interested in functions $f(X) = g(X_0)$ for some $g \in L^1(\mu)$, $\mu \in \mathscr{H}$. Thus $f(X \circ \theta_t) = g(X_t)$ and $\int f\,\mathrm{d}\xi = \int g\,\mathrm{d}\mu$.

We now state an important characterization of $\mathfrak{I}$.

Lemma 1.5.20 *Define the tail σ-field $\mathfrak{F}^\infty$ as the completion with respect to ξ of $\bigcap_{t\geq 0}\sigma(X_s : s \geq t)$. Then $\mathfrak{I} \subset \mathfrak{F}^\infty$.*

Proof Let $\mathfrak{F}_t^\infty$ denote the ξ-completion of $\sigma(X_s : s \geq t)$. If $A \in \mathfrak{I} \subset \mathfrak{F}_0^\infty$, then θ_t-invariance implies $A \in \mathfrak{F}_t^\infty$ for all t, and therefore $A \in \mathfrak{F}^\infty$. □

Lemma 1.5.21 *Let Y be a $\mathfrak{I}$-measurable, bounded random variable. Then*

(a) *$f(x) = \mathbb{E}_x[Y] := \mathbb{E}[Y \mid X_0 = x]$ satisfies $T_t f = f$, μ-a.s.;*

(b) *$M_t := f(X_t)$ is an $(\mathfrak{F}_t^X)$-martingale on $(\Omega, \mathfrak{F}, \xi)$;*

(c) *$Y = \lim_{t\to\infty} M_t$, ξ-a.s., and in $L^1(\mu)$.*

Proof Since Y is $\mathfrak{I}$-measurable, $\xi(Y \neq Y \circ \theta_t) = 0$ for all $t \geq 0$. Thus

$$\begin{aligned}
0 &= \mathbb{E}_\mu |Y - Y \circ \theta_t| \\
&= \int_S \mathbb{E}_x |Y - Y \circ \theta_t| \mu(\mathrm{d}x) \\
&\geq \int_S \left|\mathbb{E}_x [Y] - \mathbb{E}_x [Y \circ \theta_t]\right| \mu(\mathrm{d}x) \\
&= \int_S \left|\mathbb{E}_x [Y] - \mathbb{E}_x \left[\mathbb{E}\left[Y \circ \theta_t \mid \mathfrak{F}_t^X\right]\right]\right| \mu(\mathrm{d}x) \\
&= \int_S \left|\mathbb{E}_x [Y] - \mathbb{E}_x [\mathbb{E}_{X_t} [Y]]\right| \mu(\mathrm{d}x) \\
&= \int_S \left|f(x) - T_t f(x)\right| \mu(\mathrm{d}x),
\end{aligned} \tag{1.5.11}$$

which establishes (a). The fourth equality in (1.5.11) follows by the Markov property and the fact that Y is $\mathfrak{I}$-measurable and therefore $\mathfrak{F}$-measurable. It also follows that, μ-a.s., $f(X_t) = \mathbb{E}_x \left[Y \mid \mathfrak{F}_t^X\right]$. Thus $(M_t, \mathfrak{F}_t^X)$ is a regular martingale. Part (c) then follows by the convergence theorem for regular martingales [32, theorem 3.3.2, p. 50]. □

Definition 1.5.22 A set $A \in \mathscr{B}(S)$ is said to be μ-invariant under the Markov family if $P(t, x, A) = 1$ for μ-a.s $x \in A$, and $t \geq 0$. Let $\mathfrak{I}_\mu$ denote the set of all μ-invariant sets in $\mathscr{B}(S)$.

Lemma 1.5.23 *The class of μ-invariant sets $\mathfrak{I}_\mu$ is a σ-field. If A is a μ-invariant set, then $\mathbb{I}_A(X_0)$ is $\mathfrak{I}_\mu$-measurable. Moreover, if $C \in \mathfrak{I}$, then there exists $B \in \mathfrak{I}_\mu$ such that $\xi(\mathbb{I}_C \neq \mathbb{I}_B(X_0)) = 0$.*

Proof Let $A \in \mathscr{B}(S)$ be μ-invariant. Then

$$\begin{aligned}
\mu(A) &= \int_S \mu(\mathrm{d}x) P(t, x, A) \\
&= \int_A \mu(\mathrm{d}x) P(t, x, A) + \int_{A^c} \mu(\mathrm{d}x) P(t, x, A) \\
&= \mu(A) + \int_{A^c} \mu(\mathrm{d}x) P(t, x, A),
\end{aligned}$$

which shows that $P(t, x, A^c) = 1$, μ-a.e. on A^c. Thus $\mathfrak{I}_\mu$ is closed under complementation. That it is closed under countable unions is easy to prove. It clearly includes the empty set and the whole space. It follows that $\mathfrak{I}_\mu$ is a σ-field. Also

$$\xi(\mathbb{I}_A(X_0) \neq \mathbb{I}_A(X_t)) = \int_A \mu(\mathrm{d}x) P(t, x, A^c) + \int_{A^c} \mu(\mathrm{d}x) P(t, x, A) = 0,$$

which implies that $\mathbb{I}_A(X_0) \circ \theta_t = \mathbb{I}_A(X_t) = \mathbb{I}_A(X_0)$, ξ-a.s. It follows that $\mathbb{I}_A(X_0)$ is $\mathfrak{I}_\mu$-measurable. Next, let $C \in \mathfrak{I}$ and set $f(x) = \mathbb{P}_x(C)$. Then by Lemma 1.5.21

$$\mathbb{I}_C = \lim_{t \to \infty} f(X_t) \quad \xi\text{-a.s.}$$

Choose $0 < a < b < 1$, and define

$$A_b^a = \{x \in S : a \leq f(x) \leq b\}.$$

Then, since $\mathbb{I}_{A_b^a}(X_t) \to 0$, ξ-a.s. as $t \to \infty$, and X is stationary, it follows that $\mathbb{E}_\mu \left[\mathbb{I}_{A_b^a}(X_t)\right] = 0$, which in turn implies that $P(t, x, A_b^a) = 0$, μ-a.s. Therefore

$$\mu(A_b^a) = \int_S \mu(\mathrm{d}x) P(t, x, A_b^a) = 0,$$

and since a and b are arbitrary, we have

$$\mu(\{x \in S : 0 < f(x) < 1\}) = 0.$$

This shows that μ is supported on the union of the sets

$$B = \{x \in S : f(x) = 1\} \quad \text{and} \quad B' = \{x \in S : f(x) = 0\}.$$

Since $f(x) = \mathbb{I}_B(x)$ μ-a.s., invoking once more Lemma 1.5.21, we obtain

$$f(x) = \mathbb{E}_x [f(X_t)] = \mathbb{E}_x [\mathbb{I}_B(X_t)] = P(t, x, B)$$

for μ-almost all x. Thus B is a μ-invariant set and the result follows. □

Corollary 1.5.24 *The probability measure ξ is trivial on $\mathfrak{I}$ if and only if the invariant probability measure μ is trivial on $\mathfrak{I}_\mu$.*

This shows that the ergodic decomposition for a Markov process translates into a decomposition of its state space. The latter is known as the Doeblin decomposition.

1.6 Diffusion processes

The study of diffusion processes was motivated among other things by the diffusive behavior of particles in a fluid in physics. (Historically, an early motivation came from finance – see Bachelier [10].) The most famous example is of course the observation by Brown of the random motion of pollen in a liquid, which led to the discovery of the most famous diffusion of them all, the Brownian motion.

Physically, a diffusion process is an $\mathbb{R}^d$-valued process X defined by the requirements: for $\varepsilon > 0$, $t \geq 0$ and $h > 0$,

$$\begin{aligned} \mathbb{E}\left[(X_{t+h} - X_t)\,\mathbb{I}\{|X_{t+h} - X_t| \leq \varepsilon\} \;\middle|\; \mathfrak{F}_t^X\right] &= m(X_t)h + o(h), \\ \mathbb{E}\left[(X_{t+h} - X_t)(X_{t+h} - X_t)^{\mathsf{T}}\,\mathbb{I}\{|X_{t+h} - X_t| \leq \varepsilon\} \;\middle|\; \mathfrak{F}_t^X\right] &= a(X_t)h + o(h). \end{aligned} \tag{1.6.1}$$

The functions $m\colon \mathbb{R}^d \mapsto \mathbb{R}^d$ and $a\colon \mathbb{R}^d \mapsto \mathbb{R}^{d\times d}$ are called the drift vector and the diffusion matrix, respectively. The latter is seen to be nonnegative definite by virtue of being a conditional covariance. Physically, the drift gives the local (conditional) mean velocity of the particle and the diffusion matrix the local (conditional) covariance of the fluctuations. Assuming (1.6.1) and the additional technical conditions, that for $\varepsilon > 0$, $t \geq 0$ and $h > 0$, we have

$$\mathbb{P}\left(|X_{t+h} - X_t| > \varepsilon \mid \mathfrak{F}_t^X\right) = o(h),$$
$$\mathbb{E}\left[|X_{t+h} - X_t|^3 \, \mathbb{I}\{|X_{t+h} - X_t| \leq \varepsilon\} \mid \mathfrak{F}_t^X\right] = o(h),$$

one can formally derive the partial differential equations satisfied by the transition probability density $p(t, x, y)$ (i.e., the density w.r.t. the Lebesgue measure of the transition probability kernel) of this process. These are the Kolmogorov forward and backward equations:

$$\frac{\partial p}{\partial t} - \mathcal{L}^* p = 0, \quad t \in [0, T], \qquad \lim_{t \downarrow 0} p(t, x, y) = \delta_x(y),$$
$$\frac{\partial p}{\partial t} + \mathcal{L} p = 0, \quad t \in [0, T], \qquad \lim_{t \uparrow T} p(t, x, y) = \delta_y(x),$$

where $\mathcal{L}$ is the second order elliptic operator

$$\mathcal{L}f := \langle m, \nabla f \rangle + \frac{1}{2} \operatorname{tr}(a \nabla^2 f), \tag{1.6.2}$$

and $\mathcal{L}^*$ is its formal adjoint. The forward equation is also known as the Fokker–Planck equation. See Wong and Hajek [122, pp. 169–173] for a derivation.

This derivation is, of course, purely formal and one requires to fall back upon the theory of parabolic partial differential equations to study the well-posedness of these equations. For example, if m and a are bounded Lipschitz and the least eigenvalue of a is bounded away from zero, then a unique solution exists in the class of functions once continuously differentiable in t and twice continuously differentiable in (x, y), with growth in the latter variables slower than $e^{\alpha(|x|^2 + |y|^2)}$ for any $\alpha > 0$ [83]. See Bogachev *et al.* [26] for regularity of transition probabilities under weaker assumptions on the coefficients.

While this was a major development in its time, it put diffusion theory in the straightjacket of the available theory of parabolic PDEs, which in particular meant restrictions on m and a along the above lines. The semigroup theory developed by Feller and Dynkin and described in the preceding section was a major extension, which allowed for more general situations. The operator $\mathcal{L}$ above appeared as the infinitesimal generator of the associated Markov semigroup when $C^2(\mathbb{R}^d)$ was included in the domain of the generator. The latter, however, was not always true, e.g., in the important case of merely measurable m, a common situation in stochastic control as we shall see later. The crowning achievement in Markov process theory was the Stroock–Varadhan martingale formulation we described in the last section, which broadened the scope even further. Here $\mathcal{L}$ appeared as the extended generator.

The modern definition of a diffusion is as a Feller process with continuous sample paths. From this, it can be shown that its extended generator is of the above form

(1.6.2). This follows from a theorem due to Dynkin which we quote here without proof [102, pp. 258–259].

Theorem 1.6.1 *Let $\{T_t: t \geq 0\}$ be a strongly continuous Markov semigroup of linear operators $T_t: C_0(\mathbb{R}^d) \to C_0(\mathbb{R}^d)$. Suppose that the domain of its generator $\mathcal{A}$ contains $C_c^\infty(\mathbb{R}^d)$, the space of smooth functions of compact support. Then the restriction $\mathcal{L}$ of $\mathcal{A}$ to $C_c^\infty(\mathbb{R}^d)$ takes the form* (1.6.2). *Moreover, the functions $m: \mathbb{R}^d \to \mathbb{R}^d$ and $a: \mathbb{R}^d \to \mathbb{R}^{d\times d}$ are continuous and the matrix $\{a_{ij}(x)\}$ is nonnegative definite symmetric for each $x \in \mathbb{R}^d$.*

A parallel development, of great importance to this book, is Itô's formulation of diffusions as solutions of stochastic differential equations. (The scalar case was earlier handled by Doeblin, a fact that remained unknown for many years due to some unfortunate historical circumstances. This is explained in the historical summary by M. Davis and A. Etheridge [10].) This viewpoint begins with the stochastic differential equation

$$\mathrm{d}X_t = m(X_t)\,\mathrm{d}t + \sigma(X_t)\,\mathrm{d}W_t\,, \quad t \geq 0\,,$$

where σ is any nonnegative definite square-root of a and W is a standard Brownian motion in $\mathbb{R}^d$. This is to be interpreted as the *integral* equation

$$X_t = X_0 + \int_0^t m(X_s)\,\mathrm{d}s + \int_0^t \sigma(X_s)\,\mathrm{d}W_s\,, \quad t \geq 0\,.$$

The second term on the right is the *Itô stochastic integral*, which Itô introduced for this purpose. We study the solution concepts for such equations and the associated well-posedness issues in Chapter 2. We close here with the comments that while this approach departs from the purely distributional specifications (such as the semigroup or martingale approaches) in so far as it hypothesizes a "system driven by noise" on some probability space in the background, it leaves at our disposal the powerful machinery of *Itô calculus* with which one can work wonders. More of that later.

1.7 Bibliographical note

Section 1.1. For a brief account of the general theory of random processes see [32, chapter 6]. For more extensive accounts, see [47] and [66].

Sections 1.2–1.6. For the general theory of Markov processes, [45], [55], and [102, 103] are comprehensive texts, and the latter two have extensive accounts of the semigroup and martingale approaches sketched in Sections 1.3 and 1.4. An early source for the ergodic theory of Markov processes is [125]. For diffusions, the classic text by Stroock and Varadhan [115] is a must. An excellent treatment of the topic also appears in [79].

2

Controlled Diffusions

2.1 Introduction

With this chapter, we begin our exposition of ergodic control of diffusion processes, which are continuous time, continuous path Markov processes. We have to assume a certain amount of mathematical sophistication on the part of the reader. We shall limit it to a familiarity with Brownian motion, stochastic calculus and (uncontrolled) diffusions at the level of Karatzas and Shreve [76]. The few occasions we might reach beyond this, we shall provide pointers to the relevant literature.

With this introduction, we start in this chapter with basic concepts pertaining to controlled diffusions and their immediate consequences.

2.2 Solution concepts

The prototypical controlled diffusion model we consider is the $\mathbb{R}^d$-valued process $X = \{X_t : t \geq 0\}$ described by the stochastic differential equation

$$\mathrm{d}X_t = b(X_t, U_t)\,\mathrm{d}t + \sigma(X_t)\,\mathrm{d}W_t\,. \tag{2.2.1}$$

All random processes in (2.2.1) live in a complete probability space $(\Omega, \mathfrak{F}, \mathbb{P})$, and satisfy:

(i) W is a d-dimensional standard Wiener process.

(ii) The initial condition X_0 is an $\mathbb{R}^d$-valued random variable with a prescribed law ν_0, and is independent of W.

(iii) The process U takes values in a compact, metrizable set $\mathbb{U}$, and $U_t(\omega)$, $t \geq 0$, $\omega \in \Omega$ is jointly measurable in (t, ω) (in particular it has measurable sample paths). In addition, it is *non-anticipative*: For $s < t$, $W_t - W_s$ is independent of

$$\mathfrak{F}_s := \text{the completion of } \sigma(X_0, U_r, W_r : r \leq s) \text{ relative to } (\mathfrak{F}, \mathbb{P})\,. \tag{2.2.2}$$

(iv) The functions $b = [b^1, \dots, b^d]^{\mathsf{T}} : \mathbb{R}^d \times \mathbb{U} \mapsto \mathbb{R}^d$ and $\sigma = [\sigma^{ij}] : \mathbb{R}^d \mapsto \mathbb{R}^{d\times d}$ are locally Lipschitz in x, with a Lipschitz constant $K_R > 0$ depending on $R > 0$, i.e., for all $x, y \in \mathbb{R}^d$ such that $\max\{|x|, |y|\} \le R$, and $u \in \mathbb{U}$,

$$|b(x,u) - b(y,u)| + \|\sigma(x) - \sigma(y)\| \le K_R |x - y|, \tag{2.2.3}$$

where $\|\sigma\|^2 := \operatorname{tr}\left(\sigma\sigma^{\mathsf{T}}\right)$. Also, b is continuous in (x, u). Moreover, b and σ satisfy a global *linear growth condition* of the form

$$|b(x,u)|^2 + \|\sigma(x)\|^2 \le K_1(1 + |x|^2) \qquad \forall (x,u) \in \mathbb{R}^d \times \mathbb{U}. \tag{2.2.4}$$

Remark 2.2.1 Let $(\Omega, \mathfrak{F}, \mathbb{P})$ be a complete probability space, and let $\{\mathfrak{F}_t\}$ be a filtration on $(\Omega, \mathfrak{F})$ such that each $\mathfrak{F}_t$ is complete relative to $\mathfrak{F}$. Recall that a d-dimensional Wiener process $(W_t, \mathfrak{F}_t)$, or $(\mathfrak{F}_t)$-Wiener process, is an $\mathfrak{F}_t$-adapted Wiener process such that $W_t - W_s$ and $\mathfrak{F}_s$ are independent for all $t > s \ge 0$. An equivalent definition of the model for (2.2.1) starts with a d-dimensional Wiener process $(W_t, \mathfrak{F}_t)$ and requires that the control process U be $\mathfrak{F}_t$-adapted. Note then that U is necessarily non-anticipative.

A process U satisfying (iii) is called an *admissible control* and represents the largest class of controls we shall admit. This class is denoted by $\mathfrak{U}$. Intuitively, these are the controls that do not use any information about future increments of the driving "noise" W, a physically reasonable assumption.

A few remarks are in order here. To start with, the above assumptions represent a comfortable level of generality, not the greatest possible generality. We have chosen to do so in order to contain the technical overheads not central to the issues we are interested in. Secondly, the control process U enters the *drift vector b*, but not the *diffusion matrix* σ. We comment on this restriction later in this chapter. Finally, it may seem restrictive to impose the condition that σ is a square matrix. Nevertheless, there's no loss of generality in assuming that it is, as we argue later.

In integral form, (2.2.1) is written as

$$X_t = X_0 + \int_0^t b(X_s, U_s)\,\mathrm{d}s + \int_0^t \sigma(X_s)\,\mathrm{d}W_s\,. \tag{2.2.5}$$

We say that a process $X = \{X_t(\omega)\}$ is a solution of (2.2.1), if it is $\mathfrak{F}_t$-adapted, continuous in t, defined for all $\omega \in \Omega$ and $t \in [0, \infty)$ and satisfies (2.2.5) for all $t \in [0, \infty)$ at once a.s.

The second term on the right-hand side of (2.2.5) is an Itô stochastic integral. We recall here its basic properties. In general, if σ_t is an $\mathfrak{F}_t$-adapted process taking values in $\mathbb{R}^{m\times d}$, and

$$\int_0^t \|\sigma_s\|^2\,\mathrm{d}s < \infty \qquad \forall t \ge 0, \tag{2.2.6}$$

then

$$\xi_t := \int_0^t \sigma_s\,\mathrm{d}W_s \tag{2.2.7}$$

is a local martingale relative to $\{\mathfrak{F}_t\}$. If

$$\mathbb{E}\left[\int_0^t \|\sigma_s\|^2 \, \mathrm{d}s\right] < \infty \qquad \forall t \geq 0,$$

then $(\xi_t, \mathfrak{F}_t)$ is a square-integrable martingale with quadratic covariation matrix

$$\langle \xi, \xi \rangle_t = \int_0^t \sigma_s \sigma_s^{\mathsf{T}} \, \mathrm{d}s\,.$$

For $T > 0$ and an $\mathbb{R}^d$-valued process $\{Y_t : t \in [0, T]\}$ on $(\Omega, \mathfrak{F}, \mathbb{P})$ with continuous paths define

$$\|Y\|_{\infty,T} := \sup_{0 \leq t \leq T} |Y_t|\,.$$

It is well known that if $(\mathcal{X}, \|\cdot\|_{\mathcal{X}})$ is a Banach space, then

$$\|Z\|_{L^p(\mathcal{X})} := \left(\mathbb{E}\,\|Z\|_{\mathcal{X}}^p\right)^{1/p}, \qquad 1 < p < \infty\,,$$

is a complete norm on the space of (equivalence classes) of $\mathcal{X}$-valued random variables Z defined on a complete probability space $(\Omega, \mathfrak{F}, \mathbb{P})$, that satisfy $\mathbb{E}\,\|Z\|_{\mathcal{X}}^p < \infty$. Let $\mathcal{H}_T^p$ denote the subspace of processes Y defined for $t \in [0, T]$, which have continuous sample paths, i.e., take values in $C([0, T]; \mathbb{R}^d)$, and satisfy $\mathbb{E}\,\|Y\|_{\infty,T}^p < \infty$. It follows that $\mathcal{H}_T^p$ is a Banach space under the norm

$$\|Y\|_{\mathcal{H}_T^p} := \left(\mathbb{E}\,\|Y\|_{\infty,T}^p\right)^{1/p}, \qquad 1 < p < \infty\,.$$

We also let $\mathcal{H}^p$ denote the set of $\mathbb{R}^d$-valued random variables whose law has a finite p^{th} moment.

Perhaps the most important inequality in martingale theory is Doob's inequality, which asserts that if $(M_t, \mathfrak{F}_t)$ is a right-continuous martingale, and

$$A(T, \varepsilon) := \left\{\sup_{0 \leq t \leq T} |M_t| \geq \varepsilon\right\}, \qquad \varepsilon > 0\,,$$

then

$$\mathbb{P}(A(T, \varepsilon)) \leq \varepsilon^{-1}\, \mathbb{E}\left[M_T \mathbb{I}_{A(T,\varepsilon)}\right]. \tag{2.2.8}$$

Provided M is nonnegative, (2.2.8) yields

$$\|M\|_{\mathcal{H}_T^p} \leq \frac{p}{p-1} \left(\mathbb{E}\,[M_T]^p\right)^{1/p}, \qquad p > 1\,. \tag{2.2.9}$$

For ξ as in (2.2.7), suppose (2.2.6) holds and define

$$\langle \xi \rangle_t = \int_0^t \|\sigma_s\|^2 \, \mathrm{d}s\,.$$

If τ is a $\mathfrak{F}_t$-Markov time, then

$$\mathbb{E}\,[\langle \xi \rangle_\tau] \leq \mathbb{E}\left[\sup_{s < \infty} |\xi_{s \wedge \tau}|^2\right] \leq 4\,\mathbb{E}\,[\langle \xi \rangle_\tau]\,. \tag{2.2.10}$$

Moreover, for any positive numbers ε and δ,

$$\mathbb{P}\left(\sup_{s<\infty} |\xi_{s\wedge\tau}|^2 \geq \varepsilon\right) \leq \frac{\delta}{\varepsilon} + \mathbb{P}\left(\langle\xi\rangle_\tau \geq \delta\right) .$$

Also, if $\mathbb{E}\left[\langle\xi\rangle_\tau\right] < \infty$, then $(\xi_{s\wedge\tau}, \mathfrak{F}_t)$ is a square-integrable martingale for $s \geq 0$, $\lim_{s\to\infty} \xi_{s\wedge\tau} = \xi_\tau$ a.s. on the set $\{\tau < \infty\}$, and $\mathbb{E}\left[\langle\xi\rangle_\tau\right] = \mathbb{E}\left[|\xi_\tau|^2\right]$. For a proof of these results see Krylov [79, pp. 104–105].

Moreover, the Burkholder–Davis–Gundy inequalities assert that for any $p \in (0, \infty)$ there exists a positive constant C_p such that for any Markov time τ

$$\frac{1}{C_p}\,\mathbb{E}\left[\sup_{s<\infty} |\xi_{s\wedge\tau}|^p\right] \leq \mathbb{E}\left[\langle\xi\rangle_\tau^{p/2}\right] \leq C_p\,\mathbb{E}\left[\sup_{s<\infty} |\xi_{s\wedge\tau}|^p\right] . \tag{2.2.11}$$

For $p = 1$, (2.2.11) holds with $C_p = 3$ [79, pp. 160–163].

Of fundamental importance in the study of functionals of X is Itô's formula in (2.2.13) below. For $f \in C^2(\mathbb{R}^d)$ define the operator $\mathcal{L}\colon C^2(\mathbb{R}^d) \mapsto C(\mathbb{R}^d \times \mathbb{U})$ by

$$\mathcal{L}f(x, u) = \sum_{i,j} a^{ij}(x)\frac{\partial^2 f}{\partial x_i \partial x_j}(x) + \sum_i b^i(x, u)\frac{\partial f}{\partial x_i}(x) , \tag{2.2.12}$$

where

$$a := \frac{1}{2}\sigma\,\sigma^{\mathsf{T}} .$$

Then for $f \in C^2(\mathbb{R}^d)$,

$$f(X_t) = f(X_0) + \int_0^t \mathcal{L}f(X_s, U_s)\,\mathrm{d}s + M_t \quad \text{a.s.}, \tag{2.2.13}$$

where

$$M_t := \int_0^t \langle\nabla f(X_s), \sigma(X_s)\,\mathrm{d}W_s\rangle \tag{2.2.14}$$

is a local martingale.

2.2.1 Existence and uniqueness of solutions

We distinguish between the *nondegenerate* case when the least eigenvalue of $\sigma\sigma^{\mathsf{T}}$ is bounded away from zero on every compact subset of $\mathbb{R}^d$, and the *degenerate* case when it is not. More precisely, if G is a bounded open set in $\mathbb{R}^d$, we say that the controlled diffusion in (2.2.1) is nondegenerate in G if $\mathcal{L}$ is uniformly elliptic in G, i.e., if it satisfies

$$\sum_{i,j=1}^{d} a^{ij}(x)\xi_i\xi_j \geq K|\xi|^2 \qquad \forall \xi \in \mathbb{R}^d ,\ \forall x \in G ,$$

for some constant $K > 0$. We also say that the controlled diffusion is nondegenerate (in $\mathbb{R}^d$) if it is nondegenerate on every open ball B_R. Hence if the controlled diffusion is nondegenerate, using without loss of generality the parameterization in (2.2.3), it may be assumed to satisfy

$$\sum_{i,j=1}^{d} a^{ij}(x)\xi_i\xi_j \geq K_R^{-1}|\xi|^2 \qquad \forall \xi \in \mathbb{R}^d\,,\ \forall x \in B_R\,,\ \forall R > 0\,. \tag{2.2.15}$$

In either case, if W, U and X_0 are defined on a complete probability space $(\Omega, \mathfrak{F}, \mathbb{P})$ satisfying (i)–(iv), an a.s. unique solution X to (2.2.1) is guaranteed by Theorem 2.2.4 below. Theorem 2.2.2 paves the way.

Theorem 2.2.2 *Let W, U and X_0, satisfying (i)–(iii), be defined on a complete probability space $(\Omega, \mathfrak{F}, \mathbb{P})$, and let X be a solution of* (2.2.1)*. Under condition* (2.2.4)*, if $\|X_0\|_{\mathcal{H}^{2m}} = \left(\mathbb{E}\,|X_0|^{2m}\right)^{1/2m} < \infty$ with $m \in \mathbb{N}$, then for all $t > 0$,*

$$\mathbb{E}\left[|X_t|^{2m}\right]^{1/2m} \leq N_{m,t}(X_0)\,, \tag{2.2.16a}$$

$$\left\|X - X_0\right\|_{\mathcal{H}_t^{2m}} \leq 4\,(C_{2m} + t^m)^{1/2m}\, N_{m,t}(X_0)\,\sqrt{K_1 t}\,, \tag{2.2.16b}$$

where C_{2m} is the constant in (2.2.11) *and $N_{m,t}$ is given by*

$$N_{m,t}(X_0) := \left(1 + \left\|X_0\right\|_{\mathcal{H}^{2m}}\right) \mathrm{e}^{4mK_1 t}\,.$$

Moreover, if $\tau_n := \inf\{t > 0 : |X_t| > n\}$, then $\tau_n \uparrow \infty$ $\mathbb{P}$-a.s. as $n \to \infty$.

Proof Note that $\{\tau_n\}$ is a localizing sequence for the local martingale M in (2.2.14). Consequently under (2.2.4), we obtain by Dynkin's formula

$$\begin{aligned}
\mathbb{E}\left[1 + 2\,|X_{t\wedge\tau_n}|^{2m}\right] &\leq 1 + 2\,\mathbb{E}\,|X_0|^{2m} + 2\,\mathbb{E}\Biggl[\int_0^{t\wedge\tau_n} \Bigl(2m\,\langle X_s, b(X_s, U_s)\rangle \\
&\qquad\qquad + m(2m-1)\|\sigma(X_s)\|^2\Bigr)\,|X_s|^{2m-2}\,\mathrm{d}s\Biggr] \\
&\leq 1 + 2\,\mathbb{E}\,|X_0|^{2m} + 8m^2 K_1\,\mathbb{E}\left[\int_0^{t\wedge\tau_n} \left(1 + |X_s|^2\right)|X_s|^{2m-2}\,\mathrm{d}s\right] \\
&\leq 1 + 2\,\mathbb{E}\,|X_0|^{2m} + 8m^2 K_1 \int_0^t \mathbb{E}\left[1 + 2\,|X_{s\wedge\tau_n}|^{2m}\right]\mathrm{d}s\,.
\end{aligned} \tag{2.2.17}$$

Using the Gronwall inequality, (2.2.17) yields

$$\mathbb{E}\left[1 + 2\,|X_{t\wedge\tau_n}|^{2m}\right] \leq \left(1 + 2\,\mathbb{E}\,|X_0|^{2m}\right)\mathrm{e}^{8m^2 K_1 t}\,. \tag{2.2.18}$$

Therefore (2.2.16a) follows by taking limits as $n \to \infty$ in (2.2.18).

Next, by (2.2.5), employing the Burkholder–Davis–Gundy and Hölder inequalities, together with (2.2.18), we obtain after a few calculations,

$$
\begin{aligned}
\mathbb{E}\left[\sup_{0\le s\le t} |X_s - X_0|^{2m}\right] &\le 2^{2m-1}\left(\mathbb{E}\left[\int_0^t |b(X_s,U_s)|\,\mathrm{d}s\right]^{2m} + C_{2m}\,\mathbb{E}\left[\int_0^t \|\sigma(X_s)\|^2\,\mathrm{d}s\right]^m\right)\\
&\le 2^{2m-1}t^{2m-1}\,\mathbb{E}\left[\int_0^t |b(X_s,U_s)|^{2m}\,\mathrm{d}s\right]\\
&\qquad + 2^{2m-1}C_{2m}t^{m-1}\,\mathbb{E}\left[\int_0^t \|\sigma(X_s)\|^{2m}\,\mathrm{d}s\right]\\
&\le 2^{2m-1}t^{m-1}\left(t^m + C_{2m}\right)\int_0^t \mathbb{E}\left[K_1^m\left(1+|X_s|^2\right)^m\right]\mathrm{d}s\\
&\le 2^{2m-1}t^{m-1}\left(t^m + C_{2m}\right)\int_0^t 2^{m-1}K_1^m\,\mathbb{E}\left[1+|X_s|^{2m}\right]\mathrm{d}s\\
&\le 4^{2m}K_1^m t^m\left(t^m + C_{2m}\right)\mathrm{e}^{8m^2K_1t}\left(1+\mathbb{E}\,|X_0|^{2m}\right),
\end{aligned}
$$

thus proving (2.2.16b). By (2.2.16b), for any $t > 0$ and $m \in \mathbb{N}$, there exists a constant $\tilde{C}_m(t, K_1)$ such that

$$
\|X\|_{\mathcal{H}_t^{2m}} \le \tilde{C}_m(t,K_1)\left(1 + \|X_0\|_{\mathcal{H}^{2m}}\right). \tag{2.2.19}
$$

Hence, for any $t > 0$,

$$
\begin{aligned}
\mathbb{P}\left(\tau_n \le t\right) &= \mathbb{P}\left(\sup_{s\le t}|X_s| \ge n\right)\\
&\le \frac{\tilde{C}_1(t,K_1)\left(1+\|X_0\|_{\mathcal{H}^2}\right)}{n} \xrightarrow[n\to\infty]{} 0,
\end{aligned} \tag{2.2.20}
$$

which shows that $\tau_n \uparrow \infty$ $\mathbb{P}$-a.s. □

Note that in proving (2.2.16a), we have used only the Itô formula and not (2.2.5). We summarize the implications of this as a corollary, which we use later on.

Corollary 2.2.3 *Let U be given on a complete probability space $(\Omega, \mathfrak{F}, \mathbb{P})$, and suppose b and σ satisfy* (2.2.4). *If*

$$
f(X_t) - \int_0^t \mathcal{L}f(X_s, U_s)\,\mathrm{d}s
$$

is a local martingale for all $f \in C^2(\mathbb{R}^d)$, with $X_0 = x_0 \in \mathbb{R}^d$ $\mathbb{P}$-a.s., then (2.2.16a) *holds.*

Theorem 2.2.4 *Let W, U and X_0, satisfying (i)–(iii), be defined on a complete probability space $(\Omega, \mathfrak{F}, \mathbb{P})$, and suppose* (2.2.3) *and* (2.2.4) *hold. Then, provided $\mathbb{E}\,|X_0|^2 < \infty$, there exists a pathwise unique solution to* (2.2.1) *in $(\Omega, \mathfrak{F}, \mathbb{P})$.*

Proof Suppose first that b and σ are bounded, and that they satisfy (2.2.3) with some constant K. Select $T > 0$ such that $K(T + 2\sqrt{T}) \le \frac{1}{2}$, and let $\mathcal{X}$ denote the subspace of $\mathcal{H}^2_T$ of processes X which are adapted to the filtration $\{\mathfrak{F}_t\}$ defined in (2.2.2). Define the map $G\colon \mathcal{X} \to \mathcal{X}$ by

$$G(X)(t) := X_0 + \int_0^t b(X_s, U_s)\,\mathrm{d}s + \int_0^t \sigma(X_s)\,\mathrm{d}W_s\,, \quad t \in [0, T]\,.$$

Then, for $X, X' \in \mathcal{X}$, using (2.2.3),

$$\left\| \int_0^{\cdot} (b(X_s, U_s) - b(X'_s, U_s))\,\mathrm{d}s \right\|_{\mathcal{H}^2_T} \le KT\|X - X'\|_{\mathcal{H}^2_T}\,. \tag{2.2.21}$$

Similarly, by (2.2.9) and the Cauchy–Schwarz inequality

$$\begin{aligned}\left\| \int_0^{\cdot} [\sigma(X_s) - \sigma(X'_s)]\,\mathrm{d}W_s \right\|_{\mathcal{H}^2_T} &\le 2\left(\int_0^T \mathbb{E}\left\|\sigma(X_s) - \sigma(X'_s)\right\|^2 \mathrm{d}s\right)^{1/2} \\ &\le 2K\sqrt{T}\,\|X - X'\|_{\mathcal{H}^2_T}\,.\end{aligned} \tag{2.2.22}$$

Combining (2.2.21) and (2.2.22), we obtain

$$\|G(X) - G(X')\|_{\mathcal{H}^2_T} \le \tfrac{1}{2}\,\|X - X'\|_{\mathcal{H}^2_T}\,.$$

Thus the map G is a contraction and since $\mathcal{X}$ is closed, it must have a unique fixed point in $\mathcal{X}$. Therefore, (2.2.1) has a unique solution in $\mathcal{X}$. On the other hand, given a solution X on $[0, T]$ of (2.2.1), with $\mathbb{E}\,|X_0|^2 < \infty$, i.e., $X_0 \in \mathcal{H}^2$, it follows by (2.2.16b) that $\|X\|_{\mathcal{H}^2_T} < \infty$. Thus $X \in \mathcal{X}$ and therefore coincides with the unique fixed point. Repeating this argument on $[T, 2T], [2T, 3T], \dots$, completes the proof.

For b and σ not bounded, let b^n and σ^n be continuous bounded functions, which agree with b and σ on B_n, the ball of radius n centered at the origin, and satisfy (2.2.3) and (2.2.4) for each $n \in \mathbb{N}$. Let X^n denote the unique solution of (2.2.1), obtained above, with parameters (b^n, σ^n) and satisfying $X^n_0 = X_0$. Without loss of generality suppose $X_0 = x \in \mathbb{R}^d$. Define

$$\tilde{\tau}_n := \inf\,\{t > 0 : |X^n_t| \ge n\}\,.$$

We claim that

$$\mathbb{P}\left(\sup_{t<\tilde{\tau}_n\wedge\tilde{\tau}_m} |X^n_t - X^m_t| = 0\right) = 1 \qquad \forall n, m \in \mathbb{N}\,.$$

Indeed, using (2.2.3), (2.2.5) and (2.2.10) to estimate

$$f(t) := \mathbb{E}\left[\sup_{s\le t\wedge\tilde{\tau}_n\wedge\tilde{\tau}_m} |X^n_s - X^m_s|^2\right],$$

we obtain, with $\bar{\tau} := \tilde{\tau}_n \wedge \tilde{\tau}_m$,

$$f(t) \le 2\,\mathbb{E}\left[\int_0^{t\wedge\bar{\tau}} \left|b(X_s^n, U_s) - b(X_s^m, U_s)\right| \mathrm{d}s\right]^2$$

$$+ 2\,\mathbb{E}\left[\sup_{s\le t}\left(\int_0^{s\wedge\bar{\tau}} (\sigma(X_{s'}^n) - \sigma(X_{s'}^m))\,\mathrm{d}W_{s'}\right)^2\right]$$

$$\le 2t\int_0^t \mathbb{E}\left|b(X_{s\wedge\bar{\tau}}^n, U_{s\wedge\bar{\tau}}) - b(X_{s\wedge\bar{\tau}}^m, U_{s\wedge\bar{\tau}})\right|^2 \mathrm{d}s$$

$$+ 8\int_0^t \mathbb{E}\left\|\sigma(X_{s\wedge\bar{\tau}}^n) - \sigma(X_{s\wedge\bar{\tau}}^m)\right\|^2 \mathrm{d}s$$

$$\le 2(4+t)K_{n\wedge m}^2 \int_0^t f(s)\,\mathrm{d}s\,. \tag{2.2.23}$$

By (2.2.23) and the Gronwall inequality, $f(t) = 0$. By Theorem 2.2.2, $\mathbb{E}\,|X_t^n|^2$ is bounded on $t \in [0, T]$ uniformly over $n \in \mathbb{N}$, and it follows as in (2.2.20) that $\tilde{\tau}_n \uparrow \infty$, $\mathbb{P}$-a.s. By Theorem 2.2.2, for $n' > n$,

$$\mathbb{P}\left(\sup_{n'>n}\ \sup_{0\le t\le T}\left|X_t^n - X_t^{n'}\right| > 0\right) \le \mathbb{P}(\tilde{\tau}_n > T) \xrightarrow[n\to\infty]{} 0\,.$$

It follows that X^n converges w.p. 1 uniformly on $[0, T]$ to some limit X. Taking limits in

$$X_t^n = X_0^n + \int_0^t b^n(X_s^n, U_s)\,\mathrm{d}s + \int_0^t \sigma^n(X_s^n)\,\mathrm{d}W_s\,, \qquad t \in [0, T]\,,$$

we verify that X satisfies (2.2.5) on $[0, T]$, for any $T > 0$.

Lastly, we show uniqueness. Let X and $\hat{X}$ be two solutions, and let τ_n and $\hat{\tau}_n$ denote their exit times from the ball B_n, respectively. Using (2.2.3) and the Gronwall inequality, we can show as before that

$$\mathbb{E}\left[\sup_{0\le t\le T} |X_t - \hat{X}_t|^2\,\mathbb{I}\{t \le T \wedge \tau_n \wedge \hat{\tau}_n\}\right] = 0\,.$$

Thus

$$\mathbb{P}\left(\sup_{0\le t\le T} |X_t - \hat{X}_t| > 0\right) \le \mathbb{P}(\tau_n < T) + \mathbb{P}(\hat{\tau}_n < T)\,,$$

and since the probabilities on the right-hand side tend to 0 as $n \to \infty$, we obtain

$$\mathbb{P}\left(\sup_{0\le t\le T} |X_t - \hat{X}_t| = 0\right) = 1 \qquad \forall T > 0\,,$$

from which pathwise uniqueness follows. □

It is evident that $\mathcal{H}_T^2$ is a natural solution space for (2.2.1). Therefore we are going to assume by default that the initial data X_0 lies in $\mathcal{H}^2$, i.e., $\mathbb{E}\,|X_0|^2 < \infty$.

Lemma 2.2.5 *Assume* (2.2.3)–(2.2.4). *Let* $\{X^n\}$ *be a sequence of solutions of* (2.2.1) *under the same control* $U \in \mathfrak{U}$ *with initial conditions* $\{X_0^n\}$ *such that* X_0^n *converges in*

$\mathcal{H}^2$ as $n \to \infty$ to some X_0. Then, if X is the unique solution of (2.2.1) *under U and initial condition X_0, we have*

$$\|X^n - X\|_{\mathcal{H}^2_T} \xrightarrow[n\to\infty]{} 0, \quad \forall T > 0.$$

Proof Fix $n \in \mathbb{N}$, and for $i \in \{n, \infty\}$, with $X^\infty \equiv X$, define

$$X^{i,R}_t := X^i_t \, \mathbb{I}\{|X^n_0| \vee |X^\infty_0| \le R^{3/8}\}, \quad t \ge 0,$$

$$\tau^i_R := \inf \left\{t > 0 : \left|X^{i,R}_t\right| > R\right\},$$

and $\hat{X}^{i,R} := X^i - X^{i,R}$. We write

$$\left\|X^n - X^\infty\right\|_{\mathcal{H}^2_T} \le \left\|X^{n,R} - X^{\infty,R}\right\|_{\mathcal{H}^2_T} + \left\|\hat{X}^{n,R}\right\|_{\mathcal{H}^2_T} + \left\|\hat{X}^{\infty,R}\right\|_{\mathcal{H}^2_T}. \tag{2.2.24}$$

Let

$$A^n_R := \left\{|X^n_0| > R\right\} \cup \left\{|X^\infty_0| > R\right\}. \tag{2.2.25}$$

Using (2.2.19) and conditioning, we obtain

$$\begin{aligned}
\left\|\hat{X}^{n,R}\right\|^2_{\mathcal{H}^2_T} &\le \mathbb{E}\left[\mathbb{I}\{|X^n_0| > R\}\,\mathbb{E}\left[\|X^n\|^2_{\infty,T} \mid X^n_0\right]\right] \\
&\le \mathbb{E}\left[\mathbb{I}_{A^n_R}\,\mathbb{E}\left[\|X^n\|^2_{\infty,T} \mid X^n_0\right]\right] \\
&\le 2\tilde{C}^2_1\,\mathbb{E}\left[(1 + |X^n_0|^2)\,\mathbb{I}_{A^n_R}\right] \\
&= 2\tilde{C}^2_1\left(\mathbb{P}(A^n_R) + \left\|X^n_0\,\mathbb{I}_{A^n_R}\right\|^2_{\mathcal{H}^2}\right) \xrightarrow[R\to\infty]{} 0.
\end{aligned} \tag{2.2.26}$$

The analogous estimate applies for $\left\|\hat{X}^{\infty,R}\right\|_{\mathcal{H}^2_T}$. Since $\{|X^n_0|^2\}$ are uniformly integrable, the convergence in (2.2.26) is uniform in $n \in \mathbb{N}$. Let

$$f(t) := \mathbb{E}\left[\sup_{s \le t\wedge\tau^n_R\wedge\tau^\infty_R} \left|X^{n,R}_s - X^{\infty,R}_s\right|^2\right].$$

As in (2.2.23), we obtain

$$f(t) \le 3\left\|X^{n,R}_0 - X^{\infty,R}_0\right\|^2_{\mathcal{H}^2} + 3(4 + t)K_R \int_0^t f(s)\,\mathrm{d}s. \tag{2.2.27}$$

A triangle inequality yields

$$\left\|X^{n,R} - X^{\infty,R}\right\|_{\mathcal{H}^2_T} \le \sqrt{f(T)} + \left(\sum_{i=n,\infty} \mathbb{E}\left[\sup_{\tau^i_R < s \le T} \left|X^{n,R}_s - X^{\infty,R}_s\right|^2\right]\right)^{1/2}. \tag{2.2.28}$$

Using the Cauchy–Schwartz and Markov inequalities, we obtain the estimate, with

$i, j \in \{n, \infty\}$,

$$\mathbb{E}\left[\sup_{\tau_R^i < s \le T} |X_s^j|^2\right] = \mathbb{E}\left[\sup_{s \le T} |X_s^j|^2 \,\mathbb{I}\left\{s > \tau_R^i\right\}\right]$$
$$\le \sqrt{\mathbb{P}\left(\sup_{s \le T} |X_s^i| > R\right) \mathbb{E}\left[\sup_{s \le T} |X_s^j|^4\right]}$$
$$\le \frac{1}{R^2} \|X^i\|^2_{\mathcal{H}_T^4} \|X^j\|^2_{\mathcal{H}_T^4} \,. \qquad (2.2.29)$$

For any $\mathbb{R}^d$-valued random variable Y with finite second moment it holds that

$$\mathbb{E}\left[|Y|^4 \,\mathbb{I}\left\{|Y| \le R^{3/8}\right\}\right] \le \mathbb{E}\left[|Y|^4 \,\mathbb{I}\left\{|Y| \le R^{1/4}\right\}\right] + R^{3/2}\, \mathbb{P}\left(|Y| > R^{1/4}\right)$$
$$\le R\left(1 + \|Y\|^2_{\mathcal{H}^2}\right).$$

By (2.2.19) and (2.2.29), there exists a constant $\check{C} = \check{C}(T, K_1)$, such that, for all $R > 1$,

$$\left(\sum_{i=n,\infty} \mathbb{E}\left[\sup_{\tau_R^i < s \le T} \left|X_s^{n,R} - X_s^{\infty,R}\right|^2\right]\right)^{1/2} \le \frac{\check{C}}{\sqrt{R}} \left(1 + \left\|X_0^n\right\|_{\mathcal{H}^2} + \left\|X_0^\infty\right\|_{\mathcal{H}^2}\right). \qquad (2.2.30)$$

Combining (2.2.24), (2.2.25), (2.2.28), (2.2.30), and the Gronwall estimate from (2.2.27), we obtain

$$\left\|X^n - X^\infty\right\|_{\mathcal{H}_T^2} \le 2\mathrm{e}^{(6+T)TK_R} \left\|X_0^n - X_0^\infty\right\|_{\mathcal{H}^2} + \frac{\check{C}}{\sqrt{R}} \left(1 + \left\|X_0^n\right\|_{\mathcal{H}^2} + \left\|X_0^\infty\right\|_{\mathcal{H}^2}\right)$$
$$+ 2\tilde{C}_1 \left(2\sqrt{\mathbb{P}(A_R^n)} + \left\|X_0^n \,\mathbb{I}_{A_R^n}\right\|_{\mathcal{H}^2} + \left\|X_0^\infty \,\mathbb{I}_{A_R^n}\right\|_{\mathcal{H}^2}\right). \qquad (2.2.31)$$

Given $\varepsilon > 0$, we first choose $R > 0$ such that the second and third term on the right-hand side of (2.2.31) are each less than $\varepsilon/3$, and then choose $N \in \mathbb{N}$, such that the first term is also less than $\varepsilon/3$ for all $n > N$. This proves the lemma. □

Remark 2.2.6 It is evident from the proof of Lemma 2.2.5 that the estimates depend only the local Lipschitz constants K_R and the compact subset of $\mathcal{H}^2$ that $\{X_0^n\}$ lie in and is otherwise independent of the control U or other parameters in (2.2.1). Therefore the following statement is true: for every compact set $H \subset \mathcal{H}^2$ and $T > 0$, there exists a function $G_T \colon \mathbb{R}_+ \to \mathbb{R}_+$ satisfying $G_T(z) \to 0$ as $z \downarrow 0$, such that for any two solutions X and X' of (2.2.1) under the same control and with corresponding initial conditions X_0 and X_0' in H, we have

$$\|X - X'\|_{\mathcal{H}_T^2} \le G_T\left(\|X_0 - X_0'\|_{\mathcal{H}^2}\right).$$

Note also that if K_R is a constant K independent of R, then by (2.2.27) we have the stronger statement that

$$\|X - X'\|_{\mathcal{H}_T^2} \le 2\mathrm{e}^{(6+T)TK} \left\|X_0 - X_0'\right\|_{\mathcal{H}^2}, \quad \forall X_0, X_0' \in \mathcal{H}^2.$$

Remark 2.2.7 The linear growth assumption (2.2.4) guarantees that trajectories do not suffer an explosion in finite time. This assumption is quite standard but may be restrictive for some applications. Existence of solutions to (2.2.1) can be established under the weaker assumptions

$$\langle x - y, b(x, u) - b(y, u)\rangle + \|\sigma(x) - \sigma(y)\| \leq K_{|x|\vee|y|}\,|x - y|$$

and

$$\langle x, b(x, u)\rangle + \|\sigma(x)\|^2 \leq K_1(1 + |x|^2) \tag{2.2.32}$$

for all $x, y \in \mathbb{R}^d$ and $u \in \mathbb{U}$ [79]. Note that (2.2.17) used in the derivation of (2.2.16a) continues to hold for $m = 1$ if (2.2.4) is replaced by (2.2.32) and so does Corollary 2.2.3.

For $U \in \mathfrak{U}$, we use the notation $\mathbb{P}_x^U$ when we need to indicate explicitly that the process X is governed by U and starts at $X_0 = x$. The associated expectation operator is $\mathbb{E}_x^U$.

2.2.2 Feedback controls

An admissible control U is called a *feedback control* if it is progressively measurable with respect to the natural filtration $\{\mathfrak{F}_t^X\}$ of X. This is tantamount to saying that there exists a measurable map

$$v\colon [0,\infty) \times C([0,\infty); \mathbb{R}^d) \to \mathbb{U}$$

such that for each $t \geq 0$, $U_t = v_t(X)$ a.s., and is measurable with respect to $\mathfrak{F}_t^X$. In view of the latter fact, we may write $v_t(X_{[0,t]})$ in place of $v_t(X)$ by abuse of notation. It is evident that U cannot be specified a priori as in Theorem 2.2.4. Instead, one has to make sense of (2.2.1) with U_t replaced by $v_t(X_{[0,t]})$. This is not always possible and even when it is, an enlargement of the solution concept might be necessary.

In Theorem 2.2.4, we prescribed X_0, W and U on a probability space and constructed a solution X on the same space. We call this the strong formulation. Correspondingly, the equation

$$X_t = x_0 + \int_0^t b(X_s, v_s(X_{[0,s]}))\,\mathrm{d}s + \int_0^t \sigma(X_s)\,\mathrm{d}W_s \tag{2.2.33}$$

is said to have a *strong solution* if given a Wiener process $(W_t, \mathfrak{F}_t)$ on a complete probability space $(\Omega, \mathfrak{F}, \mathbb{P})$, we can construct a process X on $(\Omega, \mathfrak{F}, \mathbb{P})$, with initial value $X_0 = x_0 \in \mathbb{R}^d$, which is continuous, $\mathfrak{F}_t$-adapted, and satisfies (2.2.33) for all t at once a.s. A strong solution is called *unique* if any two such solutions X and X' agree $\mathbb{P}$-a.s. when viewed as elements of $C([0,\infty); \mathbb{R}^d)$.

Let $\{\mathfrak{F}_t^W\}$ be the filtration generated by W. It is evident that if X_t is $\mathfrak{F}_t^W$-adapted, then such a solution X is a strong solution. We say that (2.2.33) has a *weak solution* if we can find processes X and W on some probability space $(\tilde{\Omega}, \tilde{\mathfrak{F}}, \tilde{\mathbb{P}})$ such that

X_0 has the prescribed law ν_0, W is a standard Wiener process independent of X_0, and (2.2.33) holds with $W_t - W_s$ independent of $\{X_{s'}, W_{s'} : s' \le s\}$ for all $s \le t$. The weak solution is *unique* if any two weak solutions X and X', possibly defined on different probability spaces, agree in law when viewed as $C([0,\infty);\mathbb{R}^d)$-valued random variables, i.e., $\mathscr{L}(X) = \mathscr{L}(X')$.

Intuitively, a strong solution corresponds to an engineer's viewpoint, wherein one has X_0, U which, along with the noise W, are fed to a black box represented by (2.2.1) as inputs, with X the resultant output. In contrast, the weak solution corresponds to a statistician's (to be precise, a time series analyst's) viewpoint according to which one has a pair X, U to which (2.2.33) is "fitted" to extract whatever structure there is to it, with W the "residual error."

Note that with $u \in \mathbb{U}$ treated as a parameter, (2.2.12) also gives rise to an operator $\mathcal{L}^u : C^2(\mathbb{R}^d) \to C(\mathbb{R}^d)$, defined by $\mathcal{L}^u f(x) = \mathcal{L} f(x,u)$.

Weak solutions can be given a martingale characterization. We work in the canonical space $\Omega = C([0,\infty);\mathbb{R}^d)$, with $X_t(\omega) = \omega(t)$.

Definition 2.2.8 Let the feedback control v_t be fixed. A probability measure $\mathbb{P}_{x_0}$ in $\mathscr{P}(C([0,\infty);\mathbb{R}^d))$ is a solution to the *martingale problem for $\mathcal{L}^{v_t}$ started at $x_0 \in \mathbb{R}^d$*, if $\mathbb{P}_{x_0}(X_0 = x_0) = 1$ and

$$M_t := f(X_t) - \int_0^t \mathcal{L}^{v_s(X_{[0,s]})} f(X_s)\,\mathrm{d}s\,, \quad t \ge 0\,, \tag{2.2.34}$$

is a local martingale under $\mathbb{P}_{x_0}$ for all $f \in C^2(\mathbb{R}^d)$, or equivalently, if it is a square-integrable martingale under $\mathbb{P}_{x_0}$ for all $f \in C^2_c(\mathbb{R}^d)$.

We note here that (2.2.34) is a martingale if and only if for all $g \in C_b(C([0,s];\mathbb{R}^d))$, $f \in C^2_c(\mathbb{R}^d)$ and $t \ge s$, it holds that

$$\mathbb{E}\left[\left(f(X_t) - f(X_s) - \int_s^t \mathcal{L}^{v_r(X_{[0,r]})} f(X_r)\,\mathrm{d}r\right) g(X_{[0,s]})\right] = 0\,. \tag{2.2.35}$$

This can be demonstrated via a standard monotone class argument.

Theorem 2.2.9 *Let condition* (2.2.4) *hold. Then there exists a weak solution to* (2.2.33) *if and only if there exists a solution to the martingale problem for $\mathcal{L}^{v_t}$ started at x_0. Moreover, a solution to* (2.2.33) *is unique, if and only if the solution of the martingale problem for $\mathcal{L}^{v_t}$ started at x_0 is unique.*

Proof Necessity follows by (2.2.13). To prove sufficiency, let $\mathbb{P}_{x_0}$ be a solution to the martingale problem. Since $\mathbb{E}_{x_0}|X_t|^2 < \infty$ by Corollary 2.2.3, then condition (2.2.4) implies that the process M_t defined in (2.2.34) corresponding to $f(x) = |x|^2$ is a martingale. Hence, with $f(x) = x_i$, it follows that the ith coordinate of M defined by

$$M_t := X_t - \int_0^t b(X_s, v_s(X_{[0,s]}))\,\mathrm{d}s\,,$$

is a square-integrable martingale. Next, let $f(x) = x_i x_j$. Computing via (2.2.34) it follows that

$$M_t^i M_t^j - \int_0^t a^{ij}(X_s)\,\mathrm{d}s$$

is a local martingale, and therefore a martingale since both M and X are square-integrable, and condition (2.2.4) is in effect. By the representation theorem of [121], it then follows that M is of the form

$$M_t = \int_0^t \sigma(X_s)\,\mathrm{d}W_s$$

for a suitable Wiener process W, defined on a possibly augmented probability space.

If $\mathbb{P}_{x_0}$ and $\mathbb{P}'_{x_0}$ are two solutions of the martingale problem, the laws of X under $\mathbb{P}_{x_0}$ and $\mathbb{P}'_{x_0}$ must agree, by the weak uniqueness hypothesis. Thus $\mathbb{P}_{x_0} = \mathbb{P}'_{x_0}$ as elements of $\mathscr{P}(C([0,\infty);\mathbb{R}^d))$. The converse also follows in a standard fashion. □

In view of Theorem 2.2.9, we call $\mathcal{L}^u$ the *controlled extended generator* for X.

Remark 2.2.10 Recall our earlier comment on the restriction that σ be square. The later half of the proof of Theorem 2.2.9 shows that as long as we work with weak solutions (as we indeed shall), σ could be replaced by the nonnegative definite square-root of $\sigma\sigma^{\mathsf{T}}$ and thus can be taken to be square without any loss of generality. Note that the Lipschitz assumption on σ is not in general equivalent to a Lipschitz property of a. However, if the eigenvalues of $\sigma(x)$ are bounded away from zero uniformly over $x \in \mathbb{R}^d$ and it is Lipschitz continuous, then $\sqrt{\sigma\sigma^{\mathsf{T}}}$ can be chosen to be Lipschitz [115, p. 131, theorem 5.2.2]. The same is true in the degenerate case under the assumption that σ is in $C^2(\mathbb{R}^d)$ and its second partials are uniformly bounded in $\mathbb{R}^d$ [115, p. 132, theorem 5.2.3].

As mentioned in Remark 2.2.10, we shall mostly work with weak solutions. One reason is of course that it is a more natural solution concept in most situations. Further, it permits us to consider more general cases of (2.2.33). This is because existence or uniqueness of a strong solution to (2.2.33) always implies that of a weak solution, but not vice versa. See, e.g., the celebrated counterexample due to Tsirelsen in Liptser and Shiryayev [87, pp. 150–151].

The reader may still feel some apprehension in restricting the set of controls from admissible to feedback in the passage from (2.2.5) to (2.2.33), but this also causes no loss of generality as we argue later (see Theorem 2.3.4 below).

Theorem 2.2.11 *Assume* (2.2.3), (2.2.4) *and* (2.2.15). *Then equation* (2.2.33) *has a unique weak solution under any feedback control.*

Proof We first consider the case where b and σ are bounded and the diffusion is uniformly nondegenerate, or in other words, when the constant K_R in (2.2.15) satisfies $\sup_{R>0} K_R < \infty$. By Theorem 2.2.4, for $\overline{W}$ is a d-dimensional standard Wiener process independent of X_0, the equation

$$X'_t = X_0 + \int_0^t \sigma(X'_s)\,\mathrm{d}\overline{W}_s \tag{2.2.36}$$

has a unique strong solution. Let $\Omega = C([0,\infty);\mathbb{R}^d)$ with $\mathfrak{F}$ denoting its Borel σ-field, completed with respect to $\mathbb{P} = \mathscr{L}(X')$. Then the *canonical process* X defined by $X_t(\omega) = \omega(t)$ for $\omega \in \Omega$, agrees in law with X'. Let $\{\mathfrak{F}^X_t\}$ denote the natural filtration of X and $\mathbb{P}_t$ the restriction of $\mathbb{P}$ to $(\Omega, \mathfrak{F}^X_t)$ for $t \geq 0$. Define a new measure $\bar{\mathbb{P}}$ on $(\Omega, \mathfrak{F})$ as follows: For $t \geq 0$, the restriction of $\bar{\mathbb{P}}$ to $(\Omega, \mathfrak{F}^X_t)$, denoted by $\bar{\mathbb{P}}_t$, is absolutely continuous with respect to $\mathbb{P}_t$ with Radon–Nikodym derivative

$$\Lambda_t := \frac{\mathrm{d}\bar{\mathbb{P}}_t}{\mathrm{d}\mathbb{P}_t} = \exp\left(M_t(X,\overline{W};b) - \frac{1}{2}\langle M\rangle_t(X;b)\right),$$

where

$$M_t(X,\overline{W};b) := \int_0^t \langle \sigma^{-1}(X_s)\,b(X_s, v_s(X_{[0,s]})), \mathrm{d}\overline{W}_s\rangle\,,$$

$$\langle M\rangle_t(X;b) := \int_0^t \left|\sigma^{-1}(X_s)\,b(X_s, v_s(X_{[0,s]}))\right|^2 \mathrm{d}s\,.$$

Since b and σ are bounded, one can apply Novikov's criterion [74, pp. 142–144], to conclude that $(\Lambda_t, \mathfrak{F}^X_t)$ is a positive martingale with $\mathbb{E}[\Lambda_t] = 1$ for all t. This ensures that the above procedure consistently defines $\{\bar{\mathbb{P}}_t\}$, and $\bar{\mathbb{P}}$. By the Cameron–Martin–Girsanov theorem [87, p. 225], it follows that under $\bar{\mathbb{P}}$,

$$W_t := \overline{W}_t - \int_0^t \sigma^{-1}(X_s)\,b(X_s, v_s(X_{[0,s]}))\,\mathrm{d}s\,, \quad t \geq 0\,,$$

is a standard Wiener process and therefore X satisfies (2.2.33) with W. This establishes the existence of a weak solution.

To prove uniqueness, let X be a solution to (2.2.33) on some probability space $(\tilde{\Omega}, \tilde{\mathfrak{F}}, \tilde{\mathbb{P}})$. Without any loss of generality, we can take $(\tilde{\Omega}, \tilde{\mathfrak{F}}, \tilde{\mathbb{P}})$ as before, with X the canonical process. With $\{\tilde{\mathfrak{F}}^X_t\}$ denoting the natural filtration of X, we introduce a new probability measure $\tilde{\mathbb{P}}$ on $(\tilde{\Omega}, \tilde{\mathfrak{F}})$ exactly as above, but with Λ replaced by

$$\tilde{\Lambda}_t := \frac{\mathrm{d}\tilde{\mathbb{P}}_t}{\mathrm{d}\mathbb{P}_t} = \exp\left(-M_t(X,W;b) - \frac{1}{2}\langle M\rangle_t(X;b)\right), \quad t \geq 0\,,$$

with $\tilde{\mathbb{P}}_t$ being the restriction of $\tilde{\mathbb{P}}$ to $(\tilde{\Omega}, \tilde{\mathfrak{F}}^X_t)$. Using the Cameron–Martin–Girsanov theorem as above, it follows that X satisfies (2.2.36) under $\tilde{\mathbb{P}}$, with $\overline{W}$ replaced by an appropriately defined $\tilde{\mathbb{P}}$-Wiener process $\widetilde{W}$. Thus, under $\tilde{\mathbb{P}}$, X agrees in law with X' of (2.2.36). In other words, the law of X under $\mathbb{P}$ and that of X' of (2.2.36) are interconvertible through a prescribed absolutely continuous change of measure. The uniqueness of $\mathscr{L}(X)$, i.e., of the weak solution to (2.2.33), now follows from the known uniqueness of the strong, therefore weak, solution to (2.2.36).

For the general case of unbounded b and σ with linear growth and σ satisfying (2.2.15), the verification of $\mathbb{E}[\Lambda_t] = 1$ and $\mathbb{E}[\tilde{\Lambda}_t] = 1$ takes additional effort. For $n \in \mathbb{N}$, let

$$b^n(x) := \begin{cases} b(x) & \text{if } |x| \le n\,, \\ 0 & \text{if } |x| > n\,. \end{cases}$$

For each $n \in \mathbb{N}$, $\sigma^{-1} b^n$ is a bounded function and hence

$$\bar{\mathbb{P}}^n(A) := \mathbb{E}\left[\mathbb{I}_A(X_{[0,t]}) \exp\left(M_t(X, \overline{W}; b^n) - \frac{1}{2}\langle M\rangle_t(X; b^n)\right)\right],$$

where A is a Borel subset of $C([0,t]; \mathbb{R}^d)$ for $t > 0$, is the probability measure associated with the unique weak solution of

$$X^n_t = x_0 + \int_0^t b^n(X^n_s, v_s(X^n_{[0,s]}))\, \mathrm{d}s + \int_0^t \sigma(X^n_s)\, \mathrm{d}W_s\,.$$

Let $\tilde{\tau}_n = \inf\{t > 0 : |X^n_t| > n\}$. Then $X^n_{t\wedge\tilde{\tau}_n}$ is a solution of

$$X^n_{t\wedge\tilde{\tau}_n} = x_0 + \int_0^{t\wedge\tilde{\tau}_n} b(X^n_s, v_s(X^n_{[0,s]}))\, \mathrm{d}s + \int_0^{t\wedge\tilde{\tau}_n} \sigma(X^n_s)\, \mathrm{d}W_s\,.$$

Let

$$G_{R,T} := \left\{\xi \in C([0,\infty); \mathbb{R}^d) : \sup_{0\le t\le T} |\xi(t)| \le R\right\}.$$

Then $\bar{\mathbb{P}}^n(A \cap G_{R,t}) = \bar{\mathbb{P}}^{n'}(A \cap G_{R,t})$ for all Borel A, $n \wedge n' > R$ and $t > 0$. Hence the limit $\bar{\mathbb{P}}^*(G_{R,t}) = \lim_{n\to\infty} \bar{\mathbb{P}}^n(G_{R,t})$ exists for all $R > 0$, and satisfies

$$\bar{\mathbb{P}}^*(G_{R,t}) := \mathbb{E}\left[\mathbb{I}_{G_{R,t}}(X_{[0,t]}) \exp\left(M_t(X, \overline{W}; b) - \frac{1}{2}\langle M\rangle_t(X; b)\right)\right].$$

By (2.2.16b), there exists a constant $\check{C}_0 = \check{C}_0(K_1, t)$ such that

$$\bar{\mathbb{E}}^n\left[\sup_{s\le t} |X_t|^2\right] \le \check{C}_0(K_1, t)\left(1 + \|X_0\|^2_{\mathcal{H}^2}\right), \quad \forall n \in \mathbb{N},$$

which implies that $\lim_{R\to\infty} \bar{\mathbb{P}}^n(G_{R,t}) = 1$, uniformly in $n \in \mathbb{N}$. Therefore

$$\lim_{R\to\infty} \liminf_{n\to\infty} \bar{\mathbb{P}}^n(G_{R,t}) = 1\,.$$

Thus $\mathbb{E}[\Lambda_t] \ge 1$, and since by the supermartingale property the converse inequality is true, equality must hold. Uniqueness also follows as in the preceding paragraph. □

We mention here that one useful sufficient condition that guarantees $\mathbb{E}[\Lambda_t] = 1$ and $\mathbb{E}[\tilde{\Lambda}_t] = 1$ due to Portenko [96] is: For all $t > s$,

$$\mathbb{E}\left[\int_s^t \left|\sigma^{-1}(X_r)\, b(X_r, v_r(X_{[0,r]}))\right|^2 \mathrm{d}r \;\middle|\; \mathfrak{F}_s\right] \le C(t-s)^\gamma \tag{2.2.37}$$

for suitable constants $C, \gamma > 0$. Note that under the additional hypothesis that σ is uniformly nondegenerate on $\mathbb{R}^d$, condition (2.2.37) is verified by applying (2.2.16b).

A feedback control U is called a *Markov* control if it takes the form $U_t = v_t(X_t)$ for a measurable map $v\colon \mathbb{R}^d \times [0,\infty) \to \mathbb{U}$. On the other hand, if $U_t = g(X_t)$ for a measurable map $g\colon \mathbb{R}^d \to \mathbb{U}$, then U is called a *stationary Markov* control. By abuse of terminology, we often refer to the map v_t (g) as the Markov (stationary Markov) control. Existence and uniqueness of strong solutions for nondegenerate diffusions, under a Markov control was established in [116, 117, 127] for bounded data. This was subsequently extended to unbounded data under a linear growth assumption using Euler approximations in [68]. We give a bare sketch of the key steps involved in its proof.

Theorem 2.2.12 *Suppose* (2.2.3), (2.2.4) *and* (2.2.15) *hold. Then, under a Markov control* v, (2.2.33) *has a pathwise unique strong solution which is a strong Feller (and therefore strong Markov) process.*

Sketch of the proof Suppose for the time being that b and σ are bounded. Extend the definition of $\mathcal{L}^v f$ to functions f in the Sobolev space $\mathscr{W}^{2,p}_{\mathrm{loc}}(\mathbb{R}^d)$, $p \ge 2$. Fix $T > 0$. Let $\mathfrak{D}$ denote the space of functions $g \in \mathscr{W}^{1,2,p}_{\mathrm{loc}}([0,T]\times\mathbb{R}^d)$, $p \ge 2$, such that $\sup_{0\le t\le T} |g(t,x)|$ grows slower than $\exp(a|x|^2)$ for all $a > 0$. For $1 \le i \le d$, let ψ_i be the unique solution in $\mathfrak{D}$ [83, chapter 4] to the PDE

$$\frac{\partial \psi_i}{\partial t}(t,x) + \mathcal{L}^{v_t}\psi_i(t,x) = 0 \quad \text{on } (0,T)\times\mathbb{R}^d\,, \qquad \psi_i(T,x) = x_i\,.$$

Define $\Psi = [\psi_1,\dots,\psi_d]^{\mathsf{T}}\colon [0,T]\times\mathbb{R}^d \to \mathbb{R}^d$. It can be shown that, for each $t \in [0,T]$, $\Psi_t(\,\cdot\,) \equiv \Psi(t,\,\cdot\,)$ is a C^1-diffeomorphism onto its range. Intuitively (at least for small t) Ψ_t can be thought of as a perturbation (homotopy, to be precise) of the identity map. Setting $Y_t := \Psi_t(X_t)$, $t \in [0,T]$, one can apply Krylov's technique, which extends the Itô formula to functions f in the Sobolev space $\mathscr{W}^{2,p}_{\mathrm{loc}}(\mathbb{R}^d)$ (henceforth referred to as the *Itô–Krylov formula* [78, p. 122]), to show that Y satisfies the stochastic integral equation

$$Y_t = Y_0 + \int_0^t (\mathrm{d}\Psi_s\sigma)\circ\Psi_s^{-1}(Y_s)\,\mathrm{d}W_s\,, \tag{2.2.38}$$

where $\mathrm{d}\Psi_s$ and Ψ_s^{-1} are the Jacobian matrix and the inverse map of Ψ_s, respectively, and "$\circ$" stands for composition of functions. Arguments analogous to those of Theorem 2.2.4 using the locally Lipschitz property can be used to show that (2.2.38) has an a.s. unique strong solution, whence the corresponding claim for X follows from the fact that $\{\Psi_t : t \ge 0\}$ is a family of C^1-diffeomorphisms. To prove the strong Feller property, let f be a bounded measurable function on $\mathbb{R}^d$. The equation

$$\frac{\partial \varphi}{\partial t}(t,x) + \mathcal{L}^{v_t}\varphi(t,x) = 0 \quad \text{on } (0,T)\times\mathbb{R}^d\,, \qquad \varphi(T,x) = f(x)\,,$$

has a unique solution in $\mathfrak{D}$ and a straightforward application of the Itô–Krylov formula shows that $\varphi(t,x) = \mathbb{E}\,[f(X_T)\mid X_t = x]$. By Sobolev embedding, $\mathfrak{D} \subset C(\mathbb{R}^{d+1})$ and thus $\varphi(t,\,\cdot\,)$ is continuous for $t \in [0,T]$. This establishes the strong Feller property.

Suppose now that b and σ are not bounded. We approximate them via truncation by selecting for each $n \in \mathbb{N}$ a pair (b^n, σ^n) of continuous bounded functions, which agree with (b, σ) on B_n, the ball of radius n centered at the origin, and satisfy (2.2.3) and (2.2.4) for each $n \in \mathbb{N}$. Let X^n denote the unique solution of (2.2.33) with parameters (b^n, σ^n). By Theorem 2.2.2, the first exit time from B_n, satisfies $\tau_n \uparrow \infty$, a.s. By pathwise uniqueness, we have

$$\mathbb{P}\left(X^n_{t\wedge\tau_n\wedge\tau_m} = X^m_{t\wedge\tau_n\wedge\tau_m}\,,\ \forall t \in [0, T]\right) = 1\,.$$

Therefore $X^n_{t\wedge\tau_n}$ converges w.p. 1 uniformly on every interval $[0, T]$ to some X which satisfies (2.2.33) on $[0, T]$ for every $T > 0$. The strong Feller property for unbounded coefficients follows from [26, theorem 4.1], which moreover asserts that the transition probabilities have densities which are locally Hölder continuous. □

Theorem 2.2.12 shows that every measurable map $v\colon \mathbb{R}^d \times [0, \infty) \to \mathbb{U}$ gives rise to an admissible control $U_t = v_t(X_t)$. Moreover, under this control, X is a Markov process. As a converse, we have the following theorem which does not require non-degeneracy.

Theorem 2.2.13 *If X in (2.2.33) is a Markov (time-homogeneous Markov) process, then the process $U_t = v_t(X_{[0,t]})$ may be taken to be a Markov (stationary Markov) control.*

Proof Consider the first claim. For $t > s$,

$$\begin{aligned}
\mathbb{E}\left[\int_s^t b(X_r, U_r)\,\mathrm{d}r \,\Big|\, \mathfrak{F}^X_s\right] &= \mathbb{E}\left[X_t - X_s \mid \mathfrak{F}^X_s\right] \\
&= \mathbb{E}\left[X_t - X_s \mid X_s\right] \\
&= \mathbb{E}\left[\int_s^t b(X_r, U_r)\,\mathrm{d}r \,\Big|\, X_s\right]. \qquad (2.2.39)
\end{aligned}$$

Since for any continuous sample path of X the drift $b(X_t, U_t)$ is bounded over compact time intervals, it follows that as $t \to s$,

$$\frac{1}{t-s}\int_s^t b(X_r, U_r)\,\mathrm{d}r \to b(X_s, U_s) \quad \mathbb{P}\text{-a.s., a.e. in } s\,.$$

Divide the first and the last terms in (2.2.39) by $t - s$. Letting $t \to s$ and using the conditional dominated convergence theorem, we obtain

$$b(X_s, U_s) = \mathbb{E}\left[b(X_s, U_s) \mid X_s\right] \quad \mathbb{P}\text{-a.s., a.e. in } s\,,$$

where we use the fact that U_s is measurable with respect to $\mathfrak{F}^X_s$. Thus $b(X_s, U_s)$ is also measurable with respect to the σ-field generated by X_s for almost all s, where the qualification "for almost all s" can be dropped by suitably modifying U_s on a Lebesgue-null set. Then by the measurable selection theorem of [13] there exists a measurable map $v\colon [0, \infty) \times \mathbb{R}^d \to \mathbb{U}$ such that

$$b(X_s, U_s) = b(X_s, v_s(X_s))\,.$$

This proves the first claim. Now for $f \in C_c^2(\mathbb{R}^d)$, consider $\mathcal{L}^{v_t} f(x)$. Then the Markov process X has the extended generator $\mathcal{L}^{v_t}$. If it is time-homogeneous, $\mathcal{L}^{v_t}$ cannot have an explicit time-dependence. It follows that v_t can be replaced by g for some measurable $g\colon \mathbb{R}^d \to \mathbb{U}$ in (2.2.34), and hence also in the stochastic differential equation describing X. □

For the controlled process X, under $U \in \mathfrak{U}$, we adopt the notation

$$\mathbb{E}^U_{X_0}\left[f(X_{[0,\infty)})\right] := \mathbb{E}\left[f(X_{[0,\infty)}) \mid X_0\right].$$

2.3 Relaxed controls

We describe the *relaxed control* framework, originally introduced for deterministic control by Young [126]. This entails the following: The space $\mathbb{U}$ is replaced by $\mathscr{P}(\mathbb{U})$, where, as usual, $\mathscr{P}(\mathbb{U})$ denotes the space of probability measures on $\mathbb{U}$ endowed with the Prohorov topology, and b^i, $1 \le i \le d$, are replaced by

$$\bar{b}^i(x, v) = \int_{\mathbb{U}} b^i(x, u) v(\mathrm{d}u), \quad x \in \mathbb{R}^d, \quad v \in \mathscr{P}(\mathbb{U}), \quad 1 \le i \le d.$$

Note that $\bar{b}$ inherits the same continuity, linear growth and local Lipschitz (in the first argument) properties from b. The space $\mathscr{P}(\mathbb{U})$, in addition to being compact, is convex when viewed as a subset of the space of finite signed measures on $\mathbb{U}$. One may view $\mathbb{U}$ as the "original" control space and view the passage from $\mathbb{U}$ to $\mathscr{P}(\mathbb{U})$ as a "relaxation" of the problem that allows $\mathscr{P}(\mathbb{U})$-valued controls which are analogous to randomized controls in the discrete time setup. For this to be a valid relaxation in the optimization theoretic sense, it must, for a start, have the original problem embedded in it. This it does indeed, for a $\mathbb{U}$-valued control trajectory U can be identified with the $\mathscr{P}(\mathbb{U})$-valued trajectory δ_{U_t}, where δ_q denotes the Dirac measure at q. Henceforth, "control" means relaxed control, with Dirac measure-valued controls (which correspond to original $\mathbb{U}$-valued controls) being referred to as *precise* controls. The class of stationary Markov controls is denoted by $\mathfrak{U}_{\mathrm{sm}}$, and $\mathfrak{U}_{\mathrm{sd}} \subset \mathfrak{U}_{\mathrm{sm}}$ denotes the subset corresponding to precise controls. The next result, a stochastic analog of the *chattering lemma* of Young [126], further confirms that this is indeed a valid relaxation.

Theorem 2.3.1 *Let X, $U \in \mathfrak{U}$, and W satisfy* (2.2.1) *on a probability space* $(\Omega, \mathfrak{F}, \mathbb{P})$. *There exists a sequence $\{U^n\}$ of admissible precise controls on $(\Omega, \mathfrak{F}, \mathbb{P})$ such that if $\{X^n\}$ is the corresponding family of solutions to* (2.2.1), *then for each $T > 0$ and $f \in C([0, T] \times \mathbb{U})$, we have*

$$\int_0^T \int_{\mathbb{U}} f(t, u)\, U^n_t(\mathrm{d}u)\, \mathrm{d}t \to \int_0^T \int_{\mathbb{U}} f(t, u)\, U_t(\mathrm{d}u)\, \mathrm{d}t \quad \text{on } \Omega \tag{2.3.1}$$

and

$$\left\|X^n - X\right\|_{\mathcal{H}^2_T} \xrightarrow[n\to\infty]{} 0 \qquad \forall T > 0.$$

Proof For each $n \geq 1$, let

$$t_k^n := k\frac{T}{n}, \quad k = 0, \dots, n,$$
$$\mathcal{I}_i^n := [t_{i-1}^n, t_i^n), \quad i = 1, \dots, n.$$

Also, let $\{\mathbb{U}_j^n : j = 1, \dots, j_n\}$ be a finite partition of $\mathbb{U}$, satisfying $\operatorname{diam}(\mathbb{U}_j^n) \leq \frac{1}{n}$, and for each $j = 1, \dots, j_n$, select an arbitrary $u_j^n \in \mathbb{U}_j^n$. Define

$$\lambda_{ij}^n := \int_{\mathcal{I}_{i-1}^n} U_t(\mathbb{U}_j^n)\,\mathrm{d}t, \quad i = 2, \dots, n. \tag{2.3.2}$$

Then $\sum_j \lambda_{ij}^n = \frac{T}{n}$. For $i \geq 2$, subdivide each $\mathcal{I}_i^n$ further into disjoint left-closed right-open intervals $\mathcal{I}_{ij}^n$, each of length λ_{ij}^n. Let $\{U^n : n \in \mathbb{N}\}$ be a sequence of precise controls defined by

$$U_t^n = \begin{cases} \delta_{u_0} & \text{if } t \in \mathcal{I}_1^n, \\ \delta_{u_j^n} & \text{if } t \in \mathcal{I}_{ij}^n,\ i \geq 2, \end{cases} \tag{2.3.3}$$

where $u_0 \in \mathbb{U}$ is arbitrarily selected. It follows from (2.3.2)–(2.3.3) that U_t^n for $t \in I_i^n$, $i \geq 2$ is a function of $\{U_t : t \in I_{i-1}^n\}$. Hence U^n is non-anticipative.

By the uniform continuity of f on $[0, T] \times \mathbb{U}$, there exists a sequence $\{\varepsilon_n\} \subset \mathbb{R}$, satisfying $\varepsilon_n \to 0$ as $n \to \infty$, such that

$$\begin{aligned} \left|f(t,u) - f(t_i^n, u_j^n)\right| &< \varepsilon_n \qquad \forall (t,u) \in \mathcal{I}_i^n \times \mathbb{U}_j^n, \\ \sup_{u \in \mathbb{U}} \left|f(t,u) - f(t_{i-1}^n, u)\right| &< \varepsilon_n \qquad \forall t \in \mathcal{I}_i^n, \end{aligned} \tag{2.3.4}$$

for all $n \in \mathbb{N}$, $i = 1, \dots, n$, and $j = 1, \dots, j_n$. Let M_f be an upper bound of $|f|$ on $[0, T] \times \mathbb{U}$. Decomposing the integrals in (2.3.1) and using (2.3.4) in a triangle inequality, we obtain

$$\begin{aligned} &\left| \int_0^T \int_{\mathbb{U}} f(t,u) U_t^n(\mathrm{d}u)\,\mathrm{d}t - \int_0^T \int_{\mathbb{U}} f(t,u) U_t(\mathrm{d}u)\,\mathrm{d}t \right| \\ &\quad \leq \left| \int_{\mathcal{I}_1^n} f(t,u_0)\,\mathrm{d}t \right| + \left| \int_{\mathcal{I}_n^n} \int_{\mathbb{U}} f(t,u) U_t(\mathrm{d}u)\,\mathrm{d}t \right| \\ &\qquad + \sum_{i=1}^{n-1} \sum_{j=1}^{j_n} \left| \int_{\mathcal{I}_{(i+1)j}^n} f(t,u_j^n)\,\mathrm{d}t - \int_{\mathcal{I}_i^n} \int_{\mathbb{U}_j^n} f(t,u) U_t(\mathrm{d}u)\,\mathrm{d}t \right| \\ &\quad \leq 2\frac{M_f T}{n} + 2\frac{(n-1)\varepsilon_n T}{n} \\ &\qquad + \sum_{i=1}^{n-1} \sum_{j=1}^{j_n} \left| \int_{\mathcal{I}_{(i+1)j}^n} f(t_i^n, u_j^n)\,\mathrm{d}t - \int_{\mathcal{I}_i^n} \int_{\mathbb{U}_j^n} f(t_i^n, u_j^n)\, U_t(\mathrm{d}u)\,\mathrm{d}t \right|. \end{aligned} \tag{2.3.5}$$

By (2.3.2) and the definition of $\{\mathcal{I}_{ij}^n\}$, the last term in (2.3.5) vanishes identically. Hence, (2.3.1) follows.

For $R > 0$, let b^R and σ^R be continuous bounded functions which agree with b and σ on B_R, the ball of radius R centered at the origin, and satisfy (2.2.3) and (2.2.4). Let

$X^{n,R}$ denote the unique solution of (2.2.1) with parameters (b^R, σ^R) under the control U^n, satisfying $X_0^{n,R} = X_0$. Similarly $X^{\infty,R}$ denotes the corresponding solution under the control U. The solution for (2.2.1) under the unmodified coefficients b and σ, and control U^n is denoted by X^n, $n \in \mathbb{N} \cup \{\infty\}$. The argument in the proof of Lemma 2.2.5 shows that we can assume without loss of generality that the law of X_0 has compact support. Let

$$\tau_R^i := \inf\left\{t > 0 : \left|X_t^{i,R}\right| > R\right\}, \quad i \in \mathbb{N} \cup \{\infty\},$$

and

$$Z_t^{i,R} := \mathbb{I}_{[0,t]}(\tau_R^i \wedge \tau_R^\infty), \quad i \in \mathbb{N}, \quad t > 0.$$

Argue as in Theorem 2.2.4 to obtain

$$\begin{aligned}\left\|\left(X^{n,R} - X^{\infty,R}\right) Z_T^{n,R}\right\|_{\mathcal{H}_T^2} &\le K_R(T + 2\sqrt{T}) \left\|\left(X^{n,R} - X^{\infty,R}\right) Z_T^{n,R}\right\|_{\mathcal{H}_T^2} \\ &+ \left(\mathbb{E}\left[\sup_{t\in[0,T]} \left|\int_0^{t\wedge\tau_R^n\wedge\tau_R^\infty} \left(\bar{b}(X_t^{\infty,R}, U_t^n) - \bar{b}(X_t^{\infty,R}, U_t)\right) \mathrm{d}t\right|^2\right]\right)^{1/2} . \end{aligned} \tag{2.3.6}$$

By (2.3.5), it is clear that (2.3.1) holds uniformly over T in a compact set. Thus, by the dominated convergence theorem, the last term of (2.3.6) goes to zero as $n \to \infty$, from which we get that

$$\left\|\left(X^{n,R} - X^{\infty,R}\right) Z_T^{n,R}\right\|_{\mathcal{H}_T^2} \xrightarrow[n\to\infty]{} 0, \tag{2.3.7}$$

provided T is small enough that $K_R(T + 2\sqrt{T}) < 1$. However, iterating the same argument over the interval $[T, 2T]$, noting also that $\left\|X_T^{n,R} - X_T^{\infty,R}\right\|_{\mathcal{H}_T^2}$ can be made arbitrarily small if n is sufficiently large, it follows that, for any fixed $R > 0$, (2.3.7) holds for all $T > 0$.

Since X_0 has compact support, it follows by (2.2.16b) that

$$\left\{\sup_{s\le T} |X_s^n|^2 : n \in \mathbb{N} \cup \{\infty\}\right\}$$

are uniformly integrable. Therefore

$$\left\|X^n - X^{n,R} Z_T^{n,R}\right\|_{\mathcal{H}_T^2}^2 \le 2 \sum_{i\in\{n,\infty\}} \mathbb{E}\left[\sup_{s\le T} |X_s^n|^2\, \mathbb{I}\{\tau_R^i < T\}\right] \xrightarrow[R\to\infty]{} 0, \tag{2.3.8}$$

uniformly in n. Using the triangle inequality

$$\begin{aligned}\left\|X^n - X^\infty\right\|_{\mathcal{H}_T^2} &\le \left\|X^n - X^{n,R} Z_T^{n,R}\right\|_{\mathcal{H}_T^2} + \left\|X^\infty - X^{\infty,R} Z_T^{n,R}\right\|_{\mathcal{H}_T^2} \\ &\quad + \left\|\left(X^{n,R} - X^{\infty,R}\right) Z_T^{n,R}\right\|_{\mathcal{H}_T^2}, \end{aligned} \tag{2.3.9}$$

the result follows, since by (2.3.8) the first two terms on the right-hand side of (2.3.9) can be made arbitrarily small for large R uniformly in n, while, by (2.3.7), the last term can be made arbitrarily small for large n for any fixed R. □

The proof of Theorem 2.3.1 suggests a natural topology for the space of trajectories U_t, which we describe next. Let $\mathscr{U}$ denote the space of measurable maps $[0,\infty) \to \mathscr{P}(\mathbb{U})$. Let $\{f_i\}$ be a countable dense set in the unit ball of $C(\mathbb{U})$. Then $\{f_i\}$ is a convergence determining class for $\mathscr{P}(\mathbb{U})$. Let

$$\alpha_i(t) := \int_{\mathbb{U}} f_i(u) U_t(\mathrm{d}u)\,, \quad i = 1, 2, \dots$$

Then α_i has measurable paths, and $|\alpha_i(t)| \le 1$ for all $t \ge 0$. For $T > 0$, let B_T denote the space of measurable maps $[0,T] \to [-1,1]$ with the weak*-topology of $L^2[0,T]$ relativized to it. Let B denote the space of measurable maps $[0,\infty) \to [-1,1]$ with the corresponding inductive topology, i.e., the coarsest topology that renders continuous the map $B \to B_T$ that maps $x \in B$ to its restriction to B_T for every $T > 0$. Let $B^\infty = B \times B \times \cdots$ (countable product) with the product topology.

Next, note that the map $\varphi\colon \mathscr{P}(\mathbb{U}) \to [-1,1]^\infty$ defined by

$$\mu \mapsto \left(\int f_1\,\mathrm{d}\mu, \int f_2\,\mathrm{d}\mu, \dots\right)$$

is continuous, one-to-one with a compact domain, and hence is a homeomorphism onto its range. By abuse of notation, we denote the map $U \mapsto (\alpha_1, \alpha_2, \dots) \in B^\infty$ also as $\varphi\colon \mathscr{U} \to B^\infty$. We relativize the topology of B^∞ to $\varphi(\mathscr{U})$ and topologize $\mathscr{U}$ with the coarsest topology that renders $\varphi\colon \mathscr{U} \to \varphi(\mathscr{U})$ a homeomorphism.

Theorem 2.3.2 *The space $\mathscr{U}$ is compact and metrizable, hence Polish.*

Proof A standard application of the Banach–Alaoglu theorem shows that for each T, B_T is compact. It is clearly metrizable by the metric

$$d_T(x,y) = \sum_{n=1}^{\infty} 2^{-n} \left| \int_0^T e_n^T(t) x(t)\,\mathrm{d}t - \int_0^T e_n^T(t) y(t)\,\mathrm{d}t \right| \wedge 1\,,$$

where $\{e_n\}$ is a complete orthonormal basis for $L^2[0,T]$. Since the topology of B is defined inductively from those of B_T, $T > 0$, it is easy to see that it is compact. Also, it is metrizable by the metric

$$d(x,y) = \sum_{n=1}^{\infty} 2^{-n} d_n(x^n, y^n)\,,$$

where x^n and y^n are the restrictions of x and y to $[0,n]$, $n \ge 1$, respectively. Thus B^∞ is also compact metrizable. Therefore it suffices to show that $\varphi(\mathscr{U})$ is closed in B^∞. Let

$$\varphi(\mathscr{U}) \ni \alpha^n = (\alpha_1^n, \alpha_2^n, \dots) \to \alpha \in B^\infty\,.$$

Fix $T > 0$. Let $n(1) = 1$ and define $\{n(k)\}$ inductively so as to satisfy

$$\sum_{j=1}^{\infty} 2^{-j} \max_{1 \le l < k} \left| \int_0^T (\alpha_j^{n(k)}(t) - \alpha_j(t))(\alpha_j^{n(l)}(t) - \alpha_j(t))\,\mathrm{d}t \right| < \frac{1}{k}\,,$$

which is possible because $\alpha_j^n \to \alpha_j$ in B for each j. Denote by $\|\cdot\|_{2,T}$ and $\langle\cdot,\cdot\rangle$ the norm and the inner product on $L^2[0,T]$, respectively. Then, for each $j \ge 1$,

$$\left\|\frac{1}{m}\sum_{k=1}^{m}\alpha_j^{n(k)} - \alpha_j\right\|_{2,T}^2 \le \frac{4T^2}{m} + \frac{2}{m^2}\sum_{i=2}^{m}\sum_{l=1}^{i-1}\left|\langle\alpha_j^{n(i)} - \alpha_j, \alpha_j^{n(l)} - \alpha_j\rangle\right|$$
$$\le \frac{2T^2}{m^2}[2m + 2^j(m-1)] \xrightarrow[m\to\infty]{} 0\,.$$

Thus

$$\frac{1}{m}\sum_{k=1}^{m}\alpha_j^{n(k)} \to \alpha_j \quad \text{as } m \to \infty\,,$$

strongly in $L^2[0,T]$, when restricted to $[0,T]$. By a diagonal argument, we can extract a subsequence $\{m(k)\}$ of $\{m\}$ such that, for almost all t,

$$\frac{1}{m(k)}\sum_{l=1}^{m(k)}\alpha_j^{n(l)}(t) \xrightarrow[k\to\infty]{} \alpha_j(t)\,, \quad j \ge 1\,. \tag{2.3.10}$$

Define $U_t^{(k)} \in \mathscr{P}(\mathbb{U})$ by

$$\int_{\mathbb{U}} f_i(u)U_t^{(k)}(\mathrm{d}u) = \frac{1}{m(k)}\sum_{j=1}^{m(k)}\alpha_i^{n(l)}(t)\,, \quad i \ge 1\,, \ k \ge 1\,. \tag{2.3.11}$$

Since $\{f_i\}$ is dense in the unit ball of $C(\mathbb{U})$, it separates points of $\mathscr{P}(\mathbb{U})$ and therefore, $U_t^{(k)}$ is well-defined for each k. By (2.3.10) – (2.3.11), any limit point U_t of $\{U_t^{(k)}\}$ in $\mathscr{P}(\mathbb{U})$ must satisfy

$$\int_{\mathbb{U}} f_i(u)U_t(\mathrm{d}u) = \alpha_i(t)\,, \quad i \ge 1\,.$$

Thus $\alpha(t) = \varphi(U_t)$, or in other words, $\alpha(t) \in \varphi(\mathscr{P}(\mathbb{U}))$ for a.e. t, where the qualification "a.e. t" can be dropped by suitably modifying α on a Lebesgue-null set. Define $\tilde{U}_t := \varphi^{-1}(\alpha(t))$. Then $\alpha(t) = \varphi(\tilde{U}_t)$, implying that $\alpha \in \varphi(\mathscr{U})$. Thus $\varphi(\mathscr{U})$ is closed in B^∞. □

The following consequence of this topology is frequently useful:

Theorem 2.3.3 *If $U^n \to U$ in $\mathscr{U}$ and $f \in C([0,T]\times\mathbb{U})$ for some $T > 0$, then*

$$\int_0^T\int_{\mathbb{U}} f(t,u)U_t^n(\mathrm{d}u)\,\mathrm{d}t \xrightarrow[n\to\infty]{} \int_0^T\int_{\mathbb{U}} f(t,u)U_t(\mathrm{d}u)\,\mathrm{d}t\,.$$

Proof From the foregoing, the claim is clearly true for functions f taking the form $f(t,u) = g(t)f_i(u)$ for some i and some $g \in C[0,T]$. By the density of $\{f_i\}$ in the unit ball of $C(\mathbb{U})$, it follows that it is also true for f of the form $f(t,u) = g(t)f(u)$ for $g \in C[0,T]$, $f \in C(\mathbb{U})$ and hence for linear combinations of such functions. By the Stone–Weierstrass theorem, the latter are dense in $C([0,T]\times\mathbb{U})$, so the claim follows by a simple approximation argument. □

We can view the control process U of (2.2.1) as a $\mathscr{U}$-valued random variable. Using the relaxed control framework permits us to use (2.2.1) or (2.2.33) flexibly. Therefore, if one is working with the weak formulation, (2.2.1) can be replaced by (2.2.33) without any loss of generality. Conversely, given a weak solution of (2.2.33), its replica (in law) can be constructed with given W and X_0 on a prescribed $(\Omega, \mathfrak{F}, \mathbb{P})$ if an augmentation thereof is permitted. The precise forms of these statements are contained in the next theorem.

Theorem 2.3.4 *Let X_0, W be prescribed on a probability space $(\Omega, \mathfrak{F}, \mathbb{P})$.*

(a) *Let U be an admissible control on $(\Omega, \mathfrak{F}, \mathbb{P})$ and X the corresponding strong solution to* (2.2.1). *Then on a possibly augmented probability space, X also satisfies* (2.2.1) *with U and W replaced by $\tilde{U}$ and $\widetilde{W}$, respectively, where $\widetilde{W}$ is a standard Wiener process independent of X_0 and $\tilde{U}$ is a feedback control.*

(b) *Suppose $\hat{X}$, $\hat{W}$, $\hat{U}$, and $\hat{X}_0$ satisfy* (2.2.1) *on a probability space $(\hat{\Omega}, \hat{\mathfrak{F}}, \hat{\mathbb{P}})$ with $\hat{U}_t = f(t, \hat{X}_{[0,t]})$, a feedback control, and with $\mathscr{L}(W_{[0,\infty)}, X_0) = \mathscr{L}(\hat{W}_{[0,\infty)}, \hat{X}_0)$. Then by augmenting $(\Omega, \mathfrak{F}, \mathbb{P})$ if necessary, we can construct on it a process X satisfying* (2.2.1) *with prescribed W, X_0, and a feedback control $U_t = f(t, X_{[0,t]})$, so that $\mathscr{L}(X) = \mathscr{L}(\hat{X})$.*

Proof To prove part (a), let $\{f_i\}$ be a countable dense set in the unit ball of $C(\mathbb{U})$ and define $\tilde{U}$ by

$$\int f_i \, \mathrm{d}\tilde{U}_t = \mathbb{E}\left[\int f_i \, \mathrm{d}U_t \,\middle|\, \mathfrak{F}_t^X\right] \quad \mathbb{P}\text{-a.s.}, \quad t \geq 0, \quad i \geq 1,$$

taking a measurable version thereof. Write

$$X_t = X_0 + \int_0^t b(X_s, \tilde{U}_s) \, \mathrm{d}s + M_t,$$

where

$$M_t = \int_0^t \sigma(X_s) \, \mathrm{d}W_s + \int_0^t [b(X_s, U_s) - b(X_s, \tilde{U}_s)] \, \mathrm{d}s, \quad t \geq 0.$$

It is easily checked that $(M_t, \mathfrak{F}_t^X)$ is a zero mean square-integrable martingale with quadratic covariation matrix process

$$\int_0^t \sigma(X_s)\sigma^{\mathsf{T}}(X_s) \, \mathrm{d}s, \quad t \geq 0.$$

The representation theorem of Wong [121] then implies that

$$M_t = \int_0^t \sigma(X_s) \, \mathrm{d}\widetilde{W}_s, \quad t \geq 0,$$

for a d-dimensional standard Wiener process $\widetilde{W}$ defined on a possibly augmented probability space.

For part (b), let

$$Q \in \mathscr{P}(\mathbb{R}^d \times C([0,\infty); \mathbb{R}^d) \times C([0,\infty); \mathbb{R}^d))$$

denote the law of $(\hat{X}_0, \hat{W}_{[0,\infty)}, \hat{X}_{[0,\infty)})$. Disintegrate it as

$$Q(\mathrm{d}w_1, \mathrm{d}w_2, \mathrm{d}w_3) = Q_1(\mathrm{d}w_1, \mathrm{d}w_2)\, Q_2(\mathrm{d}w_3 \mid w_1, w_2)\,,$$

where

$$Q_1 = \mathscr{L}(\hat{X}_0, \hat{W}_{[0,\infty)}) = \mathscr{L}(X_0, W_{[0,\infty)})$$

and Q_2 is the regular conditional law of $\hat{X}_{[0,\infty)}$ given $(\hat{X}_0, \hat{W}_{[0,\infty)})$, defined Q_1-a.s. Augment Ω to $\breve{\Omega} = \Omega \times C([0,\infty); \mathbb{R}^d)$, let $\breve{\mathfrak{F}}$ be the product σ-field on $\breve{\Omega}$, and replace $\mathbb{P}$ by $\breve{\mathbb{P}}$, which is defined as follows: For $A \in \mathfrak{F}$, and B Borel in $C([0,\infty); \mathbb{R}^d)$,

$$\breve{\mathbb{P}}(A \times B) = \mathbb{E}\left[Q_2(B \mid X_0, W_{[0,\infty)})\, \mathbb{I}_A\right].$$

Define X on $(\breve{\Omega}, \breve{\mathfrak{F}}, \breve{\mathbb{P}})$ by

$$X_t(\omega_1, \omega_2) = \omega_2(t)\,, \quad t \in [0,\infty)\,, \quad (\omega_1, \omega_2) \in \breve{\Omega}\,.$$

The rest is routine. □

Notation 2.3.5 To facilitate the passage to relaxed controls, we adopt the following notation. For a function $g\colon \mathbb{R}^d \times \mathbb{U} \to \mathbb{R}^k$ we let $\bar{g}\colon \mathbb{R}^d \times \mathscr{P}(\mathbb{U}) \to \mathbb{R}^k$ denote its extension to relaxed controls defined by

$$\bar{g}(x,\mu) := \int_{\mathbb{U}} g(x,u)\,\mu(\mathrm{d}u)\,, \quad \mu \in \mathscr{P}(\mathbb{U})\,. \tag{2.3.12}$$

A relaxed stationary Markov control v may be viewed as a Borel measurable kernel on $\mathbb{U}$ given $\mathbb{R}^d$, and therefore we adopt the notation $v(x) = v(\mathrm{d}u \mid x)$. For any fixed $v \in \mathfrak{U}_{\mathrm{sm}}$, and g as above, $x \mapsto \bar{g}(x, v(x))$ is a Borel measurable function. In order to simplify the notation, treating v as a parameter, we define $g_v\colon \mathbb{R}^d \to \mathbb{R}^k$ by

$$g_v(x) := \bar{g}(x, v(x)) = \int_{\mathbb{U}} g(x,u)\, v(\mathrm{d}u \mid x)\,. \tag{2.3.13}$$

Also, for $v \in \mathfrak{U}_{\mathrm{sm}}$,

$$\mathcal{L}^v := a^{ij}\partial_{ij} + b^i_v \partial_i$$

denotes the extended generator of the diffusion governed by v.

That we can go back and forth between the relaxed control formulation and the precise control formulation follows from the following corollary.

Corollary 2.3.6 *If (X, U) is a solution pair of*

$$X_t = X_0 + \int_0^t \bar{b}(X_s, U_s)\,\mathrm{d}s + \int_0^t \sigma(X_s)\,\mathrm{d}W_s\,,$$

with U a relaxed control and $\mathfrak{G}_t := \mathfrak{F}_t^{X,U}$, i.e., the right-continuous completion of the natural filtration of (X, U), then, on a possibly augmented probability space, there *exists another pair (X', U'), where U' is a $\mathbb{U}$-valued control process, such that if $\tilde{U}$ is defined by*

$$\int h(u)\tilde{U}_t(\mathrm{d}u) = \mathbb{E}\left[h(U'_t) \mid \mathfrak{G}_t\right] \qquad \forall h \in C(\mathbb{U}),$$

then (X, U) and $(X', \tilde{U})$ agree in law.

Proof When $t < 0$, let $U_t = \delta_{u_0}$ for some fixed $u_0 \in \mathbb{U}$. Define U^δ, $\delta > 0$, by

$$\int_{\mathbb{U}} f(u)U^\delta_t(\mathrm{d}u) = \frac{1}{\delta}\int_{t-\delta}^t \left[\int_{\mathbb{U}} f(u)U_s(\mathrm{d}u)\right]\mathrm{d}s, \quad t \geq 0,$$

for f in a countable dense subset of $C(\mathbb{U})$. Then U^δ has continuous sample paths and is $\mathfrak{G}_t$-adapted, hence it is progressively measurable with respect to $\{\mathfrak{G}_t\}$. Since $U^\delta_t \to U_t$ as $\delta \downarrow 0$ for a.e. t, U has a progressively measurable version. Thus without loss of generality, we take U to be progressively measurable.

We may view $X_t = X_t(\omega)$ and $U_t = U_t(\omega)$ as random variables on the product probability space $(\Omega_1, \mathfrak{F}_1, e^{-t}\mathrm{d}t \times \mathbb{P})$, with $\Omega_1 := \mathbb{R}_+ \times \Omega$ and $\mathfrak{F}_1$ the product σ-field. Consider the probability space $(\Omega' := \Omega_1 \times \mathbb{U}, \mathfrak{F}', \mathbb{P}')$ where $\mathfrak{F}'$ is the progressively measurable sub σ-field of the product σ-field and $\mathbb{P}'$ is defined by: $\mathbb{P}'$ restricts to $\mathbb{P}$ under the projection $\Omega' \mapsto \Omega$ and the regular conditional law on $\mathbb{U}$ given $(t, \omega) \in \Omega_1$ is $U_t(\omega)$. Define on this space the random process U' by $U'_t = U'(t, \omega, u) = u$. By construction, $\mathbb{E}\left[f(U'_t) \mid \mathfrak{G}_t\right] = \int_{\mathbb{U}} f(u)U_t(\mathrm{d}u)$ a.s. $\forall f \in C(\mathbb{U})$, and thus we may write

$$X_t = X_0 + \int_0^t b(X_s, U'_s)\,\mathrm{d}s + \int_0^t \sigma(X_s)\,\mathrm{d}\widetilde{W}_s$$

for a suitably defined Brownian motion $\widetilde{W}$ as in Theorem 2.3.4. This proves the claim. □

2.3.1 Two technical lemmas

We present two useful facts about controlled diffusions.

The first shows that the set of controlled diffusion laws is closed under conditioning w.r.t. a stopped σ-field w.r.t. the natural filtration of the controlled process. The second establishes the tightness of laws under tight initial conditions.

Lemma 2.3.7 *Let X be a solution of* (2.2.1) *with $\{\mathfrak{F}^X_t\}$ its natural filtration and τ an $(\mathfrak{F}^X_t)$-stopping time. Then the regular conditional law of $X \circ \theta_\tau$ given $\mathfrak{F}^X_\tau$ is a.s. the law of a controlled diffusion of type* (2.2.1) *on $\{\tau < \infty\}$.*

Proof Since we are working with the weak formulation, Theorem 2.3.4 allows us to assume without any loss of generality that U is a feedback control, say $U_t = v_t(X_{[0,t]})$, $t \geq 0$, where $v\colon [0, \infty) \times C([0, \infty); \mathbb{R}^d) \to \mathscr{P}(\mathbb{U})$ is progressively measurable. The weak formulation of (2.2.1) is equivalent to (2.2.35) for $t \geq s$, $f \in C^2_c(\mathbb{R}^d)$, and

$g \in C_b(C([0,s];\mathbb{R}^d))$. But a.s. on $\{\tau < \infty\}$,

$$\mathbb{E}\left[\left(f(X_{\tau+t}) - f(X_{\tau+s}) - \int_s^t \mathcal{L}^{v_{\tau+r}(X_{[0,\tau+r]})} f(X_{\tau+r})\,\mathrm{d}r\right) g(X_{[\tau,\tau+s]}) \,\Big|\, \mathfrak{F}_\tau^X\right] = 0 .$$

From this observation and [115, p. 33, lemma 3.3], it follows that a.s. on $\{\tau < \infty\}$, the regular conditional law of $\{\tilde{X}_t\} = \{X_{\tau+t} : t \geq 0\}$ given $\mathfrak{F}_\tau^X$ is the law of a diffusion as in (2.2.1), controlled by

$$U_t = \tilde{v}_t(\tilde{X}_{[0,t]}), \quad t \geq 0 ,$$

where $\tilde{v}_t(\tilde{X}_{[0,t]}) = v_{\tau+t}(X_{[0,\tau+t]})$ with $X_{[0,\tau]}$ being treated as a fixed parameter on the right-hand side. □

Lemma 2.3.8 *Let $M_0 \in \mathscr{P}(\mathbb{R}^d)$ be tight. Then the laws of the collection*

$$\mathscr{X}_{M_0} := \{X : (X,U) \text{ solves } (2.2.1),\ \mathscr{L}(X_0) \in M_0 ,\ U \in \mathfrak{U}\}$$

are tight in $\mathscr{P}(C([0,\infty);\mathbb{R}^d))$.

Proof Since the laws of X_0 are in a tight set, for any $\varepsilon > 0$, we can find a compact set $K_\varepsilon \subset \mathbb{R}^d$ such that the probability of $\{X_0 \notin K_\varepsilon\}$ does not exceed ε under any of these laws. Thus we may consider bounded X_0 without loss of generality. Consequently, since by (2.2.16a) all moments are bounded on any interval $[0,T]$, (2.2.16b) yields

$$\mathbb{E}\left[|X_t - X_s|^4\right] \leq \hat{K}(T)(t-s)^2 , \quad 0 \leq s < t \leq T ,$$

for some $\hat{K}(T) > 0$. As shown in Billingsley [24, p. 95, theorem 12.3] this is sufficient to guarantee that

$$\sup_{X \in \mathscr{X}_{M_0}} \mathbb{P}\left(w_T(X,\delta) \geq \varepsilon\right) \xrightarrow[\delta\to 0]{} 0 \qquad \forall \varepsilon > 0 ,$$

where w_T denotes the modulus of continuity on $[0,T]$, i.e.,

$$w_T(X,\delta) := \sup_{|t-s|\leq\delta} |X_t - X_s| , \quad 0 \leq s < t \leq T .$$

Thus $\{\mathscr{L}(X) : X \in \mathscr{X}_{M_0}\}$ satisfies the hypotheses of the well-known tightness characterization of $\mathscr{P}(C([0,\infty);\mathbb{R}^d))$ [24, p. 55, theorem 8.2]. □

Corollary 2.3.9 *If $M_0 \in \mathscr{P}(\mathbb{R}^d)$ is compact, then the set of laws $\mathscr{L}(X,U)$ of the pairs (X,U) which solve* (2.2.1) *and satisfy $\mathscr{L}(X_0) \in M_0$ and $U \in \mathfrak{U}$ is also compact.*

Proof Since $\mathscr{U}$ is compact, Lemma 2.3.8 implies that the laws of (X,U) are tight. Consider the equivalent statement of the martingale formulation

$$\mathbb{E}\left[\left(f(X_t) - f(X_s) - \int_s^t \mathcal{L}^{U_r} f(X_r)\,\mathrm{d}r\right) g(X_{[0,s]}, U_{[0,s]})\right] = 0$$

for $t > s$, $f \in C_c^2(\mathbb{R}^d)$, and $g \in C_b(C([0,s];\mathbb{R}^d) \times \mathscr{U}_s)$, where $\mathscr{U}_s$ denotes the space of measurable maps $[0,s] \to \mathscr{P}(\mathbb{U})$, topologized in the same manner as $\mathscr{U}$. This equation is preserved under convergence in law. The claim then follows by Lemma 2.3.8. □

2.3.2 Remarks

We conclude this section with some remarks on the difficulties in replacing $\sigma(x)$ by $\sigma(x,u)$ in (2.2.1) (i.e., introducing explicit control dependence in the diffusion matrix).

(i) Theorem 2.2.12 established a unique strong solution for a nondegenerate diffusion with Markov controls. Suppose the control also enters σ. Because it is not reasonable in general to impose more regularity than mere measurability on the map $v\colon [0,\infty)\times\mathbb{R}^d \to \mathscr{P}(\mathbb{U})$ ($g\colon \mathbb{R}^d \to \mathscr{P}(\mathbb{U})$) that defines a Markov (stationary Markov) control, we have to deal with a stochastic differential equation with drift vector and diffusion matrix that are merely measurable, nothing more. A suitable existence result for strong solutions to these is unavailable. Weak solutions do exist in the nondegenerate case [78, pp. 86–91], but their uniqueness is not available in general [92, 107]. Parenthetically, we mention here that the martingale problem is well posed for one- and two-dimensional diffusions with bounded measurable coefficients, provided σ is positive definite on compact subsets of $\mathbb{R}^d$ [115, pp. 192-193].

Furthermore, in the nondegenerate case, if σ does not depend on U, the process X has well-behaved transition probability densities (a fact we shall often use). But if it does and U is Markov but not stationary Markov, this is not guaranteed [56].

(ii) In the passage from precise to relaxed controls, the controlled infinitesimal generator transforms as follows. For $f \in C_c^2(\mathbb{R}^d)$,

$$\mathcal{L}^u f(x) = \sum_{i,j} a^{ij}(x,u)\frac{\partial^2 f}{\partial x_i \partial x_j}(x) + \sum_i b^i(x,u)\frac{\partial f}{\partial x_i}(x)$$

gets replaced by

$$\mathcal{L}^v f(x) = \sum_{i,j}\left[\int_{\mathbb{U}} a^{ij}(x,u)v(\mathrm{d}u)\right]\frac{\partial^2 f}{\partial x_i \partial x_j}(x) + \sum_i \bar{b}^i(x,v(x))\frac{\partial f}{\partial x_i}(x)\,,$$

and not by

$$\mathcal{L}^v f(x) = \frac{1}{2}\sum_{i,j,k} \bar{\sigma}^{ik}(x,v(x))\bar{\sigma}^{jk}(x,v(x))\frac{\partial^2 f}{\partial x_i \partial x_j}(x) + \sum_i \bar{b}^i(x,v(x))\frac{\partial f}{\partial x_i}(x)\,,$$

with

$$\bar{\sigma}(x,v(x)) = \int_{\mathbb{U}} \sigma(x,u)v(\mathrm{d}u \mid x)\,.$$

This leads to problems of interpretation. We should note that the corresponding Hamilton–Jacobi–Bellman equation is fully nonlinear if the control enters in the diffusion matrix. Even if this were handled analytically, the interpretation of the stochastic differential equation would still be problematic.

2.4 A topology for Markov controls

We endow $\mathfrak{U}_{\mathrm{sm}}$ with the topology that renders it a compact metric space. We refer to it as "the" topology since, as is well known, the topology of a compact Hausdorff space has a certain rigidity and cannot be weakened or strengthened without losing the Hausdorff property or compactness, respectively [105, p. 60]. This can be accomplished by viewing $\mathfrak{U}_{\mathrm{sm}}$ as a subset of the unit ball of $L^\infty(\mathbb{R}^d, \mathfrak{M}_s(\mathbb{U}))$ under its weak*-topology, where $\mathfrak{M}_s(\mathbb{U})$ denotes the space of signed Borel measures on $\mathbb{U}$ under the weak*-topology. The space $L^\infty(\mathbb{R}^d, \mathfrak{M}_s(\mathbb{U}))$ is the dual of $L^1(\mathbb{R}^d, C(\mathbb{U}))$, and by the Banach–Alaoglu theorem the unit ball is weak*-compact. Since the space of probability measures is closed in $\mathfrak{M}_s(\mathbb{U})$, it follows that $\mathfrak{U}_{\mathrm{sm}}$ is weak*-closed in $L^\infty(\mathbb{R}^d, \mathfrak{M}_s(\mathbb{U}))$, and since it is a subset of the unit ball of the latter, it is weak*-compact. Moreover, since $L^1(\mathbb{R}^d, C(\mathbb{U}))$ is separable, the weak*-topology of its dual is metrizable. We have the following criterion for convergence in $\mathfrak{U}_{\mathrm{sm}}$.

Lemma 2.4.1 *For $v_n \to v$ in $\mathfrak{U}_{\mathrm{sm}}$ it is necessary and sufficient that*

$$\int_{\mathbb{R}^d} f(x) \int_{\mathbb{U}} g(x,u) v_n(\mathrm{d}u \mid x)\,\mathrm{d}x \xrightarrow[n\to\infty]{} \int_{\mathbb{R}^d} f(x) \int_{\mathbb{U}} g(x,u) v(\mathrm{d}u \mid x)\,\mathrm{d}x$$

for all $f \in L^1(\mathbb{R}^d) \cap L^2(\mathbb{R}^d)$ and $g \in C_b(\mathbb{R}^d \times \mathbb{U})$.

Proof Since $fg \in L^1(\mathbb{R}^d, C(\mathbb{U}))$ necessity is a direct consequence of the definition of the weak*-topology. Sufficiency also follows since $L^1(\mathbb{R}^d) \cap L^2(\mathbb{R}^d)$ is dense in $L^1(\mathbb{R}^d)$. □

Under the topology of Markov controls the solutions to (2.2.1) depend continuously on the control in the following sense:

Theorem 2.4.2 *For the class of nondegenerate controlled diffusions, the law of the controlled process under a stationary Markov control $v \in \mathfrak{U}_{\mathrm{sm}}$ and a fixed initial law ν depends continuously on v.*

Proof Without loss of generality we assume $\nu = \delta_{x_0}$ for some $x_0 \in \mathbb{R}^d$. Suppose $v_n \to v_\infty$ in $\mathfrak{U}_{\mathrm{sm}}$. Let X^n, $n \in \mathbb{N} \cup \{\infty\}$, denote the corresponding Markov processes, with $\mathscr{L}(X_0^n) = \nu$, and $\{T_t^{(n)} : t \geq 0\}$ the corresponding transition semigroup. For any $t > s > 0$, $k \in \mathbb{N}$, $f \in C_b^2(\mathbb{R}^d)$, $g \in C_b\left((\mathbb{R}^d)^k\right)$, and set of times $\{t_j : 1 \leq j \leq k\}$ satisfying

$$0 \leq t_1 < t_2 < \cdots < t_k \leq s,$$

we have

$$\begin{aligned}\mathbb{E}_\nu^{v_n}\left[\left(f(X_t^n) - T_{t-s}^{(n)} f(X_s^n)\right) g(X_{t_1}^n, \dots, X_{t_k}^n)\right] &= 0, \\ \mathbb{E}_\nu^{v_n}\left[\left(f(X_t^n) - f(X_s^n) - \int_s^t \mathcal{L}^{v_n} f(X_r^n, v_n(X_r^n))\,\mathrm{d}r\right) g(X_{t_1}^n, \dots, X_{t_k}^n)\right] &= 0.\end{aligned} \tag{2.4.1}$$

By estimates of parabolic PDEs [62, 83], the collection $\{T_{t-s}^{(n)} f : n \in \mathbb{N}\}$ is equicontinuous, and since it is bounded it is relatively compact in $C_b(\mathbb{R}^d)$. By Lemma 2.3.8,

$\{\mathscr{L}(X^n) : n \in \mathbb{N}\}$ are tight. On the other hand, $\{\mathscr{L}(U^n)\}$, where

$$U^n := \{v_n(X^n_t) : t \geq 0\},$$

are also tight since they take values in a compact space. Dropping to a subsequence if necessary, we may then suppose that

$$T^{(n)}_{t-s} f \xrightarrow[n\to\infty]{} h \in C_b(\mathbb{R}^d)$$

and

$$\mathscr{L}(X^n, U^n) \xrightarrow[n\to\infty]{} \mathscr{L}(\tilde{X}, \tilde{U})$$

for some $h \in C_b(\mathbb{R}^d)$ and $(\tilde{X}, \tilde{U})$. Taking the limit in (2.4.1) we obtain

$$\mathbb{E}^{\tilde{U}}_{\nu}\left[\left(f(\tilde{X}_t) - h(\tilde{X}_s)\right) g(\tilde{X}_{t_1}, \dots, \tilde{X}_{t_k})\right] = 0\,, \tag{2.4.2a}$$

$$\mathbb{E}^{\tilde{U}}_{\nu}\left[\left(f(\tilde{X}_t) - f(\tilde{X}_s) - \int_s^t \mathcal{L}f(\tilde{X}_r, \tilde{U}_r)\,\mathrm{d}r\right) g(\tilde{X}_{t_1}, \dots, \tilde{X}_{t_k})\right] = 0\,. \tag{2.4.2b}$$

A standard monotone class argument applied to (2.4.2a) shows that

$$\mathbb{E}^{\tilde{U}}_{\nu}\left[f(\tilde{X}_t) \mid \mathfrak{F}^{\tilde{X}}_s\right] = h(\tilde{X}_s)\,,$$

and thus $\tilde{X}$ is Markov. In turn, (2.4.2b) implies that $\tilde{X}$ is a controlled diffusion with controlled extended generator $\mathcal{L}$ and control $\tilde{U}$, which by Theorem 2.2.13 may be taken to be Markov. Also if $\{\tilde{T}_{r,y} : y \geq r \geq 0\}$ denotes the transition semigroup of $\tilde{X}$, then $h = \tilde{T}_{s,t} f$. That is

$$T^{(n)}_{t-s} f \xrightarrow[n\to\infty]{} \tilde{T}_{s,t} f\,. \tag{2.4.3}$$

By considering a countable convergence determining class of the functions f and extracting a common subsequence for (2.4.3) by a diagonal argument, we conclude that (2.4.3) holds for all $f \in C^2_b(\mathbb{R}^d)$, and all $s < t$. In particular, $\tilde{T}_{s,t}$ must be of the form $\hat{T}_{t-s}$ for a semigroup $\{\hat{T}_t : t \geq 0\}$, proving that $\tilde{X}$ is time-homogeneous Markov.

Let $p^n(t, x, y)$, $n \in \mathbb{N}$, and $\tilde{p}(t, x, y)$ denote the transition probability densities for $\{X^n\}$ and $\tilde{X}$, respectively. By Bogachev *et al.* [26, theorem 4.1], $\{p^n(t, x, \cdot)\}$ are locally bounded and Hölder equicontinuous for any $t > 0$ and $x \in \mathbb{R}^d$. Any limit point then in $C(\mathbb{R}^d)$ must coincide with $\tilde{p}$. By Scheffé's theorem [25, p. 214]

$$\left\|p^n(t, x, \cdot) - \tilde{p}(t, x, \cdot)\right\|_{L^1(\mathbb{R}^d)} \xrightarrow[n\to\infty]{} 0\,.$$

Thus, for $f \in C^2_c(\mathbb{R}^d)$,

$$\begin{aligned}\left|\int_{\mathbb{R}^d} p^n(t, x, y)\mathcal{L}^{v_n} f(y)\,\mathrm{d}y - \int_{\mathbb{R}^d} \tilde{p}(t, x, y)\mathcal{L}^{v_\infty} f(y)\,\mathrm{d}y\right| \\ \leq K_f \left\|p^n(t, x, \cdot) - \tilde{p}(t, x, \cdot)\right\|_{L^1(\mathbb{R}^d)} \\ + \left|\int_{\mathbb{R}^d} \tilde{p}(t, x, y)(\mathcal{L}^{v_n} f(y) - \mathcal{L}^{v_\infty} f(y))\,\mathrm{d}y\right|\,.\end{aligned} \tag{2.4.4}$$

Since the second term on the right-hand-side of (2.4.4) also converges to 0 as $n \to \infty$ by Lemma 2.4.1, we may take the limit as $n \to \infty$ in

$$T_t^{(n)} f(x) - T_s^{(n)} f(x) = \int_s^t T_r^{(n)} \mathcal{L}^{v_n} f(x)\,\mathrm{d}r$$
$$= \int_s^t \int_{\mathbb{R}^d} p^n(r,x,y) \mathcal{L}^{v_n} f(y)\,\mathrm{d}y\,\mathrm{d}r \qquad \forall f \in C_c^2(\mathbb{R}^d),$$

to conclude that

$$\hat{T}_t f(x) - \hat{T}_s f(x) = \int_s^t \hat{T}_r \mathcal{L}^{v_\infty} f(x)\,\mathrm{d}r \qquad \forall f \in C_c^2(\mathbb{R}^d).$$

It follows that $\tilde{X}$ is controlled by the Markov control $v_\infty \in \mathfrak{U}_{\mathrm{sm}}$, i.e., $\mathscr{L}(\tilde{X}) = \mathscr{L}(X^\infty)$. The claim follows. □

Similarly, the class of all Markov controls can be endowed with a compact metric topology by viewing it as a subset of the unit ball of $L^\infty(\mathbb{R} \times \mathbb{R}^d, \mathfrak{M}_s(\mathbb{U}))$ under its weak*-topology, and a counterpart of Theorem 2.4.2 can be proved analogously.

We make frequent use of the following convergence result.

Lemma 2.4.3 *Let $\{v_n\} \subset \mathfrak{U}_{\mathrm{sm}}$ be a sequence that converges to $v \in \mathfrak{U}_{\mathrm{sm}}$ in the topology of Markov controls, and let $\varphi_n \in \mathscr{W}^{2,p}(G)$, $p > d$, be a sequence of solutions of $\mathcal{L}^{v_n}\varphi_n = h_n$, $n \in \mathbb{N}$, on a bounded open set $G \subset \mathbb{R}^d$ with a C^2 boundary. Suppose that for some constant M, $\|\varphi_n\|_{\mathscr{W}^{2,p}(G)} \le M$ for all $n \in \mathbb{N}$, and that h_n converges weakly in $L^p(G)$ for $p > 1$, to some function h. Then any weak limit φ of φ_n in $\mathscr{W}^{2,p}(G)$ as $n \to \infty$, satisfies $\mathcal{L}^v\varphi = h$ in G.*

Proof Using the notation defined in (2.3.13), we have

$$\mathcal{L}^v\varphi - h = a^{ij}\partial_{ij}(\varphi - \varphi_n) + b^i_{v_n}\partial_i(\varphi - \varphi_n) + (b^i_v - b^i_{v_n})\partial_i\varphi - (h - h_n)\,. \tag{2.4.5}$$

Since $p > d$, by the compactness of the embedding $\mathscr{W}^{2,p}(G) \hookrightarrow C^{1,r}(\bar{G})$, $r < 1 - \frac{d}{p}$ (see Theorem A.2.15), we can select a subsequence such that $\varphi_{n_k} \to \varphi$ in $C^{1,r}(\bar{G})$. Thus $b^i_{v_n}\partial_i(\varphi - \varphi_n)$ converges to 0 in $L^\infty(G)$. By Lemma 2.4.1, and since G is bounded, $(b^i_v - b^i_{v_n})\partial_i\varphi$ converges weakly to 0, in $L^p(G)$ for any $p > 1$. The remaining two terms in (2.4.5) converge weakly to 0 in $L^p(G)$ by hypothesis. Since the left-hand side of (2.4.5) is independent of $n \in \mathbb{N}$, it solves $\mathcal{L}^v\varphi - h = 0$. □

2.5 Stability of controlled diffusions

In this section we give a brief account of the recurrence and stability properties of controlled diffusions. The term *domain* in $\mathbb{R}^d$ refers to a non-empty open connected subset of the Euclidean space $\mathbb{R}^d$. If $D \subset \mathbb{R}^d$ is a domain, we denote by $\tau(D)$, the *first exit time* of the process $\{X_t\}$ from D,

$$\tau(D) := \inf\{t > 0 : X_t \notin D\}.$$

The open ball centered at the origin in $\mathbb{R}^d$ of radius R is denoted by B_R and we often use the abbreviations $\tau_R \equiv \tau(B_R)$ and $\breve{\tau}_R \equiv \tau(B_R^c)$.

Consider (2.2.1) under a stationary Markov control $v \in \mathfrak{U}_{\mathrm{sm}}$. The controlled process is called *recurrent relative to a domain D*, or *D-recurrent*, if $\mathbb{P}_x^v(\tau(D^c) < \infty) = 1$ for all $x \in D^c$. Otherwise, it is called *transient* (relative to D). A D-recurrent process is called *positive D-recurrent* if $\mathbb{E}_x^v[\tau(D^c)] < \infty$ for all $x \in D^c$, otherwise it is called *null D-recurrent*. We refer to $\tau(D^c)$ as the *recurrence time*, or the *first hitting time* of the domain D.

A controlled process is called (positive) recurrent if it is (positive) D-recurrent for all bounded domains $D \subset \mathbb{R}^d$. For a nondegenerate controlled diffusion the recurrence properties are independent of the particular domain. Thus a nondegenerate diffusion is either recurrent or transient (see Lemma 2.6.12), and, as shown later in Theorem 2.6.10, if it is recurrent, then it is either positive or null recurrent, relative to all bounded domains.

A control $v \in \mathfrak{U}_{\mathrm{sm}}$ is called *stable*, if the associated diffusion is positive recurrent. We denote the set of such controls by $\mathfrak{U}_{\mathrm{ssm}}$, and by $\mathfrak{U}_{\mathrm{ssd}} \subset \mathfrak{U}_{\mathrm{ssm}}$ the subset of precise stable controls.

2.5.1 Stochastic Lyapunov functions

Sufficient conditions for the finiteness of the mean recurrence time can be provided via stochastic Lyapunov functions. We start with the following lemma.

Lemma 2.5.1 *Let D be a bounded C^2 domain. If there exist nonnegative φ_1, φ_2 in $C^2(\bar{D}^c)$ satisfying, for some $v \in \mathfrak{U}_{\mathrm{sm}}$,*

$$
\begin{aligned}
\mathcal{L}^v \varphi_1 &\le -1\,, \\
\mathcal{L}^v \varphi_2 &\le -2\varphi_1\,,
\end{aligned}
\tag{2.5.1}
$$

then

$$
\begin{aligned}
\mathbb{E}_x^v[\tau(D^c)] &\le \varphi_1(x)\,, \\
\mathbb{E}_x^v[\tau(D^c)^2] &\le \varphi_2(x)
\end{aligned}
\tag{2.5.2}
$$

for all $x \in D^c$.

Proof Let $\hat{\tau}_R := \tau(D^c) \wedge \tau_R$, $R > 0$. Using Dynkin's formula we obtain

$$
\mathbb{E}_x^v[\hat{\tau}_R] \le \varphi_1(x) - \mathbb{E}_x^v\left[\varphi_1(X_{\hat{\tau}_R})\right],
$$

and taking limits as $R \to \infty$, yields $\mathbb{E}_x^v[\tau(D^c)] \le \varphi_1(x)$. Similarly,

$$
\mathbb{E}_x^v\left[\int_0^{\hat{\tau}_R} 2\varphi_1(X_t)\,\mathrm{d}t\right] \le \varphi_2(x)\,,
\tag{2.5.3}
$$

and therefore, conditioning at $t \wedge \hat{\tau}_R$, we obtain

$$\begin{aligned}
\mathbb{E}^v_x[(\hat{\tau}_R)^2] &= \mathbb{E}^v_x\left[\int_0^\infty 2(\hat{\tau}_R - t)\,\mathbb{I}\{t < \hat{\tau}_R\}\,\mathrm{d}t\right] \\
&= \mathbb{E}^v_x\left[\int_0^\infty 2\,\mathbb{E}^v_x\left[(\hat{\tau}_R - t)\,\mathbb{I}\{t < \hat{\tau}_R\} \,\middle|\, \mathfrak{F}^X_{t\wedge\hat{\tau}_R}\right]\mathrm{d}t\right] \\
&= \mathbb{E}^v_x\left[\int_0^\infty 2\,\mathbb{I}\{t \wedge \hat{\tau}_R < \hat{\tau}_R\}\,\mathbb{E}^v_{X_{t\wedge\hat{\tau}_R}}[\hat{\tau}_R - t \wedge \hat{\tau}_R]\,\mathrm{d}t\right] \\
&\le \mathbb{E}^v_x\left[\int_0^\infty 2\varphi_1(X_{t\wedge\hat{\tau}_R})\,\mathbb{I}\{t < \hat{\tau}_R\}\,\mathrm{d}t\right] \\
&= \mathbb{E}^v_x\left[\int_0^{\hat{\tau}_R} 2\varphi_1(X_t)\,\mathrm{d}t\right].
\end{aligned} \tag{2.5.4}$$

Letting $R \uparrow \infty$, and combining (2.5.3)–(2.5.4), we obtain $\mathbb{E}^v_x[\tau(D^c)^2] \le \varphi_2(x)$. □

Remark 2.5.2 It is evident from the proof of Lemma 2.5.1 that if we replace v in (2.5.1) by some $U \in \mathfrak{U}$, then (2.5.2) holds for the process controlled under U. Also, if (2.5.1) holds for all $u \in \mathbb{U}$, then (2.5.2) holds uniformly over $U \in \mathfrak{U}$.

Recall that $f \in C(\mathcal{X})$, where $\mathcal{X}$ is a topological space, is called *inf-compact* if the set $\{x \in \mathcal{X} : f(x) \le \lambda\}$ is compact (or empty) for every $\lambda \in \mathbb{R}$.

Stability for controlled diffusions can be characterized with the aid of Lyapunov equations involving the operator $\mathcal{L}^u$. Consider the following sets of Lyapunov conditions, listed from the weakest to the strongest, each holding for some nonnegative, inf-compact Lyapunov function $\mathcal{V} \in C^2(\mathbb{R}^d)$:

(L2.1) For some bounded domain D

$$\mathcal{L}^u\mathcal{V}(x) \le -1 \qquad \forall (x,u) \in D^c \times \mathbb{U}\,.$$

(L2.2) There exists a nonnegative, inf-compact $h \in C(\mathbb{R}^d)$ and a constant $k_0 > 0$ satisfying

$$\mathcal{L}^u\mathcal{V}(x) \le k_0 - h(x) \qquad \forall (x,u) \in \mathbb{R}^d \times \mathbb{U}\,.$$

(L2.3) There exist positive constants k_0 and k_1 such that

$$\mathcal{L}^u\mathcal{V}(x) \le k_0 - 2k_1\mathcal{V}(x) \qquad \forall (x,u) \in \mathbb{R}^d \times \mathbb{U}\,. \tag{2.5.5}$$

As in the proof of Lemma 2.5.1, condition (L2.1) is sufficient for the finiteness of the mean recurrence times to D under all $U \in \mathfrak{U}$. The stronger condition (L2.2) guarantees the tightness of the mean empirical measures and α-discounted occupation measures, as the lemma that follows shows. Consequently, by Theorem 1.5.15, under (L2.2), the controlled diffusion has an invariant probability measure for any $v \in \mathfrak{U}_{\mathrm{sm}}$.

Lemma 2.5.3 *For* $\nu \in \mathscr{P}(\mathbb{R}^d)$ *and* $U \in \mathfrak{U}$, *define the sets of* mean empirical measures $\{\bar{\zeta}^U_{\nu,t} : t > 0\}$ *and* α-discounted occupation measures $\{\xi^U_{\nu,\alpha} : \alpha > 0\}$ *by*

$$\int_{\mathbb{R}^d \times \mathbb{U}} f \, \mathrm{d}\bar{\zeta}^U_{\nu,t} = \frac{1}{t} \int_0^t \mathbb{E}^U_\nu \left[\int_{u \in \mathbb{U}} f(X_s, u) \, U_s(\mathrm{d}u) \right] \mathrm{d}s \,,$$

and

$$\int_{\mathbb{R}^d \times \mathbb{U}} f \, \mathrm{d}\xi^U_{\nu,\alpha} = \alpha \int_0^\infty \mathrm{e}^{-\alpha s} \, \mathbb{E}^U_\nu \left[\int_{u \in \mathbb{U}} f(X_s, u) \, U_s(\mathrm{d}u) \right] \mathrm{d}s \,, \quad \alpha > 0 \,,$$

for all $f \in C_b(\mathbb{R}^d \times \mathbb{U})$, *respectively. Also, we let* $\bar{\zeta}^U_{x,t} := \bar{\zeta}^U_{\delta_x,t}$ *and similarly for* $\xi^U_{x,\alpha}$. *If (L2.2) holds, then for any* $\nu \in \mathscr{P}(\mathbb{R}^d)$ *and* $t_0 > 0$, *the families*

$$\{\bar{\zeta}^U_{\nu,t} : t \geq t_0 \,, \; U \in \mathfrak{U}\} \,, \tag{2.5.6a}$$

$$\{\xi^U_{\nu,\alpha} : \alpha > 0 \,, \; U \in \mathfrak{U}\} \tag{2.5.6b}$$

are tight. Every accumulation point $\mu \in \mathscr{P}(\mathbb{R}^d \times \mathbb{U})$ *of* $\bar{\zeta}^U_{\nu,t}$ *as* $t \to \infty$, *or of* $\xi^U_{\nu,\alpha}$ *as* $\alpha \downarrow 0$ *satisfies*

$$\int_{\mathbb{R}^d} \mathcal{L}^u f(x) \, \mu(\mathrm{d}x, \mathrm{d}u) = 0 \,, \qquad \forall f \in C^\infty_c(\mathbb{R}^d) \,. \tag{2.5.7}$$

Proof It suffices to assume that ν has compact support. With $\tau_n(t) := t \wedge \tau(B_n)$, applying Dynkin's formula we obtain

$$\mathbb{E}^U_x \left[\mathcal{V}(X_{\tau_n(t)}) \right] - \mathcal{V}(x) \leq k_0 \, \mathbb{E}^U_x [\tau_n(t)] - \mathbb{E}^U_x \left[\int_0^{\tau_n(t)} h(X_s) \, \mathrm{d}s \right] . \tag{2.5.8}$$

Letting $n \to \infty$ in (2.5.8), using monotone convergence and rearranging terms, we have that for any ball $B_R \subset \mathbb{R}^d$,

$$\begin{aligned} \left(\min_{B^c_R} h \right) \int_0^t \mathbb{E}^U_x \left[\mathbb{I}_{B^c_R}(X_s) \right] \mathrm{d}s &\leq \int_0^t \mathbb{E}^U_x \left[h(X_s) \right] \mathrm{d}s \\ &\leq k_0 t + \mathcal{V}(x) \,. \end{aligned} \tag{2.5.9}$$

By (2.5.9), integrating with respect to ν,

$$\frac{1}{t} \int_0^t \mathbb{E}^U_\nu \left[\mathbb{I}_{B^c_R}(X_s) \right] \mathrm{d}s \leq \frac{k_0 t + \int \mathcal{V} \, \mathrm{d}\nu}{t \, \min_{B^c_R} h}$$

for all $U \in \mathfrak{U}$, $t > 0$ and $x \in \mathbb{R}^d$. This implies the tightness of (2.5.6a).

To show tightness of (2.5.6b), let $\tilde{\mathcal{V}}(t, x) := \alpha \mathrm{e}^{-\alpha t} \mathcal{V}(x)$. Then by Dynkin's formula

$$\begin{aligned} \mathbb{E}^U_x \left[\tilde{\mathcal{V}}(\tau_n(t), X_{\tau_n(t)}) \right] - \alpha \mathcal{V}(x) \leq \alpha \, \mathbb{E}^U_x &\left[\int_0^{\tau_n(t)} \mathrm{e}^{-\alpha s} \left[k_0 - h(X_s) \right] \mathrm{d}s \right] \\ &- \alpha \, \mathbb{E}^U_x \left[\int_0^{\tau_n(t)} \tilde{\mathcal{V}}(s, X_s) \, \mathrm{d}s \right] , \end{aligned}$$

and rearranging terms and integrating with respect to ν,

$$\alpha\, \mathbb{E}^U_\nu\left[\int_0^{\tau_n(t)} \mathrm{e}^{-\alpha s} h(X_s)\,\mathrm{d}s\right] \le \alpha \int_{\mathbb{R}^d} \mathcal{V}(x)\,\nu(\mathrm{d}x) + k_0\,. \tag{2.5.10}$$

The result follows by taking limits first as $n \to \infty$ and then $t \to \infty$ in (2.5.10), and repeating the argument in (2.5.9).

To prove the second assertion, by Itô's formula, we obtain,

$$\frac{1}{t}\left(\mathbb{E}^U_\nu\left[f(X_t)\right] - \int_{\mathbb{R}^d} f(x)\,\nu(\mathrm{d}x)\right) = \int_{\mathbb{R}^d\times\mathbb{U}} \mathcal{L}^u f(x)\, \bar{\zeta}^U_{\nu,t}(\mathrm{d}x, \mathrm{d}u)\,,$$

and since $(x, u) \mapsto \mathcal{L}^u f(x)$ is a bounded continuous function, (2.5.7) follows by taking limits as $t \to \infty$ along a converging subsequence. An analogous argument applies to the α-discounted occupation measures. □

A probability measure μ satisfying (2.5.7) is called an *infinitesimal occupation measure.*

We now turn to the analysis of (L2.3). We need the following definition.

Definition 2.5.4 We say that (2.2.1) is *bounded in probability* under $U \in \mathfrak{U}$, if the family of measures

$$\{\mathbb{P}^U_x(X_t \in \cdot\,) : x \in K\,,\ t > 0\}$$

is tight for any compact set K.

The Lyapunov condition (L2.3) guarantees that (2.2.1) is bounded in probability uniformly in $U \in \mathfrak{U}$, as the following lemma shows.

Lemma 2.5.5 *Under (L2.3),*

$$\mathbb{E}^U_x[\mathcal{V}(X_t)] \le \frac{k_0}{2k_1} + \mathcal{V}(x)\mathrm{e}^{-2k_1 t} \qquad \forall x \in \mathbb{R}^d\,, \quad \forall U \in \mathfrak{U}\,. \tag{2.5.11}$$

In addition, if B_r is a ball in $\mathbb{R}^d$ such that

$$\mathcal{V}(x) \ge 1 + \frac{k_0}{k_1} \qquad \forall x \in B^c_r$$

and $\breve{\tau}_r := \inf\{t \ge 0 : X_t \in B_r\}$, then

$$\sup_{U\in\mathfrak{U}} \mathbb{E}^U_x[\mathrm{e}^{k_1\breve{\tau}_r}] \le \frac{k_1}{k_0 + k_1}\mathcal{V}(x) \qquad \forall x \in B^c_r\,. \tag{2.5.12}$$

Proof Let $R > 0$, and τ_R denote the first exit time of X from B_R. Applying Dynkin's formula to $f(t) = \mathrm{e}^{4k_1 t}\mathcal{V}(X_t)$, and using (2.5.5), we obtain

$$\begin{aligned}
\mathbb{E}^U_x[f(t\wedge\tau_R)] - \mathcal{V}(x) &= \mathbb{E}^U_x\left[\int_0^{t\wedge\tau_R} \mathrm{e}^{4k_1 s}[4k_1\mathcal{V}(X_s) + \mathcal{L}^{U_s}\mathcal{V}(X_s)]\,\mathrm{d}s\right]\\
&\le \mathbb{E}^U_x\left[\int_0^{t\wedge\tau_R} \mathrm{e}^{4k_1 s}[k_0 + 2k_1\mathcal{V}(X_s)]\,\mathrm{d}s\right],
\end{aligned}$$

and letting $R \to \infty$, using Fatou's lemma, yields

$$\begin{aligned}\mathbb{E}_x^U[f(t)] &\le \mathcal{V}(x) + k_0 \int_0^t \mathrm{e}^{4k_1 s}\mathrm{d}s + \int_0^t 2k_1\, \mathbb{E}_x^U[f(s)]\,\mathrm{d}s \\ &\le \mathcal{V}(x) + \frac{k_0}{4k_1}\mathrm{e}^{4k_1 t} + 2k_1 \int_0^t \mathbb{E}_x^U[f(s)]\,\mathrm{d}s\,. \end{aligned} \tag{2.5.13}$$

Let

$$g(t) := \mathcal{V}(x) + \frac{k_0}{4k_1}\mathrm{e}^{4k_1 t}.$$

Applying Gronwall's lemma to (2.5.13), yields

$$\begin{aligned}\mathbb{E}_x^U[\mathcal{V}(X_t)] &\le \mathrm{e}^{-4k_1 t}\left[g(t) + 2k_1 \int_0^t \mathrm{e}^{2k_1(t-s)} g(s)\,\mathrm{d}s\right] \\ &= \mathrm{e}^{-4k_1 t}\left[\mathcal{V}(x) + \frac{k_0}{4k_1}\mathrm{e}^{4k_1 t} + \mathcal{V}(x)\left(\mathrm{e}^{2k_1 t} - 1\right) + \frac{k_0}{4k_1}\mathrm{e}^{2k_1 t}\left(\mathrm{e}^{2k_1 t} - 1\right)\right] \\ &\le \frac{k_0}{2k_1} + \mathcal{V}(x)\mathrm{e}^{-2k_1 t}\,. \end{aligned}$$

To prove (2.5.12), let $B_R \supset B_r \cup \{x\}$. Applying Dynkin's formula, we obtain

$$\begin{aligned}\frac{k_0 + k_1}{k_1}\,\mathbb{E}_x^U\big[\mathrm{e}^{k_1(\breve{\tau}_r \wedge \tau_R)}\big] - \mathcal{V}(x) &\le \mathbb{E}_x^U\big[\mathrm{e}^{k_1(\breve{\tau}_r \wedge \tau_R)}\mathcal{V}(X_{\breve{\tau}_r \wedge \tau_R})\big] - \mathcal{V}(x) \\ &\le \mathbb{E}_x^U\left[\int_0^{\breve{\tau}_r \wedge \tau_R} \mathrm{e}^{k_1 t}(k_0 - k_1\mathcal{V}(X_t))\,\mathrm{d}t\right] \le 0\,. \end{aligned}$$

The result follows by letting $R \to \infty$. □

It follows by (2.5.12) that all moments of recurrence times to B_r are uniformly bounded. We state this as a corollary.

Corollary 2.5.6 *Suppose (L2.3) holds, and let B_r be as in Lemma 2.5.5. Then for any compact set $\Gamma \in \mathbb{R}^d$, we have*

$$\sup_{U\in\mathfrak{U}}\ \sup_{x\in\Gamma}\ \mathbb{E}_x^U[\breve{\tau}_r^m] < \infty \qquad \forall m \ge 1\,.$$

For a nondegenerate diffusion, the assertion of Corollary 2.5.6 remains true for all bounded domains $D \in \mathbb{R}^d$. This can be shown by employing the technique of the proof (b) ⇒ (a) in Theorem 2.6.10 which appears later on p. 76.

Remark 2.5.7 If instead of holding for all $\mathcal{L}^u$, $u \in \mathbb{U}$, (L2.1)–(L2.3) hold for $\mathcal{L}^v$ for some $v \in \mathfrak{U}_{\mathrm{sm}}$, then the controlled diffusion under v is positive D-recurrent, has tight mean empirical measures, or is bounded in probability, correspondingly.

2.6 Stability of nondegenerate controlled diffusions

In this section the ellipticity condition (2.2.15) is in effect. The analysis uses heavily the results from uniformly elliptic PDEs in Appendix A. Note that the assumptions on (2.2.1) use the parameter K_R, while the family of operators $\mathfrak{L}_0(\gamma)$ in Definition A.1.1 is parameterized by a function γ. It is clear that assumptions (2.2.3)–(2.2.4) imply (A.1.1b) and (A.1.1c), while the *uniform ellipticity* condition (A.1.1a) is equivalent to (2.2.15). Thus under the assumptions of the model (2.2.1) in the nondegenerate case, there exists some function $\gamma\colon (0,\infty) \to (0,\infty)$ such that $\mathcal{L}^v \in \mathfrak{L}_0(\gamma)$ for all $v \in \mathfrak{U}_{\mathrm{sm}}$. The estimates of solutions in Appendix A depend on the function γ which parameterizes the family of operators $\mathfrak{L}_0(\gamma)$. Therefore, for the class of models described by any fixed family of parameters $\{K_R : R > 0\}$, the estimates of solutions of the associated elliptic PDEs depend only on the domain in $\mathbb{R}^d$ that the solution is defined on. Keeping that in mind, and in the interest of notational economy, in the rest of this book we refrain from explicitly mentioning the dependence of the estimates of solutions on K_R or γ.

There are two important properties of solutions which are not discussed in Appendix A. These result from the linear growth condition (2.2.4), which is not imposed on the family $\mathfrak{L}_0(\gamma)$. First, as shown in (2.2.20), for any $x \in \mathbb{R}^d$,

$$\sup_{U\in\mathfrak{U}} \mathbb{P}^U_x(\tau_n \le t) \downarrow 0\,, \qquad \text{as } n \uparrow \infty\,.$$

Second, for any bounded domain D, we have

$$\mathbb{E}^U_x[\tau(D^c)] \xrightarrow[|x|\to\infty]{} \infty\,,$$

uniformly in $U \in \mathfrak{U}$. In fact a stronger statement can be made. If X^n satisfy (2.2.1) and $|X^n_0| = n$, then

$$\inf_{0\le t\le T} |X^n_t| \xrightarrow[n\to\infty]{\text{a.s.}} \infty\,, \tag{2.6.1}$$

under any $U \in \mathfrak{U}$. To prove this under (2.2.4), let $\varphi(x) := \bigl(1 + |x|^4\bigr)^{-1}$, and

$$\mathfrak{s}_n = \inf\,\{t \ge 0 : |X^n_t| \le \sqrt{n}\}\,.$$

Then for some constant $M > 0$, we have $|\mathcal{L}^u\varphi(x)| \le M\varphi(x)$ for all $u \in \mathbb{U}$, and applying Dynkin's formula

$$\begin{aligned}
\mathbb{E}^U[\varphi(X^n_{t\wedge\mathfrak{s}_n})] &= \mathbb{E}^U[\varphi(X^n_0)] + \mathbb{E}^U\left[\int_0^{t\wedge\mathfrak{s}_n} \mathcal{L}^{U_t}\varphi(X_r)\,\mathrm{d}r\right]\\
&\le (1+n^4)^{-1} + \mathbb{E}^U\left[\int_0^t \left|\mathcal{L}^{U_t}\varphi(X_{r\wedge\mathfrak{s}_n})\right|\mathrm{d}r\right]\\
&\le (1+n^4)^{-1} + M\int_0^t \mathbb{E}^U[\varphi(X^n_{r\wedge\mathfrak{s}_n})]\,\mathrm{d}r\,.
\end{aligned} \tag{2.6.2}$$

By (2.6.2),

$$\frac{1}{1+n^2}\,\mathbb{P}^U(\mathfrak{s}_n \le t) \le \mathbb{E}^U[\varphi(X^n_{t\wedge\mathfrak{s}_n})] \le \frac{\mathrm{e}^{Mt}}{1+n^4}\,, \tag{2.6.3}$$

and hence by (2.6.2)–(2.6.3),

$$\mathbb{P}^U\Big(\inf_{0\le t\le T}|X^n_t| \le \sqrt{n}\Big) \le \frac{1+n^2}{1+n^4}\mathrm{e}^{MT}\,, \tag{2.6.4}$$

from which (2.6.1) follows. Note that (2.6.4) also holds under (2.2.32).

2.6.1 Moments of recurrence times

For nondegenerate controlled diffusions sharper characterizations of the mean recurrence times are available.

Theorem 2.6.1 *Let $D \subset \mathbb{R}^d$ be a bounded C^2 domain, K a compact subset of D, and $v \in \mathfrak{U}_{\mathrm{sm}}$. Then $\varphi_1(x) = \mathbb{E}^v_x[\tau(D)]$ and $\varphi_2(x) = \mathbb{E}^v_x[\tau^2(D)]$ are the unique solutions in $\mathscr{W}^{2,p}(D) \cap \mathscr{W}^{1,p}_0(D)$, $p \in (1,\infty)$, of the Dirichlet problem*

$$\begin{aligned} \mathcal{L}^v\varphi_1 &= -1 \quad \text{in } D\,, \qquad &\varphi_1 = 0 \quad \text{on } \partial D\,,\\ \mathcal{L}^v\varphi_2 &= -2\varphi_1 \quad \text{in } D\,, \qquad &\varphi_2 = 0 \quad \text{on } \partial D\,. \end{aligned} \tag{2.6.5}$$

Moreover,

(a) $\tau(D) = \tau(\bar D)$ *and* $\tau(\bar D^c) = \tau(D^c)$, $\mathbb{P}^U$*-a.s.*, $\forall U \in \mathfrak{U}$.
(b) *There exist positive constants* $\overline{C}_k$, $\underline{C}_k$, $k = 1,2$, *depending only on D and K, such that*

$$\underline{C}_k \le \mathbb{E}^v_x\left[\tau^k(D)\right] \le \overline{C}_k \qquad \forall x \in K\,, \quad U \in \mathfrak{U}\,, \quad k = 1,2\,.$$

(c) $\mathbb{P}^U_x\left(\tau(D) > \frac{\underline{C}_1}{4}\right) > \frac{\underline{C}_1^2}{8\overline{C}_2}$ *for all* $x \in K$ *and* $U \in \mathfrak{U}$.

Proof By Theorem A.2.7, (2.6.5) has a unique solution $\varphi_i \in \mathscr{W}^{2,p}(D) \cap \mathscr{W}^{1,p}_0(D)$, $i = 1, 2$, for all $p \in (1,\infty)$. Thus this characterization is standard and can be proved using Dynkin's formula as in the proof of Lemma 2.5.1.

To show part (a) let $\{D_n : n \in \mathbb{N}\}$ be a decreasing sequence of C^2 domains such that $\cap_{n\in\mathbb{N}} D_n = \bar D$. Let φ_n be the solution of the Dirichlet problem

$$\max_{u\in\mathbb{U}} \mathcal{L}^u\varphi_n = -1 \quad \text{in } D_n\,, \qquad \varphi_n = 0 \quad \text{on } \partial D_n\,,$$

and define

$$\tilde\varphi_n(x) = \begin{cases} \varphi_n(x) & \text{for } x \in D_n\,,\\ 0 & \text{for } x \in D^c_n\,. \end{cases}$$

By Theorem A.2.7, $\tilde\varphi_n \in \mathscr{W}^{1,p}(\mathbb{R}^d)$, $p \in (1,\infty)$. Also, $\tilde\varphi_n$ is bounded in $\mathscr{W}^{1,p}(D_1)$, uniformly in $n \in \mathbb{N}$. Thus, since $\tilde\varphi_n$ is also non-increasing, it converges uniformly to a continuous function on D_1. Consequently $\varphi_n \downarrow 0$ on ∂D, uniformly in $n \in \mathbb{N}$. Then for all $U \in \mathfrak{U}$ and $x \in D$,

$$\mathbb{E}^U_x[\tau(D_n)] \le \mathbb{E}^U_x[\tau(D)] + \mathbb{E}^U_x\left[\varphi_n(X_{\tau(D)})\right],$$

from which it follows that

$$\mathbb{E}_x^U[\tau(\bar{D}) - \tau(D)] \le \mathbb{E}_x^U[\tau(D_n) - \tau(D)] \le \sup_{\partial D} \varphi_n \xrightarrow[n\to\infty]{} 0 .$$

Therefore $\mathbb{P}_x^U(\tau(D) \neq \tau(\bar{D})) = 0$. Similarly, for any ball B_R which contains $\bar{D}$, we have $\mathbb{P}_x^U\left(\tau(\bar{D}^c) \wedge \tau_R \neq \tau(D^c) \wedge \tau_R\right) = 0$, and the second assertion follows as well, by letting $R \to \infty$.

The constants $\overline{C}_k$ and $\underline{C}_k$ in part (b) are guaranteed by Theorems A.2.1 and A.2.12, respectively. Lastly, for part (c) we use the inequality

$$\mathbb{P}\left(\tau > \frac{\mathbb{E}\,\tau}{4}\right) \ge \frac{(\mathbb{E}\,\tau)^2}{8\,\mathbb{E}\,\tau^2} \tag{2.6.6}$$

in combination with the bounds in (b). The proof of (2.6.6) is as follows: We decompose $\mathbb{E}\,[\tau]$ as dictated by the partition of

$$\mathbb{R}_+ = \left[0, \tfrac{\mathbb{E}\,\tau}{4}\right] \cup \left(\tfrac{\mathbb{E}\,\tau}{4}, \tfrac{4\,\mathbb{E}\,\tau^2}{\mathbb{E}\,\tau}\right] \cup \left(\tfrac{4\,\mathbb{E}\,\tau^2}{\mathbb{E}\,\tau}, \infty\right),$$

and use upper bounds on each of the intervals to obtain

$$\begin{aligned}
\mathbb{E}\,\tau &\le \frac{\mathbb{E}\,\tau}{4} + \frac{4\,\mathbb{E}\,\tau^2}{\mathbb{E}\,\tau}\,\mathbb{P}\left(\tau > \frac{\mathbb{E}\,\tau}{4}\right) + \mathbb{E}\left[\tau\,\mathbb{I}\left\{\tau > \tfrac{4\,\mathbb{E}\,\tau^2}{\mathbb{E}\,\tau}\right\}\right] \\
&\le \frac{\mathbb{E}\,\tau}{4} + \frac{4\,\mathbb{E}\,\tau^2}{\mathbb{E}\,\tau}\,\mathbb{P}\left(\tau > \frac{\mathbb{E}\,\tau}{4}\right) + \frac{\mathbb{E}\,\tau}{4} .
\end{aligned}$$

In this derivation we use also use the inequality

$$\begin{aligned}
\left(\mathbb{E}\left[\tau\,\mathbb{I}\left\{\tau > \tfrac{4\,\mathbb{E}\,\tau^2}{\mathbb{E}\,\tau}\right\}\right]\right)^2 &\le \mathbb{E}\,\tau^2\ \mathbb{P}\left(\tau > \frac{4\,\mathbb{E}\,\tau^2}{\mathbb{E}\,\tau}\right) \\
&\le \mathbb{E}\,\tau^2 \times \frac{\mathbb{E}\,\tau^2}{\left(\frac{4\,\mathbb{E}\,\tau^2}{\mathbb{E}\,\tau}\right)^2} \\
&= \frac{(\mathbb{E}\,\tau)^2}{16} .
\end{aligned}$$

□

Remark 2.6.2 Theorem 2.6.1 (a) also holds for nonsmooth domains provided they satisfy an *exterior cone condition* [67, p. 203].

Theorem 2.6.3 *Let D be a bounded C^2 domain, and $v \in \mathfrak{U}_{\mathrm{sm}}$. Then*

(i) *If $g \in C(\partial D)$ and $h \in L^p(D)$, $p \ge d$, then*

$$\varphi(x) = \mathbb{E}_x^v[g(X_{\tau(D)})] + \mathbb{E}_x^v\left[\int_0^{\tau(D)} h(X_t)\,\mathrm{d}t\right]$$

is the unique solution in $\mathscr{W}^{2,p}_{\mathrm{loc}}(D) \cap C(\bar{D})$, of the problem $\mathcal{L}^v\varphi = -h$ in D, and $\varphi = g$ on ∂D.

(ii) *Suppose that for some nonnegative $h \in L^\infty_{\mathrm{loc}}(D^c)$,*

$$f(x) := \mathbb{E}_x^v\left[\int_0^{\tau(D^c)} h(X_t)\,\mathrm{d}t\right] \tag{2.6.7}$$

is finite at some $x_0 \in \bar{D}^c$. *Then* $f(x)$ *is finite for all* $x \in \bar{D}^c$ *and it is the minimal nonnegative solution in* $\mathscr{W}^{2,p}_{\mathrm{loc}}(\bar{D}^c) \cap C(D^c)$, $p > 1$, *of* $\mathcal{L}^v f = -h$ *in* $\bar{D}^c$, *and* $f = 0$ *on* ∂D.

Proof The first part follows by Dynkin's formula, and uniqueness is guaranteed by Theorem A.2.7. To prove (ii) let $\{R_n\}$, $n \geq 0$, be a sequence of radii, diverging to infinity, with $D \cup \{x_0\} \Subset B(R_0)$. Set $\tau_n = \tau(B_{R_n})$ and $\breve{\tau} = \tau(D^c)$. Let f_n be the solution in $\mathscr{W}^{2,p}(B_{R_n} \setminus \bar{D}) \cap \mathscr{W}^{1,p}_0(B_{R_n} \setminus \bar{D})$, $p > 1$, of the Dirichlet problem

$$\mathcal{L}^v f_n = -h \quad \text{on } B_{R_n} \setminus \bar{D}\,, \qquad f_n = 0 \quad \text{on } \partial B_{R_n} \cup \partial D\,.$$

Then

$$f_n(x) = \mathbb{E}^v_x \left[\int_0^{\breve{\tau} \wedge \tau_n} h(X_t)\, \mathrm{d}t \right], \quad x \in B_{R_n} \setminus \bar{D}\,.$$

For each $N > 0$ the functions $\psi_{n+1}(x) = f_{n+1} - f_n$, $n \geq N$, are $\mathcal{L}^v$-harmonic in $B_{R_N} \setminus \bar{D}$. It is also clear that $\psi_n > 0$ on $B_{R_n} \setminus \bar{D}$. Since

$$f_n(x_0) \leq f(x_0) < \infty\,,$$

the series of positive $\mathcal{L}^v$-harmonic functions

$$\sum_{k=N+1}^{n} \psi_k(x) = f_n(x) - f_N(x)\,, \quad x \in B_{R_N} \setminus \bar{D}\,,\ n > N$$

is bounded at x_0 and in turn, by Theorem A.2.6, it converges uniformly on $B_{R_N} \setminus \bar{D}$ to an $\mathcal{L}^v$-harmonic function ψ. Let $f = \psi + f_N$. Then f satisfies (2.6.7) on $B_{R_N} \cap D^c$, and since N is arbitrary also on $\bar{D}_c$. By the strong maximum principle (Theorem A.2.3) any nonnegative solution φ of $\mathcal{L}^v \varphi = -h$ on $\bar{D}^c$ satisfies $\varphi \geq f_n$ for all $n > 0$. Thus $\varphi \geq f$. □

Corollary 2.6.4 *Let* D *be a bounded* C^2 *domain, and* $v \in \mathfrak{U}_{\mathrm{sm}}$. *Suppose that* $\mathbb{E}^v_x[\tau^2(D^c)] < \infty$ *for some* $x \in D^c$. *Then*

$$f_1(x) := \mathbb{E}^v_x[\tau(D^c)] \quad \textit{and} \quad f_2(x) := \mathbb{E}^v_x[\tau^2(D^c)]$$

are the minimal nonnegative solutions in $\mathscr{W}^{2,p}_{\mathrm{loc}}(\bar{D}^c) \cap C(D^c)$, $p > 1$, *of*

$$\begin{aligned} \mathcal{L}^v f_1 &= -1 \quad \textit{in } \bar{D}^c\,, & f_1 &= 0 \quad \textit{on } \partial D\,, \\ \mathcal{L}^v f_2 &= -2 f_1 \quad \textit{in } \bar{D}^c\,, & f_2 &= 0 \quad \textit{on } \partial D\,. \end{aligned} \tag{2.6.8}$$

Proof If f_1 and f_2 solve (2.6.8), then by Lemma 2.5.1, $\mathbb{E}^v_x[\tau^k(D^c)] \leq f_k(x)$, $k = 1, 2$, for all $x \in D^c$. The rest follows as in Theorem 2.6.3 (ii). □

2.6.2 Invariant probability measures

We give a brief account of the ergodic behavior under stable stationary Markov controls. We show that under $v \in \mathfrak{U}_{\mathrm{ssm}}$ the controlled process X has a unique invariant probability measure. Moreover, employing a method introduced by Hasminskiĭ, we characterize the invariant probability measure via an embedded Markov chain [70], and show that it depends continuously on $v \in \mathfrak{U}_{\mathrm{ssm}}$, under the topology of Markov controls. To this end we proceed by first stating two auxiliary lemmas.

Lemma 2.6.5 *Let D_1 and D_2 be two open balls in $\mathbb{R}^d$, satisfying $D_1 \Subset D_2$. Then*

$$0 < \inf_{\substack{x\in\bar{D}_1\\ v\in\mathfrak{U}_{\mathrm{sm}}}} \mathbb{E}^v_x[\tau(D_2)] \le \sup_{\substack{x\in\bar{D}_1\\ v\in\mathfrak{U}_{\mathrm{sm}}}} \mathbb{E}^v_x[\tau(D_2)] < \infty\,, \tag{2.6.9a}$$

$$\inf_{\substack{x\in\partial D_2\\ v\in\mathfrak{U}_{\mathrm{ssm}}}} \mathbb{E}^v_x[\tau(D^c_1)] > 0\,, \tag{2.6.9b}$$

$$\sup_{x\in\partial D_2} \mathbb{E}^v_x[\tau(D^c_1)] < \infty \qquad \forall v \in \mathfrak{U}_{\mathrm{ssm}}\,, \tag{2.6.9c}$$

$$\inf_{v\in\mathfrak{U}_{\mathrm{sm}}} \inf_{x\in\Gamma} \mathbb{P}^v_x(\tau(D_2) > \tau(D^c_1)) > 0 \tag{2.6.9d}$$

for all compact sets $\Gamma \subset D_2 \setminus \bar{D}_1$.

Proof Let h be the unique solution in $\mathscr{W}^{2,p}(D_2) \cap \mathscr{W}^{1,p}_0(D_2)$, $p \ge 2$, of

$$\mathcal{L}^v h = -1 \quad \text{in } D_2\,, \qquad h = 0 \quad \text{on } \partial D_2\,.$$

By Dynkin's formula,

$$h(x) = \mathbb{E}^v_x\left[\tau(D_2)\right] \qquad \forall x \in D_2\,.$$

The positive lower bound in (2.6.9a) follows by Theorem A.2.12, while the finite upper bound follows from the weak maximum principle of Alexandroff (see Theorem A.2.1 on p. 304). Note also that (2.6.9a) is a special case of Theorem 2.6.1 (b). To establish (2.6.9b) we select an open ball $D_3 \Supset D_2$ and let φ be the solution of the Dirichlet problem

$$\mathcal{L}^v \varphi = -1 \quad \text{in } D_3 \setminus \bar{D}_1\,, \qquad \varphi = 0 \quad \text{on } \partial D_1 \cup \partial D_3\,.$$

By Theorem A.2.12,

$$\inf_{v\in\mathfrak{U}_{\mathrm{sm}}} \left(\inf_{x\in\partial D_2} \varphi(x)\right) > 0\,,$$

and the result follows since $\mathbb{E}^v_x[\tau(D^c_1)] > \varphi(x)$.

Let $n \in \mathbb{N}$ be large enough so that $D_2 \Subset B_n$, and let g_n be the solution of the Dirichlet problem

$$\mathcal{L}^v g_n = -1 \quad \text{in } B_n \setminus \bar{D}_1\,, \qquad g_n = 0 \quad \text{on } \partial B_n \cup \partial D_1\,.$$

If $x_0 \in \partial D_2$ and $v \in \mathfrak{U}_{\mathrm{ssm}}$, then $\mathbb{E}^v_{x_0}[\tau(D^c_1)] < \infty$. Since

$$g_n(x_0) = \mathbb{E}^v_{x_0}\left[\tau(D^c_1) \wedge \tau(B_n)\right] \le \mathbb{E}^v_{x_0}\left[\tau(D^c_1)\right],$$

by Harnack's inequality [67, corollary 9.25, p. 250], the increasing sequence of $\mathcal{L}^v$-harmonic functions $f_n = g_n - g_1$ is bounded locally in D_1^c, and hence approaches a limit as $n \to \infty$, which is an $\mathcal{L}^v$-harmonic function on D_1^c. Therefore $g = \lim_{n\to\infty} g_n$ is a bounded function on ∂D_2, and by monotone convergence $g(x) = \mathbb{E}_x^v[\tau(D_1^c)]$. Property (2.6.9c) follows.

Turning to (2.6.9d), let $\varphi_v(x) := \mathbb{P}_x^v(\tau(D_2) > \tau(D_1^c))$. Then φ_v is the unique solution in $\mathscr{W}^{2,p}_{\mathrm{loc}}(D_2 \setminus \bar{D}_1) \cap C(\bar{D}_2 \setminus D_1)$, $p > 1$ of

$$\mathcal{L}^v \varphi_v = 0 \quad \text{in } D_2 \setminus \bar{D}_1 , \qquad \varphi_v = 1 \quad \text{on } \partial D_2 , \qquad \varphi_v = 0 \quad \text{on } \partial D_1 .$$

It follows by Theorem A.2.8 and Theorem A.2.15 (1b) that $\{\varphi_v : v \in \mathfrak{U}_{\mathrm{sm}}\}$ is equicontinuous on $\bar{D}_2 \setminus D_1$. We argue by contradiction. If $\varphi_{v_n}(x_n) \to 0$ as $n \to \infty$ for some sequences $\{v_n\} \subset \mathfrak{U}_{\mathrm{sm}}$ and $\{x_n\} \subset \Gamma$, then Harnack's inequality implies that $\varphi_{v_n} \to 0$ uniformly over any compact subset of $D_2 \setminus \bar{D}_1$, which contradicts the equicontinuity of $\{\varphi_{v_n}\}$, since $\varphi_{v_n} = 1$ on ∂D_2. This proves the claim. □

Lemma 2.6.6 *Let D_1 and D_2 be as in Lemma 2.6.5. Let $\hat{\tau}_0 = 0$, and for $k = 0, 1, \dots$ define inductively an increasing sequence of stopping times by*

$$\begin{aligned} \hat{\tau}_{2k+1} &:= \inf\, \{t > \hat{\tau}_{2k} : X_t \in D_2^c\} , \\ \hat{\tau}_{2k+2} &:= \inf\, \{t > \hat{\tau}_{2k+1} : X_t \in D_1\} . \end{aligned} \tag{2.6.10}$$

Then $\hat{\tau}_n \uparrow \infty$, $\mathbb{P}_x^U$-a.s. for all $U \in \mathfrak{U}$.

Proof Let $U \in \mathfrak{U}$. Since $\hat{\tau}_{2n-1} \le \hat{\tau}_{2n}$, $\mathbb{P}_x^U$-a.s., it suffices to prove that for some $\varepsilon > 0$,

$$\sum_{n=0}^{\infty} \mathbb{I}\{\hat{\tau}_{2n+1} - \hat{\tau}_{2n} > \varepsilon\} = \infty \quad \mathbb{P}_x^U\text{-a.s.} \tag{2.6.11}$$

By Theorem 2.3.4, without loss of generality we may assume that U is a feedback control. The statement in (2.6.11) is then equivalent to [93, corollary, p. 151]

$$\sum_{n=0}^{\infty} \mathbb{P}_x^v \left(\hat{\tau}_{2n+1} - \hat{\tau}_{2n} > \varepsilon \mid \mathfrak{F}^X_{\hat{\tau}_{2n}}\right) = \infty \quad \mathbb{P}_x^v\text{-a.s.}$$

By Lemma 2.3.7 it is therefore enough to show that, for some $\varepsilon > 0$,

$$\inf_{U\in\mathfrak{U}} \inf_{x\in\partial D_1} \mathbb{P}_x^U(\hat{\tau}_1 > \varepsilon) > 0 . \tag{2.6.12}$$

Since Theorem 2.6.1 (c) implies (2.6.12), the result follows. □

Theorem 2.6.7 *With D_1, D_2 and $\{\hat{\tau}_n\}$ as in (2.6.10), the process $\tilde{X}_n := X_{\hat{\tau}_{2n}}$, $n \ge 1$, is a ∂D_1-valued ergodic Markov chain, under any $v \in \mathfrak{U}_{\mathrm{ssm}}$. Moreover, there exists a constant $\delta \in (0, 1)$ (which does not depend on v), such that if $\tilde{P}_v$ and $\tilde{\mu}_v$ denote the transition kernel and the stationary distribution of $\tilde{X}$ under $v \in \mathfrak{U}_{\mathrm{ssm}}$, respectively,* *then for all $x \in \partial D_1$,*

$$\begin{aligned} \big\|\tilde{P}_v^{(n)}(x,\cdot\,) - \tilde{\mu}_v(\,\cdot\,)\big\|_{\mathrm{TV}} &\le \frac{2\delta^n}{1-\delta} \qquad \forall n \in \mathbb{N}\,, \\ (1-\delta)\tilde{P}_v(x,\cdot\,) &\le \tilde{\mu}_v(\,\cdot\,)\,. \end{aligned} \tag{2.6.13}$$

Moreover, the map $v \mapsto \tilde{\mu}_v$ from $\mathfrak{U}_{\mathrm{ssm}}$ to $\mathscr{P}(\partial D_1)$ is continuous in the topology of Markov controls (see Section 2.4).

Proof The strong Markov property implies that $\{\tilde{X}_n\}_{n\in\mathbb{N}}$ is a Markov chain. Let R be large enough such that $D_2 \Subset B_R$. With $h \in C(\partial D_1)$, $h \ge 0$, let ψ be the unique solution in $\mathscr{W}^{2,p}_{\mathrm{loc}}(B_R \cap \bar{D}_1^c) \cap C(\overline{B_R \setminus D_1})$, $p > 1$, of the Dirichlet problem

$$\begin{aligned} \mathcal{L}^v\psi &= 0 \quad \text{in } B_R \cap \bar{D}_1^c\,, \\ \psi &= h \quad \text{on } \partial D_1\,, \\ \psi &= 0 \quad \text{on } \partial B_R\,. \end{aligned}$$

Then, for each $x \in \partial D_2$, the map $h \mapsto \psi(x)$ defines a continuous linear functional on $C(\partial D_1)$ and by Dynkin's formula satisfies

$$\psi(x) = \mathbb{E}_x^v\left[h(X_{\tau(D_1^c)})\,\mathbb{I}\{\tau(D_1^c) < \tau_R\}\right].$$

Therefore, by the Riesz representation theorem there exists a finite Borel measure $q_{1,R}^v(x,\cdot\,)$ on $\mathscr{B}(\partial D_1)$ such that

$$\psi(x) = \int_{\partial D_1} q_{1,R}^v(x,\mathrm{d}y)h(y)\,.$$

It is evident that for any $A \in \mathscr{B}(\partial D_1)$, $q_{1,R}^v(x,A) \uparrow q_1^v(x,A)$ as $R \to \infty$, and that $q_1^v(x,A) = \mathbb{P}_x^v(X_{\tau(D_1^c)} \in A)$. Similarly, the analogous Dirichlet problem on D_2 yields a $q_2^v(x,\cdot\,) \in \mathscr{P}(\partial D_2)$, satisfying $q_2^v(x,A) = \mathbb{P}_x^v(X_{\hat{\tau}_1} \in A)$, and by Harnack's inequality, there exists a positive constant C_H such that, for all $x, x' \in \partial D_1$ and $A \in \mathscr{B}(\partial D_2)$,

$$q_2^v(x,A) \le C_H q_2^v(x',A)\,. \tag{2.6.14}$$

Hence, the transition kernel

$$\tilde{P}_v(x,\cdot\,) = \int_{\partial D_2} q_2^v(x,\mathrm{d}y)q_1^v(y,\cdot\,)$$

of $\tilde{X}$ inherits the Harnack inequality in (2.6.14). Therefore, for any fixed $x_0 \in \partial D_1$, we have

$$\tilde{P}_v(x,\cdot\,) \ge C_H^{-1}\tilde{P}_v(x_0,\cdot\,) \qquad \forall x \in \partial D_1\,, \tag{2.6.15}$$

which implies that $\tilde{P}_v$ is a contraction under the total variation norm and satisfies

$$\left\|\int_{\partial D_1}(\mu(\mathrm{d}x) - \mu'(\mathrm{d}x))\tilde{P}_v(x,\cdot\,)\right\|_{\mathrm{TV}} \le \left(1 - C_H^{-1}\right)\|\mu - \mu'\|_{\mathrm{TV}}$$

for all μ and μ' in $\mathscr{P}(\partial D_1)$. Thus (2.6.13) holds with $\delta = (1 - C_H^{-1})$. Since the fixed point of the contraction $\tilde{P}_v$ is unique, the chain is ergodic.

We first show that the maps $v \mapsto q_k^v$, $k = 1,2$, are uniformly continuous on ∂D_2 and ∂D_1, respectively. Indeed, as described above,

$$\varphi_v(x) = \int_{\partial D_1} q_1^v(x, \mathrm{d}y)h(y)\,, \quad v \in \mathfrak{U}_{\mathrm{ssm}}\,, \quad h \in C(\partial D_1)\,,$$

is the unique bounded solution in $\mathscr{W}^{2,p}_{\mathrm{loc}}(\bar{D}_1^c) \cap C(D_1^c)$ of the problem $\mathcal{L}^v\varphi_v = 0$ in $\bar{D}_1^c$, and $\varphi_v = h$ on ∂D_1. Suppose $v_n \to v$ in $\mathfrak{U}_{\mathrm{ssm}}$ as $n \to \infty$. If G is a C^2 bounded domain such that $\partial D_2 \subset G \Subset \bar{D}_1^c$, then by Lemma 2.4.3 every subsequence of $\{\varphi_{v_n}\}$ contains a further subsequence $\{\varphi_n : n \in \mathbb{N}\}$, which converges weakly as $n \to \infty$ in $\mathscr{W}^{2,p}(G)$, $p > 1$, to some $\mathcal{L}^v$-harmonic function. Since G is arbitrary, this limit yields a function $\tilde{\varphi} \in \mathscr{W}^{2,p}_{\mathrm{loc}}(\bar{D}_1^c) \cap C(D_1^c)$ that satisfies $\mathcal{L}^v\tilde{\varphi} = 0$ on $\bar{D}_1^c$ and $\tilde{\varphi} = h$ on ∂D_1. By uniqueness $\tilde{\varphi} = \varphi_v$. Since the convergence is uniform on compact sets, we have $\sup_{\partial D_2} |\varphi_{v_n} - \varphi_v| \to 0$ as $n \to \infty$. This implies that $v \mapsto q_1^v$ is uniformly continuous on ∂D_2. Similarly, $v \mapsto q_2^v$ is uniformly continuous on ∂D_1. Thus their composition $v \mapsto \tilde{P}_v(x, \cdot\,)$ is uniformly continuous on $x \in \partial D_1$. On the other hand, if $\{v_n\} \subset \mathfrak{U}_{\mathrm{ssm}}$ is any sequence converging to $v \in \mathfrak{U}_{\mathrm{ssm}}$ as $n \to \infty$, then since $\mathscr{P}(\partial D_1)$ is compact, there exists a further subsequence also denoted as $\{v_n\}$ along which $\tilde{\mu}_{v_n} \to \tilde{\mu} \in \mathscr{P}(\partial D_1)$. Thus, by the uniform convergence of

$$\int_{\partial D_1} \tilde{P}_{v_n}(x, \mathrm{d}y) f(y) \xrightarrow[n\to\infty]{} \int_{\partial D_1} \tilde{P}_v(x, \mathrm{d}y) f(y)$$

for any $f \in C(\partial D_1)$, we obtain

$$\tilde{\mu}(\cdot\,) = \lim_{n\to\infty} \tilde{\mu}_{v_n}(\cdot\,) = \lim_{n\to\infty} \int_{\partial D_1} \tilde{\mu}_{v_n}(\mathrm{d}x)\tilde{P}_{v_n}(x, \cdot\,) = \int_{\partial D_1} \tilde{\mu}(\mathrm{d}x)\tilde{P}_v(x, \cdot\,)\,,$$

and by uniqueness $\tilde{\mu} = \tilde{\mu}_v$. Thus $v \mapsto \tilde{\mu}_v$ from $\mathfrak{U}_{\mathrm{ssm}}$ to $\mathscr{P}(\partial D_1)$ is continuous. □

Remark 2.6.8 Uniqueness of $\tilde{\mu}_v$ also follows from the fact that since ∂D_1 is compact, the chain under $v \in \mathfrak{U}_{\mathrm{ssm}}$ has a non-empty compact convex set $\mathcal{M}_v \subset \mathscr{P}(\partial D_1)$ of invariant probability measures, the extreme points of which correspond to ergodic measures. If $\mu \in \mathcal{M}_v$, then $\mu(\cdot\,) = \int_{\partial D_1} \mu(\mathrm{d}x)\tilde{P}_v(x, \cdot\,)$, and thus if μ' is another element of $\mathcal{M}_v$, we obtain $\mu \le C_H\mu'$, and vice versa, which implies that μ and μ' are mutually absolutely continuous with respect to each other. Since any two distinct ergodic measures are mutually singular (see Theorem 1.5.5), μ and μ' must coincide.

The proof of Theorem 2.6.7 shows that $\tilde{P}_v$ is a contraction on $\mathscr{P}(\partial D_1)$ under the total variation norm topology. We now exhibit a metric for the weak topology of $\mathscr{P}(\partial D_1)$ under which $\tilde{P}_v$ is a contraction, and the contraction constant does not depend on v. This supplies an alternate proof that the unique fixed point $\tilde{\mu}_v$ is continuous in v.

The Wasserstein metric

$$\rho_w(\mu_1, \mu_2) := \inf \left(\mathbb{E}\left|\Xi_1 - \Xi_2\right|^2\right)^{1/2},$$

where the infimum is over all pairs of random variables (Ξ_1, Ξ_2) with $\mathscr{L}(\Xi_i) = \mu_i$, $i = 1,2$, is a complete metric on $\left\{\mu \in \mathscr{P}(\mathbb{R}^d) : \int |x|^2 \mu(\mathrm{d}x) < \infty\right\}$ [50, 98]. Define the family of *bounded Lipschitz metrics* $\{\delta_p : p > 0\}$ on $\mathscr{P}(\partial D_1)$ by

$$\delta_p(\mu,\mu') := \sup\left\{\int_{\partial D_1} f(x)(\mu-\mu')(\mathrm{d}x) : f \in C^{0,1}(\partial D_1),\ \mathrm{Lip}(f) \le p\right\},$$

where "Lip" stands for the Lipschitz constant. It is well known that δ_1 is equivalent to the Wasserstein metric on $\mathscr{P}(\partial D_1)$.

Let $\tilde{P}_v(x,f) := \int_{\partial D_1} \tilde{P}_v(x,\mathrm{d}y)f(y)$. By (2.6.15), for all $f \in C(\partial D_1)$ we have

$$\left|\tilde{P}_v(x,f) - \tilde{P}_v(y,f)\right| \le \left(1 - C_H^{-1}\right)\mathrm{span}(f) \qquad \forall x,y \in \partial D_1\,.$$

In other words, the map $f \mapsto \tilde{P}_v(\cdot\,,f)$ is a span contraction. Note that δ_p remains unchanged if we restrict the family of f in its definition to

$$\mathscr{F} := \left\{f \in C^{0,1}(\partial D_1) : \mathrm{Lip}(f) \le p,\ \min_{\partial D_1} f = 0\right\}.$$

Note also that $\tilde{P}_v(\cdot\,,f)$ is well defined on D_1 and satisfies $\tilde{P}_v(x\,,f) = \mathbb{E}_x^v[h_f(X_{\tau(D_2)})]$, where $h_f(y) = \mathbb{E}_y^v[f(X_{\tau(D_1^c)})]$. Thus $\sup_{x\in D_1} \tilde{P}_v(x,f) \le \mathrm{span}(f)$ for all $f \in \mathscr{F}$, and it follows by Lemma A.2.5 that for some constant K_0,

$$\left\|\tilde{P}_v(\cdot\,,f)\right\|_{\mathscr{W}^{2,p}(D_1)} \le K_0\,\mathrm{span}(f) \qquad \forall v \in \mathfrak{U}_{\mathrm{ssm}}\,, \quad \forall f \in \mathscr{F}\,.$$

Then, by the compact embedding of $\mathscr{W}^{2,p}(D_1) \hookrightarrow C^{1,r}(\bar{D}_1)$, for $p > \frac{d}{1-r}$, we have

$$\mathrm{Lip}(\tilde{P}_v(\cdot\,,f)) \le C_1\,\mathrm{span}(f)\,, \quad f \in C(\partial D_1)\,, \tag{2.6.16}$$

for some constant C_1 that does not depend on v. Let

$$M_0 := \sup\,\{\mathrm{span}(f) : f \in \mathscr{F}\} < \infty\,.$$

By (2.6.16) and the span contraction property, if we define

$$p_n := (C_1 M_0)^{-1}(1 - C_H^{-1})^{1-n}\,, \quad n \in \mathbb{N}\,,$$

then

$$\begin{aligned}
\mathrm{Lip}(\tilde{P}_v^n(\cdot\,,f)) &= \mathrm{Lip}\left(\tilde{P}_v(\cdot\,,\tilde{P}_v^{n-1}(\cdot\,,f))\right)\\
&\le C_1\,\mathrm{span}(\tilde{P}_v^{n-1}(\cdot\,,f))\\
&\le C_1\left(1 - C_H^{-1}\right)^{n-1}\mathrm{span}(f)\\
&\le \frac{1}{p_n} \qquad \forall f \in \mathscr{F}\,, \quad \forall v \in \mathfrak{U}_{\mathrm{ssm}}\,.
\end{aligned} \tag{2.6.17}$$

Therefore, by (2.6.17) and the fact that $\delta_p(\mu,\mu') = p\,\delta_1(\mu,\mu')$ for all $p > 0$, we obtain

$$\begin{aligned}
\delta_1(\mu\tilde{P}_v^n, \mu'\tilde{P}_v^n) &= \frac{1}{p_n}\,\delta_{p_n}(\mu\tilde{P}_v^n, \mu'\tilde{P}_v^n)\\
&\le \frac{1}{p_n}\,\delta_1(\mu,\mu')\,.
\end{aligned} \tag{2.6.18}$$

Define

$$\tilde{\delta}(\mu,\mu') := \sup_{v\in\mathfrak{U}_{\mathrm{ssm}}} \sum_{k=0}^{\infty} \delta_1(\mu\tilde{P}_v^k, \mu'\tilde{P}_v^k)\,.$$

By (2.6.18),

$$\tilde{\delta}(\mu,\mu') \le (1 + C_1 M_0 C_H)\,\delta_1(\mu,\mu')\,.$$

Since $\tilde{\delta}(\mu,\mu') \ge \delta_1(\mu,\mu')$, it follows that $\tilde{\delta}$ and δ_1 are equivalent. On the other hand,

$$\begin{aligned}\tilde{\delta}(\mu\tilde{P}_v, \mu'\tilde{P}_v) &= \tilde{\delta}(\mu,\mu') - \delta_1(\mu,\mu')\\ &\le \left(1 - (1 + C_1 M_0 C_H)^{-1}\right)\tilde{\delta}(\mu,\mu')\,,\end{aligned}$$

so that $\tilde{P}_v$ is a contraction relative to $\tilde{\delta}$. Therefore, if $\tilde{\mu}_v$ and $\tilde{\mu}_{v'}$ are the unique fixed points of $\tilde{P}_v$ and $\tilde{P}_{v'}$, respectively, we obtain

$$\begin{aligned}\tilde{\delta}(\mu_{v'},\mu_v) &= \tilde{\delta}(\mu_{v'}\tilde{P}_{v'}, \mu_v\tilde{P}_v)\\ &\le \tilde{\delta}(\mu_{v'}\tilde{P}_{v'}, \mu_v\tilde{P}_{v'}) + \tilde{\delta}(\mu_v\tilde{P}_{v'}, \mu_v\tilde{P}_v)\\ &\le \left(1 - (1 + C_1 M_0 C_H)^{-1}\right)\tilde{\delta}(\mu_{v'},\mu_v) + \tilde{\delta}(\mu_v\tilde{P}_{v'}, \mu_v\tilde{P}_v)\,,\end{aligned}$$

which implies that

$$\tilde{\delta}(\mu_{v'},\mu_v) \le (1 + C_1 M_0 C_H)\,\tilde{\delta}(\mu_v\tilde{P}_{v'}, \mu_v\tilde{P}_v)\,. \tag{2.6.19}$$

Since $\{\tilde{P}_v(\,\cdot\,, f) : f \in \mathscr{F}\,,\ v \in \mathfrak{U}_{\mathrm{ssm}}\}$ is a family of uniformly continuous functions and $v \mapsto \tilde{P}_v(\,\cdot\,, f)$ is continuous for each $f \in \mathscr{F}$, the continuity of $v \mapsto \tilde{\mu}_v$ in $\mathscr{P}(\partial D_1)$ follows by (2.6.19).

Theorem 2.6.9 *Let $\{\hat{\tau}_n\}$ be as defined in* (2.6.10)*, and let $\tilde{\mu}_v$ denote the unique stationary probability distribution of $\{\tilde{X}_n\}$, under $v \in \mathfrak{U}_{\mathrm{ssm}}$. Define $\eta_v \in \mathscr{P}(\mathbb{R}^d)$ by*

$$\int_{\mathbb{R}^d} f\,\mathrm{d}\eta_v = \frac{\int_{\partial D_1} \mathbb{E}_x^v\left[\int_0^{\hat{\tau}_2} f(X_t)\,\mathrm{d}t\right]\tilde{\mu}_v(\mathrm{d}x)}{\int_{\partial D_1} \mathbb{E}_x^v\,[\hat{\tau}_2]\,\tilde{\mu}_v(\mathrm{d}x)}\,, \qquad f \in C_b(\mathbb{R}^d)\,.$$

Then η_v is the unique invariant probability measure of X, under $v \in \mathfrak{U}_{\mathrm{ssm}}$.

Proof Define the measure μ_v by

$$\int_{\mathbb{R}^d} g(x)\mu_v(\mathrm{d}x) = \int_{\partial D_1} \mathbb{E}_x^v\left[\int_0^{\hat{\tau}_2} g(X_t)\,\mathrm{d}t\right]\tilde{\mu}_v(\mathrm{d}x) \qquad \forall g \in C_b(\mathbb{R}^d)\,. \tag{2.6.20}$$

For any $s \ge 0$ and $f \in C_b(\mathbb{R}^d)$, we have

$$\begin{aligned}\mathbb{E}_x^v\left[\int_0^{\hat{\tau}_2} \mathbb{E}_{X_t}^v[f(X_s)]\mathrm{d}t\right] &= \mathbb{E}_x^v\left[\int_0^{\infty} \mathbb{I}\{t < \hat{\tau}_2\}\,\mathbb{E}_x^v[f(X_{s+t}) \mid \mathfrak{F}_t^X]\,\mathrm{d}t\right]\\ &= \mathbb{E}_x^v\left[\int_0^{\infty} \mathbb{E}_x^v\left[\mathbb{I}\{t < \hat{\tau}_2\} f(X_{s+t}) \;\middle|\; \mathfrak{F}_t^X\right]\mathrm{d}t\right]\\ &= \mathbb{E}_x^v\left[\int_0^{\hat{\tau}_2} f(X_{s+t})\,\mathrm{d}t\right].\end{aligned} \tag{2.6.21}$$

Since $\tilde{\mu}_v$ is the stationary probability distribution of $\{\tilde{X}_n\}$, we obtain

$$\int_{\partial D_1} \mathbb{E}_x^v \left[\int_{\hat{\tau}_2}^{s+\hat{\tau}_2} f(X_t)\, \mathrm{d}t \right] \tilde{\mu}_v(\mathrm{d}x) = \int_{\partial D_1} \mathbb{E}_x^v \left[\mathbb{E}_{X_{\hat{\tau}_2}}^v \int_0^s f(X_t)\, \mathrm{d}t \right] \tilde{\mu}_v(\mathrm{d}x)$$
$$= \int_{\partial D_1} \mathbb{E}_x^v \left[\int_0^s f(X_t)\, \mathrm{d}t \right] \tilde{\mu}_v(\mathrm{d}x)\,. \qquad (2.6.22)$$

By (2.6.22) and (2.6.20), we have

$$\int_{\partial D_1} \mathbb{E}_x^v \left[\int_0^{\hat{\tau}_2} f(X_{s+t})\, \mathrm{d}t \right] \tilde{\mu}_v(\mathrm{d}x) = \int_{\partial D_1} \mathbb{E}_x^v \left[\int_0^{s+\hat{\tau}_2} f(X_t)\, \mathrm{d}t - \int_0^s f(X_t)\, \mathrm{d}t \right] \tilde{\mu}_v(\mathrm{d}x)$$
$$= \int_{\partial D_1} \mathbb{E}_x^v \left[\int_0^{s+\hat{\tau}_2} f(X_t)\, \mathrm{d}t - \int_{\hat{\tau}_2}^{s+\hat{\tau}_2} f(X_t)\, \mathrm{d}t \right] \tilde{\mu}_v(\mathrm{d}x)$$
$$= \int_{\partial D_1} \mathbb{E}_x^v \left[\int_0^{\hat{\tau}_2} f(X_t)\, \mathrm{d}t \right] \tilde{\mu}_v(\mathrm{d}x)$$
$$= \int_{\mathbb{R}^d} f(x) \mu_v(\mathrm{d}x)\,. \qquad (2.6.23)$$

Thus, first applying (2.6.20) with $g(x) := \mathbb{E}_x^v[f(X_s)]$, and next (2.6.21) followed by (2.6.23), we obtain

$$\int_{\mathbb{R}^d} \mathbb{E}_x^v[f(X_s)] \mu_v(\mathrm{d}x) = \int_{\partial D_1} \mathbb{E}_x^v \left[\int_0^{\hat{\tau}_2} \mathbb{E}_{X_t}^v[f(X_s)]\, \mathrm{d}t \right] \tilde{\mu}_v(\mathrm{d}x)$$
$$= \int_{\partial D_1} \mathbb{E}_x^v \left[\int_0^{\hat{\tau}_2} f(X_{s+t})\, \mathrm{d}t \right] \tilde{\mu}_v(\mathrm{d}x)$$
$$= \int_{\mathbb{R}^d} f(x) \mu_v(\mathrm{d}x)\,.$$

Hence, μ_v is an invariant measure for X, and its normalization $\eta_v := \frac{\mu_v}{\mu_v(\mathbb{R}^d)}$ yields an invariant probability measure.

By Theorem A.3.5, the resolvent $Q_\alpha(x, \mathrm{d}y)$ of X controlled by v is equivalent to the Lebesgue measure $\mathfrak{m}$ for all $x \in \mathbb{R}^d$, and takes the form $Q_\alpha(x, \mathrm{d}y) = q_\alpha(x, y)\mathfrak{m}(\mathrm{d}y)$. If η_v^1 and η_v^2 are two invariant probability measures for X then, for all $A \in \mathscr{B}(\mathbb{R}^d)$,

$$\eta_v^i(A) = \int_A \left(\int_{\mathbb{R}^d} \eta_v^i(\mathrm{d}x) q_\alpha(x, y) \right) \mathfrak{m}(\mathrm{d}y)\,, \quad i = 1, 2\,. \qquad (2.6.24)$$

It follows from (2.6.24) that η_v^1 and η_v^2 are mutually absolutely continuous with respect to the Lebesgue measure and hence with respect to each other. However, if η_v^1 and η_v^2 are distinct ergodic measures they are singular with respect to each other. Hence, $\eta_v^1 = \eta_v^2$. □

We conclude this section with some useful equivalent criteria for stability.

Theorem 2.6.10 *For a nondegenerate diffusion the following are equivalent:*

(a) $v \in \mathfrak{U}_{\mathrm{sm}}$ *is stable;*

(b) $\mathbb{E}^v_x[\tau(G^c)] < \infty$ *for some bounded domain* $G \subset \mathbb{R}^d$ *and all* $x \in \mathbb{R}^d$*;*

(c) $\mathbb{E}^v_{x_0}[\tau(G^c)] < \infty$ *for some bounded domain* $G \subset \mathbb{R}^d$ *and some* $x_0 \in \mathbb{R}^d \setminus \bar{G}$*;*

(d) *there exists a unique invariant probability measure;*

(e) *for any* $g \in C(\mathbb{R}^d, [0, 1])$*, satisfying* $g(0) = 0$ *and* $\lim_{|x|\to\infty} g(x) = 1$*, there exists a nonnegative, inf-compact function* $\mathcal{V} \in \mathscr{W}^{2,p}_{\mathrm{loc}}(\mathbb{R}^d)$*,* $p \geq 2$*, and a constant* $\varrho \in (0, 1)$ *which solve*

$$\mathcal{L}^v\mathcal{V}(x) + g(x) - \varrho = 0\,; \tag{2.6.25}$$

(f) *there exist a nonnegative, inf-compact function* $\mathcal{V} \in \mathscr{W}^{2,p}_{\mathrm{loc}}(\mathbb{R}^d)$*, a constant* $\varepsilon > 0$ *and a compact set* $K \subset \mathbb{R}^d$*, such that*

$$\mathcal{L}^v\mathcal{V}(x) \leq -\varepsilon \qquad \forall x \notin K\,. \tag{2.6.26}$$

Proof We show (c) ⇒ (b) ⇒ (a) ⇒ (d) ⇒ (e) ⇒ (f) ⇒ (c).

(c) ⇒ (b): This follows by Theorem 2.6.3 (ii).

(b) ⇒ (a): Let G_1 be any bounded domain. Observe that if $\mathbb{E}^v_x[\tau(G^c)] < \infty$, then $\mathbb{E}^v_x[\tau(\tilde{G}^c)] < \infty$ for any domain $\tilde{G} \Supset G \cup G_1$. Therefore it is enough to show that $\mathbb{E}^v_x[\tau(G^c)] < \infty$ implies $\mathbb{E}^v_x[\tau(G^c_0)] < \infty$ for any domain $G_0 \Subset G$. Let D_1 and D_2 be open balls in $\mathbb{R}^d$ such that $D_2 \Supset D_1 \Supset G$ and define the sequence of stopping times $\{\hat{\tau}_k\}$, $k \in \mathbb{N}$, as in Lemma 2.6.6, and $\hat{\tau}_0 := \min\{t \geq 0 : X_t \in D_1\}$. By hypothesis $\mathbb{E}^v_x[\hat{\tau}_0] < \infty$, and hence by Lemma 2.6.5, $\mathbb{E}^v_x[\hat{\tau}_k] < \infty$ for all $k \in \mathbb{N}$. As in Lemma 2.6.6, $\hat{\tau}_k \uparrow \infty$, $\mathbb{P}^v_x$-a.s. Let $Q = D_2 \setminus \bar{G}_0$. Then

$$\varphi(x) = \mathbb{P}^v_x(X_{\tau(Q)} \in \partial D_2)\,, \quad x \in Q$$

is the unique solution in $\mathscr{W}^{2,p}_{\mathrm{loc}}(Q) \cap C(\bar{Q})$, $p \geq 2$, of the Dirichlet problem

$$\begin{aligned} \mathcal{L}^v\varphi &= 0 \quad \text{in } Q\,,\\ \varphi &= 0 \quad \text{on } \partial G_0\,,\\ \varphi &= 1 \quad \text{on } \partial D_2\,. \end{aligned}$$

By the strong maximum principle φ cannot have a local maximum in Q. Therefore

$$p_0 := \sup_{x\in\partial D_1} \mathbb{P}^v_x(X_{\tau(Q)} \in \partial D_2) < 1\,.$$

By the strong Markov property, for $k \in \mathbb{N}$,

$$\mathbb{P}^v_x(\tau(G^c_0) > \hat{\tau}_{2k}) \leq p_0\, \mathbb{P}^v_x(\tau(G^c_0) > \hat{\tau}_{2k-2}) \leq \cdots \leq p_0^k\,.$$

Thus, for $x \in \partial D_1$,

$$
\begin{aligned}
\mathbb{E}_x^v[\tau(G_0^c)] &\le \sum_{k=1}^{\infty} \mathbb{E}_x^v\Big[\hat\tau_{2k}\,\mathbb{I}\big\{\hat\tau_{2k-2} < \tau(G_0^c) < \hat\tau_{2k}\big\}\Big] \\
&= \mathbb{E}_x^v[\hat\tau_0] + \sum_{k=1}^{\infty}\sum_{\ell=1}^{k} \mathbb{E}_x^v\Big[(\hat\tau_{2\ell} - \hat\tau_{2\ell-2})\,\mathbb{I}\big\{\hat\tau_{2k-2} < \tau(G_0^c) < \hat\tau_{2k}\big\}\Big] \\
&= \mathbb{E}_x^v[\hat\tau_0] + \sum_{\ell=1}^{\infty}\sum_{k=l}^{\infty} \mathbb{E}_x^v\Big[(\hat\tau_{2\ell} - \hat\tau_{2\ell-2})\,\mathbb{I}\big\{\hat\tau_{2k-2} < \tau(G_0^c) < \hat\tau_{2k}\big\}\Big] \\
&= \mathbb{E}_x^v[\hat\tau_0] + \sum_{\ell=1}^{\infty} \mathbb{E}_x^v\Big[(\hat\tau_{2\ell} - \hat\tau_{2\ell-2})\,\mathbb{I}\big\{\hat\tau_{2\ell-2} < \tau(G_0^c)\big\}\Big] \\
&\le \mathbb{E}_x^v[\hat\tau_0] + \sum_{k=1}^{\infty} p_0^{k-1} \sup_{x\in\partial D_1} \mathbb{E}_x^v[\hat\tau_2] \\
&= \mathbb{E}_x^v\left[\tau(G^c)\right] + \frac{\sup_{x\in\partial D_1} \mathbb{E}_x^v[\hat\tau_2]}{1-p_0}\,.
\end{aligned}
$$

(a) $\Rightarrow$ (d): This follows from Theorem 2.6.9.

(d) $\Rightarrow$ (e): This is a special case of the HJB equation under the near-monotone costs analyzed in Section 3.6.

(e) $\Rightarrow$ (f): This follows from the fact that $K = \{x \in \mathbb{R}^d : g(x) \le \varrho + \varepsilon\}$ is compact for $\varepsilon > 0$ sufficiently small.

(f) $\Rightarrow$ (c): Let G be an open bounded set and $B_R \supset G \cup \{x\}$. By Dynkin's theorem,

$$\mathbb{E}_x^v[\mathcal{V}(X_{\tau(G^c)\wedge\tau_R})] - \mathcal{V}(x) \le -\varepsilon\,\mathbb{E}_x^v[\tau(G^c)\wedge\tau_R]\,.$$

Taking limits as $R \to \infty$ and using Fatou's lemma, we obtain

$$\mathbb{E}_x^v[\tau(G^c)] \le \frac{1}{\varepsilon}\Big(\mathcal{V}(x) - \inf_{\mathbb{R}^d} \mathcal{V}\Big)\,.$$

This completes the proof. □

Remark 2.6.11 In (2.6.25) g may be selected so that $\lim_{|x|\uparrow\infty} g(x) = \infty$, provided it is integrable with respect to the invariant probability measure η_v under $v \in \mathfrak{U}_{\mathrm{ssm}}$, in which case $\varrho = \int g\,\mathrm{d}\eta_v$ and need not be in $(0, 1)$. Similarly (2.6.26) may be replaced by $\lim_{|x|\uparrow\infty} \mathcal{L}^v\mathcal{V}(x) = -\infty$.

2.6.3 A characterization of recurrence

If the diffusion under $v \in \mathfrak{U}_{\mathrm{sm}}$ is recurrent, then μ_v defined in (2.6.20) is a Radon measure, and hence it is a σ-finite invariant measure. Showing this amounts to proving that for any compact set $K \subset \mathbb{R}^d$, it holds that

$$\mathbb{E}_x^v\left[\int_0^{\hat\tau_2} \mathbb{I}_K(X_t)\,\mathrm{d}t\right] < \infty \qquad \forall x \in \partial D_1\,. \tag{2.6.27}$$

Since the portion of the integral in (2.6.27) over $[0, \hat{\tau}_1]$ has clearly finite mean, (2.6.27) follows from Lemma 2.6.13 below.

As also stated earlier, if a nondegenerate diffusion under $v \in \mathfrak{U}_{\mathrm{sm}}$ is D-recurrent relative to a bounded domain D, then it is recurrent. This is the result of the following lemma, which also provides a useful characterization.

Lemma 2.6.12 *Let D be an open ball in $\mathbb{R}^d$. The process X under $v \in \mathfrak{U}_{\mathrm{sm}}$ is D-recurrent if and only if the Dirichlet problem $\mathcal{L}^v\varphi = 0$ in $\bar{D}^c$, $\varphi = f$ on ∂D has a unique bounded solution φ_f for all $f \in C(\partial D)$. Moreover, if X under $v \in \mathfrak{U}_{\mathrm{sm}}$ is D-recurrent for some bounded domain $D \subset \mathbb{R}^d$, then it is recurrent.*

Proof For a nonnegative $f \in C(\partial D)$, let $\hat{\varphi}_f$ denote the minimal nonnegative solution in $\mathscr{W}^{2,p}_{\mathrm{loc}}(\bar{D}^c) \cap C(D^c)$ of the Dirichlet problem $\mathcal{L}^v\varphi = 0$ in $\bar{D}^c$, $\varphi = f$ on ∂D. Using Dynkin's formula it is straightforward to verify that

$$\hat{\varphi}_f(x) = \mathbb{E}^v_x\bigl[f(X_{\tau(D^c)})\,\mathbb{I}\{\tau(D^c) < \infty\}\bigr], \qquad x \in D^c. \tag{2.6.28}$$

With $\mathbf{1}$ denoting the function identically equal to 1, if there exists a unique bounded solution φ_f of this Dirichlet problem, then we must have $\hat{\varphi}_{\mathbf{1}} = \mathbf{1}$, and by (2.6.28), $\mathbb{P}^v_x(\tau(D^c) < \infty) = 1$ for all $x \in D^c$. Suppose there are two distinct such bounded solutions for some $f \in C(\partial D)$. Their difference is then some nonzero bounded solution ψ with zero boundary condition, which without loss of generality satisfies $|\psi(x)| < 1$ for all $x \in D^c$, and $\psi(x') > 0$ for some $x' \in \bar{D}^c$. It follows that $\varphi_{\mathbf{1}} = \mathbf{1} - \psi$ is a nonnegative solution, with boundary condition $f = \mathbf{1}$, satisfying $\varphi_{\mathbf{1}}(x') < 1$. Hence, the minimal nonnegative solution satisfies $\hat{\varphi}_{\mathbf{1}}(x') < 1$, and it follows by (2.6.28) that X is transient relative to D.

Next, suppose X is D-recurrent. By arguing as in (b) $\Rightarrow$ (a) in the proof of Theorem 2.6.10, it suffices to show that it is also G-recurrent for every bounded C^2 domain $G \Subset D$. Let $\hat{\varphi}$ be the minimal nonnegative solution of $\mathcal{L}^v\varphi = 0$ in $\bar{G}^c$, $\varphi = 1$ on ∂G. Suppose $\hat{\varphi}$ is not constant, otherwise $\hat{\varphi} = 1$ and X is G-recurrent. Clearly $\hat{\varphi} \le 1$ on $\bar{G}^c$. Let $B_R \Supset D$ such that $\hat{\varphi} \not\equiv 1$ on $B_R \setminus D$. Choose $\bar{x} \in \partial B_R$ such that $\hat{\varphi}(\bar{x}) = \min_{\partial B_R} \hat{\varphi}$. By the strong maximum principle $\min_{\partial D} \hat{\varphi} > \hat{\varphi}(\bar{x})$. However since X is D-recurrent,

$$\hat{\varphi}(\bar{x}) = \mathbb{E}_{\bar{x}}[\hat{\varphi}(X_{\tau(D^c)})] \ge \min_{\partial D} \hat{\varphi},$$

and we reach a contradiction. □

We continue with a useful technical lemma.

Lemma 2.6.13 *Let $D \subset \mathbb{R}^d$ be a bounded domain and $G \subset \mathbb{R}^d$ a compact set. Define*

$$\xi^v_{D,G}(x) := \mathbb{E}^v_x\left[\int_0^{\tau(D^c)} \mathbb{I}_G(X_t)\,\mathrm{d}t\right].$$

Then

(i) $\sup_{v\in\mathfrak{U}_{\mathrm{sm}}} \sup_{x\in\bar{D}^c} \xi^v_{D,G}(x) < \infty$;

(ii) *if X is recurrent under* $v \in \mathfrak{U}_{\mathrm{sm}}$*, then* $\xi^v_{D,G}$ *is the unique bounded solution in* $\mathscr{W}^{2,p}_{\mathrm{loc}}(\bar{D}^c) \cap C(D^c)$ *of the Dirichlet problem* $\mathcal{L}^v \xi = -\mathbb{I}_G$ *in* D^c *and* $\xi = 0$ *on* ∂D*;*

(iii) *if* $\mathcal{U} \subset \mathfrak{U}_{\mathrm{sm}}$ *is a closed set of controls under which X is recurrent, the map* $(v, x) \mapsto \xi^v_{D,G}(x)$ *is continuous on* $\mathcal{U} \times \bar{D}^c$*.*

Proof Without loss of generality we may suppose that D is a ball centered at the origin. Let $\hat{\tau} := \tau(D^c)$ and define

$$Z_t := \int_0^t \mathbb{I}_G(X_s)\,\mathrm{d}s\,, \quad t \ge 0\,.$$

Select $R > 0$ such that $D \cup G \subset B_R$. Using the strong Markov property, since $\mathbb{I}_G = 0$ on B^c_R, we obtain

$$\begin{aligned}
&\mathbb{E}^v_x[Z_{\hat{\tau}}] \le \sup_{x' \in \partial B_R} \mathbb{E}^v_{x'}[Z_{\hat{\tau}}] \qquad \forall x \in B^c_R\,,\\
&\mathbb{E}^v_x[Z_{\hat{\tau}}] \le \mathbb{E}^v_x[Z_{\hat{\tau}\wedge\tau_R}] + \sup_{x' \in \partial B_R} \mathbb{E}^v_{x'}[Z_{\hat{\tau}}] \qquad \forall x \in B_R \cap D^c\,.
\end{aligned} \tag{2.6.29}$$

Since, by (2.6.9a),

$$\sup_{v \in \mathfrak{U}_{\mathrm{sm}}} \sup_{x \in B_R \cap D^c} \mathbb{E}^v_x[Z_{\hat{\tau}\wedge\tau_R}] \le \sup_{v \in \mathfrak{U}_{\mathrm{sm}}} \sup_{x \in B_R \cap D^c} \mathbb{E}^v_x[\hat{\tau} \wedge \tau_R] < \infty\,, \tag{2.6.30}$$

it suffices by (2.6.29), to exhibit a uniform bound for $\mathbb{E}^v_x[Z_{\hat{\tau}}]$ on ∂B_R. Let $R' > R$. By (2.6.9d), for some constant $\beta < 1$, we have

$$\sup_{v \in \mathfrak{U}_{\mathrm{sm}}} \sup_{x \in \partial B_R} \mathbb{P}^v_x(\hat{\tau} \ge \tau_{R'}) < \beta\,.$$

Set $\hat{\tau}(t) = t \wedge \hat{\tau}$. By conditioning first at $\tau_{R'}$, and using the fact that $\mathbb{I}_G = 0$ on B^c_R and (2.6.29), we obtain, for $x \in \partial B_R$,

$$\begin{aligned}
\mathbb{E}^v_x\left[Z_{\hat{\tau}(t)}\right] &= \mathbb{E}^v_x\left[Z_{\hat{\tau}(t)}\,\mathbb{I}\{\hat{\tau}(t) < \tau_{R'}\}\right] + \mathbb{E}^v_x\left[Z_{\hat{\tau}(t)}\,\mathbb{I}\{\hat{\tau}(t) \ge \tau_{R'}\}\right]\\
&\le \mathbb{E}^v_x\left[Z_{\hat{\tau}\wedge\tau_{R'}}\right] + \left(\sup_{x \in \partial B_R} \mathbb{P}^v_x(\hat{\tau}(t) \ge \tau_{R'})\right)\left(\sup_{x \in \partial B_{R'}} \mathbb{E}^v_x\left[Z_{\hat{\tau}(t)}\right]\right)\\
&\le \mathbb{E}^v_x\left[Z_{\hat{\tau}\wedge\tau_{R'}}\right] + \beta \sup_{x \in \partial B_R} \mathbb{E}^v_x\left[Z_{\hat{\tau}(t)}\right]\,.
\end{aligned} \tag{2.6.31}$$

By (2.6.30) and (2.6.31), for all $v \in \mathfrak{U}_{\mathrm{sm}}$,

$$\begin{aligned}
\sup_{x \in \partial B_R} \mathbb{E}^v_x\left[Z_{\hat{\tau}(t)}\right] &\le (1-\beta)^{-1} \sup_{x \in \partial B_R} \mathbb{E}^v_x\left[Z_{\hat{\tau}\wedge\tau_{R'}}\right]\\
&\le (1-\beta)^{-1} \sup_{v \in \mathfrak{U}_{\mathrm{sm}}} \sup_{x \in B_{R'} \cap D^c} \mathbb{E}^v_x[\hat{\tau} \wedge \tau_{R'}] < \infty\,.
\end{aligned} \tag{2.6.32}$$

Taking limits as $t \to \infty$ in (2.6.32), using monotone convergence, (i) follows.

Next we prove (ii). With $R > 0$ such that $B_R \Supset D$, let φ_R be the unique solution in $\mathscr{W}^{2,p}(B_R \cap \bar{D}^c) \cap \mathscr{W}^{1,p}_0(\bar{B}_R \cap D^c)$, $p > 1$, of the Dirichlet problem

$$\mathcal{L}^v \varphi_R = -\mathbb{I}_G \quad \text{in } B_R \cap \bar{D}^c\,, \qquad \varphi_R = 0 \quad \text{on } \partial B_R \cup \partial D\,.$$

By Dynkin's formula, φ_R is dominated by $\xi^v_{D,G}$, and since it is nondecreasing in R, it converges, uniformly over compact subsets of $\mathbb{R}^d$, to some $\varphi \in \mathscr{W}^{2,p}_{\mathrm{loc}}(\bar{D}^c) \cap C_b(D^c)$, as $R \uparrow \infty$. The function φ solves

$$\mathcal{L}^v \varphi = -\mathbb{I}_G \quad \text{in } \bar{D}^c\,, \qquad \varphi = 0 \quad \text{on } \partial D\,. \tag{2.6.33}$$

Since by hypothesis $\mathbb{P}^v(\tau(D^c) < \infty) = 1$, an application of Dynkin's formula yields $\varphi = \xi^v_{D,G}$. Hence $\xi^v_{D,G}$ is a bounded solution of (2.6.33). Suppose φ' is another bounded solution of (2.6.33). Then $\varphi - \varphi'$ is $\mathcal{L}^v$-harmonic in $\bar{D}^c$ and equals zero on ∂D. By Lemma 2.6.12, $\varphi - \varphi'$ must be identically zero on D^c. Uniqueness follows.

To show (iii), let $v_n \to v$ in $\mathcal{U}$. By Lemmas A.2.5 and 2.4.3, every subsequence of $\{\xi^{v_n}_{D,G}\}$ contains a further subsequence converging weakly in $\mathscr{W}^{2,p}(D')$, $p > 1$, over any bounded domain $D' \Subset \bar{D}^c$ to some ψ satisfying $\mathcal{L}^v\psi = -\mathbb{I}_G$ in D^c. By uniqueness of the solution of the Dirichlet problem this limit must be $\xi^v_{D,G}$, and since convergence is uniform on compact subsets of $\bar{D}^c$, continuity of $(v, x) \mapsto \xi^v_{D,G}(x)$ follows. □

2.6.4 Infinitesimally invariant probability measures

We start with a key technical lemma. Recall that under $v \in \mathfrak{U}_{\mathrm{sm}}$, X is a strong Feller process. Denote by $T^v_t : C_b(\mathbb{R}^d) \to C_b(\mathbb{R}^d)$ the operator

$$T^v_t f(x) := \mathbb{E}^v_x[f(X_t)]\,, \quad t \ge 0\,.$$

This is the associated transition semigroup with infinitesimal generator $\mathcal{L}^v$, whose domain in denoted by $\mathscr{D}(\mathcal{L}^v)$.

Lemma 2.6.14 *A probability measure $\eta \in \mathscr{P}(\mathbb{R}^d)$ is invariant for X under $v \in \mathfrak{U}_{\mathrm{sm}}$, if and only if*

$$\int_{\mathbb{R}^d} \mathcal{L}^v f(x)\,\eta(\mathrm{d}x) = 0 \qquad \forall f \in \mathscr{D}(\mathcal{L}^v)\,. \tag{2.6.34}$$

Proof If η is invariant for X, and X_0 has law η, then so does X_t for all $t \ge 0$. Then, for $f \in \mathscr{D}(\mathcal{L}^v)$, using the notation

$$\mathbb{E}^v_\eta[f(X_t)] := \int_{\mathbb{R}^d} \mathbb{E}^v_x[f(X_t)]\,\eta(\mathrm{d}x)\,,$$

we have

$$\begin{aligned} 0 = \mathbb{E}^v_\eta[f(X_t)] - \mathbb{E}^v_\eta[f(X_0)] &= \mathbb{E}^v_\eta\left[\int_0^t \mathcal{L}^v f(X_s)\,\mathrm{d}s\right] \\ &= \int_0^t \mathbb{E}^v_\eta[\mathcal{L}^v f(X_s)]\,\mathrm{d}s \\ &= t\int_{\mathbb{R}^d} \mathcal{L}^v f(x)\,\eta(\mathrm{d}x)\,, \end{aligned} \tag{2.6.35}$$

which implies (2.6.34). Now suppose η satisfies (2.6.34). We have

$$T^v_t f(x) - f(x) = \int_0^t T^v_s(\mathcal{L}^v f)(x)\,\mathrm{d}s \qquad \forall t \ge 0\,. \tag{2.6.36}$$

Note that if $f \in \mathscr{D}(\mathcal{L}^v)$, then $T^v_s f \in \mathscr{D}(\mathcal{L}^v)$ for all $s \geq 0$. Therefore

$$\int_{\mathbb{R}^d} \mathcal{L}^v(T^v_s f)(x)\,\eta(\mathrm{d}x) = 0 \qquad \forall s \geq 0\,,\ \forall f \in \mathscr{D}(\mathcal{L}^v)\,. \tag{2.6.37}$$

Hence, integrating (2.6.36) with respect to η over $\mathbb{R}^d$, applying Fubini's theorem, and using (2.6.37) together with the property

$$T^v_s(\mathcal{L}^v f) = \mathcal{L}^v(T^v_s f) \qquad \forall f \in \mathscr{D}(\mathcal{L}^v)\,,$$

we obtain

$$\int_{\mathbb{R}^d} T^v_t f(x)\,\eta(\mathrm{d}x) = \int_{\mathbb{R}^d} f(x)\,\eta(\mathrm{d}x) \qquad \forall f \in \mathscr{D}(\mathcal{L}^v)\,. \tag{2.6.38}$$

Since $\mathcal{L}^v$ is the generator of a strongly continuous semigroup on $C_b(\mathbb{R}^d)$, $\mathscr{D}(\mathcal{L}^v)$ is dense in $C_b(\mathbb{R}^d)$. Thus (2.6.38) holds for all $f \in C_b(\mathbb{R}^d)$, and it follows that η is an invariant probability measure for X. □

If an invariant probability measure η is absolutely continuous with respect to the Lebesgue measure, then its density ψ is a generalized solution to the adjoint equation given by

$$(\mathcal{L}^v)^*\psi(x) = \sum_{i=1}^d \frac{\partial}{\partial x_i}\left(\sum_{j=1}^d a^{ij}(x)\frac{\partial \psi}{\partial x_j}(x) + \hat{b}^i_v(x)\psi(x)\right) = 0\,, \tag{2.6.39}$$

where

$$\hat{b}^i_v = \sum_{j=1}^d \frac{\partial a^{ij}}{\partial x_j} - b^i_v\,.$$

A probability measure μ is called *infinitesimally invariant* for $\mathcal{L}^v$, $v \in \mathfrak{U}_{\mathrm{sm}}$, if

$$\int_{\mathbb{R}^d} \mathcal{L}^v f(x)\,\mu(\mathrm{d}x) = 0 \qquad \forall f \in C^\infty_c(\mathbb{R}^d)\,,$$

which is also denoted as $(\mathcal{L}^v)^*\mu = 0$. As we show later in Theorem 5.3.4, which applies to a more general model, if μ is infinitesimally invariant for $\mathcal{L}^v$, then μ possesses a density $\psi \in \mathscr{W}^{1,p}_{\mathrm{loc}}(\mathbb{R}^d)$ for all $p > 1$ (see also [26]). It is also shown by Bogachev *et al.* [27] that there exists an infinitesimally invariant probability measure for $\mathcal{L}^v$, provided there exists a Lyapunov function $\mathcal{V} \in C^2(\mathbb{R}^d)$ satisfying (compare with Theorem 2.6.10)

$$\lim_{|x|\to\infty} \mathcal{V}(x) = \infty \quad \text{and} \quad \lim_{|x|\to\infty} \mathcal{L}^v\mathcal{V}(x) = -\infty\,.$$

Theorem 2.6.16 below enables us to replace $\mathscr{D}(\mathcal{L}^v)$ in Lemma 2.6.14 with the space of functions in $C^2(\mathbb{R}^d)$ with compact support. Moreover, it asserts that invariant probability measures of nondegenerate positive-recurrent diffusions possess a density which is in $\mathscr{W}^{1,p}_{\mathrm{loc}}(\mathbb{R}^d)$ for all $p > 1$.

We need the following definition.

Definition 2.6.15 Let $\mathscr{C}$ denote a countable dense subset of $C_0^2(\mathbb{R}^d)$ with compact supports, where as usual $C_0^2(\mathbb{R}^d)$ denotes the Banach space of functions $f : \mathbb{R}^d \to \mathbb{R}$ that are twice continuously differentiable and they and their derivatives up to second order vanish at infinity, with the norm

$$\|f\|_{C^2} := \sup_{x\in\mathbb{R}^d} \left\{ |f(x)| + \sum_i \left|\tfrac{\partial f}{\partial x_i}(x)\right| + \sum_{i,j} \left|\tfrac{\partial^2 f}{\partial x_i \partial x_j}(x)\right| \right\}.$$

Theorem 2.6.16 *A Borel probability measure η on $\mathbb{R}^d$ is an invariant measure for the process associated with $\mathcal{L}^v$, if and only if*

$$\int_{\mathbb{R}^d} \mathcal{L}^v f(x)\,\eta(\mathrm{d}x) = 0 \qquad \forall f \in \mathscr{C}. \tag{2.6.40}$$

Moreover, if η satisfies (2.6.40), *then it has a density $\psi \in \mathscr{W}^{1,p}_{\mathrm{loc}}(\mathbb{R}^d)$ with respect to the Lebesgue measure and for any $R > 0$ there exist positive constants $k_1(v, R)$ and $k_2(v, R)$ such that*

$$k_1(v, R) \le \psi(x) \le k_2(v, R) \qquad \forall x \in B_R. \tag{2.6.41}$$

Proof Applying Itô's formula for $f \in \mathscr{C}$ yields (2.6.35), and necessity follows. Under the linear growth assumption (2.2.4), Proposition 1.10 (c) and Remark 1.11 (i) in Stannat [109], combined with Theorem 5.3.4, which applies to a more general model, assert that η has a density and is an invariant measure for the diffusion process. The bounds in (2.6.41) are obtained by applying Harnack's inequality for the divergence form equation $(\mathcal{L}^v)^*\psi = 0$ [67, theorem 8.20, p. 199]. □

2.7 The discounted and ergodic control problems

Let $c : \mathbb{R}^d \times \mathbb{U} \to \mathbb{R}$ be a continuous function bounded from below. In accordance with the relaxed control framework, we define

$$\bar{c}(x, v) = \int_{\mathbb{U}} c(x, u)v(\mathrm{d}u), \quad x \in \mathbb{R}^d, \quad v \in \mathscr{P}(\mathbb{U}).$$

The function c serves as the *running cost*. Before we introduce the ergodic control problem, we review the infinite horizon discounted control problem and the associated dynamic programming principle. This plays an important role in developing the dynamic programming principle for ergodic control, which is obtained from the discounted problem via an asymptotic analysis known as the *vanishing discount* limit.

We define the infinite horizon α-discounted cost

$$J^U_\alpha(x) := \mathbb{E}^U_x \left[\int_0^\infty \mathrm{e}^{-\alpha t}\bar{c}(X_t, U_t)\,\mathrm{d}t \right], \quad x \in \mathbb{R}^d, \quad U \in \mathfrak{U}, \tag{2.7.1}$$

where $\alpha > 0$ is called the *discount factor*. The infinite horizon *α-discounted control problem* seeks to minimize (2.7.1) over all $U \in \mathfrak{U}$. This prompts the definition

$$V_\alpha(x) := \inf_{U\in\mathfrak{U}} J^U_\alpha(x)$$

of the *α-discounted value function* V_α.

The following theorem, also known as the *principle of optimality*, characterizes the α-discounted control problem.

Theorem 2.7.1 *Suppose that $J_\alpha^{\hat{U}}(x) < \infty$ for each $x \in \mathbb{R}^d$ for some control $\hat{U} \in \mathfrak{U}$. Then for each initial law, there exists an optimal U. Also, V_α satisfies the dynamic programming principle*

$$V_\alpha(x) = \min_{U \in \mathfrak{U}} \mathbb{E}_x^U \left[\int_0^{t \wedge \tau} e^{-\alpha s} \bar{c}(X_s, U_s)\, ds + e^{-\alpha(t \wedge \tau)} V_\alpha(X_{t \wedge \tau}) \right] \qquad \forall t \geq 0, \qquad (2.7.2)$$

for all $x \in \mathbb{R}^d$ and all $(\mathfrak{F}_t^X)$-stopping times τ. Moreover, the minimum on the right-hand side of (2.7.2) *is attained at an optimal control.*

Proof If c is bounded, then $J_\alpha^U(x)$ is a bounded linear functional of the set A_x of laws $\mathscr{L}(X, U)$ corresponding to the solutions of (2.2.1) with $X_0 = x$. Otherwise, consider the truncation $c_N = c \wedge N$, with $N \in \mathbb{N}$. Since lower semicontinuity is preserved under limits of increasing sequences of such functions, it follows that $J_\alpha^U(x)$ is a lower semicontinuous functional of the law of (X, U), and since the set A_x is compact by Corollary 2.3.9, the infimum is attained.

Let (X^*, U^*) be an optimal pair with $X_0^* = x$. By Theorem 2.3.4 we may assume that U^* is a feedback control of the form $U_t^* = f^*(t, X_{[0,t]})$. Then

$$V_\alpha(x) = \mathbb{E}_x^{U^*} \left[\int_0^\infty e^{-\alpha t} \bar{c}(X_t^*, f^*(t, X_{[0,t]}))\, dt \right]$$

by definition. Hence, for any $(\mathfrak{F}_t^{X^*})$-stopping time τ, using Lemma 2.3.7, we obtain

$$\begin{aligned} V_\alpha(x) &= \mathbb{E}_x^{U^*} \Bigg[\int_0^{t \wedge \tau} e^{-\alpha s} \bar{c}(X_s^*, f^*(s, X_{[0,s]}))\, ds \\ &\qquad + e^{-\alpha(t \wedge \tau)}\, \mathbb{E}_x^{U^*} \left[\int_{t \wedge \tau}^\infty e^{-\alpha(s - t \wedge \tau)} \bar{c}(X_s^*, f^*(s, X_{[0,s]}))\, ds \,\middle|\, \mathfrak{F}_{t \wedge \tau}^{X^*} \right] \Bigg] \\ &\geq \mathbb{E}_x^{U^*} \left[\int_0^{t \wedge \tau} e^{-\alpha s} \bar{c}(X_s^*, f^*(s, X_{[0,s]}))\, ds + e^{-\alpha(t \wedge \tau)} V_\alpha(X_{t \wedge \tau}^*) \right] \\ &\geq \inf_{U \in \mathfrak{U}} \mathbb{E}_x^U \left[\int_0^{t \wedge \tau} e^{-\alpha s} \bar{c}(X_s, U_s)\, ds + e^{-\alpha(t \wedge \tau)} V_\alpha(X_{t \wedge \tau}) \right]. \end{aligned} \qquad (2.7.3)$$

Let $(\tilde{X}, \tilde{U})$ be such that $\tilde{X}_0 = x$ and $\tilde{U} \circ \theta_{t \wedge \tau}$ is conditionally independent of $\mathfrak{F}_{t \wedge \tau}^{\tilde{X}}$ given $\tilde{X}_{t \wedge \tau}$, and optimal for the initial condition $\tilde{X}_{t \wedge \tau}$. That it is possible to construct such a control follows by arguments similar to those used in the proof of Theorem 2.3.4 (b) (see also a more general construction in Lemma 6.4.10, p. 238). We have

$$\begin{aligned} V_\alpha(x) &\leq \mathbb{E}_x^{\tilde{U}} \left[\int_0^\infty e^{-\alpha t} \bar{c}(\tilde{X}_t, \tilde{U}_t)\, dt \right] \\ &= \mathbb{E}_x^{\tilde{U}} \left[\int_0^{t \wedge \tau} e^{-\alpha s} \bar{c}(\tilde{X}_s, \tilde{U}_s)\, ds + e^{-\alpha(t \wedge \tau)} V_\alpha(\tilde{X}_{t \wedge \tau}) \right]. \end{aligned} \qquad (2.7.4)$$

Since $\{\tilde{U}_s : 0 \leq s \leq t \wedge \tau\}$ is unconstrained except for the non-anticipativity requirement, taking the infimum over such controls in (2.7.4) we obtain

$$V_\alpha(x) \leq \inf_{\tilde{U} \in \mathfrak{U}} \mathbb{E}_x^{\tilde{U}} \left[\int_0^{t\wedge\tau} e^{-\alpha s} \bar{c}(\tilde{X}_s, \tilde{U}_s)\, ds + e^{-\alpha(t\wedge\tau)} V_\alpha(\tilde{X}_{t\wedge\tau}) \right]. \tag{2.7.5}$$

By (2.7.3) and (2.7.5) equality must hold and (2.7.2) follows. The last assertion is then evident by (2.7.3). □

Corollary 2.7.2 (The martingale dynamic programming principle) *The process*

$$e^{-\alpha t} V_\alpha(X_t) + \int_0^t e^{-\alpha s} \bar{c}(X_s, U_s)\, ds\,, \quad t \geq 0\,,$$

is an $(\mathfrak{F}_t^X)$*-submartingale and it is a martingale if and only if* (X, U) *is an optimal pair.*

Proof For any a.s. bounded $(\mathfrak{F}_t^X)$-stopping time τ, letting $t \to \infty$ in (2.7.2), we obtain

$$V_\alpha(x) \leq \mathbb{E}_x^U \left[\int_0^{\tau} e^{-\alpha s} \bar{c}(X_s, U_s)\, ds + e^{-\alpha \tau} V_\alpha(X_\tau) \right],$$

with equality if and only if (X, U) is optimal. The claim follows from a standard characterization of submartingales and martingales, respectively. □

The first part of Theorem 2.7.1 can be improved to assert the existence of time-homogeneous Markov control which is optimal, by using a procedure known as Krylov selection. This is presented in Section 6.7 for a fairly general class of Markov processes.

The ergodic control problem, in its *almost sure* (or *pathwise*) formulation, seeks to a.s. minimize over all admissible $U \in \mathfrak{U}$

$$\limsup_{t\to\infty} \frac{1}{t} \int_0^t \bar{c}(X_s, U_s)\, ds\,. \tag{2.7.6}$$

A weaker, *average* formulation seeks to minimize

$$\limsup_{t\to\infty} \frac{1}{t} \int_0^t \mathbb{E}^U [\bar{c}(X_s, U_s)]\, ds\,. \tag{2.7.7}$$

The analysis of the ergodic control problem is closely tied with stability. This is partly because stability is a desirable property for control systems. In the case of the average formulation, we seek an admissible control U^* which attains the infimum of (2.7.7) over $\mathfrak{U}$, while at the same time the mean empirical measures $\{\bar{\zeta}_{x,t}^{U^*}\}$ are kept tight. Problems where this is not possible are deemed pathological and are uninteresting. Suppose then that the mean empirical measures are tight under some $U^* \in \mathfrak{U}$. By Lemma 2.5.3, with $\mathscr{G}_0 \subset \mathscr{P}(\mathbb{R}^d \times \mathbb{U})$ denoting the class of infinitesimal occupation measures defined in (2.5.7), we have

$$\liminf_{t\to\infty} \frac{1}{t} \int_0^t \mathbb{E}_x^U [\bar{c}(X_s, U_s)]\, ds \geq \inf_{\mu \in \mathscr{G}_0} \int_{\mathbb{R}^d \times \mathbb{U}} c(x, u)\, \mu(dx, du)\,. \tag{2.7.8}$$

Clearly, a sufficient condition for U^* to be optimal with respect to the average criterion, is that $\lim_{t\to\infty} \bar{\zeta}_{x,t}^{U^*} \to \mu^* \in \mathscr{P}(\mathbb{R}^d \times \mathbb{U})$, and that μ^* attains the infimum of the right-hand side of (2.7.8).

At the same time, of special importance is the class of stationary Markov controls $\mathfrak{U}_{\mathrm{sm}}$, and in particular its subclass of precise controls $\mathfrak{U}_{\mathrm{sd}}$ because of ease of implementation. One question that arises then is whether every infinitesimal occupation measure is attainable as a limit of mean empirical measures as $t \to \infty$ over some admissible control, and a second question is whether this is also possible if we consider only stationary Markov controls. For nondegenerate diffusions both questions are answered affirmatively. In fact these questions have already been answered by Theorem 2.6.16, which we restate in Chapter 3 in a different form. In the degenerate case, the first question has essentially an affirmative answer, and this is in fact true for a much more general model which we investigate in Chapter 6.

2.8 Bibliographical note

Section 2.2. An excellent exposition of stochastic calculus and allied topics can be found in [76]. Some other books on these topics are [102, 103, 74, 87].

Sections 2.3–2.4. Here we follow [28, 29].

Sections 2.5–2.6. These are mostly extensions of results in [69, 70] to the controlled case.

Section 2.7. An early, albeit a bit dated, exposition of controlled diffusions appears in [58]. Our exposition is along the lines of [28]. For a more analytic approach, see [14, 78].

3

Nondegenerate Controlled Diffusions

3.1 Introduction

In this chapter, we study controlled diffusions whose infinitesimal generator is locally uniformly elliptic, in other words, it satisfies (2.2.15). Intuitively, this implies that the noise (i.e., the driving Wiener process) has d-degrees of freedom, thus affecting all components in an even manner.

Section 3.2 develops a convex analytic framework, while Section 3.3 introduces a notion of stability that is uniform over the set of stationary Markov controls. These results are used in Section 3.4 which studies the existence of an optimal strategy. Section 3.5 reviews the α-discounted control problem. Sections 3.6 and 3.7 are devoted to dynamic programming and existence of solutions to the Hamilton–Jacobi–Bellman (HJB) equation. The last section specializes to one-dimensional diffusions.

3.2 Convex analytic properties

In this section we develop a convex analytic framework for the study of nondegenerate controlled diffusions. Recall that $\mathfrak{U}_{\mathrm{sm}}$ stands for the class of stationary Markov controls, and $\mathfrak{U}_{\mathrm{ssm}} \subset \mathfrak{U}_{\mathrm{sm}}$ is the subclass of stable controls. We first introduce the concept of ergodic occupation measures which is essentially the class of stationary mean empirical measures for the joint state and control process under controls in $\mathfrak{U}_{\mathrm{ssm}}$ and we show that this collection is identical to the class of infinitesimal empirical measures. Ergodic occupation measures form a closed, convex subset of $\mathscr{P}(\mathbb{R}^d \times \mathbb{U})$ and a key result in this section is to establish that its extreme points correspond to the class of precise controls $\mathfrak{U}_{\mathrm{ssd}}$. Therefore, for nondegenerate diffusions, the ergodic control problem may be restricted to this class.

3.2.1 Ergodic occupation measures

By Theorem 1.5.18 (Birkhoff's ergodic theorem), if $v \in \mathfrak{U}_{\mathrm{ssm}}$, then

$$\lim_{T\to\infty} \frac{1}{T}\int_0^T c_v(X_t)\,\mathrm{d}t = \int_{\mathbb{R}^d}\int_{\mathbb{U}} c(x,u)v(\mathrm{d}u \mid x)\,\eta_v(\mathrm{d}x) \quad \text{a.s.,} \tag{3.2.1}$$

where c_v is as in Notation 2.3.5 on p. 53. This motivates the following definition.

Definition 3.2.1 For $v \in \mathfrak{U}_{\mathrm{ssm}}$ let $\eta_v \in \mathscr{P}(\mathbb{R}^d)$ the associated unique invariant probability measure. Define the *ergodic occupation measure* $\pi_v \in \mathscr{P}(\mathbb{R}^d \times \mathbb{U})$ by

$$\pi_v(\mathrm{d}x, \mathrm{d}u) = \eta_v(\mathrm{d}x)\,v(\mathrm{d}u \mid x)\,.$$

For notational convenience, we often use the abbreviated notation $\pi_v = \eta_v \circledast v$. Also, we denote the set of all ergodic occupation measures by $\mathscr{G}$ and the set of associated invariant probability measures by

$$\mathscr{H} := \{\eta \in \mathscr{P}(\mathbb{R}^d) : \eta \circledast v \in \mathscr{G}\,,\ \text{for some } v \in \mathfrak{U}_{\mathrm{ssm}}\}\,.$$

Thus, by (3.2.1), the ergodic control problem over $\mathfrak{U}_{\mathrm{ssm}}$ is equivalent to a linear optimization problem over $\mathscr{G}$. We first give a characterization of $\mathscr{G}$ and also show that it is closed and convex. Then we study the extreme points of $\mathscr{G}$ and show that these correspond to controls in $\mathfrak{U}_{\mathrm{ssd}}$. This is the basic framework of the convex analytic approach to the ergodic control problem whose main aim is the existence results in Section 3.4. First we show that for nondegenerate diffusions every infinitesimal occupation measure (see p. 63) is an ergodic occupation measure and vice versa.

Lemma 3.2.2 *A probability measure $\pi \in \mathscr{P}(\mathbb{R}^d \times \mathbb{U})$ is an ergodic occupation measure if and only if,*

$$\int_{\mathbb{R}^d\times\mathbb{U}} \mathcal{L}^u f(x)\,\pi(\mathrm{d}x, \mathrm{d}u) = 0 \qquad \forall f \in \mathscr{C}\,, \tag{3.2.2}$$

or equivalently, if

$$\int_{\mathbb{R}^d} \mathcal{L}^v f(x)\,\eta_v(\mathrm{d}x) = 0 \qquad \forall f \in \mathscr{C}\,, \tag{3.2.3}$$

where $v \in \mathfrak{U}_{\mathrm{sm}}$ and $\eta_v \in \mathscr{P}(\mathbb{R}^d)$ correspond to the decomposition $\pi = \eta_v \circledast v$.

Proof This easily follows from Theorem 2.6.16. □

Lemma 3.2.3 *The set of ergodic occupation measures $\mathscr{G}$ is a closed and convex subset of $\mathscr{P}(\mathbb{R}^d \times \mathbb{U})$.*

Proof If a sequence $\{\pi_n\}$ in $\mathscr{P}(\mathbb{R}^d \times \mathbb{U})$ satisfies (3.2.2) and converges to π as $n \to \infty$, then π also satisfies (3.2.2). Hence, $\mathscr{G}$ is closed. To show that $\mathscr{G}$ is convex, let $\pi_{v_i} = \eta_{v_i} \circledast v_i$, $i = 1, 2$, be two elements of $\mathscr{G}$ and $\delta \in (0, 1)$. Set $\eta = \delta\eta_{v_1} + (1-\delta)\eta_{v_2}$. Noting that η_{v_1} and η_{v_2} are both absolutely continuous with respect to η we define

$$v(\mathrm{d}u \mid x) := \delta\,\frac{\mathrm{d}\eta_{v_1}}{\mathrm{d}\eta}(x)\,v_1(\mathrm{d}u \mid x) + (1-\delta)\,\frac{\mathrm{d}\eta_{v_2}}{\mathrm{d}\eta}(x)\,v_2(\mathrm{d}u \mid x)\,,$$

where $\frac{\mathrm{d}\eta_{v_i}}{\mathrm{d}\eta}$ denotes the Radon–Nikodym derivative. It follows that

$$\eta \circledast v = \delta\,(\eta_{v_1} \circledast v_1) + (1-\delta)\,(\eta_{v_2} \circledast v_2)\,.$$

Also, for $f \in C_b^2(\mathbb{R}^d)$,

$$\mathcal{L}^v f(x) = \delta\,\frac{\mathrm{d}\eta_{v_1}}{\mathrm{d}\eta}(x)\,\mathcal{L}^{v_1} f(x) + (1-\delta)\,\frac{\mathrm{d}\eta_{v_1}}{\mathrm{d}\eta}(x)\,\mathcal{L}^{v_2} f(x)\,,$$

and hence, by (3.2.3),

$$\int_{\mathbb{R}^d} \mathcal{L}^v f\,\mathrm{d}\eta = \delta \int_{\mathbb{R}^d} \mathcal{L}^{v_1} f\,\mathrm{d}\eta_{v_1} + (1-\delta)\int_{\mathbb{R}^d} \mathcal{L}^{v_2} f\,\mathrm{d}\eta_{v_2} = 0\,,$$

proving that $\eta \circledast v \in \mathscr{G}$. □

3.2.2 Continuity and compactness of invariant measures

We start with some important continuity results.

Lemma 3.2.4 *Let K be a compact subset of $\mathscr{G}$. Define*

$$\mathscr{H}(K) = \{\eta \in \mathscr{P}(\mathbb{R}^d) : \eta \circledast v \in K\,,\ \text{for some } v \in \mathfrak{U}_{\mathrm{ssm}}\}\,. \tag{3.2.4}$$

Let $\varphi[\eta]$ denote the density of $\eta \in \mathscr{H}$ and define

$$\Phi(K) := \{\varphi[\eta] : \eta \in \mathscr{H}(K)\}\,. \tag{3.2.5}$$

Then

(a) *for every $R > 0$ there exists a constant $C_H = C_H(R)$ such that*

$$\varphi(x) \le C_H \varphi(y) \qquad \forall \varphi \in \Phi(K)\,,\ \forall x, y \in B_R\,;$$

(b) *there exists $R_0 > 0$ such that for all $R > R_0$, with $|B_R|$ denoting the volume of $B_R \subset \mathbb{R}^d$, we have*

$$\frac{1}{2C_H|B_R|} \le \inf_{B_R}\ \varphi \le \sup_{B_R}\ \varphi \le \frac{C_H}{|B_R|} \qquad \forall \varphi \in \Phi(K)\,; \tag{3.2.6}$$

(c) *there exists a constant $C_1 = C_1(R, K) > 0$ and $a_1 > 0$ such that*

$$|\varphi(x) - \varphi(y)| \le C_1 |x-y|^{a_1} \qquad \forall \varphi \in \Phi(K)\,,\ \forall x, y \in B_R\,.$$

Proof Part (a) is nothing more than Harnack's inequality for the divergence form equation in (2.6.39) [67, theorem 8.20, p. 199]. Since, by part (a),

$$C_H \inf_{B_R}\ \varphi \ge \sup_{B_R}\ \varphi\,,$$

choosing R_0 large enough so that $\eta(B_{R_0}) > \frac{1}{2}$ for all $\eta \in \mathscr{H}(K)$, we obtain

$$C_H\,|B_R| \inf_{B_R}\ \varphi \ge |B_R| \sup_{B_R}\ \varphi \ge \int_{B_R} \varphi(x)\,\mathrm{d}x > \frac{1}{2}\,,$$

and

$$\frac{|B_R|}{C_H} \sup_{B_R} \varphi \le |B_R| \inf_{B_R} \varphi \le \int_{B_R} \varphi(x)\,\mathrm{d}x < 1\,.$$

The local Hölder continuity in part (c) follows from standard estimates for solutions of (2.6.39) [67, theorem 8.24, p. 202]. □

Lemma 3.2.5 *Let K be a compact subset of $\mathscr{G}$. The set $\mathscr{H}(K)$ defined in* (3.2.4) *is compact in $\mathscr{P}(\mathbb{R}^d)$ under the total variation norm topology.*

Proof First we show that $\mathscr{H}(K)$ is closed in $\mathscr{P}(\mathbb{R}^d)$. Suppose $\{\eta_{v_n}\}$, $n \in \mathbb{N}$, is a sequence in $\mathscr{H}(K)$ converging to $\hat{\eta} \in \mathscr{P}(\mathbb{R}^d)$ under the total variation norm. Let $\pi_n = \eta_{v_n} \circledast v_n$ be the corresponding sequence of ergodic occupation measures in K. Since K is compact, π_n converges along a subsequence, which is also denoted by $\{\pi_n\}$, to some $\hat{\pi} = \eta_{\hat{v}} \circledast \hat{v}$. It follows that $\eta_{v_n} \to \eta_{\hat{v}}$ as $n \to \infty$, and hence $\hat{\eta} = \eta_{\hat{v}}$. By Lemma 3.2.4, the collection $\Phi(K)$, defined in (3.2.5), is equibounded and Hölder equicontinuous on bounded subdomains of $\mathbb{R}^d$. Therefore any sequence in $\Phi(K)$ contains a subsequence which converges uniformly on compact sets. Let $\varphi_n \to \hat{\varphi}$ be such a convergent sequence in $\Phi(K)$ and $\{\eta_n\}$ the corresponding invariant probability measures. Since K is compact, the family $\mathscr{H}(K)$ is tight, and hence the associated densities $\Phi(K)$ are uniformly integrable. It follows that $\{\varphi_n\}$ converges in $L^1(\mathbb{R}^d)$ as well. Therefore $\int \hat{\varphi} = 1$, $\hat{\varphi} \ge 0$, and for $h \in C_b(\mathbb{R}^d)$,

$$\int_{\mathbb{R}^d} h(x)\,\varphi_n(x)\,\mathrm{d}x \xrightarrow[n\to\infty]{} \int_{\mathbb{R}^d} h(x)\,\hat{\varphi}(x)\,\mathrm{d}x\,.$$

This implies $\eta_n \to \hat{\eta}$ in $\mathscr{P}(\mathbb{R}^d)$ and, since $\varphi_n \to \hat{\varphi}$ in $L^1(\mathbb{R}^d)$, also in total variation. Since $\mathscr{H}(K)$ is closed in $\mathscr{P}(\mathbb{R}^d)$, it follows that $\hat{\eta} \in \mathscr{H}(K)$. This completes the proof. □

For a subset $\mathcal{U} \subset \mathfrak{U}_{\mathrm{ssm}}$, let $\mathscr{H}_{\mathcal{U}}$ and $\mathscr{G}_{\mathcal{U}}$ denote the set of associated invariant probability measures and ergodic occupation measures respectively. In the lemma which follows we show that if $\mathscr{H}_{\mathcal{U}}$ is tight then the map $v \mapsto \eta_v$ from $\bar{\mathcal{U}}$[1] to $\mathscr{H}_{\bar{\mathcal{U}}}$ is continuous under the total variation norm topology. Since $\bar{\mathcal{U}}$ is compact, it follows that $\mathscr{H}_{\bar{\mathcal{U}}}$ is compact in the total variation norm topology.

Lemma 3.2.6 *Let $\mathcal{U} \subset \mathfrak{U}_{\mathrm{ssm}}$ and suppose $\mathscr{H}_{\mathcal{U}}$ is tight. Then*

(a) *the map $v \mapsto \eta_v$ from $\bar{\mathcal{U}}$ to $\mathscr{H}_{\bar{\mathcal{U}}}$ is continuous under the total variation norm topology of $\mathscr{H}$;*

(b) *the map $v \mapsto \pi_v$ from $\bar{\mathcal{U}}$ to $\mathscr{G}_{\bar{\mathcal{U}}} \subset \mathscr{P}(\mathbb{R}^d \times \mathbb{U})$ is continuous.*

[1] As usual, $\bar{\mathcal{U}}$ denotes the closure of $\mathcal{U}$ in the topology of Markov controls.

Proof Let $\{v_n\}$ be a sequence in $\mathcal{U}$ which converges (under the topology of Markov controls) to $v^* \in \bar{\mathcal{U}}$. Then, for all $g \in C_b(\mathbb{R}^d \times \mathbb{U})$ and $h \in L^1(\mathbb{R}^d)$,

$$\int_{\mathbb{R}^d\times\mathbb{U}} g(x,u)\,h(x)\,(v_n(\mathrm{d}u \mid x) - v^*(\mathrm{d}u \mid x))\,\mathrm{d}x \xrightarrow[n\to\infty]{} 0\,. \tag{3.2.7}$$

By the tightness assumption and Prohorov's theorem, $\mathscr{H}_{\mathcal{U}}$ is relatively compact in $\mathscr{P}(\mathbb{R}^d)$, and thus $\{\eta_{v_n}\}$ has a limit point $\eta^* \in \mathscr{P}(\mathbb{R}^d)$. Passing to a subsequence (which we also denote by $\{\eta_{v_n}\}$) converging to this limit point, we conclude, as in the proof of Lemma 3.2.5, that $\eta_{v_n} \to \eta^*$ in total variation, and that the associated densities $\varphi_{v_n} \to \varphi^*$ in $L^1(\mathbb{R}^d)$ (and also converge uniformly on compact subsets of $\mathbb{R}^d$). Using the notation for relaxed controls introduced in (2.3.12), let

$$\bar{g}(x, v(x)) := \int_{\mathbb{U}} g(x,u)\,v(\mathrm{d}u \mid x)\,, \qquad v \in \mathfrak{U}_{\mathrm{sm}}\,. \tag{3.2.8}$$

We use the triangle inequality

$$\begin{aligned}
\biggl|\int_{\mathbb{R}^d} \bar{g}(x, v_n(x))\,\eta_{v_n}(\mathrm{d}x) &- \int_{\mathbb{R}^d} \bar{g}(x, v^*(x))\,\eta^*(\mathrm{d}x)\biggr| \\
&\le \biggl|\int_{\mathbb{R}^d} \bar{g}(x, v_n(x))(\varphi_{v_n}(x) - \varphi^*(x))\,\mathrm{d}x\biggr| \\
&\quad + \biggl|\int_{\mathbb{R}^d} \bar{g}(x, v_n(x))\,\varphi^*(x)\,\mathrm{d}x - \int_{\mathbb{R}^d} \bar{g}(x, v^*(x))\,\varphi^*(x)\,\mathrm{d}x\biggr|\,.
\end{aligned} \tag{3.2.9}$$

Since $\varphi_{v_n} \to \varphi^*$ in $L^1(\mathbb{R}^d)$, the first term on the right-hand side of (3.2.9) converges to zero as $n \to \infty$, and so does the second term by (3.2.7). Hence, by (3.2.9),

$$0 = \int_{\mathbb{R}^d} \mathcal{L}^{v_n} f(x)\,\eta_{v_n}(\mathrm{d}x) \xrightarrow[n\to\infty]{} \int_{\mathbb{R}^d} \mathcal{L}^{v^*} f(x)\,\eta^*(\mathrm{d}x) \qquad \forall f \in \mathscr{C}\,,$$

implying $\eta^* = \eta_{v^*} \in \mathscr{H}_{\bar{\mathcal{U}}}$, by Theorem 2.6.16, and this establishes (a). Since

$$\int_{\mathbb{R}^d} \bar{g}(x, v(x))\,\eta_v(\mathrm{d}x) = \int_{\mathbb{R}^d\times\mathbb{U}} g(x,u)\,\pi_v(\mathrm{d}x, \mathrm{d}u)\,,$$

(3.2.9) also implies (b). □

3.2.3 Extreme points of the set of invariant measures

We now turn to the analysis of the set of extreme points of $\mathscr{G}$. As seen in the proof of Lemma 3.2.3, if $\pi_{v_i} \in \mathscr{G}$, $i = 1, 2, 3$, satisfy $\pi_{v_1} = \delta\,\pi_{v_2} + (1-\delta)\,\pi_{v_3}$ for some $\delta > 0$, then with $\pi_{v_i} = \eta_{v_i} \circledast v_i$, we have

$$v_1 = \delta\,\frac{\mathrm{d}\eta_{v_2}}{\mathrm{d}\eta_{v_1}}\,v_2 + (1-\delta)\,\frac{\mathrm{d}\eta_{v_3}}{\mathrm{d}\eta_{v_1}}\,v_3\,.$$

Since each η_{v_i} and the Lebesgue measure on $\mathbb{R}^d$ are mutually absolutely continuous, the Radon–Nikodym derivatives $\frac{\mathrm{d}\eta_{v_i}}{\mathrm{d}\eta_{v_1}}$, $i = 2, 3$ are positive a.e. in $\mathbb{R}^d$. It follows that if some pair (v_i, v_j) agree a.e. on some Borel set A of positive Lebesgue measure, then all three v_i's must agree a.e. on A. Note that this also implies that if some pair (v_i, v_j) differ a.e. on A then all pairs differ a.e. on A.

Lemma 3.2.7 *Let $A \subset \mathscr{B}(\mathbb{R}^d)$ be a bounded set of positive Lebesgue measure. Suppose $v_1, v_2 \in \mathfrak{U}_{\mathrm{sm}}$ agree on A^c and differ a.e. on A and that for some $v_0 \in \mathfrak{U}_{\mathrm{ssm}}$ and $\gamma \in \mathcal{B}(\mathbb{R}^d)$, satisfying*

$$\inf_{x \in A} \min\{\gamma(x), 1 - \gamma(x)\} \geq \gamma_A > 0,$$

it holds that

$$v_0(\cdot \mid x) = \gamma(x)\, v_1(\cdot \mid x) + (1 - \gamma(x))\, v_2(\cdot \mid x).$$

Then there exist $\hat{v}_1, \hat{v}_2 \in \mathfrak{U}_{\mathrm{ssm}}$, which agree a.e. on A^c and differ a.e. on A, such that

$$\pi_{v_0} = \tfrac{1}{2}(\pi_{\hat{v}_1} + \pi_{\hat{v}_2}). \tag{3.2.10}$$

In particular, π_{v_0} is not an extreme point of $\mathscr{G}$.

Proof Let $R > 0$ be such that $A \subset B_R$. For $v_0 \in \mathfrak{U}_{\mathrm{ssm}}$, define

$$\mathfrak{U}(v_0, R) := \{v \in \mathfrak{U}_{\mathrm{sm}} : v(x) = v_0(x),\ \text{if } |x| > R\}.$$

Since any $v \in \mathfrak{U}(v_0, R)$ agrees with v_0 on B_R^c, we have

$$\mathbb{E}_x^v\left[\tau(B_R^c)\,\mathbb{I}\{\tau(B_R^c) > t\}\right] = \mathbb{E}_x^{v_0}\left[\tau(B_R^c)\,\mathbb{I}\{\tau(B_R^c) > t\}\right] \qquad \forall v \in \mathfrak{U}(v_0, R),$$

and it follows from Lemma 3.3.4 (ix), which appears later on p. 97, that $\mathfrak{U}(v_0, R)$ are uniformly stable. In particular, v_1 and v_2 are stable. We are going to establish that there exists $\delta \in (0, 1)$, and $\hat{v} \in \mathfrak{U}(v_0, R)$ such that

$$\pi_{v_0} = \delta\, \pi_{v_1} + (1 - \delta)\, \pi_{\hat{v}}.$$

Then the hypotheses of the lemma imply that $v_1 \neq \hat{v}$, a.e. on A. Hence, if we select $\pi_{\hat{v}_1}$ and $\pi_{\hat{v}_2}$ as

$$\pi_{\hat{v}_1} = \delta\, \pi_{v_1} + (1 - \delta)\, \pi_{v_0}, \qquad \pi_{\hat{v}_2} = (1 - \delta)\, \pi_{\hat{v}} + \delta\, \pi_{v_0},$$

then (3.2.10) holds, and by the remarks in the paragraph before the lemma, $\hat{v}_1$ and $\hat{v}_2$ differ a.e. on A and agree a.e. on A^c. In view of the proof of Lemma 3.2.3, it suffices to show that

$$v_0(\cdot \mid x) = \delta \frac{\mathrm{d}\eta_{v_1}}{\mathrm{d}\eta_0}(x)\, v_1(\cdot \mid x) + (1 - \delta) \frac{\mathrm{d}\eta_{\hat{v}}}{\mathrm{d}\eta_0}(x)\, \hat{v}(\cdot \mid x),$$

where $\eta_0 = \delta\, \eta_{v_1} + (1 - \delta)\, \eta_{\hat{v}}$, or equivalently, expressed in terms of the associated densities, that

$$v_0(\cdot \mid x) = \frac{\delta\, \varphi_{v_1}(x)\, v_1(\cdot \mid x) + (1 - \delta)\, \varphi_{\hat{v}}(x)\, \hat{v}(\cdot \mid x)}{\delta\, \varphi_{v_1}(x) + (1 - \delta)\, \varphi_{\hat{v}}(x)}. \tag{3.2.11}$$

Since $\mathfrak{U}(v_0, R)$ is closed, it follows by Lemma 3.3.4 (viii) that

$$K_0 := \{\pi_v : v \in \mathfrak{U}(v_0, R)\}$$

is a compact subset of $\mathscr{G}$. Thus, by Lemma 3.2.5, $\mathscr{H}(K_0)$ is compact in $\mathscr{P}(\mathbb{R}^d)$, under the total variation norm topology. Also, by Lemma 3.2.4 (b), for some positive constants $\underline{\delta}$ and $\overline{\delta}$,

$$0 < \underline{\delta} \le \varphi_v(x) \le \overline{\delta} \quad \text{for } |x| \le R\,, \quad \forall v \in \mathfrak{U}(v_0, R)\,. \tag{3.2.12}$$

Let

$$\delta = \frac{\underline{\delta}\,\gamma_A}{\underline{\delta}\,\gamma_A + \overline{\delta}\,(1-\gamma_A)}\,.$$

For $v \in \mathfrak{U}(v_0, R)$, define w by

$$w(\cdot \mid x) := v_0(\cdot \mid x) + \frac{\delta\,\varphi_{v_1}(x)}{(1-\delta)\,\varphi_v(x)}\bigl(v_0(\cdot \mid x) - v_1(\cdot \mid x)\bigr)\,. \tag{3.2.13}$$

We claim that $w \in \mathfrak{U}(v_0, R)$. To prove the claim, first note that since $v_1 = v_0$ on A^c, (3.2.13) implies that $w(\cdot \mid x) = v_0(\cdot \mid x)$ for all $x \in A^c$. On the other hand, by (3.2.12), $\overline{\delta}\varphi_v(x) \ge \underline{\delta}\varphi_{v_1}(x)$ on A for all $v \in \mathfrak{U}(v_0, R)$, which implies

$$\delta\,\varphi_{v_1}(x) + (1-\delta)\,\varphi_v(x) \ge \delta\,\gamma_A^{-1}\varphi_{v_1}(x) \qquad \forall x \in A\,. \tag{3.2.14}$$

Moreover, by the hypotheses of the lemma,

$$v_0(\cdot \mid x) - \gamma_A v_1(\cdot \mid x) \ge \gamma_A v_2(\cdot \mid x) \qquad \forall x \in A\,. \tag{3.2.15}$$

Thus, by (3.2.13)–(3.2.15),

$$\begin{aligned} w(\cdot \mid x) &\ge \frac{\delta\,\varphi_{v_1}(x)}{(1-\delta)\,\varphi_v(x)}\bigl(\gamma_A^{-1} v_0(\cdot \mid x) - v_1(\cdot \mid x)\bigr) \\ &\ge \frac{\delta\,\varphi_{v_1}(x)}{(1-\delta)\,\varphi_v(x)}\, v_2(\cdot \mid x) \\ &\ge 0 \qquad \forall x \in A\,. \end{aligned}$$

It also follows by (3.2.13) that $w(\mathbb{U} \mid x) = 1$ for all $x \in \mathbb{R}^d$. This finishes the proof of the claim. Therefore (3.2.13) defines a map $\eta_v \mapsto \eta_w$ on $\mathscr{H}(K_0)$ which we denote by G. Since $\mathscr{H}(K_0)$ is compact in the total variation norm topology and convex, then if we show that the map G is continuous in the total variation norm topology, it follows by Schauder's fixed point theorem that G has a fixed point $\eta_{\hat{v}} \in \mathscr{H}(K_0)$, and the corresponding $\hat{v} \in \mathfrak{U}(v_0, R)$ satisfies (3.2.11) by construction. Thus in order to complete the proof it remains to show that the map $\eta_v \mapsto \eta_w$ in (3.2.13) is continuous. Let $\{\eta_{v_n}\}$, $n \in \mathbb{N}$, be a sequence in $\mathscr{H}(K_0)$, converging to $\eta_{v^*} \in \mathscr{H}(K_0)$, and denote by w_n and w^* the corresponding elements of $\mathfrak{U}(v_0, R)$ defined by (3.2.13). By the proof of Lemma 3.2.5, $\varphi_{v_n} \to \varphi_{v^*}$ uniformly on compact sets and also in $L^1(\mathbb{R}^d)$. Therefore, by employing (3.2.13), and using the notation in (3.2.8), we deduce that, for all $g \in C_b(\mathbb{R}^d \times \mathbb{U})$ and $h \in L^1(\mathbb{R}^d)$,

$$\int_{\mathbb{R}^d} \bar{g}(x, w_n)\,h(x)\,\mathrm{d}x \xrightarrow[n\to\infty]{} \int_{\mathbb{R}^d} \bar{g}(x, w^*)\,h(x)\,\mathrm{d}x\,. \tag{3.2.16}$$

By (3.2.16), $w_n \to w^*$ in $\mathfrak{U}(v_0, R)$. Since $\mathscr{H}(K_0) = \mathscr{H}(\mathfrak{U}(v_0, R))$ is compact, then by Lemma 3.2.6, $\eta_{w_n} \to \eta_{w^*}$ in total variation, and the proof is complete. □

Lemmas 3.2.3 and 3.2.7 yield the following theorem.

Theorem 3.2.8 *The set $\mathscr{G}$ is closed and convex, and its extreme points correspond to precise controls $v \in \mathfrak{U}_{\mathrm{ssd}}$.*

Proof Suppose $\pi_{v_0} \in \mathscr{G}_e$ and $v_0 \notin \mathfrak{U}_{\mathrm{ssd}}$. Then v_0 can be expressed pointwise as a strict convex combination of some v_1 and v_2 in $\mathfrak{U}_{\mathrm{sm}}$ which differ on a set of positive Lebesgue measure. If these are not in $\mathfrak{U}_{\mathrm{ssm}}$, they can be modified to match v_0 outside some ball of radius $R > 0$. Thus v_0 satisfies the hypotheses of Lemma 3.2.7 and the result follows. □

Let $\bar{\mathbb{R}}^d$ denote the one-point compactification of $\mathbb{R}^d$. We view $\mathbb{R}^d \subset \bar{\mathbb{R}}^d$ via the natural imbedding. Similarly, $\mathscr{P}(\mathbb{R}^d \times \mathbb{U})$ is viewed as a subset of $\mathscr{P}(\bar{\mathbb{R}}^d \times \mathbb{U})$. Let $\bar{\mathscr{G}}$ denote the closure of $\mathscr{G}$ in $\mathscr{P}(\bar{\mathbb{R}}^d \times \mathbb{U})$. Also, $\mathscr{G}_e$ ($\bar{\mathscr{G}}_e$) denote the set of extreme points of $\mathscr{G}$ ($\bar{\mathscr{G}}$).

Lemma 3.2.9 *It holds that $\mathscr{G}_e \subset \bar{\mathscr{G}}_e$ and any $\pi \in \mathscr{G}$ is the barycenter of a probability measure supported on $\mathscr{G}_e$.*

Proof We first show that $\mathscr{G}_e \subset \bar{\mathscr{G}}_e$. If not, let $\pi \in \mathscr{G}_e \setminus \bar{\mathscr{G}}_e$. Then π must be a strict convex combination of two distinct elements of $\bar{\mathscr{G}}$, at least one of which must assign strictly positive probability to $\{\infty\} \times \mathbb{U}$. But then $\pi(\{\infty\} \times \mathbb{U}) > 0$, a contradiction. This proves the first part of the lemma. If $\mathscr{G}$ is compact, then $\mathscr{G}_e \neq \varnothing$ and $\mathscr{G}_e = \bar{\mathscr{G}}_e$, and the proof of the second part follows directly from Choquet's theorem. If $\mathscr{G}$ is not compact, applying Choquet's theorem to $\bar{\mathscr{G}}$ we deduce that $\bar{\mathscr{G}}_e \neq \varnothing$ and each $\pi \in \mathscr{G}$ is the barycenter of a probability measure ξ on $\bar{\mathscr{G}}_e$. If $\xi(\bar{\mathscr{G}}_e \setminus \mathscr{G}_e) > 0$, we must have $\pi(\{\infty\} \times \mathbb{U}) > 0$, a contradiction. Thus $\xi(\mathscr{G}_e) = 1$. □

Corollary 3.2.10 *If $\pi \in \mathscr{G} \mapsto \int c\,\mathrm{d}\pi$ attains its minimum, then this is attained at some π_v, with $v \in \mathfrak{U}_{\mathrm{ssd}}$.*

Proof This follows directly from Theorem 3.2.8 and Lemma 3.2.9. □

Lemma 3.2.11 *For each $\pi \in \bar{\mathscr{G}}$, there exist $\pi' \in \mathscr{G}$, $\pi'' \in \mathscr{P}(\{\infty\} \times \mathbb{U})$ and $\delta \in [0, 1]$ such that, for all $B \in \mathscr{B}(\bar{\mathbb{R}}^d \times \mathbb{U})$,*

$$\pi(B) = \delta\,\pi'(B \cap (\mathbb{R}^d \times \mathbb{U})) + (1 - \delta)\,\pi''(B \cap (\{\infty\} \times \mathbb{U}))\,. \tag{3.2.17}$$

Proof If $\pi(\mathbb{R}^d \times \mathbb{U}) = 1$, then $\pi \in \mathscr{G}$ and (3.2.17) holds for $\delta = 1$ and $\pi' = \pi$. If $\pi(\mathbb{R}^d \times \mathbb{U}) = 0$ then (3.2.17) holds for $\delta = 0$ and $\pi'' = \pi$. Suppose $\pi(\mathbb{R}^d \times \mathbb{U}) \in (0, 1)$. Set $\delta = \pi(\mathbb{R}^d \times \mathbb{U})$ and

$$\pi'(B) = \frac{\pi(B \cap (\mathbb{R}^d \times \mathbb{U}))}{\delta}\,, \qquad \pi''(B) = \frac{\pi(B \cap (\{\infty\} \times \mathbb{U}))}{1 - \delta}\,. \tag{3.2.18}$$

Then (3.2.17) holds for π' and π'', as defined in (3.2.18). Let $\{\pi_n : n \in \mathbb{N}\} \subset \mathscr{G}$ be such that $\pi_n \to \pi$. By Lemma 3.2.2,

$$\int_{\mathbb{R}^d \times \mathbb{U}} \mathcal{L}^u f(x)\, \pi_n(\mathrm{d}x, \mathrm{d}u) = 0 \qquad \forall f \in \mathscr{C}\,. \tag{3.2.19}$$

Letting $n \to \infty$ in (3.2.19), we obtain

$$\int_{\mathbb{R}^d \times \mathbb{U}} \mathcal{L}^u f(x)\, \pi'(\mathrm{d}x, \mathrm{d}u) = 0 \qquad \forall f \in \mathscr{C}\,.$$

Therefore, by Lemma 3.2.2, $\pi' \in \mathscr{G}$. □

3.3 Uniform recurrence properties

Recall that $v \in \mathfrak{U}_{\mathrm{sm}}$ is called stable if $\mathbb{E}^v_x[\tau(G^c)] < \infty$ for all $x \in \mathbb{R}^d$, and for all bounded domains $G \subset \mathbb{R}^d$. By Theorem 2.6.9, the process X under a stable control v has a unique invariant probability measure η_v. We turn our attention to ergodic properties that are uniform with respect to controls in $\mathfrak{U}_{\mathrm{ssm}}$.

3.3.1 Uniform boundedness of mean recurrence times

The next theorem establishes a uniform bound for a certain class of functionals of the controlled process, when $\mathfrak{U}_{\mathrm{sm}} = \mathfrak{U}_{\mathrm{ssm}}$. Recall the notation introduced in (2.3.13) on p. 53.

Theorem 3.3.1 *Suppose $\mathfrak{U}_{\mathrm{sm}} = \mathfrak{U}_{\mathrm{ssm}}$ and let $h \in C(\mathbb{R}^d \times \mathbb{U})$ be a nonnegative function. If for some bounded domain D and some $x_0 \in D^c$*

$$\mathbb{E}^v_{x_0}\left[\int_0^{\tau(D^c)} h_v(X_t)\,\mathrm{d}t\right] < \infty \qquad \forall v \in \mathfrak{U}_{\mathrm{ssm}}\,,$$

then for any open ball $B \subset \mathbb{R}^d$ and any compact set $\Gamma \subset \bar{B}^c$, we have

$$\sup_{v \in \mathfrak{U}_{\mathrm{ssm}}} \sup_{x \in \Gamma} \mathbb{E}^v_x\left[\int_0^{\tau(B^c)} h_v(X_t)\,\mathrm{d}t\right] < \infty\,. \tag{3.3.1}$$

Proof Set

$$\beta^v_x[\tau] := \mathbb{E}^v_x\left[\int_0^{\tau} h_v(X_t)\,\mathrm{d}t\right].$$

Note that by Theorem 2.6.3 (ii), $\beta^v_x[\tau(D^c)]$ is finite for all $x \in D^c$ and $v \in \mathfrak{U}_{\mathrm{sm}}$, and it also follows as in the proof of (b) ⇒ (a) of Theorem 2.6.10 that $\beta^v_x[\tau(B^c)] < \infty$ for any open ball $B \subset \mathbb{R}^d$, $x \in B^c$ and $v \in \mathfrak{U}_{\mathrm{sm}}$. We argue by contradiction. If (3.3.1) does not hold, then there exists a sequence $\{v_n\} \subset \mathfrak{U}_{\mathrm{ssm}}$, an open ball $B \subset \mathbb{R}^d$, and a compact set $\Gamma \subset \bar{B}^c$ such that $\sup_{x \in \Gamma} \beta^{v_n}_x[\tau(B^c)] \to \infty$ as $n \to \infty$. Then, by Harnack's inequality, for all compact sets $\Gamma \subset \bar{B}^c$, it holds that

$$\inf_{x \in \Gamma} \beta^{v_n}_x[\tau(B^c)] \to \infty\,. \tag{3.3.2}$$

Suppose that for some open ball $G \subset \mathbb{R}^d$ and a compact set $K \subset \bar{G}^c$,

$$\sup_{n\in\mathbb{N}} \inf_{x\in K} \beta_x^{v_n}\left[\tau(G^c)\right] < \infty\,.$$

It then follows as in the proof of (b) ⇒ (a) of Theorem 2.6.10 that

$$\sup_{n\in\mathbb{N}} \inf_{x\in\Gamma} \beta_x^{v_n}\left[\tau(B^c)\right] < \infty\,,$$

which contradicts (3.3.2). Therefore (3.3.2) holds for any open ball $B \subset \mathbb{R}^d$ and any compact set $\Gamma \subset \bar{B}^c$.

Fix a ball G_0, and let $\Gamma \subset \bar{G}_0^c$. Select $v_0 \in \mathfrak{U}_{\mathrm{ssm}}$ such that $\inf_{x\in\Gamma} \beta_x^{v_0}\left[\tau(G_0^c)\right] > 2$. Let G_1 be an open ball such that $\Gamma \cup G_0 \Subset G_1$, and satisfying

$$\beta_x^{v_0}\left[\tau(G_0^c)\right] \le 2\beta_x^{v_0}\left[\tau(G_0^c) \wedge \tau(G_1)\right] \qquad \forall x \in \Gamma\,.$$

This is always possible since

$$\beta_x^{v_0}\left[\tau(G_0^c) \wedge \tau_R)\right] \uparrow \beta_x^{v_0}\left[\tau(G_0^c)\right] \quad \text{as } R \to \infty\,,$$

uniformly on Γ. Select any open ball $\tilde{G}_1 \Supset G_1$, and let

$$p_1 := \inf_{v\in\mathfrak{U}_{\mathrm{ssm}}} \inf_{x\in\Gamma} \mathbb{P}_x^v(\tau(\tilde{G}_1) < \tau(G_0^c))\,. \tag{3.3.3}$$

By (2.6.9d), $p_1 > 0$. By (3.3.2), we may select $v_1 \in \mathfrak{U}_{\mathrm{ssm}}$ such that

$$\inf_{x\in\partial\tilde{G}_1} \beta_x^{v_1}\left[\tau(G_1^c)\right] > 8p_1^{-1}\,. \tag{3.3.4}$$

Define

$$\check{v}_1(x) := \begin{cases} v_0(x) & \text{for } x \in G_1\,, \\ v_1(x) & \text{for } x \in G_1^c\,. \end{cases}$$

It follows by (3.3.3) and (3.3.4) that

$$\inf_{x\in\Gamma} \beta_x^{\check{v}_1}\left[\tau(G_0^c)\right] \ge \left(\inf_{x\in\Gamma} \mathbb{P}_x^{\check{v}_1}(\tau(\tilde{G}_1) < \tau(G_0^c))\right)\left(\inf_{x\in\partial\tilde{G}_1} \beta_x^{v_1}\left[\tau(G_1^c)\right]\right) \ge 8\,. \tag{3.3.5}$$

Therefore there exists an open ball $G_2 \Supset G_1$ satisfying

$$\beta_x^{\check{v}_1}\left[\tau(G_0^c) \wedge \tau(G_2)\right] > 4\,.$$

We proceed inductively as follows. Suppose $\check{v}_{k-1} \in \mathfrak{U}_{\mathrm{ssm}}$ and G_k an open ball in $\mathbb{R}^d$ are such that

$$\beta_x^{\check{v}_{k-1}}\left[\tau(G_0^c) \wedge \tau(G_k)\right] > 2^k\,.$$

First pick any open ball $\tilde{G}_k \Supset G_k$, and then select $v_k \in \mathfrak{U}_{\mathrm{ssm}}$ satisfying

$$\inf_{x\in\partial\tilde{G}_k} \beta_x^{v_k}[\tau(G_k^c)] > 2^{k+2}\Bigl(\inf_{v\in\mathfrak{U}_{\mathrm{ssm}}} \inf_{x\in\Gamma} \mathbb{P}_x^v(\tau(\tilde{G}_k) < \tau(G_0^c))\Bigr)^{-1}\,.$$

This is always possible by (3.3.2). Proceed by defining the concatenated control

$$\check{v}_k(x) := \begin{cases} \check{v}_{k-1}(x) & \text{for } x \in G_k\,, \\ v_k(x) & \text{for } x \in G_k^c\,. \end{cases}$$

It follows as in (3.3.5) that

$$\inf_{x \in \Gamma} \beta_x^{\check{v}_k} \left[\tau(G_0^c) \right] > 2^{k+2}\,.$$

Subsequently choose $G_{k+1} \Supset \tilde{G}_k$ such that

$$\inf_{x \in \Gamma} \beta_x^{\check{v}_k} \left[\tau(G_0^c) \wedge \tau(G_{k+1}) \right] > \frac{1}{2} \inf_{x \in \Gamma} \beta_x^{\check{v}_k} \left[\tau(G_0^c) \right],$$

thus yielding

$$\beta_x^{\check{v}_k} \left[\tau(G_0^c) \wedge \tau(G_{k+1}) \right] > 2^{k+1}\,. \tag{3.3.6}$$

By construction, each $\check{v}_k$ agrees with $\check{v}_{k-1}$ on G_k. It is also evident that the sequence $\{\check{v}_k\}$ converges to a control $v^* \in \mathfrak{U}_{\mathrm{ssm}}$, which agrees with $\check{v}_k$ on G_k for each $k \ge 1$, and hence, by (3.3.6),

$$\inf_{x \in \Gamma} \beta_x^{v^*} \left[\tau(G_0^c) \wedge \tau(G_k) \right] > 2^k \qquad \forall k \in \mathbb{N}\,.$$

Thus $\beta_x^{v^*} \left[\tau(G_0^c) \right] = \infty$, contradicting the original hypothesis. □

Theorem 3.3.1 implies that if $\mathfrak{U}_{\mathrm{sm}} = \mathfrak{U}_{\mathrm{ssm}}$, then $\mathbb{E}_x^v[\tau(G^c)]$ is bounded uniformly in $v \in \mathfrak{U}_{\mathrm{ssm}}$ for every fixed non-empty open set G and $x \in G^c$ (Theorem 3.3.1). We call this property *uniform positive recurrence*.

Now, let $D_1 \Subset D_2$ be two fixed open balls in $\mathbb{R}^d$, and let $\hat{\tau}_2$ be as defined in Theorem 2.6.7. Let $h \in C_b(\mathbb{R}^d \times \mathbb{U})$ be a nonnegative function and define

$$\Phi_R^v(x) := \mathbb{E}_x^v \left[\int_0^{\tau(D_1^c)} \mathbb{I}_{B_R^c}(X_t)\, h_v(X_t)\, \mathrm{d}t \right], \quad x \in \partial D_2\,, \quad v \in \mathfrak{U}_{\mathrm{ssm}}\,. \tag{3.3.7}$$

Choose $R_0 > 0$ such that $B_{R_0} \Supset D_2$. Then, provided $R > R_0$, $\Phi_R^v(x)$ satisfies $\mathcal{L}^v \Phi_R^v = 0$ in $B_{R_0} \cap \bar{D}_1^c$, and by Harnack's inequality there exists a constant C_H, independent of $v \in \mathfrak{U}_{\mathrm{ssm}}$, such that $\Phi_R^v(x) \le C_H \Phi_R^v(y)$ for all $x, y \in \partial D_2$ and $v \in \mathfrak{U}_{\mathrm{sm}}$. Harnack's inequality also holds for the function $x \mapsto \mathbb{E}_x^v[\hat{\tau}_2]$ on ∂D_1 (for this we apply Theorem A.2.13). Also, by Lemma 2.6.5, for some constant $C_0 > 0$, we have

$$\inf_{v \in \mathfrak{U}_{\mathrm{ssm}}} \inf_{x \in \partial D_2} \mathbb{E}_x^v[\tau(D_1^c)] \ge C_0 \sup_{v \in \mathfrak{U}_{\mathrm{ssm}}} \sup_{x \in \partial D_1} \mathbb{E}_x^v[\tau(D_2)]\,.$$

Consequently, using these estimates and applying Theorem 2.6.9 with $f = h_v$, we obtain positive constants k_1 and k_2, which depend only on D_1, D_2, and R_0, such that, for all $R > R_0$ and $x \in \partial D_2$,

$$k_1 \int_{B_R^c \times \mathbb{U}} h\, \mathrm{d}\pi_v \le \frac{\Phi_R^v(x)}{\inf_{x \in \partial D_2} \mathbb{E}_x^v\left[\tau(D_1^c)\right]} \le k_2 \int_{B_R^c \times \mathbb{U}} h\, \mathrm{d}\pi_v \qquad \forall v \in \mathfrak{U}_{\mathrm{ssm}}\,. \tag{3.3.8}$$

Similarly, applying Theorem 2.6.9 with $f = \mathbb{I}_{D_1}$ along with the Harnack inequalities for the maps $x \mapsto \mathbb{E}_x^v[\tau(D_1^c)]$ on ∂D_2 and $x \mapsto \mathbb{E}_x^v[\tau(D_2^c)]$ on ∂D_1, there exists positive constants k_3 and k_3', which depend only on D_1 and D_2, such that

$$\begin{aligned}
\eta_v(D_1) \sup_{x\in\partial D_2} \mathbb{E}_x^v[\tau(D_1^c)] &\le k_3 \inf_{x\in\partial D_1} \mathbb{E}_x^v[\tau(D_2)] \qquad \forall v \in \mathfrak{U}_{\mathrm{ssm}}\,, \\
\eta_v(D_2) \sup_{x\in\partial D_2} \mathbb{E}_x^v[\tau(D_1^c)] &\ge k_3' \sup_{x\in\partial D_1} \mathbb{E}_x^v[\tau(D_2)] \qquad \forall v \in \mathfrak{U}_{\mathrm{ssm}}\,.
\end{aligned} \tag{3.3.9}$$

We obtain the following useful variation of Theorem 3.3.1.

Corollary 3.3.2 *Suppose $\mathfrak{U}_{\mathrm{sm}} = \mathfrak{U}_{\mathrm{ssm}}$ and that $h \in C(\mathbb{R}^d \times \mathbb{U})$ is integrable with respect to all $\pi \in \mathscr{G}$. Then $\sup_{\pi\in\mathscr{G}} \int |h|\,\mathrm{d}\pi < \infty$.*

Proof Since by hypothesis $\int |h|\,\mathrm{d}\pi_v < \infty$ for all $v \in \mathfrak{U}_{\mathrm{ssm}}$, (3.3.8) together with Theorem 3.3.1 imply that $\Phi_R^v(x) < \infty$ for all $v \in \mathfrak{U}_{\mathrm{ssm}}$, $x \in \partial D_2$. Therefore applying Theorem 3.3.1 once more, we obtain $\sup_{v\in\mathfrak{U}_{\mathrm{ssm}}} \Phi_R^v(x) < \infty$, and the result follows by the left-hand inequality of (3.3.8). □

3.3.2 Uniform stability

We define uniform stability as follows.

Definition 3.3.3 A collection of stationary Markov controls $\mathcal{U} \in \mathfrak{U}_{\mathrm{ssm}}$ is called uniformly stable if the set $\{\eta_v : v \in \mathcal{U}\}$ is tight.

Let $\mathcal{U} \subset \mathfrak{U}_{\mathrm{ssm}}$ and recall that $\mathscr{G}_{\mathcal{U}}$ and $\mathscr{H}_{\mathcal{U}}$ denote the corresponding ergodic occupation and invariant probability measures, respectively. The following lemma provides some important equivalences of uniform stability over $v \in \mathcal{U}$.

Lemma 3.3.4 *Let $\mathcal{U}$ be an arbitrary subset of $\mathfrak{U}_{\mathrm{ssm}}$. The following statements are equivalent (with $h \in C(\mathbb{R}^d \times \mathbb{U}; \mathbb{R}_+)$ an inf-compact function which is Lipschitz continuous in its first argument, uniformly in the second, and which is common to* (i)–(iv)*).*

(i) *For some open ball $D \subset \mathbb{R}^d$ and some $x \in \bar{D}^c$,*

$$\sup_{v\in\mathcal{U}} \mathbb{E}_x^v\left[\int_0^{\tau(D^c)} h_v(X_t)\,\mathrm{d}t\right] < \infty\,.$$

(ii) *For all open balls $D \subset \mathbb{R}^d$ and compact sets $\Gamma \subset \mathbb{R}^d$,*

$$\sup_{v\in\mathcal{U}} \sup_{x\in\Gamma} \mathbb{E}_x^v\left[\int_0^{\tau(D^c)} h_v(X_t)\,\mathrm{d}t\right] < \infty\,.$$

(iii) *A uniform bound holds:*

$$\sup_{v\in\mathcal{U}} \int_{\mathbb{R}^d\times\mathbb{U}} h(x,u)\,\pi_v(\mathrm{d}x,\mathrm{d}u) < \infty\,. \tag{3.3.10}$$

(iv) *Provided* $\mathcal{U} = \mathfrak{U}_{\mathrm{sm}}$, *there exist a nonnegative, inf-compact* $\mathcal{V} \in C^2(\mathbb{R}^d)$ *and a constant* k_0 *satisfying*

$$\mathcal{L}^u \mathcal{V}(x) \le k_0 - h(x,u) \qquad \forall (x,u) \in \mathbb{R}^d \times \mathbb{U}\,.$$

(v) *Provided* $\mathcal{U} = \mathfrak{U}_{\mathrm{sm}}$, *for any compact set* $K \subset \mathbb{R}^d$ *and* $t_0 > 0$, *the family of mean empirical measures*

$$\left\{\bar{\zeta}^U_{x,t} : x \in K\,,\ t \ge t_0\,,\ U \in \mathfrak{U}\right\},$$

defined in Lemma 2.5.3 on p. 62, is tight.

(vi) $\mathscr{H}_{\mathcal{U}}$ *is tight.*

(vii) $\mathscr{G}_{\mathcal{U}}$ *is tight.*

(viii) $\mathscr{G}_{\bar{\mathcal{U}}}$ *is compact.*

(ix) *For some open ball* $D \subset \mathbb{R}^d$ *and* $x \in \bar{D}^c$, $\{(\tau(D^c), \mathbb{P}^v_x) : v \in \mathcal{U}\}$ *are uniformly integrable, i.e.,*

$$\sup_{v\in\mathcal{U}} \mathbb{E}^v_x\left[\tau(D^c)\,\mathbb{I}\{\tau(D^c) > t\}\right] \downarrow 0 \quad \text{as } t \uparrow \infty\,.$$

(x) *The family* $\{(\tau(D^c), \mathbb{P}^v_x) : v \in \mathcal{U}\,,\ x \in \Gamma\}$ *is uniformly integrable for all open balls* $D \subset \mathbb{R}^d$ *and compact sets* $\Gamma \subset \mathbb{R}^d$.

Proof It is evident that (ii) ⇒ (i) and (x) ⇒ (ix). Since $\mathbb{U}$ is compact, (vi) ⇔ (vii). By Prohorov's theorem, (viii) ⇒ (vii). With $D_1 \Subset D_2$ any two open balls in $\mathbb{R}^d$, we apply (3.3.8) and (3.3.9). Letting $D = D_1$, (i) ⇒ (iii) follows by (3.3.8). It is easy to show that (iii) ⇒ (vii). Therefore, since under (iii) $\mathscr{H}_{\mathcal{U}}$ is tight, (3.2.6) implies that $\inf_{v\in\mathcal{U}} \eta_v(D_1) > 0$. In turn, by (2.6.9a) and (3.3.9),

$$\sup_{v\in\mathcal{U}} \sup_{x\in\partial D_2} \mathbb{E}^v_x[\tau(D^c_1)] \le k_3 \frac{\sup_{v\in\mathcal{U}} \sup_{x\in\partial D_1} \mathbb{E}^v_x[\tau(D_2)]}{\inf_{v\in\mathcal{U}} \eta_v(D_1)} < \infty\,. \tag{3.3.11}$$

Consequently, by applying (3.3.8) and Lemma 2.6.13 (i), (iii) ⇒ (ii) follows. Also, (iv) ⇒ (v) can be demonstrated by using the arguments in the proof of Lemma 2.5.3. We proceed by proving the implications (vi) ⇒ (iii) ⇒ (iv) ⇒ (i), (ix) ⇒ (vii), and (v) ⇒ (viii) ⇒ (x).

(vi) ⇒ (iii): Let

$$\hat{h}_v(x) := \left(\eta_v(B^c_{|x|})\right)^{-1/2}, \quad v \in \mathcal{U}\,, \quad x \in \mathbb{R}^d\,,$$

and define $h := \inf_{v\in\mathcal{U}} \hat{h}_v$. Note that $\int_{\mathbb{R}^d} \hat{h}_v \, \mathrm{d}\eta_v = 2$. Indeed, since

$$\eta_v(\{x : \hat{h}_v(x) > t\}) = \begin{cases} t^{-2} & \text{if } t \in [1,\infty)\,, \\ 1 & \text{otherwise}, \end{cases}$$

we have

$$\int_{\mathbb{R}^d} \hat{h}_v(x)\,\eta_v(\mathrm{d}x) = \int_0^\infty \eta_v(\{x : \hat{h}_v(x) > t\})\,\mathrm{d}t = 1 + \int_1^\infty t^{-2}\,\mathrm{d}t = 2\,.$$

Thus (3.3.10) holds. Since $\mathscr{H}_{\mathcal{U}}$ is tight, $\sup_{v\in\mathcal{U}} \eta_v(B^c_{|x|}) \to 0$ as $|x| \to \infty$, and thus h is inf-compact. Next, we show that h is locally Lipschitz continuous. Let $R > 0$ and $x, x' \in B_R$. Then, with $g(x) := \eta_v(B^c_{|x|})$,

$$\left|\hat{h}_v(x) - \hat{h}_v(x')\right| = \frac{|g(x) - g(x')|}{\sqrt{g(x)g(x')}\left(\sqrt{g(x)} + \sqrt{g(x')}\right)}. \qquad (3.3.12)$$

By (3.2.6), the denominator of (3.3.12) is uniformly bounded away from zero on B_R, while the numerator has the upper bound $\left(\sup_{B_R} \varphi_v\right)\left||B_{|x|}| - |B_{|x'|}|\right|$, where φ_v is the density of η_v. Therefore, by Lemma 3.2.6 and (3.3.12), $(x, v) \mapsto \hat{h}_v(x)$ is continuous in $\mathbb{R}^d \times \bar{\mathcal{U}}$ and locally Lipschitz in the first argument. Since $\bar{\mathcal{U}}$ is compact, local Lipschitz continuity of h follows.

(iii) ⇒ (iv): We borrow some results from Section 3.5. By Theorem 3.5.3 on p. 108, the Dirichlet problem

$$\max_{u\in\mathbb{U}} \left[\mathcal{L}^u f_r(x) + h(x, u)\right] = 0, \quad x \in B_r \setminus \bar{D}_1,$$
$$f_r\big|_{\partial D_1 \cap \partial B_r} = 0$$

has a solution $f_r \in C^{2,s}(\bar{B}_r \setminus D_1)$, $s \in (0, 1)$. Let $v_r \in \mathfrak{U}_{\mathrm{sd}}$ satisfy $\mathcal{L}^{v_r} f_r(x) + h_{v_r}(x) = 0$ for all $x \in \mathbb{R}^d$. Recall the definition of $\xi^v_{D,G}$ in Lemma 2.6.13. Then using (3.3.7) and (3.3.8), with $r > R > R_0$, and since under the hypothesis of (iii) equation (3.3.11) holds, we obtain

$$\begin{aligned} f_r(x) &= \mathbb{E}^{v_r}_x\left[\int_0^{\tau(D_1^c)\wedge\tau_r} h_v(X_t)\,\mathrm{d}t\right] \\ &\le \Big(\sup_{B_R\times\mathbb{U}} h\Big)\xi^{v_r}_{D_1,B_R}(x) + \Phi^{v_r}_R(x) \\ &\le \Big(\sup_{B_R\times\mathbb{U}} h\Big)\xi^{v_r}_{D_1,B_R}(x) + k_2' \int_{\mathbb{R}^d} h\,\mathrm{d}\pi_{v_r}, \quad x \in \partial D_2, \end{aligned}$$

for some constant $k_2' > 0$ that depends only on D_1, D_2, and R_0. Therefore, by (iii) and Lemma 2.6.13, f_r is bounded above, and since it is monotone in r, it converges by Lemma 3.5.4, as $r \to \infty$, to some $\mathcal{V} \in C^2(D_1^c)$ satisfying

$$\mathcal{L}^u\mathcal{V}(x) \le -h(x, u) \qquad \forall (x, u) \in \bar{D}_1^c \times \mathbb{U}.$$

It remains to extend $\mathcal{V}$ to a smooth function. This can be accomplished, for instance, by selecting $D_4 \Supset D_3 \Supset D_1$, and with ψ any smooth function that equals zero on D_3 and $\psi = 1$ on D_4^c, to define $\tilde{\mathcal{V}} = \psi\mathcal{V}$ on D_1^c and $\tilde{\mathcal{V}} = 0$ on D_1. Then $\mathcal{L}^u\tilde{\mathcal{V}} \le -h$ on D_4^c, for all $u \in \mathbb{U}$, and since $|\mathcal{L}^u\tilde{\mathcal{V}}|$ is bounded in $\bar{D}_4$, uniformly in $u \in \mathbb{U}$, (iv) follows.

(iv) ⇒ (i): Let D be an open ball such that $h(x, u) \ge 2k_0$ for all $x \in D^c$ and $u \in \mathbb{U}$. By Dynkin's formula, for any $R > 0$ and $v \in \mathfrak{U}_{\mathrm{sm}}$,

$$\mathbb{E}^v_x\left[\int_0^{\tau(D^c)\wedge\tau_R} (h_v(X_t) - k_0)\,\mathrm{d}t\right] \le \mathcal{V}(x) \qquad \forall x \in \bar{D}^c. \qquad (3.3.13)$$

Since $h \le 2(h - k_0)$ on D^c, the result follows by taking limits as $R \to \infty$ in (3.3.13).

(ix) ⇒ (vii): Using (3.3.7) and (3.3.8) with $h \equiv 1$, we obtain, for any $t_0 \ge 0$ and $R > R_0$,

$$\begin{aligned}\pi_v(B_R^c \times \mathbb{U}) &\le k_1' \, \mathbb{E}_x^v \left[\int_0^{\tau(D_1^c)} \mathbb{I}_{B_R^c}(X_t)\, \mathrm{d}t \right] \\ &\le k_1' t_0 \, \mathbb{P}_x^v(\tau_R \le t_0) + k_1' \, \mathbb{E}_x^v \left[\tau(D_1^c) \, \mathbb{I}\{\tau(D_1^c) \ge t_0\} \right], \quad x \in \partial D_2 ,\end{aligned}$$

for some constant $k_1' > 0$ that depends only on D_1, D_2 and R_0. By (ix), we can select t_0 large enough so that the second term on the right-hand side is as small as desired, uniformly in $v \in \mathcal{U}$ and $x \in \partial D_2$. By (2.2.20), for any fixed $t_0 > 0$,

$$\sup_{v \in \mathfrak{U}_{\mathrm{sm}}} \sup_{x \in \partial D_2} \mathbb{P}_x^v(\tau_R \le t_0) \xrightarrow[R \to \infty]{} 0 ,$$

and (vii) follows.

(v) ⇒ (viii): Since the mean empirical measures are tight, their closure is compact by Prohorov's theorem. Then, by Lemma 2.5.3, every accumulation point of a sequence of mean empirical measures is an ergodic occupation measure. Also, if $v \in \mathfrak{U}_{\mathrm{ssm}}$, then $\bar{\zeta}_{x,t}^v$ converges as $t \to \infty$ to π_v. Therefore tightness implies that the set of accumulation points as $t \to \infty$ of sequences of mean empirical measures is precisely the set of ergodic occupation measures, and hence, being closed, $\mathscr{G}$ is compact.

(viii) ⇒ (x): Let $D = D_1$ and, without loss of generality, let also $\Gamma = \partial D_2$. Then (3.3.8) implies

$$\sup_{v \in \mathcal{U}} \sup_{x \in \partial D_2} \mathbb{E}_x^v \left[\int_0^{\tau(D_1^c)} \mathbb{I}_{B_R^c}(X_t)\, \mathrm{d}t \right] \xrightarrow[R \to \infty]{} 0 . \tag{3.3.14}$$

Given any sequence $\{(v_n, x_n)\} \subset \mathcal{U} \times \partial D_2$ converging to some $(v, x) \in \bar{\mathcal{U}} \times \partial D_2$, Lemma 2.6.13 (iii) asserts that

$$\mathbb{E}_{x_n}^{v_n} \left[\int_0^{\tau(D_1^c)} \mathbb{I}_{B_R}(X_t)\, \mathrm{d}t \right] \xrightarrow[n \to \infty]{} \mathbb{E}_x^v \left[\int_0^{\tau(D_1^c)} \mathbb{I}_{B_R}(X_t)\, \mathrm{d}t \right] \tag{3.3.15}$$

for all R such that $D_2 \Subset B_R$. By (3.3.14) and (3.3.15), we obtain

$$\mathbb{E}_{x_n}^{v_n}[\tau(D_1^c)] \to \mathbb{E}_x^v[\tau(D_1^c)] \quad as\ n \to \infty ,$$

and (x) follows. □

3.4 Existence of optimal controls

In this section we establish the existence of an optimal control in $\mathfrak{U}_{\mathrm{ssd}}$.

Based on Theorem 3.2.8 and Lemmas 3.2.9 and 3.2.11 we formulate the ergodic control problem as a convex program. Note that for $v \in \mathfrak{U}_{\mathrm{ssm}}$ and $\pi_v = \eta_v \circledast v$,

$$\lim_{T \to \infty} \frac{1}{T} \int_0^T \int_{\mathbb{U}} c(X_s, u)\, v(\mathrm{d}u \mid X_s)\, \mathrm{d}s = \int_{\mathbb{R}^d \times \mathbb{U}} c\, \mathrm{d}\pi \quad \text{a.s.}$$

This suggests the following convex programming problem

$$\text{minimize} \quad \int_{\mathbb{R}^d \times \mathrm{U}} c \, \mathrm{d}\pi$$
$$\text{over} \quad \pi \in \mathscr{G} .$$

For $\varepsilon \geq 0$, $\pi^*_\varepsilon \in \mathscr{G}$ is called *ε-optimal* if it satisfies

$$\varrho^* := \inf_{\pi \in \mathscr{G}} \int_{\mathbb{R}^d \times \mathrm{U}} c \, \mathrm{d}\pi \leq \int_{\mathbb{R}^d \times \mathrm{U}} c \, \mathrm{d}\pi^*_\varepsilon \leq \varrho^* + \varepsilon .$$

When $\varepsilon = 0$, we refer to $\pi^*_0 \in \mathscr{G}$ simply as optimal, and also denote it by π^*.

Lemma 3.4.1 *For $\varepsilon \geq 0$, if there exists an ε-optimal $\pi^*_\varepsilon \in \mathscr{G}$, then there exists an ε-optimal $\hat{\pi}^*_\varepsilon \in \mathscr{G}_e$.*

Proof By Lemma 3.2.9, there exists $\Psi \in \mathscr{P}(\mathscr{G}_e)$ such that

$$\varrho^* \leq \int_{\mathbb{R}^d \times \mathrm{U}} c \, \mathrm{d}\pi^*_\varepsilon = \int_{\mathscr{G}_e} \left(\int_{\mathbb{R}^d \times \mathrm{U}} c(x, u) \, \pi(\mathrm{d}x, \mathrm{d}u) \right) \Psi(\mathrm{d}\pi) \leq \varrho^* + \varepsilon .$$

Thus, for some $\hat{\pi}^*_\varepsilon \in \operatorname{supp}(\Psi) \subset \mathscr{G}_e$,

$$\varrho^* \leq \int_{\mathbb{R}^d \times \mathrm{U}} c \, \mathrm{d}\hat{\pi}^*_\varepsilon \leq \varrho^* + \varepsilon . \qquad \square$$

We now turn to the question of existence of an optimal $\pi^* \in \mathscr{G}$. In general, this is not the case as the following counterexample shows. Let $c(x) = \mathrm{e}^{-|x|^2}$. Then for every $v \in \mathfrak{U}_{\mathrm{ssm}}$ the ergodic cost is positive a.s., while for every unstable Markov control the ergodic cost equals 0 a.s., thus making the latter optimal. We focus on problems where stability and optimality are not at odds, and for this reason we impose two alternate sets of hypotheses: (a) a condition on the cost function that penalizes unstable behavior, and (b) the assumption that all stationary Markov controls are uniformly stable.

Assumption 3.4.2 The running cost function c is *near-monotone*, i.e.,

$$\liminf_{|x| \to \infty} \min_{u \in \mathrm{U}} c(x, u) > \varrho^* . \tag{3.4.1}$$

Assumption 3.4.3 The set $\mathfrak{U}_{\mathrm{sm}}$ of stationary Markov controls is uniformly stable (see Definition 3.3.3).

Remark 3.4.4 It may seem at first that (3.4.1) cannot be verified unless ϱ^* is known. However there are two important cases where (3.4.1) always holds. The first is the case where $\min_{u \in \mathrm{U}} c(x, u)$ is inf-compact and $\varrho^* < \infty$. The second covers problems in which $c(x, u) = c(x)$ does not depend on u and $c(x) < \lim_{|z| \to \infty} c(z)$ for all $x \in \mathbb{R}^d$. In particular for $d = 1$, if $c(x, u)$ is independent of u and is monotonically (strictly) increasing with $|x|$, then (3.4.1) holds; hence the terminology near-monotone.

Theorem 3.4.5 *Under either Assumption 3.4.2 or 3.4.3, the map $\pi \mapsto \int c \, \mathrm{d}\pi$ attains its minimum in $\mathscr{G}$ at some $\pi^* \in \mathscr{G}_e$.*

Proof Note that the map $\pi \mapsto \int c\,\mathrm{d}\pi$ is lower semicontinuous. Thus, since under Assumption 3.4.3, Lemma 3.3.4 asserts that $\mathcal{G}$ is compact, this map attains a minimum in $\mathcal{G}$, and hence, by Lemma 3.4.1 in $\mathcal{G}_e$. Note then that there exists a $v^* \in \mathfrak{U}_{\mathrm{ssd}}$ such that

$$\varrho^* = \int_{\mathbb{R}^d} c_{v^*}(x)\,\eta_{v^*}(\mathrm{d}x)\,. \tag{3.4.2}$$

Next consider Assumption 3.4.2. Let $\{\pi_n : n \in \mathbb{N}\} \subset \mathcal{G}$ such that $\int c\,\mathrm{d}\pi_n \downarrow \varrho^*$. Viewing $\mathcal{G}$ as a subset of $\mathcal{P}(\bar{\mathbb{R}}^d \times \mathbb{U})$, select a subsequence of $\{\pi_n\}$ converging to some $\hat{\pi} \in \mathcal{P}(\bar{\mathbb{R}}^d \times \mathbb{U})$. By Lemma 3.2.11, there exist $\pi' \in \mathcal{G}$, $\pi'' \in \mathcal{P}(\{\infty\} \times \mathbb{U})$ and $\delta \in [0,1]$ satisfying (3.2.17) for all $B \in \mathcal{B}(\bar{\mathbb{R}}^d \times \mathbb{U})$. By (3.4.1), there exists $R > 0$ and $\varepsilon > 0$ such that

$$\inf_{u \in \mathbb{U}} \{c(x,u) : |x| \geq R\} \geq \varrho^* + \varepsilon\,.$$

Define a sequence $c_n : \mathbb{R}^d \times \mathbb{U} \to \mathbb{R}$, for $n \in \mathbb{N}$, by

$$c_n(x,u) := \begin{cases} c(x,u) & \text{if } |x| < R+n\,, \\ \varrho^* + \varepsilon & \text{if } |x| \geq R+n\,. \end{cases} \tag{3.4.3}$$

Observe that the functions c_n defined in (3.4.3) are lower semicontinuous. Then, for any $m \in \mathbb{N}$, we obtain by letting $c_m(\infty) = \varrho^* + \varepsilon$

$$\varrho^* \geq \liminf_{n\to\infty} \int c_m\,\mathrm{d}\pi_n \geq \delta \int c_m\,\mathrm{d}\pi' + (1-\delta)(\varrho^* + \varepsilon)\,. \tag{3.4.4}$$

Letting $m \to \infty$ in (3.4.4), we obtain

$$\begin{aligned} \varrho^* &\geq \delta \int c\,\mathrm{d}\pi' + (1-\delta)(\varrho^* + \varepsilon) \\ &\geq \delta\varrho^* + (1-\delta)(\varrho^* + \varepsilon)\,. \end{aligned}$$

Thus $\delta = 1$ and $\int c\,\mathrm{d}\pi' = \varrho^*$. By Corollary 3.2.10, there exists $v^* \in \mathfrak{U}_{\mathrm{ssd}}$ such that (3.4.2) holds. □

We now wish to establish that $v^* \in \mathfrak{U}_{\mathrm{ssd}}$ satisfying (3.4.2) is optimal among all admissible controls $\mathfrak{U}$. For any admissible control $U \in \mathfrak{U}$ we define the process ζ^U_t of (random) empirical measures as a $\mathcal{P}(\mathbb{R}^d \times \mathbb{U})$-valued process satisfying, for all $f \in C_b(\mathbb{R}^d \times \mathbb{U})$,

$$\int_{\mathbb{R}^d \times \mathbb{U}} f\,\mathrm{d}\zeta^U_t = \frac{1}{t} \int_0^t \int_{\mathbb{U}} f(X_s,u) U_s(\mathrm{d}u)\,\mathrm{d}s\,, \quad t > 0\,, \tag{3.4.5}$$

where X denotes the solution of the diffusion in (2.2.1).

Lemma 3.4.6 *Almost surely, every limit point $\hat{\zeta} \in \mathcal{P}(\bar{\mathbb{R}}^d \times \mathbb{U})$ as $t \to \infty$, of the process ζ_t^U defined in* (3.4.5) *takes the form*

$$\hat{\zeta} = \delta\zeta' + (1-\delta)\zeta'' , \quad \delta \in [0,1] , \tag{3.4.6}$$

with $\zeta' \in \mathcal{G}$ and $\zeta''(\{\infty\} \times \mathbb{U}) = 1$. An identical claim holds for the mean empirical measures without the need of the qualification "almost surely."

Proof Clearly $\hat{\zeta}$ can be decomposed as in (3.4.6) for some $\zeta' \in \mathcal{P}(\mathbb{R}^d \times \mathbb{U})$. For $f \in \mathcal{C}$, we have

$$\frac{f(X_t) - f(X_0)}{t} = \frac{1}{t}\int_0^t \mathcal{L}^{U_s} f(X_s)\,\mathrm{d}s + \frac{1}{t}\int_0^t \langle \nabla f(X_s), \sigma(X_s)\,\mathrm{d}W_s\rangle . \tag{3.4.7}$$

Let

$$M_t := \int_0^t \langle \nabla f(X_s), \sigma(X_s)\,\mathrm{d}W_s\rangle , \quad t \geq 0 .$$

Then M_t is a zero mean, square-integrable martingale with continuous paths, whose quadratic variation process is given by

$$\langle M\rangle_t = \int_0^t \left|\sigma^{\mathsf{T}}(X_s)\nabla f(X_s)\right|^2 \mathrm{d}s , \quad t \geq 0 .$$

By a random time change argument, for some suitably defined one-dimensional Brownian motion $\mathbb{B}$, we obtain

$$M_t = \mathbb{B}(\langle M\rangle_t) , \quad t \geq 0 .$$

Let $\langle M\rangle_\infty := \lim_{t\to\infty} \langle M\rangle_t$. On the set $\{\langle M\rangle_\infty = \infty\}$, we have

$$\lim_{t\to\infty} \frac{\mathbb{B}(\langle M\rangle_t)}{\langle M\rangle_t} = \lim_{t\to\infty} \frac{\mathbb{B}(t)}{t} = 0 \quad \text{a.s.,}$$

and since f has compact support, $\langle M\rangle_t \in \mathcal{O}(t)$, implying that

$$\limsup_{t\to\infty} \frac{\langle M\rangle_t}{t} < \infty \quad \text{a.s.}$$

On the other hand, on the set $\{\langle M\rangle_\infty < \infty\}$,

$$\lim_{t\to\infty} \frac{\mathbb{B}(\langle M\rangle_t)}{\langle M\rangle_t} < \infty \quad \text{a.s.}$$

and

$$\lim_{t\to\infty} \frac{\langle M\rangle_t}{t} = 0 \quad \text{a.s.}$$

Therefore, in either case,

$$\lim_{t\to\infty} \frac{\mathbb{B}(\langle M\rangle_t)}{t} = \lim_{t\to\infty} \frac{\mathbb{B}(t)}{t} = 0 \quad \text{a.s.} \tag{3.4.8}$$

Equation (3.4.8) implies that the last term in (3.4.7) tends to zero a.s., as $n \to \infty$. The left-hand side of (3.4.7) also tends to zero as $n \to \infty$, since f is bounded. Since $\mathcal{C}$ is countable, these limits are zero outside a null set, for any $f \in \mathscr{C}$. By the definition of ζ_t^U, we have

$$\frac{1}{t}\int_0^t \mathcal{L}^{U_s} f(X_s)\,\mathrm{d}s = \frac{1}{t}\int_0^t \int_{\mathbb{U}} \mathcal{L}^u f(X_s) U_s(\mathrm{d}u)\,\mathrm{d}s = \int_{\mathbb{R}^d\times\mathbb{U}} \mathcal{L}^u f(x)\,\zeta_t^U(\mathrm{d}x,\mathrm{d}u)\,.$$

Thus any limit point $\hat{\zeta}$ of $\{\zeta_t^U\}$ as in (3.4.6) must satisfy, whenever $\delta > 0$,

$$\int_{\mathbb{R}^d\times\mathbb{U}} \mathcal{L}^u f(x)\,\zeta'(\mathrm{d}x,\mathrm{d}u) = 0 \qquad \forall f \in \mathscr{C}\,,$$

and thus, by Lemma 3.2.2, $\zeta' \in \mathscr{G}$. If $\delta = 0$, ζ' in (3.4.6) can be arbitrarily selected, so it may be chosen in $\mathscr{G}$. The claim for the mean empirical measures follows analogously. □

Theorem 3.4.7 *Under either Assumption 3.4.2 or Assumption 3.4.3 together with the added hypothesis that $\{\zeta_t^U : t \geq 0\}$ is a.s. tight in $\mathscr{P}(\mathbb{R}^d \times \mathbb{U})$, we have*

$$\liminf_{t\to\infty} \frac{1}{t}\int_0^t \int_{\mathbb{U}} c(X_s,u) U_s(\mathrm{d}u)\,\mathrm{d}s \geq \min_{v\in\mathfrak{U}_{\mathrm{ssd}}} \int_{\mathbb{R}^d\times\mathbb{U}} c\,\mathrm{d}\pi_v \quad a.s.$$

and

$$\liminf_{t\to\infty} \frac{1}{t}\int_0^t \mathbb{E}_x^U\left[c(X_s,U_s)\right]\mathrm{d}s \geq \min_{v\in\mathfrak{U}_{\mathrm{ssd}}} \int_{\mathbb{R}^d\times\mathbb{U}} c\,\mathrm{d}\pi_v\,,$$

for any $U \in \mathfrak{U}$.

Proof Let $\{\zeta_t^U\}$ be defined by (3.4.5) with $U \in \mathfrak{U}$. First, suppose Assumption 3.4.2 holds. Consider a sample point such that the result of Lemma 3.4.6 holds, and let $\hat{\zeta}$ be a limit point of ζ_t^U as $t \to \infty$. Then, for some sequence $\{t_\ell\}$ such that $t_\ell \uparrow \infty$, we have $\zeta_{t_\ell}^U \to \hat{\zeta}$. Let $\{c_n : n \in \mathbb{N}\}$ be the sequence defined in (3.4.3). Then

$$\begin{aligned}\liminf_{\ell\to\infty} \int_{\mathbb{R}^d\times\mathbb{U}} c\,\mathrm{d}\zeta_{t_\ell}^U &\geq \int_{\mathbb{R}^d\times\mathbb{U}} c_n\,\mathrm{d}\hat{\zeta}\\ &\geq \delta\int_{\mathbb{R}^d\times\mathbb{U}} c_n\,\mathrm{d}\zeta' + (1-\delta)(\varrho^*+\varepsilon)\,,\end{aligned}$$

and letting $n \to \infty$, we obtain

$$\begin{aligned}\liminf_{\ell\to\infty} \int_{\mathbb{R}^d\times\mathbb{U}} c\,\mathrm{d}\zeta_{t_\ell}^U &\geq \delta\int_{\mathbb{R}^d\times\mathbb{U}} c\,\mathrm{d}\zeta' + (1-\delta)(\varrho^*+\varepsilon)\\ &\geq \delta\varrho^* + (1-\delta)(\varrho^*+\varepsilon)\,.\end{aligned}$$

The analogous inequality holds for the mean empirical measures $\bar{\zeta}_{x,t}^U$. This completes the first part of the proof. Suppose now Assumption 3.4.3 holds and $\{\zeta_t^U\}$ are a.s. tight in $\mathscr{P}(\mathbb{R}^d \times \mathbb{U})$. Decomposing the arbitrary limit point $\hat{\zeta}$ of $\{\zeta_t^U\}$ as in (3.4.6), it follows that $\delta = 1$, and the result is immediate, since $\zeta' \in \mathscr{G}$. □

Remark 3.4.8 Note that Theorem 3.4.7 establishes a much stronger optimality of v^*, viz., the most "pessimistic" pathwise cost under v^* is no worse than the most "optimistic" pathwise cost under any other admissible control.

The a.s. tightness of the family $\{\zeta_t^U\}$, in the stable case, is guaranteed under the hypothesis that the hitting times of bounded domains have finite second moments bounded uniformly over all admissible controls $U \in \mathfrak{U}$. A Lyapunov condition that enforces this property is the following.

Assumption 3.4.9 There exist nonnegative $\mathcal{V}_1$ and $\mathcal{V}_2$ in $C^2(\mathbb{R}^d)$ which satisfy, outside some bounded domain D,

$$\mathcal{L}^u\mathcal{V}_1 \le -1\,, \qquad \mathcal{L}^u\mathcal{V}_2 \le -\mathcal{V}_1$$

for all $u \in \mathbb{U}$.

It is evident from Lemma 3.3.4 that Assumption 3.4.9 implies uniform stability. Employing Dynkin's formula as in the proof of Lemma 2.5.1 shows that the second moment of $\tau(D^c)$ is uniformly bounded over all admissible $U \in \mathfrak{U}$.

Corollary 3.4.10 *Let Assumption 3.4.9 hold. Then $\mathcal{G}$ is compact, and for any bounded domain D and compact set $\Gamma \subset \mathbb{R}^d$,*

$$\sup_{U\in\mathfrak{U}}\ \sup_{x\in\Gamma}\ \mathbb{E}_x^U\bigl[\tau(D^c)^2\bigr] < \infty\,.$$

Theorem 3.4.11 *Under Assumption 3.4.9, the empirical measures $\{\zeta_t^U : t \ge 0\}$ are tight a.s., under any $U \in \mathfrak{U}$.*

Proof Recall the definition of $\{\hat{\tau}_n\}$ in (2.6.10). Since by Assumption 3.4.9,

$$\sup_{x\in\partial D_2}\ \sup_{U\in\mathfrak{U}}\ \mathbb{E}_x^U\left[\int_0^{\tau(D_1^c)} \mathcal{V}_1(X_t)\,\mathrm{d}t\right] < \infty\,,$$

and $\mathcal{V}_1$ is inf-compact, and also $\inf_{x\in\partial D_2}\ \inf_{U\in\mathfrak{U}}\ \mathbb{E}_x^U\left[\tau(D_1^c)\right] > 0$, it follows that set $\hat{\mathcal{H}}$ of $\eta \in \mathcal{P}(\mathbb{R}^d)$ defined by

$$\int_{\mathbb{R}^d} f\,\mathrm{d}\eta = \frac{\mathbb{E}_{X_0}^U\left[\int_0^{\hat{\tau}_2} f(X_t)\,\mathrm{d}t\right]}{\mathbb{E}_{X_0}^U\left[\hat{\tau}_2\right]} \qquad \forall f \in C_b(\mathbb{R}^d)$$

for $U \in \mathfrak{U}$, and with the law of X_0 supported on ∂D_1, is tight.

Let $\{f_n\} \subset C(\mathbb{R}^d; [0,1])$ be a collection of maps, satisfying $f_n(x) = 0$ for $|x| \le n$, and $f_n(x) = 1$ for $|x| \ge n+1$. Then, for any $\varepsilon > 0$, we can find $N_\varepsilon \ge 1$ such that $\int_{\mathbb{R}^d} f_n\,\mathrm{d}\eta < \varepsilon$ for all $n \ge N_\varepsilon$ and $\eta \in \hat{\mathcal{H}}$. By Corollary 3.4.10, we can use the strong law of large numbers for martingales to conclude that, for all $f \in C_b(\mathbb{R}^d)$,

$$\frac{1}{n}\int_0^{\hat{\tau}_{2n}} f(X_t)\,\mathrm{d}t - \frac{1}{n}\sum_{m=0}^{n-1}\mathbb{E}_{X_0}^U\left[\int_{\hat{\tau}_{2m}}^{\hat{\tau}_{2m+2}} f(X_t)\,\mathrm{d}t \;\Big|\; \mathfrak{F}_{\hat{\tau}_{2m}}^X\right] \xrightarrow{\text{a.s.}} 0\,, \tag{3.4.9}$$

and

$$\frac{1}{n}\Big(\hat{\tau}_{2n} - \sum_{m=0}^{n-1} \mathbb{E}^U_{X_0}\big[\hat{\tau}_{2m+2} - \hat{\tau}_{2m} \mid \mathfrak{F}^X_{\hat{\tau}_{2m}}\big]\Big) \xrightarrow{\text{a.s.}} 0\,. \tag{3.4.10}$$

Therefore, since $\mathbb{E}^U_{X_0}\big[\hat{\tau}_{2m+2} - \hat{\tau}_{2m} \mid \mathfrak{F}^X_{\hat{\tau}_{2m}}\big]$ is bounded away from zero uniformly in $U \in \mathfrak{U}$ and $m \in \mathbb{N}$ by Lemma 2.3.7 and Theorem 2.6.1 (b), it follows by (3.4.10) that $\liminf_{n\to\infty} \hat{\tau}_{2n}$ is bounded away from zero a.s. Hence, subtracting the left-hand sides of (3.4.9) and (3.4.10) from the numerator and denominator, respectively, of

$$\frac{n^{-1}\int_0^{\hat{\tau}_{2n}} f_N(X_t)\,\mathrm{d}t}{n^{-1}\hat{\tau}_{2n}}$$

and taking limits, we obtain

$$\begin{aligned}
\limsup_{n\to\infty} \frac{1}{\hat{\tau}_{2n}} \int_0^{\hat{\tau}_{2n}} f_N(X_t)\,\mathrm{d}t &= \limsup_{n\to\infty} \frac{\frac{1}{n}\int_0^{\hat{\tau}_{2n}} f_N(X_t)\,\mathrm{d}t}{\frac{\hat{\tau}_{2n}}{n}} \\
&= \limsup_{n\to\infty} \frac{\frac{1}{n}\sum_{m=0}^{n-1} \mathbb{E}^U_{X_0}\Big[\int_{\hat{\tau}_{2m}}^{\hat{\tau}_{2m+2}} f_N(X_t)\,\mathrm{d}t \,\Big|\, \mathfrak{F}^X_{\hat{\tau}_{2m}}\Big]}{\frac{1}{n}\sum_{m=0}^{n-1} \mathbb{E}^U_{X_0}\big[\hat{\tau}_{2m+2} - \hat{\tau}_{2m} \mid \mathfrak{F}^X_{\hat{\tau}_{2m}}\big]} \\
&\le \varepsilon \quad \text{a.s.,} \qquad \forall N \ge N_\varepsilon\,.
\end{aligned}$$

Let $\varkappa(t)$ denote the number of cycles completed at time t, i.e.,

$$\varkappa(t) = \max\,\{k : t > \hat{\tau}_{2k}\}\,.$$

It is evident that $\varkappa(t) \uparrow \infty$ a.s. as $t \uparrow \infty$. By by Theorem 2.6.1 (b),

$$\sup_m \mathbb{E}^U_{X_0}\big[\hat{\tau}_{2m+2} - \hat{\tau}_{2m} \mid \mathfrak{F}^X_{\hat{\tau}_{2m}}\big] < \infty \quad \text{a.s.}$$

Therefore, $\frac{\hat{\tau}_{2m+2} - \hat{\tau}_{2m}}{\hat{\tau}_{2m}} \to 0$ a.s. by (3.4.10). Using this property together with the fact that $\hat{\tau}_{2\varkappa(t)} \le t < \hat{\tau}_{2\varkappa(t)+2}$ we obtain

$$\begin{aligned}
\limsup_{t\to\infty} \frac{1}{t}\int_0^t f_N(X_s)\,\mathrm{d}s &\le \limsup_{t\to\infty} \frac{1}{\hat{\tau}_{2\varkappa(t)}}\int_0^{\hat{\tau}_{2\varkappa(t)+2}} f_N(X_s)\,\mathrm{d}s \\
&\le \limsup_{n\to\infty} \frac{1}{\hat{\tau}_{2n}}\int_0^{\hat{\tau}_{2n+2}} f_N(X_s)\,\mathrm{d}s \\
&= \limsup_{n\to\infty} \frac{1}{\hat{\tau}_{2n}}\int_0^{\hat{\tau}_{2n}} f_N(X_s)\,\mathrm{d}s \\
&\le \varepsilon \qquad \forall N \ge N_\varepsilon\,,
\end{aligned}$$

thus establishing the a.s. tightness of $\{\zeta^U_t\}$. □

3.5 The discounted control problem

In this section, we study the Hamilton–Jacobi–Bellman (HJB) equation for the discounted problem, which is in turn used to study the ergodic problem via the vanishing discount limit. In view of Corollary 3.2.10, for the rest of this chapter we work only with precise controls.

Provided that the cost function is locally Hölder continuous in x, we can obtain C^2 solutions for the HJB equation. In order to avoid keeping track of the Hölder constants we assume local Lipschitz continuity for c.

Assumption 3.5.1 The cost function $c\colon \mathbb{R}^d \times \mathbb{U} \to \mathbb{R}_+$ is continuous and locally Lipschitz in its first argument uniformly in $u \in \mathbb{U}$. More specifically, for some function $K_c\colon \mathbb{R}_+ \to \mathbb{R}_+$,

$$\bigl|c(x,u) - c(y,u)\bigr| \le K_c(R)|x-y| \qquad \forall x, y \in B_R\,,\ \forall u \in \mathbb{U}\,,$$

and all $R > 0$. We denote the class of such functions by $\mathfrak{C}$.

Let $\alpha > 0$ be a constant, which we refer to as the *discount factor*. For an admissible control $U \in \mathfrak{U}$, we define the *α-discounted cost* by

$$J_\alpha^U(x) := \mathbb{E}_x^U\left[\int_0^\infty \mathrm{e}^{-\alpha t}\bar{c}(X_t, U_t)\,\mathrm{d}t\right],$$

and we let

$$V_\alpha(x) := \inf_{U \in \mathfrak{U}} J_\alpha^U(x)\,. \tag{3.5.1}$$

3.5.1 Quasilinear elliptic operators

Hamilton–Jacobi–Bellman equations that are of interest to us involve *quasilinear operators* of the form

$$\begin{aligned} S\psi(x) &:= a^{ij}(x)\,\partial_{ij}\psi(x) + \inf_{u\in\mathbb{U}} \hat{b}(x,u,\psi) \\ \hat{b}(x,u,\psi) &:= b^i(x,u)\,\partial_i\psi(x) - \lambda\,\psi(x) + c(x,u)\,. \end{aligned} \tag{3.5.2}$$

We suitably parameterize families of quasilinear operators of this form as follows.

Definition 3.5.2 For a nondecreasing function $\gamma\colon (0,\infty) \to (0,\infty)$ we denote by $\mathfrak{Q}(\gamma)$ the class of operators of the form (3.5.2), with $\lambda \in \mathbb{R}_+$ and whose coefficients b^i and c belong to $C(\mathbb{R}^d \times \mathbb{U})$, satisfy (A.1.1a)–(A.1.1b), and for all $x, y \in B_R$,

$$\max_{u\in\mathbb{U}} \Bigl\{\max_i \bigl|b^i(x,u) - b^i(y,u)\bigr| + \bigl|c(x,u) - c(y,u)\bigr|\Bigr\} \le \gamma(R)|x-y|$$

and

$$\sum_{i,j=1}^d |a^{ij}(x)| + \sum_{i=1}^d \max_{u\in\mathbb{U}} |b^i(x,u)| + \max_{u\in\mathbb{U}} |c(x,u)| \le \gamma(R)\,.$$

The Dirichlet problem for quasilinear equations is more involved than the linear case. Here we investigate the existence of solutions to the problem

$$S\psi(x) = 0 \quad \text{in } D\,, \qquad \psi = 0 \quad \text{on } \partial D \tag{3.5.3}$$

for a sufficiently smooth bounded domain D. We follow the approach of Gilbarg and Trudinger [67, section 11.2], which utilizes the Leray–Schauder fixed point theorem, to obtain the following result.

Theorem 3.5.3 *Let D be a bounded $C^{2,1}$ domain in $\mathbb{R}^d$. Then the Dirichlet problem in* (3.5.3) *has a solution in $C^{2,r}(\bar{D})$, $r \in (0,1)$ for any $S \in \mathfrak{Q}(\gamma)$.*

Proof Let

$$S_\delta\psi(x) := a^{ij}\partial_{ij}\psi(x) + \delta \inf_{u\in\mathbb{U}} \hat{b}(x,u,\psi)\,, \qquad \delta \in [0,1]\,.$$

Then according to Gilbarg and Trudinger [67, theorem 11.4, p. 281] it is enough to show that there exist constants $\rho > 0$ and M such that any $C^{2,r}(\bar{D})$ solution of the family of Dirichlet problems $S_\delta\psi = 0$ in D, $\psi = 0$ on ∂D, $\delta \in [0,1]$, satisfies

$$\|\psi\|_{C^{1,\rho}(\bar{D})} < M\,. \tag{3.5.4}$$

Suppose ψ is such a solution. Then, for each fixed u, the map

$$x \mapsto \hat{b}(x,u,\psi)$$

belongs to $C^{0,1}(\bar{D})$. If $v\colon \mathbb{R}^d \to \mathbb{U}$ is a measurable selector from the minimizer $\operatorname{Arg\,min}_{u\in\mathbb{U}} \hat{b}(x,u,\psi)$, then ψ satisfies the linear problem

$$\mathcal{L}_\delta\psi(x) = -\delta\, c(x,v(x)) \quad \text{in } D\,, \qquad \psi = 0 \quad \text{on } \partial D\,,$$

where

$$\mathcal{L}_\delta\psi(x) = a^{ij}(x)\,\partial_{ij}\psi(x) + \delta\left(b^i(x,v(x))\,\partial_i\psi(x) - \lambda\,\psi(x)\right).$$

Hence $\mathcal{L} \in \mathfrak{L}(\gamma)$ (see Definition A.1.1), and by (A.2.1), ψ satisfies

$$\sup_D \psi \le C_a|D|^{1/d} \sup_{D\times\mathbb{U}} |c|\,. \tag{3.5.5}$$

Applying Theorem A.2.7 for the linear problem, with $p \equiv 2d$, we obtain by (A.2.4)

$$\|\psi\|_{\mathscr{W}^{2,2d}(D)} \le C_0'\,\|\mathcal{L}\psi\|_{L^{2d}(D)}\,. \tag{3.5.6}$$

Since C_0' and the bound in (3.5.5) are independent of ψ, then by combining (3.5.5)–(3.5.6) and using the compactness of the embedding $\mathscr{W}^{2,2d}(D) \hookrightarrow C^{1,r}(\bar{D})$, $r < \frac{1}{2}$, asserted in Theorem A.2.15 (2b), the estimate in (3.5.4) follows, and the proof is complete. □

We conclude this section with a useful convergence result.

Lemma 3.5.4 *Let D be a bounded C^2 domain. Suppose $\{h_n\} \subset L^p(D)$, for $p > 1$, and $\{\psi_n\} \subset \mathscr{W}^{2,p}(D)$ are a pair of sequences of functions satisfying the following:*

(a) *for some* $S \in \mathfrak{Q}(\gamma)$, $S\psi_n = h_n$ *in* D *for all* $n \in \mathbb{N}$;

(b) *for some constant* M, $\|\psi_n\|_{\mathscr{W}^{2,p}(D)} \le M$ *for all* $n \in \mathbb{N}$;

(c) h_n *converges in* $L^p(D)$ *to some function* h.

Then there exists $\psi \in \mathscr{W}^{2,p}(D)$ *and a sequence* $\{n_k\} \subset \mathbb{N}$ *such that* $\psi_{n_k} \to \psi$ *in* $\mathscr{W}^{1,p}(D)$ *as* $k \to \infty$, *and*

$$S\psi = h \quad \text{in } D. \tag{3.5.7}$$

If in addition $p > d$, *then* $\psi_{n_k} \to \psi$ *in* $C^{1,r}(D)$ *for any* $r < 1 - \frac{d}{p}$.

If $h \in C^{0,\rho}(D)$, *for some* $\rho > 0$, *then* $\psi \in C^{2,\rho}(D)$.

Proof By the weak compactness of $\{\varphi : \|\varphi\|_{\mathscr{W}^{2,p}(D)} \le M\}$ and the compactness of the imbedding $\mathscr{W}^{2,p}(D) \hookrightarrow \mathscr{W}^{1,p}(D)$, we can select $\psi \in \mathscr{W}^{2,p}(D)$ and $\{n_k\}$ such that $\psi_{n_k} \to \psi$, weakly in $\mathscr{W}^{2,p}(D)$ and strongly in $\mathscr{W}^{1,p}(D)$ as $k \to \infty$. The inequality

$$\left|\inf_{u\in\mathbb{U}} \hat{b}(x,u,\psi) - \inf_{u\in\mathbb{U}} \hat{b}(x,u,\psi')\right| \le \sup_{u\in\mathbb{U}} \left|\hat{b}(x,u,\psi) - \hat{b}(x,u,\psi')\right| \tag{3.5.8}$$

shows that $\inf_{u\in\mathbb{U}} \hat{b}(\,\cdot\,,u,\psi_{n_k})$ converges in $L^p(D)$. Since, by weak convergence,

$$\int_D g(x)\,\partial_{ij}\psi_{n_k}(x)\,\mathrm{d}x \xrightarrow[k\to\infty]{} \int_D g(x)\,\partial_{ij}\psi(x)\,\mathrm{d}x$$

for all $g \in L^{\frac{p}{p-1}}(D)$ and $h_n \to h$ in $L^p(D)$, we obtain

$$\begin{aligned}\int_D g(x)(S\psi(x) - h(x))\,\mathrm{d}x &= \lim_{k\to\infty} \int_D g(x)(S\psi_{n_k}(x) - h_{n_k}(x))\,\mathrm{d}x \\ &= 0\end{aligned}$$

for all $g \in L^{\frac{p}{p-1}}(D)$. Thus the pair (ψ, h) satisfies (3.5.7).

If $p > d$, the compactness of the embedding $\mathscr{W}^{2,p}(D) \hookrightarrow C^{1,r}(\bar{D})$, $r < 1 - \frac{d}{p}$, allows us to select the subsequence such that $\psi_{n_k} \to \psi$ in $C^{1,r}(\bar{D})$. The inequality (3.5.8) shows that $\inf_{u\in\mathbb{U}} \hat{b}(\,\cdot\,,u,\psi_{n_k})$ converges uniformly in D, and the inequality

$$\left|\inf_{u\in\mathbb{U}} \hat{b}(x,u,\psi) - \inf_{u\in\mathbb{U}} \hat{b}(y,u,\psi)\right| \le \sup_{u\in\mathbb{U}} \left|\hat{b}(x,u,\psi) - \hat{b}(y,u,\psi)\right| \tag{3.5.9}$$

implies that the limit is in $C^{0,r}(D)$.

If $h \in C^{0,\rho}(D)$, then by Lemma A.2.10, $\psi \in \mathscr{W}^{2,p}(D)$ for all $p > 1$. Using the continuity of the embedding $\mathscr{W}^{2,p}(D) \hookrightarrow C^{1,r}(\bar{D})$ for $r \le 1 - \frac{d}{p}$, and (3.5.9), we conclude that $\inf_{u\in\mathbb{U}} \hat{b}(\,\cdot\,,u,\psi) \in C^{0,r}$ for all $r < 1$. Thus ψ satisfies

$$a^{ij}\partial_{ij}\psi \in C^{0,\rho}(D),$$

and it follows by Theorem A.2.9 that $\psi \in C^{2,\rho}(D)$. □

Remark 3.5.5 If we replace $S \in \mathfrak{Q}(\gamma)$ with $\mathcal{L} \in \mathfrak{L}(\gamma)$ in Lemma 3.5.4, all the assertions of the lemma other than the last sentence follow. The proof is identical.

3.5.2 The HJB equation for the discounted problem

The *optimal α-discounted cost* V_α can be characterized as a solution of a HJB equation. This is the subject of the next theorem.

Theorem 3.5.6 *Suppose c satisfies Assumption 3.5.1 and is bounded on $\mathbb{R}^d$. Then V_α defined in* (3.5.1) *is the unique solution in $C^2(\mathbb{R}^d) \cap C_b(\mathbb{R}^d)$ of*

$$\min_{u \in \mathbb{U}} \left[\mathcal{L}^u V_\alpha(x) + c(x,u)\right] = \alpha V_\alpha(x)\,. \tag{3.5.10}$$

Moreover, $v \in \mathfrak{U}_{\mathrm{sm}}$ is α-discounted optimal if and only if v a.e. realizes the pointwise infimum in (3.5.10)*, i.e., if and only if*

$$\sum_{i=1}^{d} b_v^i(x) \frac{\partial V_\alpha}{\partial x_i}(x) + c_v(x) = \min_{u \in \mathbb{U}} \left[\sum_{i=1}^{d} b^i(x,u) \frac{\partial V_\alpha}{\partial x_i}(x) + c(x,u)\right], \tag{3.5.11}$$

a.e. $x \in \mathbb{R}^d$.

Proof For $\alpha > 0$ and $R > 0$, consider the Dirichlet problem

$$\begin{aligned} \min_{u \in \mathbb{U}} \left[\mathcal{L}^u \varphi(x) + c(x,u)\right] &= \alpha \varphi(x)\,, \quad x \in B_R\,, \\ \varphi\,\big|_{\partial B_R} &= 0\,. \end{aligned} \tag{3.5.12}$$

By Theorem 3.5.3, (3.5.12) has a solution $\varphi_R \in C^{2,r}(\bar{B}_R)$, $r \in (0,1)$. Let $v_R \in \mathfrak{U}_{\mathrm{sd}}$ satisfy $\mathcal{L}^{v_R}\varphi_R(x) + c_{v_R}(x) = \alpha\varphi_R(x)$ for all $x \in \mathbb{R}^d$. By Lemma A.3.2,

$$\varphi_R(x) = \mathbb{E}_x^{v_R}\left[\int_0^{\tau_R} \mathrm{e}^{-\alpha t} c_{v_R}(X_t)\,\mathrm{d}t\right], \quad x \in B_R\,.$$

On the other hand, for any $u \in \mathbb{U}$, $\mathcal{L}^u\varphi_R(x) + c(x,u) \ge \alpha\varphi_R(x)$, and this yields

$$\varphi_R(x) \le \mathbb{E}_x^U\left[\int_0^{\tau_R} \mathrm{e}^{-\alpha t} \bar{c}(X_t, U_t)\,\mathrm{d}t\right], \quad x \in B_R\,,\ U \in \mathfrak{U}\,.$$

Therefore

$$\varphi_R(x) = \inf_{U \in \mathfrak{U}} \mathbb{E}_x^U\left[\int_0^{\tau_R} \mathrm{e}^{-\alpha t} \bar{c}(X_t, U_t)\,\mathrm{d}t\right], \quad x \in B_R\,.$$

It is clear that $\varphi_R \le V_\alpha$ and that φ_R is nondecreasing in R. If $R' > 2R$, Lemma A.2.5 asserts that for any $p \in (1,\infty)$ there is a constant C_0, which is independent of R', such that

$$\begin{aligned} \left\|\varphi_{R'}\right\|_{\mathscr{W}^{2,p}(B_R)} &\le C_0\left(\left\|\varphi_{R'}\right\|_{L^p(B_{2R})} + \left\|\mathcal{L}^{v_{R'}}\varphi_{R'} - \alpha\varphi_{R'}\right\|_{L^p(B_{2R})}\right) \\ &\le C_0\left(\left\|V_\alpha\right\|_{L^p(B_{2R})} + \left\|c_{v_{R'}}\right\|_{L^p(B_{2R})}\right) \\ &\le C_0\left(\left\|V_\alpha\right\|_{L^p(B_{2R})} + K_c(2R)\left|B_{2R}\right|^{1/p}\right). \end{aligned}$$

Thus by Lemma 3.5.4, $\varphi_{R'} \uparrow \psi$ along a subsequence as $R' \to \infty$, where $\psi \in C^{2,r}(B_R)$ for any $r \in (0,1)$ and satisfies

$$\min_{u \in \mathbb{U}} \left[\mathcal{L}^u \psi(x) + c(x,u)\right] = \alpha \psi(x), \quad x \in B_R. \tag{3.5.13}$$

Since R was arbitrary, $\varphi_{R'}$ converges on $\mathbb{R}^d$ to some $\psi \in C^2(\mathbb{R}^d) \cap C_b(\mathbb{R}^d)$, satisfying (3.5.13).

Let $\tilde{v} \in \mathfrak{U}_{\mathrm{sd}}$ be such that

$$\min_{u \in \mathbb{U}} \left[\mathcal{L}^u \psi(x) + c(x,u)\right] = \mathcal{L}^{\tilde{v}} \psi(x) + c_{\tilde{v}}(x) \quad \text{a.e. } x \in \mathbb{R}^d. \tag{3.5.14}$$

For any pair $(\psi, \tilde{v})$, with $\psi \in C^2(\mathbb{R}^d) \cap C_b(\mathbb{R}^d)$ satisfying (3.5.13) and $\tilde{v} \in \mathfrak{U}_{\mathrm{sd}}$ satisfying (3.5.14), using Itô's formula, we obtain

$$\begin{aligned} \psi(x) &= \mathbb{E}_x^{\tilde{v}} \left[\int_0^\infty \mathrm{e}^{-\alpha t} c_{\tilde{v}}(X_t)\, \mathrm{d}t\right] \\ &= \inf_{U \in \mathfrak{U}} \mathbb{E}_x^U \left[\int_0^\infty \mathrm{e}^{-\alpha t} \bar{c}(X_t, U_t)\, \mathrm{d}t\right]. \end{aligned} \tag{3.5.15}$$

However, (3.5.15) implies that $\psi = V_\alpha$ and that $\tilde{v}$ is α-discounted optimal. It remains to show that if $\hat{v} \in \mathfrak{U}_{\mathrm{sd}}$ is α-discounted optimal then it satisfies (3.5.11). Indeed, if $\hat{v} \in \mathfrak{U}_{\mathrm{sd}}$ is α-discounted optimal then

$$V_\alpha(x) = \mathbb{E}_x^{\hat{v}} \left[\int_0^\infty \mathrm{e}^{-\alpha t} c_{\hat{v}}(X_t)\, \mathrm{d}t\right].$$

Therefore, by Corollary A.3.6, V_α satisfies

$$\mathcal{L}^{\hat{v}} V_\alpha + c_{\hat{v}} = \alpha V_\alpha \quad \text{in } \mathbb{R}^d. \tag{3.5.16}$$

Let

$$h(x) := \mathcal{L}^{\hat{v}} V_\alpha(x) + c_{\hat{v}}(x) - \min_{u \in \mathbb{U}} \left[\mathcal{L}^u V_\alpha(x) + c(x,u)\right], \quad x \in \mathbb{R}^d. \tag{3.5.17}$$

For $R > 0$, let $\hat{v}' \in \mathfrak{U}_{\mathrm{sd}}$ be a control that attains the minimum in (3.5.17) in B_R, and agrees with $\hat{v}$ in B_R^c. By (3.5.16)–(3.5.17), with $h_R := h\,\mathbb{I}_{B_R}$,

$$\mathcal{L}^{\hat{v}'} V_\alpha - \alpha V_\alpha = -c_{\hat{v}'} - h_R \quad \text{in } \mathbb{R}^d.$$

We claim that $h \in L^\infty(B_R)$. Indeed, since V_α and $c_{\hat{v}}$ are bounded by Lemma A.2.5 and (3.5.16), it follows that V_α is bounded in $\mathscr{W}^{2,p}(B_R)$ for any $p \in (1,\infty)$. Thus by the continuity of the embedding $\mathscr{W}^{2,p}(B_R) \hookrightarrow C^{1,r}(B_R)$ for $p > d$ and $r \le 1 - \frac{d}{p}$ (Theorem A.2.15), ∇V_α is bounded in B_R. In turn, by (3.5.17),

$$h(x) = \sum_{i=1}^d \left(b^i_{\hat{v}}(x) - b^i_{\hat{v}'}(x)\right) \frac{\partial V_\alpha}{\partial x_i} + \left(c_{\hat{v}}(x) - c_{\hat{v}'}(x)\right),$$

and it follows that h is bounded in B_R.

Thus, by Corollary A.3.6,

$$
\begin{aligned}
V_\alpha(x) &= \mathbb{E}_x^{\hat{v}'}\left[\int_0^\infty \mathrm{e}^{-\alpha t} c_{\hat{v}'}(X_t)\,\mathrm{d}t\right] + \mathbb{E}_x^{\hat{v}'}\left[\int_0^\infty \mathrm{e}^{-\alpha t} h_R(X_t)\,\mathrm{d}t\right] \\
&\geq V_\alpha(x) + \int_0^\infty \mathrm{e}^{-\alpha t}\,\mathbb{E}_x^{\hat{v}'}\left[h_R(X_t)\right]\mathrm{d}t\,, \quad x \in \mathbb{R}^d\,.
\end{aligned}
\tag{3.5.18}
$$

By Theorem A.3.5 and (3.5.18), we obtain $\|h\|_{L^1(B_R)} = 0$. Since $R > 0$ was arbitrary, $h = 0$ a.e. in $\mathbb{R}^d$, and (3.5.17) shows that $\hat{v}$ satisfies (3.5.11). □

Definition 3.5.7 Let $V \in C^2(\mathbb{R}^d)$. We say that a function $v\colon \mathbb{R}^d \to \mathscr{P}(\mathbb{U})$ is a *measurable selector from the minimizer* $\min_{u\in\mathbb{U}}\ [\mathcal{L}^u V(x) + c(x,u)]$ if v is Borel measurable and satisfies

$$
\mathcal{L}^v V(x) + c_v(x) = \min_{u\in\mathbb{U}}\ \left[\mathcal{L}^u V(x) + c(x,u)\right]\,.
$$

In other words, $v \in \mathfrak{U}_{\mathrm{sm}}$. Since the map $(x,u) \mapsto b^i(x,u)\partial_i V(x) + c(x,u)$ is continuous and locally Lipschitz in its first argument, it is well known that there always exists such a selector $v \in \mathfrak{U}_{\mathrm{sd}}$ and that $x \mapsto b^i_v(x)\partial_i V(x) + c_v(x)$ is locally Lipschitz continuous.

Remark 3.5.8 If c is not bounded, then V_α still satisfies (3.5.10), provided of course that V_α is finite. Proceed as in the proof of Theorem 3.5.6. Observe that in establishing (3.5.13) we did not use the hypothesis that c is bounded. Let $v \in \mathfrak{U}_{\mathrm{sd}}$ be a measurable selector from the minimizer in (3.5.13). Then for any $R > 0$,

$$
\varphi_R(x) \leq \mathbb{E}_x^v\left[\int_0^{\tau_R} \mathrm{e}^{-\alpha t} c_v(X_t)\,\mathrm{d}t\right] \leq \psi(x)\,, \quad x \in B_R\,. \tag{3.5.19}
$$

Taking limits as $R \to \infty$ in (3.5.19), since $\varphi_R \uparrow \psi$, we obtain $\psi = J^v_\alpha$. On the other hand, since $\varphi_R \leq V_\alpha$, $\psi \leq V_\alpha$. Thus $\psi = V_\alpha$. Since, by the strong maximum principle, any nonnegative solution ψ of (3.5.13), satisfies $\varphi_R(x) \leq \psi$, it follows that V_α is the minimal nonnegative solution of (3.5.10). The rest of the conclusions of Theorem 3.5.6 also hold.

3.6 The HJB equation under a near-monotone assumption

In this section, we study the Hamilton–Jacobi–Bellman (HJB) equation under Assumption 3.4.2 and characterize the optimal stationary Markov control in terms of its solution. For a stable stationary Markov control v, let

$$
\varrho_v := \int_{\mathbb{R}^d} c_v(x)\,\eta_v(\mathrm{d}x)\,,
$$

where η_v is the unique invariant probability measure corresponding to v, and

$$
\varrho^* := \inf_{v\in\mathfrak{U}_{\mathrm{ssd}}} \varrho_v\,. \tag{3.6.1}
$$

In view of the existence results in Section 3.4, the infimum in (3.6.1) is attained in $\mathfrak{U}_{\mathrm{ssd}}$ and the minimizing control is optimal.

We follow the vanishing discount approach and derive the HJB equation for the ergodic criterion by taking the limit of the HJB equation for the discounted criterion as the discount factor approaches zero. The intuition is guided by the fact that, under the near-monotone assumption, all accumulation points as $\alpha \downarrow 0$ of the α-discounted occupation measures $\{\xi^{v_\alpha}_{x,\alpha}\}$, defined in Lemma 2.5.3, corresponding to a family $\{v_\alpha : \alpha > 0\} \subset \mathfrak{U}_{\mathrm{sm}}$ of α-discounted optimal controls, are optimal ergodic occupation measures. This of course implies that all accumulation points of α-discounted optimal stationary Markov controls as $\alpha \downarrow 0$ are stable controls. To prove this claim, suppose that $(\mu, \hat{v}) \in \mathscr{P}(\bar{\mathbb{R}}^d \times \mathbb{U}) \times \mathfrak{U}_{\mathrm{sm}}$ is an accumulation point of $\{(\xi^{v_\alpha}_{x,\alpha}, v_\alpha) : \alpha \in (0,1)\}$ as $\alpha \downarrow 0$. Decomposing $\mu = \delta\mu' + (1-\delta)\mu''$, with $\mu' \in \mathscr{P}(\mathbb{R}^d \times \mathbb{U})$ and $\mu'' \in \mathscr{P}(\{\infty\} \times \mathbb{U})$, it follows by the near-monotone hypothesis that $\delta > 0$. Using Itô's formula, we have

$$\mathrm{e}^{-\alpha t}\, \mathbb{E}^{v_\alpha}_x \left[f(X_t)\right] - f(x) = \int_0^t \mathrm{e}^{-\alpha s}\, \mathbb{E}^{v_\alpha}_x \left[\mathcal{L}^{v_\alpha} f(X_s) - \alpha f(X_s)\right] \mathrm{d}s \quad \forall f \in \mathscr{C}. \tag{3.6.2}$$

Taking limits in (3.6.2) as $t \to \infty$, we obtain

$$\int_{\mathbb{R}^d \times \mathbb{U}} \left(\mathcal{L}^u f(z) - \alpha f(z)\right) \xi^{v_\alpha}_{x,\alpha}(\mathrm{d}z, \mathrm{d}u) + \alpha f(x) = 0\,,$$

and letting $\alpha \downarrow 0$ along the convergent subsequence yields $\int \mathcal{L}^u f(x,u)\,\mu'(\mathrm{d}x, \mathrm{d}u) = 0$ for all $f \in \mathscr{C}$, which implies that $\mu' \in \mathscr{G}$. Therefore $\int c\,\mathrm{d}\mu' \ge \varrho^*$. At the same time one can show, using a Tauberian theorem, that $\limsup_{\alpha\downarrow 0} \int c\,\mathrm{d}\xi^{v_\alpha}_{x,\alpha} \le \varrho^*$, and it follows as in the proof of Theorem 3.4.5 that $\delta = 1$, thus establishing the claim.

Observe that when the discount factor vanishes, i.e., when $\alpha \to 0$, typically $V_\alpha(x) \to \infty$ for each $x \in \mathbb{R}^d$. For the asymptotic analysis we fix a point, say $0 \in \mathbb{R}^d$, and define

$$\bar{V}_\alpha(x) := V_\alpha(x) - V_\alpha(0)\,.$$

Then, by (3.5.10),

$$\min_{u\in\mathbb{U}} \left[\mathcal{L}^u \bar{V}_\alpha(x) + c(x,u)\right] = \alpha \bar{V}_\alpha(x) + \alpha V_\alpha(0)\,. \tag{3.6.3}$$

Provided the family $\{\bar{V}_\alpha : \alpha > 0\}$ is equicontinuous and uniformly bounded on compact subsets of $\mathbb{R}^d$ and $\alpha V_\alpha(0)$ is also bounded for $\alpha > 0$, taking limits in (3.6.3), and with V and ϱ denoting the limits of $\bar{V}_\alpha$ and $\alpha V_\alpha(0)$, respectively, we obtain

$$\min_{u\in\mathbb{U}} \left[\mathcal{L}^u V(x) + c(x,u)\right] = \varrho\,. \tag{3.6.4}$$

This equation is referred to as the HJB equation for the ergodic control problem. This asymptotic analysis is carried out in detail under both Assumptions 3.4.2 and 3.4.3. The scalar ϱ is the optimal cost and a measurable selector from the minimizer in (3.6.4) yields an optimal control.

Note that if $\varrho_v < \infty$ for some $v \in \mathfrak{U}_{\mathrm{ssm}}$, then integrating J^v_α with respect to η_v and using Fubini's theorem we obtain $\int J^v_\alpha \, \mathrm{d}\eta_v = \alpha^{-1}\varrho_v$. It follows that since η_v has positive density, J^v_α must be a.e. finite, and being continuous, it must be finite at each $x \in \mathbb{R}^d$ (see Theorem A.3.7).

3.6.1 Technical results

We proceed with some technical lemmas that play a crucial role in the analysis. For $\varrho > \inf_{\mathbb{R}^d \times \mathbb{U}} c$, define

$$\mathcal{K}(\varrho) := \{x \in \mathbb{R}^d : \min_{u \in \mathbb{U}} c(x,u) \le \varrho\}.$$

Lemma 3.6.1 *Let $\alpha > 0$. Suppose $v \in \mathfrak{U}_{\mathrm{sm}}$ and $\hat{v} \in \mathfrak{U}_{\mathrm{ssm}}$ are such that $\varrho_{\hat{v}} < \infty$, and*

$$J^v_\alpha(x) \le J^{\hat{v}}_\alpha(x) \qquad \forall x \in \mathbb{R}^d, \tag{3.6.5a}$$

$$\varrho_{\hat{v}} < \liminf_{|x|\to\infty} \min_{u \in \mathbb{U}} c(x,u). \tag{3.6.5b}$$

Then

$$\inf_{\mathcal{K}(\varrho_{\hat{v}})} J^v_\alpha = \inf_{\mathbb{R}^d} J^v_\alpha \le \frac{\varrho_{\hat{v}}}{\alpha}.$$

Proof Since $\mathbb{U}$ is compact, the map $x \mapsto \min_{u\in\mathbb{U}} c(x,u)$ is continuous. Therefore $\mathcal{K}(\varrho_{\hat{v}})$ is closed, and by (3.6.5b) it is also bounded, hence compact. In fact, (3.6.5b) implies that $\mathcal{K}(\varrho_{\hat{v}} + \varepsilon)$ is bounded and hence compact for $\varepsilon > 0$ sufficiently small. Integrating both sides of the inequality in (3.6.5a) with respect to $\eta_{\hat{v}}$, and using Fubini's theorem, we obtain

$$\inf_{\mathbb{R}^d} J^v_\alpha \le \inf_{\mathbb{R}^d} J^{\hat{v}}_\alpha \le \int_{\mathbb{R}^d} J^{\hat{v}}_\alpha(x)\,\eta_{\hat{v}}(\mathrm{d}x) = \frac{\varrho_{\hat{v}}}{\alpha}. \tag{3.6.6}$$

Choose $\varepsilon > 0$ such that $\mathcal{K}(\varrho_{\hat{v}} + 2\varepsilon)$ is compact and let $x' \in \mathbb{R}^d \setminus \mathcal{K}(\varrho_{\hat{v}} + 2\varepsilon)$ be an arbitrary point. Select $R > 0$ such that

$$B_R \supset K(\varrho_{\hat{v}} + 2\varepsilon) \cup \{x'\},$$

and let $\tau' = \tau(\mathcal{K}^c(\varrho_{\hat{v}} + \varepsilon))$, and as usual, $\tau_R \equiv \tau(B_R)$. Applying the strong Markov property relative to $\tau' \wedge \tau_R$, letting $R \to \infty$, and making use of (3.6.6), we obtain

$$\begin{aligned} J^v_\alpha(x') &\ge \frac{\varrho_{\hat{v}} + \varepsilon}{\alpha} \mathbb{E}^v_{x'}\left[1 - \mathrm{e}^{-\alpha\tau'}\right] + \mathbb{E}^v_{x'}\left[\mathrm{e}^{-\alpha\tau'}\right] \inf_{\mathbb{R}^d} J^v_\alpha \\ &\ge \inf_{\mathbb{R}^d} J^v_\alpha + \frac{\varepsilon}{\alpha} \mathbb{E}^v_{x'}\left[1 - \mathrm{e}^{-\alpha\tau'}\right]. \end{aligned} \tag{3.6.7}$$

The inequality

$$\mathbb{E}^v_{x'}\left[1 - \mathrm{e}^{-\alpha\tau'}\right] \ge \left(1 - \mathrm{e}^{-\alpha C}\right) \mathbb{P}^v_{x'}(\tau' > C),$$

which is valid for any constant $C > 0$ together with Theorem 2.6.1 (c) show that $\mathbb{E}^v_{x'}[1 - \mathrm{e}^{-\alpha\tau'}]$ is bounded away from zero in $\mathcal{K}^c(\varrho_{\hat{v}} + 2\varepsilon)$. Therefore (3.6.6) and (3.6.7) imply that J^v_α attains its minimum in $\mathcal{K}(\varrho_{\hat{v}} + \varepsilon)$ for arbitrarily small $\varepsilon > 0$, thus completing the proof. □

Definition 3.6.2 Let $\kappa\colon (0,\infty) \to (0,\infty)$ be a positive function that serves as a parameter. Let $\Gamma(\kappa, R_0)$, $R_0 > 0$, denote the class of all pairs (φ, α), with $\alpha \in [0,1]$ and φ a nonnegative function that belongs to $\mathscr{W}^{2,p}_{\mathrm{loc}}(\mathbb{R}^d)$, $p \in (1,\infty)$, and satisfies

$$\begin{aligned} \mathcal{L}^v\varphi - \alpha\varphi &= g \quad \text{in } \mathbb{R}^d\,, \\ \inf_{B_{R_0}} \varphi &= \inf_{\mathbb{R}^d} \varphi\,, \end{aligned} \tag{3.6.8}$$

for some $v \in \mathfrak{U}_{\mathrm{sm}}$ and $g \in L^\infty_{\mathrm{loc}}(\mathbb{R}^d)$, such that $\|g\|_{L^\infty(B_R)} \le \kappa(R)$ for all $R > 0$.

Lemma 3.6.3 *For each $R > R_0$ and $p > 1$, there exist constants $C_\Gamma(R)$ and $\tilde{C}_\Gamma(R,p)$, which depend only on d, κ and R_0, such that, for all $(\varphi,\alpha) \in \Gamma(\kappa, R_0)$,*

$$\operatorname*{osc}_{B_{2R}} \varphi \le C_\Gamma(R)\left(1 + \alpha \inf_{B_{R_0}} \varphi\right) \tag{3.6.9a}$$

$$\|\varphi - \varphi(0)\|_{\mathscr{W}^{2,p}(B_R)} \le \tilde{C}_\Gamma(R,p)\left(1 + \alpha \inf_{B_{R_0}} \varphi\right). \tag{3.6.9b}$$

Proof Let $(\varphi,\alpha) \in \Gamma(\kappa, R_0)$ satisfy (3.6.8). By Lemma A.2.10, $\varphi \in \mathscr{W}^{2,p}_{\mathrm{loc}}(\mathbb{R}^d)$ for all $p > 1$. Fix $R > R_0$ and let

$$\tilde{g} := \alpha(g - 2\kappa(4R))\,, \qquad \tilde{\varphi} := 2\kappa(4R) + \alpha\varphi\,.$$

Then $\tilde{g} < 0$ in B_{4R}, and $\tilde{\varphi}$ satisfies

$$\mathcal{L}^v\tilde{\varphi}(x) - \alpha\tilde{\varphi}(x) = \tilde{g}(x) \quad \text{in } B_{4R}\,.$$

We obtain

$$\begin{aligned} \|\tilde{g}\|_{L^\infty(B_{4R})} &\le \alpha\left(2\kappa(4R) + \|g\|_{L^\infty(B_{4R})}\right) \\ &\le 3\alpha\left(2\kappa(4R) - \|g\|_{L^\infty(B_{4R})}\right) \\ &\le 3|B_{4R}|^{-1}\|\tilde{g}\|_{L^1(B_{4R})}\,. \end{aligned}$$

Hence $\tilde{g} \in \mathfrak{K}(3, B_{4R})$ (see Definition A.2.11). Therefore, by Theorem A.2.13 and (3.6.8), there exists a constant $\tilde{C}_H$ such that

$$\sup_{B_{3R}} \tilde{\varphi} \le \tilde{C}_H \inf_{B_R} \tilde{\varphi}\,,$$

which implies that

$$\alpha \sup_{B_{3R}} \varphi \le \tilde{C}_H\left(2\kappa(4R) + \alpha \inf_{B_R} \varphi\right). \tag{3.6.10}$$

Let $\psi \in \mathscr{W}^{2,p}_{\mathrm{loc}}(B_{3R}) \cap C(\bar{B}_{3R})$ be the solution of the Dirichlet problem

$$\mathcal{L}^v\psi = 0 \quad \text{in } B_{3R}\,, \qquad \psi = \varphi \quad \text{on } \partial B_{3R}\,.$$

Then

$$\mathcal{L}^v(\varphi - \psi) = \alpha\varphi + g \quad \text{in } B_{3R}\,,$$

and by Theorem A.2.1 and (3.6.10), there exists a constant $\hat{C}_a$ such that

$$\sup_{B_{3R}} |\varphi - \psi| \le \hat{C}_a(1 + \alpha\varphi(\hat{x}))\,, \tag{3.6.11}$$

where $\hat{x} \in B_{R_0}$ is any point where φ attains its infimum. Since ψ is $\mathcal{L}^v$-harmonic in B_{3R}, it cannot have a minimum in B_{3R}, and as a result the function $\psi(x) - \varphi(\hat{x})$ is a nonnegative $\mathcal{L}^v$-harmonic function in B_{3R}. By Harnack's inequality (Theorem A.2.4) and (3.6.11), for some constant C_H,

$$\begin{aligned}\psi(x) - \varphi(\hat{x}) &\le C_H(\psi(\hat{x}) - \varphi(\hat{x}))\\ &\le C_H\hat{C}_a(1 + \alpha\varphi(\hat{x})) \qquad \forall x \in B_{2R}\,.\end{aligned} \tag{3.6.12}$$

Thus by (3.6.11) and (3.6.12),

$$\begin{aligned}\operatorname*{osc}_{B_{2R}} \varphi &\le \sup_{B_{2R}} (\varphi - \psi) + \sup_{B_{2R}} \psi - \varphi(\hat{x})\\ &\le \hat{C}_a(1 + C_H)(1 + \alpha\varphi(\hat{x}))\,.\end{aligned} \tag{3.6.13}$$

Let $\bar{\varphi} := \varphi - \varphi(0)$. Then

$$\mathcal{L}^v\bar{\varphi} - \alpha\bar{\varphi} = g + \alpha\varphi(0) \quad \text{in } B_{3R}\,.$$

Applying Lemma A.2.5, with $D = B_{2R}$ and $D' = B_R$, relative to the operator $\mathcal{L}^v - \alpha$, we obtain

$$\begin{aligned}\|\bar{\varphi}\|_{\mathscr{W}^{2,p}(B_R)} &\le C_0\left(\|\bar{\varphi}\|_{L^p(B_{2R})} + \|\mathcal{L}^v\bar{\varphi} - \alpha\bar{\varphi}\|_{L^p(B_{2R})}\right)\\ &\le C_0\,|B_{2R}|^{1/p}\left(\operatorname*{osc}_{B_{2R}} \varphi + \kappa(2R) + \alpha\varphi(0)\right).\end{aligned} \tag{3.6.14}$$

Since

$$\varphi(0) \le \operatorname*{osc}_{B_{R_0}} \varphi + \inf_{B_{R_0}} \varphi\,,$$

using (3.6.13) in (3.6.14) yields (3.6.9b). □

Suppose that $\{(\varphi_n, \alpha_n) : n \in \mathbb{N}\} \subset \Gamma(\kappa, R_0)$ is a sequence satisfying

$$\left|\inf_{B_{R_0}} \{\alpha_n\varphi_n\}\right| \le M_0 \qquad \forall n \in \mathbb{N}\,, \tag{3.6.15}$$

for some $M_0 \in \mathbb{R}$. Set $\bar{\varphi}_n = \varphi_n - \varphi_n(0)$. By Lemma 3.6.3, $\{\bar{\varphi}_n\}$ is bounded in $\mathscr{W}^{2,p}(B_R)$. The Sobolev imbedding theorem then implies that $\{\bar{\varphi}_n\}$ contains a subsequence which converges uniformly on compact subsets of $\mathbb{R}^d$. We combine Lemmas 3.5.4 and 3.6.3 to obtain useful convergence criteria for α-discounted value functions and the associated HJB equations.

Lemma 3.6.4 *Suppose* $\{(\varphi_n, \alpha_n)\}_{n\in\mathbb{N}} \subset \Gamma(\kappa, R_0)$, *for some* $R_0 > 0$, *satisfies* (3.6.15) *for some* $M_0 > 0$, *and* $\alpha_n \to 0$ *as* $n \to \infty$. *Moreover, suppose that either of the two pairs of conditions (A1)–(A2) or (B1)–(B2) hold.*

(A1) $\mathcal{L}^v\varphi_n - \alpha_n\varphi_n = g_n$ *in* $\mathbb{R}^d$ *for some* $v \in \mathfrak{U}_{\mathrm{sm}}$.

(A2) $g_n \to g$ *in* $L^\infty(B_R)$ *for every* $R > 0$.

(B1) $\min_{u\in\mathbb{U}} [\mathcal{L}^u\varphi_n(x) + c_n(x,u)] = \alpha_n\varphi_n(x) + h_n(x)$, $x \in \mathbb{R}^d$.

(B2) $c_n(\cdot\,,u) \to c(\cdot\,,u)$ *and* $h_n \to h$ *in* $C^{0,1}(B_R)$ *for every* $R > 0$, *uniformly over* $u \in \mathbb{U}$.

Define $\bar{\varphi}_n := \varphi_n - \varphi_n(0)$. *Then there exists a subsequence* $\{\bar{\varphi}_{k_n}\}$ *which converges in* $C^{1,r}(B_R)$, $r < 1$, *to some function* ψ *for every* $R > 0$, *and such that* $\alpha\varphi_{k_n}(0)$ *converges to some constant* $\varrho \le M_0$. *If (A1) – (A2) hold, then* $\psi \in \mathscr{W}^{2,p}_{\mathrm{loc}}(\mathbb{R}^d)$ *for any* $p \in (1,\infty)$, *and satisfies*

$$\mathcal{L}^v\psi = g + \varrho \quad \text{in } \mathbb{R}^d\,. \tag{3.6.16}$$

On the other hand, under (B1)–(B2), $\psi \in C^{2,r}_{\mathrm{loc}}(\mathbb{R}^d)$ *for any* $r < 1$, *and satisfies*

$$\min_{u\in\mathbb{U}} [\mathcal{L}^u\psi(x) + c(x,u)] = \varrho + h(x)\,. \tag{3.6.17}$$

In either case, we have

$$\inf_{B_{R_0}} \psi = \inf_{\mathbb{R}^d} \psi \ge -C_\Gamma(R_0)(1 + M_0)\,, \tag{3.6.18}$$

where C_Γ *is as in Lemma 3.6.3.*

Proof Since

$$\alpha_n\varphi_n(0) \le \alpha_n \operatorname*{osc}_{B_{R_0}} \varphi_n + \alpha_n \inf_{B_{R_0}} \varphi_n\,, \tag{3.6.19}$$

it follows by (3.6.9a) and (3.6.15) that the sequence $\{\alpha_n\varphi_n(0)\}$ is bounded. Select any subsequence $\{\hat{k}_n\} \subset \mathbb{N}$ over which it converges. By (3.6.9a), (3.6.15), and (3.6.19), the limit does not exceed M_0. By (3.6.9b) and the assumptions of the lemma, the sequence $\{\bar{\varphi}_{\hat{k}_n}\}$ satisfies the hypotheses of Lemma 3.5.4 on each ball B_R (see also Remark 3.5.5). Convergence over some subsequence $\{k_n\} \subset \{\hat{k}_n\}$, on each bounded domain, then follows by Lemma 3.5.4 and the limit satisfies (3.6.16) or (3.6.17), accordingly. The standard diagonal procedure yields convergence in $\mathbb{R}^d$. Next, for each $x \in \mathbb{R}^d$,

$$\begin{aligned}
\psi(x) &= \lim_{n\to\infty} \bar{\varphi}_{k_n}(x)\\
&\ge -\limsup_{n\to\infty} \Big(\varphi_{k_n}(0) - \inf_{\mathbb{R}^d} \varphi_{k_n}\Big) + \liminf_{n\to\infty} \Big(\varphi_{k_n}(x) - \inf_{\mathbb{R}^d} \varphi_{k_n}\Big)\\
&\ge -\limsup_{n\to\infty} \Big(\operatorname*{osc}_{B_{R_0}} \varphi_{k_n}\Big),
\end{aligned}$$

and the inequality in (3.6.18) follows by (3.6.9a) and (3.6.15). Lastly, the equality in (3.6.18) easily follows by taking limits as $n \to \infty$ in

$$\psi(x) - \inf_{B_{R_0}} \psi \ge \psi(x) - \bar{\varphi}_{k_n}(x) + \Big(\varphi_{k_n}(x) - \inf_{B_{R_0}} \varphi_{k_n}\Big) + \inf_{B_{R_0}} (\bar{\varphi}_{k_n} - \psi)\,,$$

and utilizing the uniform convergence of φ_{k_n} on B_{R_0}, to obtain

$$\psi(x) - \inf_{B_{R_0}} \psi \geq \liminf_{n\to\infty} \left(\varphi_{k_n}(x) - \inf_{B_{R_0}} \varphi_{k_n}\right) \geq 0 \qquad \forall x \in \mathbb{R}^d\,,$$

thus completing the proof. □

Let $v_\alpha \in \mathfrak{U}_{\mathrm{sm}}$ be an optimal control for the α-discounted criterion. As we have shown in Section 3.4, under the near-monotone hypothesis there exists $v^* \in \mathfrak{U}_{\mathrm{ssd}}$ which is optimal. Clearly $V_\alpha = J_\alpha^{v_\alpha} \leq J_\alpha^{v^*}$ for all $\alpha > 0$. Thus by Lemma 3.6.1,

$$\inf_{\mathcal{K}(\varrho^*)} V_\alpha = \inf_{\mathbb{R}^d} V_\alpha \leq \frac{\varrho^*}{\alpha} \qquad \forall \alpha > 0\,. \tag{3.6.20}$$

By (3.6.20), Lemma 3.6.3, and the remark in the paragraph preceding Lemma 3.6.4, it follows that $\{V_\alpha\}$ is equicontinuous on bounded subsets of $\mathbb{R}^d$, when α lies in a bounded interval. These properties of the optimal α-discounted cost function under the near-monotone hypothesis are summarized in the corollary that follows.

Corollary 3.6.5 *Under Assumption 3.4.2,* (3.6.20) *holds. In addition, for any $R > 0$ and $\alpha_0 > 0$, the family $\{V_\alpha : \alpha < \alpha_0\}$ is equicontinuous on B_R and there exists a constant $\delta_0 = \delta_0(R)$ such that*

$$V_\alpha(x) < \frac{\varrho^*}{\alpha} + \delta_0 \qquad \forall x \in B_R\,,\ \forall \alpha \in (0, \alpha_0)\,.$$

3.6.2 The HJB equation

The basic theorem characterizing optimal controls with respect to the ergodic criterion via the HJB equation follows.

Theorem 3.6.6 *Let Assumptions 3.4.2 and 3.5.1 hold. There exists $V \in C^2(\mathbb{R}^d)$ and a constant $\varrho \in \mathbb{R}$ such that*

$$\min_{u\in\mathbb{U}} \left[\mathcal{L}^u V(x) + c(x,u)\right] = \varrho \qquad \forall x \in \mathbb{R}^d\,, \tag{3.6.21a}$$

$$\varrho \leq \varrho^*\,, \quad V(0) = 0 \quad \textit{and} \quad \inf_{\mathbb{R}^d} V > -\infty\,. \tag{3.6.21b}$$

Proof Assume at first that c is bounded in $\mathbb{R}^d \times \mathbb{U}$. By Theorem 3.5.6, V_α satisfies (3.5.10). Select R_0 such that $\mathcal{K}(\varrho^*) \subset B_{R_0}$. By (3.6.20), the hypotheses (B1)–(B2) of Lemma 3.6.4 are met for any sequence (α_n, V_{α_n}) with $\alpha_n \to 0$. Thus there exists a pair (V, ϱ) satisfying (3.6.21a) – (3.6.21b), and, moreover, V is the uniform limit (in each bounded domain of $\mathbb{R}^d$) of $V_{\alpha_n} - V_{\alpha_n}(0)$ over some sequence $\alpha_n \to 0$, while ϱ is the corresponding limit of $\alpha_n V_{\alpha_n}(0)$.

Now drop the assumption that the cost function is bounded. We construct a sequence of bounded near-monotone cost functions $c_n \colon \mathbb{R}^d \times \mathbb{U} \to \mathbb{R}_+$, $n \in \mathbb{N}$, as follows. Let $\varepsilon_0 > 0$ satisfy

$$\lim_{|x|\to\infty} \min_{u\in\mathbb{U}} c(x,u) > \varrho^* + \varepsilon_0\,.$$

With $\mathfrak{d}(x,B)$ denoting the Euclidean distance of the point $x \in \mathbb{R}^d$ from the set $B \subset \mathbb{R}^d$, and $B_n \subset \mathbb{R}^d$ denoting the ball of radius n, we define

$$c_n(x,u) = \frac{\mathfrak{d}(x, B_{n+1}^c)\, c(x,u) + \mathfrak{d}(x, B_n)\,(\varrho^* + \varepsilon_0)}{\mathfrak{d}(x, B_n) + \mathfrak{d}(x, B_{n+1}^c)}\,.$$

Note that each c_n is Lipschitz in x, and by the near-monotone property of c, we have $c_n \le c_{n+1} \le c$ for all n larger than some finite n_0. In addition, $c_n \to c$. Observe also that c_n has a continuous extension in $\bar{\mathbb{R}}^d \times \mathbb{U}$ by letting $c_n(\infty) = \varrho^* + \varepsilon_0$. Let ϱ_n^* be the optimal ergodic cost relative to c_n. We first establish that $\varrho_n^* \to \varrho^*$ as $n \to \infty$. Since c_n is monotone, $\varrho_n^* \uparrow \hat{\varrho}$. Let π_n, $n \in \mathbb{N}$, denote the ergodic occupation measure corresponding to an optimal $v_n \in \mathfrak{U}_{\mathrm{ssm}}$ relative to c_n. Let $\hat{\pi} \in \mathscr{P}(\bar{\mathbb{R}}^d \times \mathbb{U})$ be any limit point of $\{\pi_n\}$, and denote again by $\{\pi_n\}$ a subsequence converging to it. By Lemma 3.2.11,

$$\hat{\pi} = \delta\,\hat{\pi}' + (1-\delta)\,\hat{\pi}''$$

for some $\hat{\pi}' \in \mathscr{G}$, $\hat{\pi}'' \in \mathscr{P}(\{\infty\} \times \mathbb{U})$, and $\delta \in [0,1]$. Since $c_n \le c_{n+k} \le c$ for all $n \ge n_0$, we have

$$\int_{\mathbb{R}\times\mathbb{U}} c_n \,\mathrm{d}\pi_{n+k} \le \int_{\mathbb{R}\times\mathbb{U}} c_{n+k}\,\mathrm{d}\pi_{n+k} \le \varrho^*\,. \tag{3.6.22}$$

Taking limits as $k \to \infty$ in (3.6.22) and using the fact that c_n is continuous in $\bar{\mathbb{R}}^d \times \mathbb{U}$ and $\pi_n \to \hat{\pi} \in \mathscr{P}(\bar{\mathbb{R}}^d \times \mathbb{U})$, we obtain

$$\delta \int_{\mathbb{R}\times\mathbb{U}} c_n \,\mathrm{d}\hat{\pi}' + (1-\delta)(\varrho^* + \varepsilon_0) \le \hat{\varrho} \le \varrho^*\,. \tag{3.6.23}$$

Next, taking limits in (3.6.23) as $n \to \infty$, and using monotone convergence,

$$\delta \int_{\mathbb{R}\times\mathbb{U}} c \,\mathrm{d}\hat{\pi}' + (1-\delta)(\varrho^* + \varepsilon_0) \le \hat{\varrho} \le \varrho^*\,. \tag{3.6.24}$$

Since by the definition of ϱ^*, $\int c\,\mathrm{d}\hat{\pi}' \ge \varrho^*$, (3.6.24) implies, all at once, that $\delta = 1$, $\int c\,\mathrm{d}\hat{\pi} = \varrho^*$ and $\hat{\varrho} = \varrho^*$. This concludes the proof that $\varrho_n^* \to \varrho^*$.

Continuing, let $(V_n, \varrho_n) \in C^2(\mathbb{R}^d)$ be the solution to the HJB equation corresponding to the bounded cost c_n, satisfying $\inf_{\mathbb{R}^d} V_n > -\infty$ and $V_n(0) = 0$, which was constructed in the first part of the proof. By (3.6.18) of Lemma 3.6.4,

$$\inf_{B_{R_0}} V_n = \inf_{\mathbb{R}^d} V_n \ge -C_\Gamma(R_0)(1+\varrho^*)$$

(in fact, V_n attains its infimum in $\mathcal{K}(\varrho^*)$). Also, V_n satisfies assumption (B2) of Lemma 3.6.4 with $\alpha \equiv 0$ and $h_n \equiv \varrho_n$. Observe that (3.6.15) holds trivially if $\alpha_n = 0$. Therefore the family $\{(V_n, 0)\}$ belongs to $\Gamma(\kappa, R_0)$ and satisfies the hypotheses (B1)–(B2) of Lemma 3.6.4. Convergence of $V_n - V_n(0)$ (over some subsequence) to a limit satisfying (3.6.21a) follows. Note that the constant ϱ arises here as a limit point of $\{\varrho_n\}$, and hence $\varrho \le \varrho^*$. Also, V is bounded below in $\mathbb{R}^d$ by (3.6.18), and thus (3.6.21b) holds. This completes the proof. □

Remark 3.6.7 Equation (3.6.21a) is usually referred to as the HJB equation for the ergodic control problem. Note that (3.6.21a) is under-determined, in the sense that it is a single equation in two variables, V and ϱ. The function V is called the *cost potential* and for an appropriate solution pair (V, ϱ) of (3.6.21a) ϱ is the optimal ergodic cost ϱ^*, as we show later.

Lemma 3.6.8 *Let Assumption 3.4.2 hold. Let $(V, \varrho) \in C^2(\mathbb{R}^d) \times \mathbb{R}$ be a solution of (3.6.21a), and suppose that V is bounded below in $\mathbb{R}^d$. Let $v \in \mathfrak{U}_{\mathrm{sm}}$ satisfy, a.e. in $\mathbb{R}^d$,*

$$\min_{u \in \mathbb{U}} \left[\sum_{i=1}^{d} b^i(x,u) \frac{\partial V}{\partial x_i}(x) + c(x,u) \right] = \sum_{i=1}^{d} b^i_v(x) \frac{\partial V}{\partial x_i}(x) + c_v(x)\,. \tag{3.6.25}$$

Then

(i) *v is stable;*

(ii) *$\varrho \geq \varrho^*$;*

(iii) *if $\varrho = \varrho^*$, then v is optimal.*

Proof Since V satisfies $\mathcal{L}^v V \leq -\varepsilon$ for some $\varepsilon > 0$ outside some compact set and is bounded below, Theorem 2.6.10 (f) asserts that v is stable.

By (3.6.21a) and (3.6.25), for $R > 0$ and $|x| < R$,

$$\mathbb{E}^v_x[V(X_{t\wedge\tau_R})] - V(x) = \mathbb{E}^v_x\left[\int_0^{t\wedge\tau_R} [\varrho - c_v(X_s)]\,\mathrm{d}s\right].$$

Therefore

$$\mathbb{E}^v_x\left[\int_0^{t\wedge\tau_R} [\varrho - c_v(X_s)]\,\mathrm{d}s\right] \geq \inf_{y\in\mathbb{R}^d} V(y) - V(x)\,.$$

Since v is stable, letting $R \to \infty$, we have

$$\varrho t - \mathbb{E}^v_x\left[\int_0^t c_v(X_s)\,\mathrm{d}s\right] \geq \inf_{y\in\mathbb{R}^d} V(y) - V(x)\,.$$

Next, dividing by t and letting $t \to \infty$, we obtain

$$\begin{aligned} \varrho &\geq \limsup_{t\to\infty} \frac{1}{t}\, \mathbb{E}^v_x\left[\int_0^t c_v(X_s)\,\mathrm{d}s\right] \\ &\geq \int_{\mathbb{R}^d} c_v(x)\,\eta_v(\mathrm{d}x) = \varrho_v \geq \varrho^*\,. \end{aligned}$$

Thus, if $\varrho = \varrho^*$, the third claim follows. □

It follows from Lemma 3.6.8 that ϱ^* is the smallest value of ϱ for which (3.6.21a) admits a solution V that is bounded below in $\mathbb{R}^d$. It is also the case that, provided (V, ϱ) satisfies (3.6.21a) – (3.6.21b), (3.6.25) is a sufficient condition for optimality. This condition is also necessary, as we prove later. We also show that (3.6.21a) has a unique solution (V, ϱ^*), such that V is bounded below in $\mathbb{R}^d$. Therefore there exists a unique solution satisfying (3.6.21a) – (3.6.21b). We need the following stochastic representation of V.

Lemma 3.6.9 *Let* $\tilde{\mathfrak{U}}_{\mathrm{ssd}} := \{v \in \mathfrak{U}_{\mathrm{ssd}} : \varrho_v < \infty\}$, *and* $\breve{\tau}_r = \tau(B_r^c)$, $r > 0$. *Then, under Assumption 3.4.2,*

(a) *if* $(V, \varrho) \in C^2(\mathbb{R}^d) \times \mathbb{R}$ *is any pair satisfying* (3.6.21a)–(3.6.21b), *then*

$$V(x) \geq \limsup_{r \downarrow 0} \inf_{v \in \mathfrak{U}_{\mathrm{ssd}}} \mathbb{E}_x^v \left[\int_0^{\breve{\tau}_r} (c_v(X_t) - \varrho^*) \, \mathrm{d}t \right], \quad x \in \mathbb{R}^d; \tag{3.6.26}$$

(b) *if in addition to satisfying* (3.6.21a)–(3.6.21b), V *is the limit of* $V_{\alpha_n} - V_{\alpha_n}(0)$ *over some sequence* $\alpha_n \to 0$, *then*

$$V(x) = \lim_{r \downarrow 0} \inf_{v \in \tilde{\mathfrak{U}}_{\mathrm{ssd}}} \mathbb{E}_x^v \left[\int_0^{\breve{\tau}_r} (c_v(X_t) - \varrho^*) \, \mathrm{d}t \right], \quad x \in \mathbb{R}^d. \tag{3.6.27}$$

Proof For $x \in \mathbb{R}^d$, choose $R > r > 0$ such that $r < |x| < R$. Let $v^* \in \mathfrak{U}_{\mathrm{sm}}$ satisfy (3.6.25). By (3.6.21b) and Lemma 3.6.8 (ii), $\varrho = \varrho^*$. Using (3.6.21a) and Dynkin's formula, we obtain

$$\begin{aligned} V(x) = \mathbb{E}_x^{v^*} \Bigg[\int_0^{\breve{\tau}_r \wedge \tau_R} & (c_{v^*}(X_t) - \varrho^*) \, \mathrm{d}t + V(X_{\breve{\tau}_r}) \mathbb{I}\{\breve{\tau}_r < \tau_R\} \\ & + V(X_{\tau_R}) \mathbb{I}\{\breve{\tau}_r \geq \tau_R\} \Bigg]. \end{aligned} \tag{3.6.28}$$

Since by Lemma 3.6.8, v^* is stable, $\mathbb{E}_x^{v^*}[\breve{\tau}_r] < \infty$. In addition, V is bounded below in $\mathbb{R}^d$. Thus

$$\liminf_{R \to \infty} \mathbb{E}_x^{v^*} [V(X_{\tau_R}) \mathbb{I}\{\breve{\tau}_r \geq \tau_R\}] \geq 0 \qquad \forall x \in \mathbb{R}^d.$$

Hence, letting $R \to \infty$ in (3.6.28), and using Fatou's lemma, we obtain

$$\begin{aligned} V(x) &\geq \mathbb{E}_x^{v^*} \left[\int_0^{\breve{\tau}_r} (c_{v^*}(X_t) - \varrho^*) \, \mathrm{d}t + V(X_{\breve{\tau}_r}) \right] \\ &\geq \inf_{v \in \tilde{\mathfrak{U}}_{\mathrm{ssd}}} \mathbb{E}_x^v \left[\int_0^{\breve{\tau}_r} (c_v(X_t) - \varrho^*) \, \mathrm{d}t \right] + \inf_{\partial B_r} V. \end{aligned}$$

Next, letting $r \to 0$ and using the fact that $V(0) = 0$, yields

$$V(x) \geq \limsup_{r \downarrow 0} \inf_{v \in \tilde{\mathfrak{U}}_{\mathrm{ssd}}} \mathbb{E}_x^v \left[\int_0^{\breve{\tau}_r} (c_v(X_t) - \varrho^*) \, \mathrm{d}t \right].$$

Now suppose $V = \lim_{n \to \infty} (V_{\alpha_n} - V_{\alpha_n}(0))$. We need to prove the reverse inequality. We proceed as follows. Let $v_\alpha \in \mathfrak{U}_{\mathrm{sd}}$ be an α-discounted optimal control. For $v \in \tilde{\mathfrak{U}}_{\mathrm{ssd}}$, define the admissible control $U \in \mathfrak{U}$ by

$$U_t = \begin{cases} v & \text{if } t \leq \breve{\tau}_r \wedge \tau_R, \\ v_\alpha & \text{otherwise.} \end{cases}$$

Since U is in general sub-optimal for the α-discounted criterion, using the strong Markov property, we obtain

$$\begin{aligned} V_\alpha(x) &\le \mathbb{E}_x^U\left[\int_0^\infty e^{-\alpha t} c(X_t, U_t)\, dt\right] \\ &= \mathbb{E}_x^v\left[\int_0^{\breve{\tau}_r \wedge \tau_R} e^{-\alpha t} c_v(X_t)\, dt + e^{-\alpha(\breve{\tau}_r \wedge \tau_R)} V_\alpha(X_{\breve{\tau}_r \wedge \tau_R})\right]. \end{aligned} \tag{3.6.29}$$

Since $v \in \tilde{\mathfrak{U}}_{\mathrm{ssd}}$, we have $\inf_{\mathbb{R}^d} \{\alpha J_\alpha^v\} \le \varrho_v < \infty$. Hence, by Theorem A.3.7, J_α^v is a.e. finite. Since $V_\alpha \le J_\alpha^v$, we have, by Remark A.3.8,

$$\mathbb{E}_x^v\left[e^{-\alpha \tau_R} V_\alpha(X_{\tau_R}) \mathbb{I}\{\breve{\tau}_r \ge \tau_R\}\right] \le \mathbb{E}_x^v[e^{-\alpha \tau_R} J_\alpha^v(X_{\tau_R})] \xrightarrow[R\to\infty]{} 0. \tag{3.6.30}$$

Decomposing the term $e^{-\alpha(\breve{\tau}_r \wedge \tau_R)} V_\alpha(X_{\breve{\tau}_r \wedge \tau_R})$ in (3.6.29), then taking limits as $R \to \infty$ applying (3.6.30), and subtracting $V_\alpha(0)$ from both sides of the inequality, we obtain

$$\begin{aligned} V_\alpha(x) - V_\alpha(0) &\le \mathbb{E}_x^v\left[\int_0^{\breve{\tau}_r} e^{-\alpha t} c_v(X_t)\, dt + e^{-\alpha \breve{\tau}_r} V_\alpha(X_{\breve{\tau}_r}) - V_\alpha(0)\right] \\ &= \mathbb{E}_x^v\left[\int_0^{\breve{\tau}_r} e^{-\alpha t}(c_v(X_t) - \varrho^*)\, dt\right] + \mathbb{E}_x^v\left[V_\alpha(X_{\breve{\tau}_r}) - V_\alpha(0)\right] \\ &\quad + \mathbb{E}_x^v\left[\alpha^{-1}(1 - e^{-\alpha \breve{\tau}_r})[\varrho^* - \alpha V_\alpha(X_{\breve{\tau}_r})]\right]. \end{aligned} \tag{3.6.31}$$

By Theorem 3.6.6 and Lemma 3.6.8 (ii), we have $\lim_{\alpha \downarrow 0} \alpha V_\alpha(0) = \varrho = \varrho^*$. Hence

$$\sup_{B_r} \left|\varrho^* - \alpha V_\alpha\right| \to 0 \quad \text{as } \alpha \to 0$$

by Lemma 3.6.3. We also have the bound

$$\mathbb{E}_x^v\left[\alpha^{-1}(1 - e^{-\alpha \breve{\tau}_r})\right] \le \mathbb{E}_x^v[\breve{\tau}_r] \qquad \forall \alpha > 0,$$

and hence letting $\alpha \to 0$ along the subsequence $\{\alpha_n\}$ in (3.6.31), yields

$$V(x) \le \mathbb{E}_x^v\left[\int_0^{\breve{\tau}_r} (c_v(X_t) - \varrho^*)\, dt + V(X_{\breve{\tau}_r})\right].$$

Since $V(0) = 0$,

$$\lim_{r \downarrow 0} \sup_{v \in \tilde{\mathfrak{U}}_{\mathrm{ssd}}} \mathbb{E}_x^v[V(X_{\breve{\tau}_r})] = 0.$$

Therefore, we obtain

$$V(x) \le \liminf_{r \downarrow 0} \inf_{v \in \tilde{\mathfrak{U}}_{\mathrm{ssd}}} \mathbb{E}_x^v\left[\int_0^{\breve{\tau}_r} (c_v(X_t) - \varrho^*)\, dt\right],$$

and the proof is complete. □

Lemma 3.6.9 implies, in particular, that $V_\alpha - V_\alpha(0)$ has a unique limit as $\alpha \to 0$. Next we show that the solution (V, ϱ) of (3.6.21a) which satisfies (3.6.21b) is unique, and that (3.6.25) is necessary for optimality.

Theorem 3.6.10 *Suppose that Assumptions 3.4.2 and 3.5.1 hold. Then there exists a unique pair* (V, ϱ) *in* $\mathscr{W}^{2,p}_{\mathrm{loc}}(\mathbb{R}^d) \times \mathbb{R}$, $1 < p < \infty$, *satisfying* (3.6.21a) – (3.6.21b). *Also, a stable stationary Markov control* v *is optimal if and only if it satisfies* (3.6.25) *a.e.*

Proof Suppose $(\tilde{V}, \tilde{\varrho}) \in \mathscr{W}^{2,p}_{\mathrm{loc}}(\mathbb{R}^d) \times \mathbb{R}$, $1 < p < \infty$, satisfy (3.6.21a) – (3.6.21b). Let $\tilde{v}$ be a stationary Markov control satisfying, a.e. in $\mathbb{R}^d$,

$$\min_{u \in \mathbb{U}} \left[\sum_{i=1}^{d} b^i(x,u) \frac{\partial \tilde{V}}{\partial x_i}(x) + c(x,u) \right] = \sum_{i=1}^{d} b^i_{\tilde{v}}(x) \frac{\partial \tilde{V}}{\partial x_i}(x) + c_{\tilde{v}}(x)\,. \tag{3.6.32}$$

By Lemma 3.6.8, $\tilde{v}$ is stable and $\tilde{\varrho} = \varrho^*$. Let (V, ϱ^*) be the solution pair obtained in Theorem 3.6.6. By Lemma 3.6.9, $\tilde{V}$ satisfies (3.6.26), and V satisfies (3.6.27). Thus $V - \tilde{V} \le 0$ in $\mathbb{R}^d$. By (3.6.21a), $\mathcal{L}^{\tilde{v}}(V - \tilde{V}) \ge 0$. Since $V(0) - \tilde{V}(0) = 0$, then by the strong maximum principle, $V = \tilde{V}$ a.e. in $\mathbb{R}^d$. This establishes uniqueness. To complete the proof, it suffices to show that if $\tilde{v} \in \mathfrak{U}_{\mathrm{ssd}}$ is optimal, then (3.6.25) holds. By Corollary 3.6.5, the α-discounted cost $J^{\tilde{v}}_\alpha = V_\alpha$ satisfies

$$\inf_{\mathcal{K}(\varrho^*)} J^{\tilde{v}}_\alpha = \inf_{\mathbb{R}^d} J^{\tilde{v}}_\alpha \le \frac{\varrho^*}{\alpha} \qquad \forall \alpha > 0\,. \tag{3.6.33}$$

Thus $J^{\tilde{v}}_\alpha$ is a.e. finite, and Theorem A.3.7 asserts that it is a solution to

$$\mathcal{L}^{\tilde{v}} J^{\tilde{v}}_\alpha + c_{\tilde{v}} = \alpha J^{\tilde{v}}_\alpha \quad \text{in } \mathbb{R}^d\,. \tag{3.6.34}$$

The hypotheses (A1)–(A2) of Lemma 3.6.4 are clearly satisfied for $(J^{\tilde{v}}_\alpha, \alpha)$. Observe that (3.6.34) conforms to case (A1) of Lemma 3.6.4 with $g_n \equiv c_{\tilde{v}}$. Hence, over some subsequence $\{\alpha_n\}$,

$$J^{\tilde{v}}_{\alpha_n} - J^{\tilde{v}}_{\alpha_n}(0) \to \tilde{V}\,, \qquad \alpha_n J^{\tilde{v}}_{\alpha_n}(0) \to \tilde{\varrho}\,,$$

and the pair $(\tilde{V}, \tilde{\varrho}) \in \mathscr{W}^{2,p}_{\mathrm{loc}}(\mathbb{R}^d) \times \mathbb{R}^d$, for any $p \in (1, \infty)$, and satisfies

$$\mathcal{L}^{\tilde{v}} \tilde{V} + c_{\tilde{v}} = \tilde{\varrho} \quad \text{in } \mathbb{R}^d \quad \text{and} \quad \inf_{\mathbb{R}^d} \tilde{V} > -\infty\,. \tag{3.6.35}$$

By (3.6.33), $\tilde{\varrho} \le \varrho^*$. Also since $\tilde{V}$ is bounded below, following the argument in the proof of Lemma 3.6.8 we obtain $\tilde{\varrho} \ge \varrho_{\tilde{v}}$. Thus $\tilde{\varrho} = \varrho^*$. Now we use the method in the proof of the second part of Lemma 3.6.9. Replacing V_α and v with $J^{\tilde{v}}_\alpha$ and $\tilde{v}$, respectively, in (3.6.29), the inequality becomes equality, and following the same steps, we obtain

$$\tilde{V}(x) = \lim_{r \downarrow 0} \mathbb{E}^{\tilde{v}}_x \left[\int_0^{\breve{\tau}_r} (c_{\tilde{v}}(X_t) - \varrho^*)\, \mathrm{d}t \right], \quad x \in \mathbb{R}^d\,. \tag{3.6.36}$$

If V denotes the solution obtained in Theorem 3.6.6, then by Lemma 3.6.9 and (3.6.36), $V - \tilde{V} \le 0$. Also, by (3.6.21a) and (3.6.35), $\mathcal{L}^{\tilde{v}}(V - \tilde{V}) \ge 0$, and we conclude as earlier that $V = \tilde{V}$ a.e. in $\mathbb{R}^d$. Thus $\tilde{V}$ satisfies (3.6.21a) and $\tilde{v}$ satisfies (3.6.32). This completes the proof. □

Remark 3.6.11 The restricted (i.e., under the condition $\varrho \le \varrho^*$) uniqueness obtained in Theorem 3.6.10 cannot be improved upon in general. In Section 3.8 we give an example which exhibits a continuum of solutions of (3.6.21a) for $\varrho > \varrho^*$.

3.7 The stable case

In this section we study the ergodic control problem in the stable case, i.e., under the assumption that $\mathfrak{U}_{\mathrm{sm}}$ is uniformly stable. Much of the analysis relies on some sharp equicontinuity estimates for the resolvents of the process under stable controls which are obtained in Theorem 3.7.4. Theorem 3.7.6 extends these results even further: without assuming that all stationary Markov controls are stable it asserts that as long as $\varrho^* < \infty$, then the family $\{V_\alpha - V_\alpha(0) : \alpha \in (0,1)\}$ is equicontinuous and equibounded. One approach to the ergodic control problem is to express the running cost functional as the difference of two near-monotone functions and then utilize the results obtained from the study of the near-monotone case [28, 39, 65]. This approach limits the study to bounded running costs. The equicontinuity results in Theorem 3.7.4 and Theorem 3.7.6 facilitate a direct treatment of the stable case which extends to unbounded running costs (see Theorems 3.7.11 and 3.7.14).

3.7.1 A lemma on control Lyapunov functions

We consider the following growth assumption on the running cost c.

Assumption 3.7.1 The cost c satisfies

$$\sup_{v\in\mathfrak{U}_{\mathrm{ssm}}} \int_{B_R^c\times\mathbb{U}} (1 + c(x,u))\, \pi_v(\mathrm{d}x, \mathrm{d}u) \xrightarrow[R\to\infty]{} 0\,. \tag{3.7.1}$$

Assumption 3.7.1 simply states that $1 + c$ is uniformly integrable with respect to $\{\pi_v : v \in \mathfrak{U}_{\mathrm{ssm}}\}$. Note that if $c \in C_b(\mathbb{R}^d \times \mathbb{U})$, Lemma 3.3.4 asserts that (3.7.1) is equivalent to uniform stability of $\mathfrak{U}_{\mathrm{ssm}}$.

Let $h \in C(\mathbb{R}^d)$ be a positive function. We denote by $\mathscr{O}(h)$ the set of functions $f \in C(\mathbb{R}^d)$ having the property

$$\limsup_{|x|\to\infty} \frac{|f(x)|}{h(x)} < \infty\,, \tag{3.7.2}$$

and by $\mathfrak{o}(h)$ the subset of $\mathscr{O}(h)$ over which the limit in (3.7.2) is zero. We extend this definition to functions on $C(\mathbb{R}^d \times \mathbb{U})$ as follows: For $h \in C(\mathbb{R}^d \times \mathbb{U})$, with $h > 0$,

$$g \in \mathfrak{o}(h) \iff \limsup_{|x|\to\infty} \sup_{u\in\mathbb{U}} \frac{|g(x,u)|}{h(x,u)} = 0\,,$$

and analogously for $\mathscr{O}(h)$.

Recall that $\breve{\tau}_r \equiv \tau(B_r^c)$, $r > 0$, and the definition of $\mathfrak{C}$ in Assumption A3.5.1 on p. 107.

Lemma 3.7.2 *There exist a constant $k_0 > 0$, and a pair of nonnegative, inf-compact functions $(\mathcal{V}, h) \in C^2(\mathbb{R}^d) \times \mathfrak{C}$ satisfying $1 + c \in \mathfrak{o}(h)$ and such that*

$$\mathcal{L}^u \mathcal{V}(x) \le k_0 - h(x,u), \qquad \forall (x,u) \in \mathbb{R}^d \times \mathbb{U} \tag{3.7.3}$$

if and only if (3.7.1) *holds. Moreover,*

(i) *for any $r > 0$,*

$$x \mapsto \sup_{v \in \mathfrak{U}_{\mathrm{ssm}}} \mathbb{E}_x^v \left[\int_0^{\breve{\tau}_r} (1 + c_v(X_t))\, \mathrm{d}t \right] \in \mathfrak{o}(\mathcal{V}); \tag{3.7.4}$$

(ii) *if $\varphi \in \mathfrak{o}(\mathcal{V})$, then for any $x \in \mathbb{R}^d$,*

$$\lim_{t\to\infty} \sup_{v \in \mathfrak{U}_{\mathrm{ssm}}} \frac{1}{t}\, \mathbb{E}_x^v [\varphi(X_t)] = 0, \tag{3.7.5}$$

and

$$\lim_{R\to\infty} \mathbb{E}_x^v [\varphi(X_{t\wedge\tau_R})] = \mathbb{E}_x^v [\varphi(X_t)] \qquad \forall v \in \mathfrak{U}_{\mathrm{ssm}}, \ \forall t \ge 0. \tag{3.7.6}$$

Proof To show sufficiency, let

$$\check{c}_v(x) := 1 + \int_{\mathbb{U}} c(x,u)\, v(\mathrm{d}u \mid x), \quad v \in \mathfrak{U}_{\mathrm{ssm}}.$$

A theorem of Abel states that if $\sum_n a_n$ is a convergent series of positive terms, and if $r_n := \sum_{k\ge n} a_k$ are its remainders, then $\sum_n r_n^{-\lambda} a_n$ converges for all $\lambda \in (0,1)$. There is a counterpart of this theorem for integrals. Thus, if we define

$$\breve{g}(r) := \left(\sup_{v \in \mathfrak{U}_{\mathrm{ssm}}} \int_{B_r^c} \check{c}_v(x)\, \eta_v(\mathrm{d}x) \right)^{-1/2}, \quad r > 0,$$

it follows by (3.7.1) that

$$\int_{\mathbb{R}^d} \check{c}_v(x)\, \breve{g}^\beta(|x|)\, \eta_v(\mathrm{d}x) < \infty \qquad \forall v \in \mathfrak{U}_{\mathrm{ssm}}, \quad \forall \beta \in [0,2). \tag{3.7.7}$$

Let

$$h(x,u) := (1 + c(x,u))\, \breve{g}(|x|).$$

By (3.7.1), $x \mapsto \breve{g}(|x|)$ is inf-compact, and it is straightforward to verify, using an estimate analogous to (3.3.12), that it is also locally Lipschitz. Adopting the notation in (2.3.13), we have $h_v(x) = \check{c}_v(x)\, \breve{g}(|x|)$. Since by (3.7.7), the function $\check{c}_v(x)\, \breve{g}^\beta(|x|)$, $\beta \in [0,2)$, is integrable with respect to η_v for any $v \in \mathfrak{U}_{\mathrm{sm}}$, it follows by Corollary 3.3.2 that the integrals are uniformly bounded, or in other words, that

$$\sup_{v \in \mathfrak{U}_{\mathrm{ssm}}} \int_{\mathbb{R}^d} h_v(x)\, \breve{g}^\beta(|x|)\, \eta_v(\mathrm{d}x) < \infty \qquad \forall \beta \in [0,1).$$

It then follows by Lemma 3.3.4 that there exists a nonnegative, inf-compact function $\mathcal{V} \in C^2(\mathbb{R}^d)$ and a constant $k_0 > 0$ which satisfy (3.7.3). To show necessity, suppose that (3.7.3) holds. By Lemma 3.3.4, the bound in (3.3.10) holds, and since we have $1 + c \in \mathfrak{o}(h)$, (3.7.1) follows. Next we prove (i). Define

$$\tilde{g}(r) := \min_{B_r^c} \check{g}(|x|) = \min_{B_r^c \times \mathbb{U}} \frac{h(x,u)}{1 + c(x,u)}, \qquad r > 0.$$

Since $1 + c \in \mathfrak{o}(h)$, we obtain $\tilde{g}(r) \to \infty$ as $r \to \infty$. With $R > 0$ large enough so that $x \in B_R$, applying Dynkin's formula to (3.7.3), we have

$$\mathbb{E}_x^v[\mathcal{V}(X_{\check{\tau}_r \wedge \tau_R})] - \mathcal{V}(x) \le \mathbb{E}_x^v\left[\int_0^{\check{\tau}_r \wedge \tau_R} [k_0 - h_v(X_t)]\,\mathrm{d}t\right] \tag{3.7.8}$$

for all $v \in \mathfrak{U}_{\mathrm{ssm}}$. Therefore

$$\mathbb{E}_x^v\left[\int_0^{\check{\tau}_r \wedge \tau_R} h_v(X_t)\,\mathrm{d}t\right] \le \mathcal{V}(x) + k_0\,\mathbb{E}_x^v[\check{\tau}_r \wedge \tau_R]\,. \tag{3.7.9}$$

It is straightforward to show that $\sup_{v \in \mathfrak{U}_{\mathrm{ssm}}} \mathbb{E}_x^v[\check{\tau}_r] \in \mathscr{O}(\mathcal{V})$. Indeed, if $R_0 > 0$ is large enough such that $h_v(x) - k_0 \ge 1$ for all $x \in B_{R_0}^c$, then using the strong Markov property and (3.7.8), we obtain

$$\begin{aligned}
\mathbb{E}_x^v[\check{\tau}_r] &\le \mathbb{E}_x^v\left[\int_0^{\check{\tau}_{R_0}} [h_v(X_t) - k_0]\,\mathrm{d}t\right] + \sup_{y \in \partial B_{R_0}} \mathbb{E}_y^v[\check{\tau}_r] \\
&\le \mathcal{V}(x) + \sup_{v \in \mathfrak{U}_{\mathrm{ssm}}} \sup_{y \in \partial B_{R_0}} \mathbb{E}_y^v[\check{\tau}_r]
\end{aligned}$$

and the claim follows by Theorem 3.3.1. Therefore, taking limits as $R \to \infty$ in (3.7.9), we have

$$\sup_{v \in \mathfrak{U}_{\mathrm{ssm}}} \mathbb{E}_x^v\left[\int_0^{\check{\tau}_r} h_v(X_t)\,\mathrm{d}t\right] \in \mathscr{O}(\mathcal{V})\,. \tag{3.7.10}$$

For each $x \in B_r^c$ and $v \in \mathfrak{U}_{\mathrm{ssm}}$, select the maximal radius $\rho_v(x)$ satisfying

$$\mathbb{E}_x^v\left[\int_0^{\check{\tau}_r} \mathbb{I}_{B_{\rho_v(x)}}(X_t)\check{c}_v(X_t)\,\mathrm{d}t\right] \le \frac{1}{2}\,\mathbb{E}_x^v\left[\int_0^{\check{\tau}_r} \check{c}_v(X_t)\,\mathrm{d}t\right]. \tag{3.7.11}$$

Thus, by (3.7.10)–(3.7.11),

$$\begin{aligned}
\mathbb{E}_x^v\left[\int_0^{\check{\tau}_r} \check{c}_v(X_t)\,\mathrm{d}t\right] &\le 2\,\mathbb{E}_x^v\left[\int_0^{\check{\tau}_r} \mathbb{I}_{B^c_{\rho_v(x)}}(X_t)\check{c}_v(X_t)\,\mathrm{d}t\right] \\
&\le \frac{2}{\tilde{g}(\rho_v(x))}\,\mathbb{E}_x^v\left[\int_0^{\check{\tau}_r} \mathbb{I}_{B^c_{\rho_v(x)}}(X_t)\check{c}_v(X_t)\check{g}(|X_t|)\,\mathrm{d}t\right] \\
&\le \frac{2}{\tilde{g}(\rho_v(x))}\,\mathbb{E}_x^v\left[\int_0^{\check{\tau}_r} h_v(X_t)\,\mathrm{d}t\right] \\
&\in \mathscr{O}\left(\frac{\mathcal{V}}{\tilde{g} \circ \rho_v}\right).
\end{aligned} \tag{3.7.12}$$

Since for any fixed ball B_ρ the function

$$x \mapsto \sup_{v \in \mathfrak{U}_{\mathrm{ssm}}} \mathbb{E}_x^v\left[\int_0^{\check{\tau}_r} \mathbb{I}_{B_\rho}(X_t)\check{c}_v(X_t)\,\mathrm{d}t\right]$$

is bounded on B_r^c by Lemma 2.6.13 (i), whereas the function on the left-hand side of (3.7.12) grows unbounded as $|x| \to \infty$ uniformly in $v \in \mathfrak{U}_{\mathrm{ssm}}$, it follows that

$$\liminf_{|x|\to\infty} \inf_{v\in\mathfrak{U}_{\mathrm{ssm}}} \rho_v(x) \to \infty\,.$$

Therefore, (3.7.4) follows from (3.7.12).

We turn now to part (ii). Applying Itô's formula and Fatou's lemma, we obtain from (3.7.3) that

$$\mathbb{E}_x^v[\mathcal{V}(X_t)] \le k_0 t + \mathcal{V}(x) \qquad \forall v \in \mathfrak{U}_{\mathrm{ssm}}\,. \tag{3.7.13}$$

If φ is $\mathfrak{o}(\mathcal{V})$, then there exists $\check{f}\colon \mathbb{R}_+ \to \mathbb{R}_+$ satisfying $\check{f}(R) \to \infty$ as $R \to \infty$, such that

$$\mathcal{V}(x) \ge |\varphi(x)|\, \check{f}(|x|)\,.$$

Define

$$R(t) := t \wedge \inf\left\{|x| : |\varphi(x)| \ge \sqrt{t}\right\}, \quad t \ge 0\,.$$

Then, by (3.7.13),

$$\begin{aligned}
\mathbb{E}_x^v\left[|\varphi(X_t)|\right] &\le \mathbb{E}_x^v\left[\left|\varphi(X_t)\right| \mathbb{I}_{B_{R(t)}}(X_t)\right] + \frac{\mathbb{E}_x^v[\mathcal{V}(X_t)\,\mathbb{I}_{B^c_{R(t)}}(X_t)]}{\check{f}(R(t))} \\
&\le \sqrt{t} + \frac{k_0 t + \mathcal{V}(x)}{\check{f}(R(t))}\,,
\end{aligned} \tag{3.7.14}$$

and taking the supremum over $v \in \mathfrak{U}_{\mathrm{ssm}}$ in (3.7.14), then dividing by t, and taking limits as $t \to \infty$, (3.7.5) follows.

To prove (3.7.6), first write

$$\mathbb{E}_x^v\left[\varphi(X_{t\wedge\tau_R})\right] = \mathbb{E}_x^v\left[\varphi(X_t)\,\mathbb{I}\{t < \tau_R\}\right] + \mathbb{E}_x^v\left[\varphi(X_{\tau_R})\,\mathbb{I}\{t \ge \tau_R\}\right]. \tag{3.7.15}$$

By (3.7.3), $\mathbb{E}_x^v[\mathcal{V}(X_{t\wedge\tau_R})] \le k_0 t + \mathcal{V}(x)$. Therefore

$$\mathbb{E}_x^v\left[\varphi(X_{\tau_R})\,\mathbb{I}\{t \ge \tau_R\}\right] \le [k_0 t + \mathcal{V}(x)] \sup_{x\in\partial B_R} \frac{\varphi(x)}{\mathcal{V}(x)}\,,$$

and since $\varphi \in \mathfrak{o}(\mathcal{V})$, this shows that the second term on the right-hand side of (3.7.15) vanishes as $R \to \infty$. Since $|\varphi(X_t)| \le M\mathcal{V}(X_t)$ for some constant $M > 0$, Fatou's lemma yields

$$\begin{aligned}
\mathbb{E}_x^v\left[\varphi(X_t)\right] &\le \varliminf_{R\to\infty} \mathbb{E}_x^v\left[\varphi(X_t)\,\mathbb{I}\{t < \tau_R\}\right] \\
&\le \varlimsup_{R\to\infty} \mathbb{E}_x^v\left[\varphi(X_t)\,\mathbb{I}\{t < \tau_R\}\right] \le \mathbb{E}_x^v\left[\varphi(X_t)\right],
\end{aligned}$$

thus obtaining (3.7.6). □

Following the proof of Lemma 3.7.2, we obtain the following useful corollary:

Corollary 3.7.3 *Suppose $v \in \mathfrak{U}_{\mathrm{ssm}}$ and $\varrho_v < \infty$. For some fixed $r > 0$ define*

$$\varphi(x) := \mathbb{E}_x^v \left[\int_0^{\breve{\tau}_r} (1 + c_v(X_t))\, \mathrm{d}t \right], \quad x \in \mathbb{R}^d .$$

Then

$$\lim_{t\to\infty} \frac{1}{t}\, \mathbb{E}_x^v \left[\varphi(X_t)\right] = 0 \qquad \forall x \in \mathbb{R}^d ,$$

and for any $t \ge 0$,

$$\lim_{R\to\infty} \mathbb{E}_x^v \left[\varphi(X_{t\wedge\tau_R})\right] = \mathbb{E}_x^v \left[\varphi(X_t)\right] \qquad \forall v \in \mathfrak{U}_{\mathrm{ssm}} ,\ \forall x \in \mathbb{R}^d .$$

3.7.2 Equicontinuity of resolvents

We next show that, for a stable control $v \in \mathfrak{U}_{\mathrm{ssm}}$, the resolvents J_α^v are bounded in $\mathscr{W}^{2,p}(B_R)$ uniformly in α for any $R > 0$. Assumption 3.7.1 is not used in these results.

Theorem 3.7.4 *There exists a constant C_0 depending only on the radius $R > 0$, such that, for all $v \in \mathfrak{U}_{\mathrm{ssm}}$ and $\alpha \in (0, 1)$,*

$$\left\| J_\alpha^v - J_\alpha^v(0) \right\|_{\mathscr{W}^{2,p}(B_R)} \le \frac{C_0}{\eta_v(B_{2R})} \left(\frac{\varrho_v}{\eta_v(B_{2R})} + \sup_{B_{4R}\times\mathbb{U}} c \right) \tag{3.7.16a}$$

$$\sup_{B_R} \alpha J_\alpha^v \le C_0 \left(\frac{\varrho_v}{\eta_v(B_R)} + \sup_{B_{2R}\times\mathbb{U}} c \right) . \tag{3.7.16b}$$

Proof Let

$$\hat{\tau} = \inf\, \{t > \tau(B_{2R}) : X_t \in B_R\} .$$

For $x \in \partial B_R$, we have

$$\begin{aligned} J_\alpha^v(x) &= \mathbb{E}_x^v \left[\int_0^{\hat{\tau}} \mathrm{e}^{-\alpha t} c_v(X_t)\, \mathrm{d}t + \mathrm{e}^{-\alpha\hat{\tau}} J_\alpha^v(X_{\hat{\tau}}) \right] \\ &= \mathbb{E}_x^v \left[\int_0^{\hat{\tau}} \mathrm{e}^{-\alpha t} c_v(X_t)\, \mathrm{d}t + J_\alpha^v(X_{\hat{\tau}}) - (1 - \mathrm{e}^{-\alpha\hat{\tau}}) J_\alpha^v(X_{\hat{\tau}}) \right] . \end{aligned} \tag{3.7.17}$$

Let $\tilde{P}_x(A) = \mathbb{P}_x^v(X_{\hat{\tau}} \in A)$. By Theorem 2.6.7, there exists $\delta \in (0, 1)$ depending only on R, such that

$$\left\| \tilde{P}_x - \tilde{P}_y \right\|_{\mathrm{TV}} \le 2\delta \qquad \forall x, y \in \partial B_R .$$

Therefore

$$\left| \mathbb{E}_x^v \left[J_\alpha^v(X_{\hat{\tau}})\right] - \mathbb{E}_y^v \left[J_\alpha^v(X_{\hat{\tau}})\right] \right| \le \delta \operatorname*{osc}_{\partial B_R} J_\alpha^v \qquad \forall x, y \in \partial B_R . \tag{3.7.18}$$

Thus (3.7.17) and (3.7.18) yield

$$\begin{aligned} \operatorname*{osc}_{\partial B_R} J_\alpha^v \le \frac{1}{1-\delta} \sup_{x\in\partial B_R} \mathbb{E}_x^v &\left[\int_0^{\hat{\tau}} \mathrm{e}^{-\alpha t} c_v(X_t)\, \mathrm{d}t \right] \\ &+ \frac{1}{1-\delta} \sup_{x\in\partial B_R} \mathbb{E}_x^v \left[(1 - \mathrm{e}^{-\alpha\hat{\tau}}) J_\alpha^v(X_{\hat{\tau}}) \right] . \end{aligned} \tag{3.7.19}$$

Next, we bound the terms on the right-hand side of (3.7.19). First,

$$\mathbb{E}^v_x\Big[(1-\mathrm{e}^{-\alpha\hat{\tau}})J^v_\alpha(X_{\hat{\tau}})\Big] \le \mathbb{E}^v_x\Big[\alpha^{-1}(1-\mathrm{e}^{-\alpha\hat{\tau}})\Big] \sup_{x\in\partial B_R} \alpha J^v_\alpha(x)$$
$$\le \Big(\sup_{\partial B_R} \alpha J^v_\alpha\Big)\mathbb{E}^v_x[\hat{\tau}] \qquad \forall x\in\partial B_R\,. \tag{3.7.20}$$

Let

$$M(R) := \sup_{B_R\times\mathbb{U}} c\,, \quad R>0\,.$$

By Lemma A.2.10, the function

$$\varphi_\alpha = \frac{M(2R)}{\alpha} + J^v_\alpha$$

is in $\mathscr{W}^{2,p}_{\mathrm{loc}}(\mathbb{R}^d)$ for all $p>1$ and satisfies

$$\mathcal{L}^v\varphi_\alpha(x) - \alpha\varphi_\alpha(x) = -c_v(x) - M(2R) \qquad \forall x\in B_{2R}\,, \tag{3.7.21}$$

and thus

$$M(2R) \le \big|(\mathcal{L}^v-\alpha)\varphi_\alpha(x)\big| \le 2M(2R) \qquad \forall x\in B_{2R}\,. \tag{3.7.22}$$

By (3.7.22),

$$\big\|(\mathcal{L}^v-\alpha)\varphi_\alpha\big\|_{L^\infty(B_{2R})} \le 2|B_{2R}|^{-1}\big\|(\mathcal{L}^v-\alpha)\varphi_\alpha\big\|_{L^1(B_{2R})}\,. \tag{3.7.23}$$

Hence $\varphi_\alpha \in \mathfrak{K}(2,B_{2R})$ (see Definition A.2.11), and by Theorem A.2.13, there exists a constant $\tilde{C}_H$ depending only on R, such that

$$\varphi_\alpha(x) \le \tilde{C}_H\varphi_\alpha(y) \qquad \forall x,y\in B_R \text{ and } \alpha\in(0,1)\,. \tag{3.7.24}$$

Integrating with respect to η_v, and using Fubini's theorem, we obtain

$$\int_{\mathbb{R}^d} \alpha J^v_\alpha(x)\,\eta_v(\mathrm{d}x) = \varrho_v \qquad \forall v\in\mathfrak{U}_{\mathrm{ssm}}\,. \tag{3.7.25}$$

By (3.7.25), $\inf_{B_R}\{\alpha J^v_\alpha\} \le \frac{\varrho_v}{\eta_v(B_R)}$, and (3.7.24) yields

$$\sup_{B_R} \alpha J^v_\alpha \le \tilde{C}_H\left(M(2R) + \frac{\varrho_v}{\eta_v(B_R)}\right), \tag{3.7.26}$$

thus proving (3.7.16b). On the other hand,

$$\psi_\alpha(x) = \mathbb{E}^v_x\left[\int_0^{\hat{\tau}} \mathrm{e}^{-\alpha t}(M(2R)+c_v(X_t))\,\mathrm{d}t\right]$$

also satisfies (3.7.21)–(3.7.23), in B_{2R}, and therefore (3.7.24) holds for ψ_α. Thus using the bound

$$\inf_{x\in\partial B_R} \mathbb{E}^v_x\left[\int_0^{\hat{\tau}} (M(2R)+c_v(X_t))\,\mathrm{d}t\right] \le (M(2R)+\varrho_v) \sup_{x\in\partial B_R} \mathbb{E}^v_x[\hat{\tau}]\,,$$

we obtain

$$\sup_{x\in\partial B_R} \mathbb{E}_x^v\left[\int_0^{\hat\tau} e^{-\alpha t} c_v(X_t)\,dt\right] \le \tilde C_H(M(2R)+\varrho_v) \sup_{x\in\partial B_R} \mathbb{E}_x^v[\hat\tau]\,. \tag{3.7.27}$$

By (3.7.19), (3.7.20), (3.7.26) and (3.7.27),

$$\operatorname*{osc}_{\partial B_R} J_\alpha^v \le \frac{2\tilde C_H}{1-\delta}\left(M(2R)+\frac{\varrho_v}{\eta_v(B_R)}\right) \sup_{x\in\partial B_R} \mathbb{E}_x^v[\hat\tau]\,. \tag{3.7.28}$$

Applying Theorem A.2.13 to the $\mathcal{L}^v$-superharmonic function $x \mapsto \mathbb{E}_x^v[\hat\tau]$, we obtain

$$\sup_{x\in\partial B_R} \mathbb{E}_x^v[\hat\tau] \le \tilde C'_H \inf_{x\in\partial B_R} \mathbb{E}_x^v[\hat\tau] \tag{3.7.29}$$

for some constant $\tilde C'_H > 0$. By (2.6.9a), (3.7.29), and the estimate

$$\inf_{x\in\partial B_R} \mathbb{E}_x^v[\hat\tau] \le \frac{1}{\eta_v(B_R)} \sup_{x\in\partial B_R} \mathbb{E}_x^v[\tau(B_{2R})]\,,$$

which is obtained from Theorem 2.6.9, we have

$$\sup_{x\in\partial B_R} \mathbb{E}_x^v[\hat\tau] \le \frac{\tilde C_1}{\eta_v(B_R)} \tag{3.7.30}$$

for some positive constant $\tilde C_1 = \tilde C_1(R)$.

Next, we use the expansion

$$J_\alpha^v(x) = \mathbb{E}_x^v\left[\int_0^{\tau_R} e^{-\alpha t} c_v(X_t)\,dt + J_\alpha^v(X_{\tau_R}) - (1-e^{-\alpha\tau_R})J_\alpha^v(X_{\tau_R})\right] \qquad \forall x \in B_R\,,$$

together with (3.7.20) to obtain

$$\operatorname*{osc}_{B_R} J_\alpha^v \le \sup_{x\in B_R} \mathbb{E}_x^v\left[\int_0^{\tau_R} e^{-\alpha t} c_v(X_t)\,dt\right] + \operatorname*{osc}_{\partial B_R} J_\alpha^v + \left(\sup_{\partial B_R} \alpha J_\alpha^v\right) \sup_{x\in B_R} \mathbb{E}_x^v[\tau_R]\,. \tag{3.7.31}$$

By Theorem A.2.1, there exists a constant $\tilde C'_1 = \tilde C'_1(R)$ such that $\mathbb{E}_x^v[\tau_R] \le \tilde C'_1$ for all $x \in B_R$, and thus the first term on the right-hand side of (3.7.31) can be bounded by

$$\sup_{x\in B_R} \mathbb{E}_x^v\left[\int_0^{\tau_R} e^{-\alpha t} c_v(X_t)\,dt\right] \le \tilde C'_1 \sup_{B_R\times\mathbb{U}} c \qquad \forall v \in \mathfrak{U}_{\mathrm{sm}}\,.$$

For second and third terms on the right-hand side of (3.7.31) we use the bounds in (3.7.26), (3.7.28) and (3.7.30). Therefore, (3.7.31) yields

$$\operatorname*{osc}_{B_R} J_\alpha^v \le \frac{\tilde C_2}{\eta_v(B_R)}\left(M(2R)+\frac{\varrho_v}{\eta_v(B_R)}\right) \tag{3.7.32}$$

for some constant $\tilde C_2 = \tilde C_2(R)$.

Let $\bar\varphi_\alpha := J_\alpha^v - J_\alpha^v(0)$. Then

$$\mathcal{L}^v\bar\varphi_\alpha - \alpha\bar\varphi_\alpha = -c_v + \alpha J_\alpha^v(0)\,, \quad \text{in } B_{2R}\,.$$

Applying Lemma A.2.5, with $D = B_{2R}$ and $D' = B_R$, relative to the operator $\mathcal{L}^v - \alpha$, we obtain, for some positive constant $\tilde{C}_3 = \tilde{C}_3(R)$,

$$\|\bar{\varphi}_\alpha\|_{\mathscr{W}^{2,p}(B_R)} \le \tilde{C}_3 \left(\|\bar{\varphi}_\alpha\|_{L^p(B_{2R})} + \|\mathcal{L}^v \bar{\varphi}_\alpha - \alpha\bar{\varphi}_\alpha\|_{L^p(B_{2R})}\right)$$
$$\le \tilde{C}_3 |B_{2R}|^{1/p} \left(\operatorname*{osc}_{B_{2R}} J^v_\alpha + M(2R) + \sup_{B_{2R}} \alpha J^v_\alpha\right),$$

and the bound in (3.7.16a) follows by (3.7.26) and (3.7.32). □

The bounds in (3.7.16) along with Theorem 3.3.1 imply that if $\mathfrak{U}_{\mathrm{sm}} = \mathfrak{U}_{\mathrm{ssm}}$, then as long as $\varrho_v < \infty$ for all $v \in \mathfrak{U}_{\mathrm{sm}}$, the functions $J^v_\alpha - J^v_\alpha(0)$ are bounded in $\mathscr{W}^{2,p}(B_R)$ on any ball B_R, uniformly in α and $v \in \mathfrak{U}_{\mathrm{ssm}}$. Next we relax the assumption that $\mathfrak{U}_{\mathrm{sm}} = \mathfrak{U}_{\mathrm{ssm}}$. We assume only that $\varrho^* < \infty$, and this of course implies that there exists some $\hat{v} \in \mathfrak{U}_{\mathrm{ssm}}$ such that $\varrho_{\hat{v}} < \infty$. We show that the α-discounted value functions $\{V_\alpha\}$ are bounded in $\mathscr{W}^{2,p}(B_R)$ on any ball B_R uniformly in $\alpha \in (0,1)$.

We first need a technical lemma. In this lemma we use the Markov transition kernel $\tilde{P}_v$, for $v \in \mathfrak{U}_{\mathrm{ssm}}$, defined in Theorem 2.6.7, relative to the domains $D_1 = B_R$ and $D_2 = B_{2R}$.

Lemma 3.7.5 *Let $v \in \mathfrak{U}_{\mathrm{ssm}}$ and $R > 0$ be fixed, and $\tilde{P}_v(x, A)$ for $A \in \mathscr{B}(\partial B_R)$ be the transition kernel of the Markov chain on ∂B_R defined by,*

$$\tilde{P}_v(x, f) \equiv \int_{\partial B_R} \tilde{P}_v(x, \mathrm{d}y) f(y) := \mathbb{E}^v_x \left[\mathbb{E}^v_{X_{\tau_{2R}}} [f(X_{\breve{\tau}_R})]\right], \quad x \in \partial B_R,$$

for $f \in C(\partial B_R)$. Also let

$$\overline{m}_\alpha := \max_{\partial B_R} V_\alpha \quad \textit{and} \quad \underline{m}_\alpha := \min_{\partial B_R} V_\alpha.$$

Then there exists a constant $\tilde{\delta} \in (0,1)$ depending only on R such that

$$\tilde{P}_v(x, V_\alpha) - \underline{m}_\alpha \le \max\left\{1, \tilde{\delta}\,(\overline{m}_\alpha - \underline{m}_\alpha)\right\} \qquad \forall \alpha \in (0,1). \tag{3.7.33}$$

Proof For $z \in \partial B_R$ and $r < R$, let $\tilde{B}_r(z) := \partial B_R \cap B_r(z)$, where $B_r(z) \subset \mathbb{R}^d$ denotes the ball of radius r centered at z. We first show that for any $r > 0$ there exists $\varepsilon = \varepsilon(r) > 0$ such that

$$\tilde{P}_v(x, \tilde{B}_r(z)) > \varepsilon(r) \qquad \forall x, z \in \partial B_R. \tag{3.7.34}$$

Since $\tilde{P}_v(x, \cdot) \ge C_H^{-1} \tilde{P}_v(x', \cdot)$ for all $x, x' \in \partial B_R$ by (2.6.15), it is enough to prove (3.7.34) for some fixed x. Let $\psi_z \colon \partial B_R \to \mathbb{R}_+$ be a smooth function supported on $\tilde{B}_r(z)$ and such that $\psi_z > 0$ on $\tilde{B}_r(z)$ and $\|\psi_z\|_\infty = 1$. For $z' \ne z$, the function $\psi_{z'}$ is simply the translation of ψ_z on ∂B_R. Hence $\|\psi_{z'} - \psi_z\|_\infty \to 0$ as $z' \to z$. Consider the map $z \mapsto \tilde{P}_v(x, \psi_z)$. This is the composition of the following continuous maps:

(i) $\partial B_R \ni z \mapsto \psi_z \in C(\partial B_R)$.
(ii) $C(\partial B_R) \ni f \mapsto \mathbb{E}^v_\cdot [f(X_{\breve{\tau}_R})] \in C(B_{2R})$ (this is continuous by Lemma 2.6.13 (ii)).
(iii) $C(B_{2R}) \ni g \mapsto \mathbb{E}^v_\cdot [g(X_{\tau_{2R}})] \in C(B_R)$.

It is clear that $\tilde{P}_v(x, \psi_z) > 0$ for all $z \in \partial B_R$, and hence it follows that

$$\min_{z\in\partial B_R} \tilde{P}_v(x, \psi_z) > 0 .$$

On the other hand, $\tilde{P}_v(x, \tilde{B}_r(z)) \geq \tilde{P}_v(x, \psi_z)$, so (3.7.34) holds.

If $\overline{m}_\alpha - \underline{m}_\alpha < 1$, then since $\tilde{P}_v(x, V_\alpha) < \overline{m}_\alpha$, (3.7.33) clearly holds. Suppose then $\overline{m}_\alpha - \underline{m}_\alpha \geq 1$. In this case it is enough to show that

$$\tilde{P}_v(x, V_\alpha) - \underline{m}_\alpha \leq \tilde{\delta}(\overline{m}_\alpha - \underline{m}_\alpha) \qquad \forall \alpha \in (0, 1) . \tag{3.7.35}$$

By the linearity of the operator $\tilde{P}_v$, equation (3.7.35) can be written in the equivalent scaled form

$$\tilde{P}_v(x, \varphi_\alpha) \leq \tilde{\delta} \qquad \forall \alpha \in (0, 1) , \qquad \text{where} \quad \varphi_\alpha(x) := \frac{V_\alpha(x) - \underline{m}_\alpha}{\overline{m}_\alpha - \underline{m}_\alpha} .$$

In this scaling $\varphi_\alpha \colon \partial B_R \to [0, 1]$, and maps onto its range. Note that φ_α satisfies

$$\mathcal{L}^v \varphi_\alpha - \alpha \varphi_\alpha = -\frac{c_v - \alpha \underline{m}_\alpha}{\overline{m}_\alpha - \underline{m}_\alpha} . \tag{3.7.36}$$

Also, since $\alpha \underline{m}_\alpha \leq \alpha V_\alpha(x) \leq \alpha J^v_\alpha(x)$ for all $x \in \partial B_R$, the term $\alpha \underline{m}_\alpha$ is bounded uniformly in $\alpha \in (0, 1)$ by (3.7.26). By (3.7.36) and Lemma A.2.5 there exists a constant $\hat{C}$ such that $\|\varphi_\alpha\|_{\mathscr{W}^{2,p}(B_R)} \leq \hat{C}$ for all $\alpha \in (0, 1)$. In particular, $\{\varphi_\alpha : \alpha \in (0, 1)\}$ is an equi-Lipschitzean family. Hence, since φ_α is onto $[0, 1]$, there exists $r_0 > 0$ such that

$$\varphi_\alpha(x) \leq \frac{1}{2} \qquad \forall x \in \tilde{B}_{r_0}(z) , \quad \forall \alpha \in (0, 1) ,$$

for some $z \in \partial B_R$ which depends on α. Therefore, by (3.7.34) we have

$$\begin{aligned}\tilde{P}_v(x, \varphi_\alpha) &\leq (1 - \varepsilon(r_0)) + \frac{1}{2}\varepsilon(r_0) \\ &= 1 - \frac{\varepsilon(r_0)}{2} =: \tilde{\delta} ,\end{aligned}$$

and the proof is complete. □

As mentioned earlier, the next theorem does not assume $\mathfrak{U}_{\mathrm{sm}} = \mathfrak{U}_{\mathrm{ssm}}$.

Theorem 3.7.6 *Suppose $\varrho_{\hat{v}} < \infty$ for some $\hat{v} \in \mathfrak{U}_{\mathrm{ssm}}$. There exists a constant $\tilde{C}_0 > 0$ depending only on the radius $R > 0$, such that, for all $\alpha \in (0, 1)$, we have*

$$\left\|V_\alpha - V_\alpha(0)\right\|_{\mathscr{W}^{2,p}(B_R)} \leq \frac{\tilde{C}_0}{\eta_{\hat{v}}(B_{2R})} \left(\frac{\varrho_{\hat{v}}}{\eta_{\hat{v}}(B_{2R})} + \sup_{B_{4R}\times\mathbb{U}} c \right) , \tag{3.7.37a}$$

$$\sup_{B_R} \alpha V_\alpha \leq \tilde{C}_0 \left(\frac{\varrho_{\hat{v}}}{\eta_{\hat{v}}(B_R)} + \sup_{B_{2R}\times\mathbb{U}} c \right) . \tag{3.7.37b}$$

Proof Since $V_\alpha \le J_\alpha^{\hat{v}}$, (3.7.37b) follows by (3.7.26). To prove (3.7.37a) we define the admissible control $U \in \mathfrak{U}$ by

$$U_t = \begin{cases} \hat{v} & \text{if } t \le \hat{\tau}\,, \\ v_\alpha & \text{otherwise,} \end{cases}$$

where $\hat{\tau}$ is as in the proof of Theorem 3.7.4, and $v_\alpha \in \mathfrak{U}_{\mathrm{sm}}$ is an α-discounted optimal control. Since U is in general sub-optimal for the α-discounted criterion, we have

$$V_\alpha(x) \le \mathbb{E}_x^{\hat{v}}\left[\int_0^{\hat{\tau}} \mathrm{e}^{-\alpha t} c_{\hat{v}}(X_t)\,\mathrm{d}t + \mathrm{e}^{-\alpha\hat{\tau}} V_\alpha(X_{\hat{\tau}})\right]. \tag{3.7.38}$$

To simplify the notation, we define

$$\Psi_\alpha := \sup_{x\in\partial B_R} \mathbb{E}_x^{\hat{v}}\left[\int_0^{\hat{\tau}} \mathrm{e}^{-\alpha t} c_{\hat{v}}(X_t)\right]\mathrm{d}t + \sup_{x\in\partial B_R} \mathbb{E}_x^{\hat{v}}\left[(1-\mathrm{e}^{-\alpha\hat{\tau}})J_\alpha^{\hat{v}}(X_{\hat{\tau}})\right].$$

We then expand (3.7.38) as in (3.7.17), strengthen the inequality using the fact that $\alpha V_\alpha \le \alpha J_\alpha^{\hat{v}}$, subtract $\underline{m}_\alpha$ from both sides, and employ Lemma 3.7.5 to obtain

$$\begin{aligned} V_\alpha(x) - \underline{m}_\alpha &\le \mathbb{E}_x^{\hat{v}}\left[V_\alpha(X_{\hat{\tau}})\right] - \underline{m}_\alpha + \Psi_\alpha \\ &\le \max\{1, \tilde{\delta}(\overline{m}_\alpha - \underline{m}_\alpha)\} + \Psi_\alpha\,. \end{aligned} \tag{3.7.39}$$

We evaluate (3.7.39) at a point $\hat{x} = \hat{x}(\alpha) \in \partial B_R$ such that $V_\alpha(\hat{x}) = \overline{m}_\alpha$, i.e., where it attains its maximum, to obtain

$$\overline{m}_\alpha - \underline{m}_\alpha \le \max\left\{\frac{1}{\tilde{\delta}}, \frac{\Psi_\alpha}{1-\tilde{\delta}}\right\}.$$

The estimates for Ψ_α derived in the proof of Theorem 3.7.4 provide a bound for $\operatorname{osc}_{\partial B_R} V_\alpha \equiv \overline{m}_\alpha - \underline{m}_\alpha$. We combine this with estimates for the other two terms in the expansion

$$\operatorname*{osc}_{B_R} V_\alpha \le \sup_{x\in B_R} \mathbb{E}_x^{v_\alpha}\left[\int_0^{\tau_R} \mathrm{e}^{-\alpha t} c_{\hat{v}}(X_t)\,\mathrm{d}t\right] + \operatorname*{osc}_{\partial B_R} V_\alpha + \left(\sup_{\partial B_R} \alpha V_\alpha\right) \sup_{x\in B_R} \mathbb{E}_x^{v_\alpha}[\tau_R]$$

to obtain a bound for $\operatorname{osc}_{B_R} V_\alpha$ similar to the one in (3.7.32). Finally, the bound in (3.7.37a) is obtained by repeating the steps in the last paragraph of the proof of Theorem 3.7.4. □

3.7.3 The HJB equation in the stable case

We now embark on the study of the HJB equation in the stable case via the vanishing discount approach. As seen in Theorem 3.4.7, the existence of a stationary Markov control which is optimal with respect to the pathwise ergodic criterion (2.7.6) requires the tightness of the empirical measures ζ_t^U defined in (3.4.5). The hypotheses imposed in this section are weaker than Assumption 3.4.9 which guarantees tightness of the family $\{\zeta_t^U\}$ (see Theorem 3.4.11). As a result, in this section we study optimality with respect to the average formulation of the ergodic criterion in (2.7.7), and we agree to call a control $U \in \mathfrak{U}$ *average-cost optimal* if it attains the minimum of (2.7.7) over all admissible controls.

On the other hand, uniform stability implies the tightness of the mean empirical measures $\{\bar{\zeta}^U_{x,t}\}$ by Lemma 3.3.4, which in turn implies that every limit point of $\bar{\zeta}^U_{x,t}$ as $t \to \infty$ lies in $\mathscr{G}$. Therefore

$$\begin{aligned}\liminf_{t\to\infty} \frac{1}{t}\int_0^t \mathbb{E}^U_x[\bar{c}(X_s, U_s)]\,\mathrm{d}s &= \liminf_{t\to\infty} \int_{\mathbb{R}^d\times\mathbb{U}} c(z,u)\,\bar{\zeta}^U_{x,t}(\mathrm{d}z,\mathrm{d}u)\\ &\ge \min_{v\in\mathfrak{U}_{\mathrm{ssm}}} \int_{\mathbb{R}^d\times\mathbb{U}} c\,\mathrm{d}\pi_v \qquad \forall U \in \mathfrak{U}\,,\end{aligned}$$

and it follows that, provided $\mathfrak{U}_{\mathrm{sm}}$ is uniformly stable, a stationary Markov control which is average-cost optimal in the class $\mathfrak{U}_{\mathrm{ssm}}$ is also average-cost optimal over all admissible controls. It is worth noting that if $\mathfrak{U}_{\mathrm{sm}}$ is uniformly stable, then we obtain a stronger form of optimality, namely that

$$\varrho^* \le \inf_{U\in\mathfrak{U}} \left(\liminf_{T\to\infty} \frac{1}{T}\int_0^T \mathbb{E}^U_x[\bar{c}(X_t, U_t)]\,\mathrm{d}t\right). \tag{3.7.40}$$

We start by showing some fairly general properties of the vanishing discount limit. We need the following definition.

Definition 3.7.7 Let $\tilde{\mathfrak{U}}_{\mathrm{ssm}} := \{v \in \mathfrak{U}_{\mathrm{ssm}} : \varrho_v < \infty\}$. Define

$$\Psi^v(x;\varrho) := \lim_{r\downarrow 0} \mathbb{E}^v_x\left[\int_0^{\check{\tau}_r} (c_v(X_t) - \varrho)\,\mathrm{d}t\right], \quad v \in \tilde{\mathfrak{U}}_{\mathrm{ssm}}\,,$$

$$\Psi^*(x;\varrho) := \liminf_{r\downarrow 0} \inf_{v\in\tilde{\mathfrak{U}}_{\mathrm{ssm}}} \mathbb{E}^v_x\left[\int_0^{\check{\tau}_r} (c_v(X_t) - \varrho)\,\mathrm{d}t\right],$$

provided the limits exist and are finite.

In the next lemma we do not assume (3.7.1), nor do we assume that $\mathfrak{U}_{\mathrm{sm}} = \mathfrak{U}_{\mathrm{ssm}}$.

Lemma 3.7.8 *The following hold:*

(i) *Suppose $\varrho^* < \infty$. Then for each sequence $\alpha_n \downarrow 0$ there exists a further subsequence also denoted as $\{\alpha_n\}$, a function $V \in C^2(\mathbb{R}^d)$ and a constant $\varrho \in \mathbb{R}$ such that, as $n \to \infty$,*

$$\alpha_n V_{\alpha_n}(0) \to \varrho \qquad \textit{and} \qquad \bar{V}_{\alpha_n} := V_{\alpha_n} - V_{\alpha_n}(0) \to V\,,$$

uniformly on compact subsets of $\mathbb{R}^d$. The pair (V, ϱ) satisfies

$$\min_{u\in\mathbb{U}} [\mathcal{L}^u V(x) + c(x,u)] = \varrho\,, \quad x \in \mathbb{R}^d\,. \tag{3.7.41}$$

Moreover,

$$V(x) \le \Psi^*(x;\varrho)\,, \qquad \varrho \le \varrho^*\,,$$

and for any $r > 0$, we have

$$V(x) \ge -\varrho^* \limsup_{\alpha\downarrow 0} \mathbb{E}^{v_\alpha}_x[\check{\tau}_r] - \sup_{B_r} V \qquad \forall x \in B^c_r\,, \tag{3.7.42}$$

where $\{v_\alpha : 0 < \alpha < 1\} \subset \mathfrak{U}_{\mathrm{sm}}$ is a collection of α-discounted optimal controls.

(ii) *If* $\hat{v} \in \mathfrak{U}_{\mathrm{ssm}}$ *and* $\varrho_{\hat{v}} < \infty$, *then there exist* $\hat{V} \in \mathscr{W}^{2,p}_{\mathrm{loc}}(\mathbb{R}^d)$, *for any* $p > 1$, *and* $\hat{\varrho} \in \mathbb{R}$, *satisfying*

$$\mathcal{L}^{\hat{v}}\hat{V} + c_{\hat{v}} = \hat{\varrho} \quad \text{in } \mathbb{R}^d ,$$

and such that $\alpha J^{\hat{v}}_\alpha(0) \to \hat{\varrho}$ *and* $J^{\hat{v}}_\alpha - J^{\hat{v}}_\alpha(0) \to \hat{V}$, *as* $\alpha \downarrow 0$, *uniformly on compact subsets of* $\mathbb{R}^d$. *Moreover,*

$$\hat{V}(x) = \Psi^{\hat{v}}(x;\hat{\varrho}) \quad \text{and} \quad \hat{\varrho} = \varrho_{\hat{v}} .$$

Proof By Theorem 3.7.6, $\alpha V_\alpha(0)$ are bounded, and $\bar{V}_\alpha = V_\alpha - V_\alpha(0)$ are bounded in $\mathscr{W}^{2,p}(B_R)$, $p > 1$, uniformly in $\alpha \in (0,1)$. Therefore we start with (3.5.10) and applying Lemma 3.5.4 we deduce that $\bar{V}_{\alpha_n}$ converges uniformly on any bounded domain, along some sequence $\alpha_n \downarrow 0$, to $V \in C^2(\mathbb{R}^d)$ satisfying (3.7.41), with ϱ being the corresponding limit of $\alpha_n V_{\alpha_n}(0)$.

We first show $\varrho \le \varrho^*$. Let $v_\varepsilon \in \mathfrak{U}_{\mathrm{ssm}}$ be an ε-optimal control and select $R \ge 0$ large enough such that $\eta_{v_\varepsilon}(B_R) \ge 1 - \varepsilon$. Since $V_\alpha \le J^{v_\varepsilon}_\alpha$, by integrating with respect to η_{v_ε} and using Fubini's theorem, we obtain

$$\Bigl(\inf_{B_R} V_\alpha\Bigr)\eta_{v_\varepsilon}(B_R) \le \int_{\mathbb{R}^d} V_\alpha(x)\,\eta_{v_\varepsilon}(\mathrm{d}x) \le \int_{\mathbb{R}^d} J^{v_\varepsilon}_\alpha(x)\,\eta_{v_\varepsilon}(\mathrm{d}x) \le \frac{\varrho^* + \varepsilon}{\alpha} .$$

Thus

$$\inf_{B_R} V_\alpha \le \frac{(\varrho^* + \varepsilon)}{\alpha(1-\varepsilon)} ,$$

and since $V_\alpha(0) - \inf_{B_R} V_\alpha$ is bounded uniformly in $\alpha \in (0,1)$, we obtain

$$\varrho \le \limsup_{\alpha \downarrow 0} \ \alpha V_\alpha(0) \le \frac{(\varrho^* + \varepsilon)}{(1-\varepsilon)} .$$

Since ε was arbitrary, $\varrho \le \varrho^*$.

Let $v_\alpha \in \mathfrak{U}_{\mathrm{sm}}$ be an α-discounted optimal control. For $v \in \tilde{\mathfrak{U}}_{\mathrm{ssm}}$, define the admissible control $U \in \mathfrak{U}$ by

$$U_t = \begin{cases} v & \text{if } t \le \breve{\tau}_r \wedge \tau_R , \\ v_\alpha & \text{otherwise.} \end{cases}$$

Since U is in general sub-optimal for the α-discounted criterion, using the strong Markov property relative to the stopping time $\breve{\tau}_r \wedge \tau_R$, we obtain for $x \in B_R \setminus \bar{B}_r$,

$$\begin{aligned} V_\alpha(x) &\le \mathbb{E}^U_x\left[\int_0^\infty \mathrm{e}^{-\alpha t}\bar{c}(X_t, U_t)\,\mathrm{d}t\right] \\ &= \mathbb{E}^v_x\left[\int_0^{\breve{\tau}_r \wedge \tau_R} \mathrm{e}^{-\alpha t} c_v(X_t)\,\mathrm{d}t + \mathrm{e}^{-\alpha(\breve{\tau}_r \wedge \tau_R)} V_\alpha(X_{\breve{\tau}_r \wedge \tau_R})\right] . \end{aligned} \tag{3.7.43}$$

Since $v \in \tilde{\mathfrak{U}}_{\mathrm{ssm}}$, applying Fubini's theorem, $\int_{\mathbb{R}^d} \alpha J^v_\alpha(x)\,\eta_v(\mathrm{d}x) = \varrho_v < \infty$. Hence, J^v_α is a.e. finite, and by Theorem A.3.7 and Remark A.3.8, since $V_\alpha \le J^v_\alpha$, we have

$$\mathbb{E}^v_x\Bigl[\mathrm{e}^{-\alpha\tau_R} V_\alpha(X_{\tau_R})\,\mathbb{I}\{\breve{\tau}_r \ge \tau_R\}\Bigr] \le \mathbb{E}^v_x[\mathrm{e}^{-\alpha\tau_R} J^v_\alpha(X_{\tau_R})] \xrightarrow[R\to\infty]{} 0 \tag{3.7.44}$$

for all $v \in \tilde{\mathfrak{U}}_{\mathrm{ssm}}$. Decomposing the term $\mathrm{e}^{-\alpha(\check{\tau}_r \wedge \tau_R)} V_\alpha(X_{\check{\tau}_r \wedge \tau_R})$ in (3.7.43), then taking limits as $R \to \infty$, applying (3.7.44) and monotone convergence, and subtracting $V_\alpha(0)$ from both sides of the inequality, we obtain

$$\bar{V}_\alpha(x) \le \mathbb{E}_x^v \left[\int_0^{\check{\tau}_r} \mathrm{e}^{-\alpha t} [c_v(X_t) - \alpha V_\alpha(0)] \, \mathrm{d}t \right] + \mathbb{E}_x^v \left[\mathrm{e}^{-\alpha \check{\tau}_r} \bar{V}_\alpha(X_{\check{\tau}_r}) \right] \tag{3.7.45}$$

for all $v \in \tilde{\mathfrak{U}}_{\mathrm{ssm}}$. We write (3.7.45) in the form

$$\begin{aligned} \bar{V}_\alpha(x) &\le \mathbb{E}_x^v \left[\int_0^{\check{\tau}_r} \mathrm{e}^{-\alpha t} c_v(X_t) \, \mathrm{d}t + \mathrm{e}^{-\alpha \check{\tau}_r} V_\alpha(X_{\check{\tau}_r}) - V_\alpha(0) \right] \\ &= \mathbb{E}_x^v \left[\int_0^{\check{\tau}_r} \mathrm{e}^{-\alpha t} (c_v(X_t) - \varrho) \, \mathrm{d}t \right] + \mathbb{E}_x^v \left[V_\alpha(X_{\check{\tau}_r}) - V_\alpha(0) \right] \\ &\qquad + \mathbb{E}_x^v \left[\alpha^{-1} (1 - \mathrm{e}^{-\alpha \check{\tau}_r}) [\varrho - \alpha V_\alpha(X_{\check{\tau}_r})] \right]. \end{aligned} \tag{3.7.46}$$

Since $|\varrho - \alpha V_\alpha(0)| \to 0$ as $\alpha \to 0$, and $V_\alpha - V_\alpha(0)$ is bounded on compact sets uniformly in $\alpha \in (0, 1)$ by Theorem 3.7.6, we obtain $\sup_{B_r} |\varrho - \alpha V_\alpha| \to 0$ as $\alpha \to 0$. Also,

$$\mathbb{E}_x^v \left[\alpha^{-1} (1 - \mathrm{e}^{-\alpha \check{\tau}_r}) \right] \le \mathbb{E}_x^v [\check{\tau}_r] \, .$$

Thus, letting $\alpha \to 0$ along the subsequence $\{\alpha_n\}$, (3.7.46) yields

$$V(x) \le \mathbb{E}_x^v \left[\int_0^{\check{\tau}_r} (c_v(X_t) - \varrho) \, \mathrm{d}t + V(X_{\check{\tau}_r}) \right] \qquad \forall v \in \tilde{\mathfrak{U}}_{\mathrm{ssm}} \, . \tag{3.7.47}$$

Since $V(0) = 0$,

$$\limsup_{r \downarrow 0} \sup_{v \in \tilde{\mathfrak{U}}_{\mathrm{ssm}}} \mathbb{E}_x^v [V(X_{\check{\tau}_r})] = 0 \, .$$

Therefore

$$V(x) \le \liminf_{r \downarrow 0} \inf_{v \in \mathfrak{U}_{\mathrm{ssm}}} \mathbb{E}_x^v \left[\int_0^{\check{\tau}_r} (c_v(X_t) - \varrho) \, \mathrm{d}t \right] .$$

Thus $V(x) \le \Psi^*(x; \varrho)$. In order to derive the lower bound for $\bar{V}$, we let $v = v_\alpha$ to obtain

$$\begin{aligned} \bar{V}_\alpha(x) &= \mathbb{E}_x^{v_\alpha} \left[\int_0^{\check{\tau}_r} \mathrm{e}^{-\alpha t} (c_{v_\alpha}(X_t) - \varrho) \, \mathrm{d}t \right] + \mathbb{E}_x^{v_\alpha} \left[\bar{V}_\alpha(X_{\check{\tau}_r}) \right] \\ &\qquad + \mathbb{E}_x^{v_\alpha} \left[\alpha^{-1} (1 - \mathrm{e}^{-\alpha \check{\tau}_r}) [\varrho - \alpha V_\alpha(X_{\check{\tau}_r})] \right], \quad x \in B_r^c . \end{aligned} \tag{3.7.48}$$

It follows by (3.7.48) that

$$\begin{aligned} \bar{V}_\alpha(x) &\ge -\varrho \, \mathbb{E}_x^{v_\alpha} [\check{\tau}_r] - \sup_{B_r} \bar{V}_\alpha - \left(\sup_{B_r} |\varrho - \alpha V_\alpha| \right) \mathbb{E}_x^{v_\alpha} [\check{\tau}_r] \\ &\ge -\left(\varrho^* + \sup_{B_r} |\varrho - \alpha V_\alpha| \right) \sup_{v \in \mathfrak{U}_{\mathrm{ssm}}} \mathbb{E}_x^v [\check{\tau}_r] - \sup_{B_r} \bar{V}_\alpha \qquad \forall x \in B_r^c . \end{aligned} \tag{3.7.49}$$

Taking limits as $\alpha \to 0$ in (3.7.49) along the convergent subsequence yields (3.7.42). This completes the proof of (i).

If $\varrho_{\hat{v}} < \infty$, then using the bounds in Theorem 3.7.4 along with Lemma 3.5.4 and Remark 3.5.5 it follows that $J^{\hat{v}}_{\alpha_n} - J^{\hat{v}}_{\alpha_n}(0)$ and $\alpha_n J^{\hat{v}}_{\alpha_n}(0)$ converge along some sequence $\alpha_n \to 0$ to $\hat{V}$ and $\hat{\varrho}$, respectively, satisfying $\mathcal{L}^{\hat{v}}\hat{V} + c_{\hat{v}} = \hat{\varrho}$. Since (3.7.45) and (3.7.46) hold with equality if we replace V_α with $J^{\hat{v}}_{\alpha_n}$, ϱ with $\hat{\varrho}$, and v with $\hat{v}$, letting $\alpha_n \to 0$, we obtain

$$\hat{V}(x) = \mathbb{E}^{\hat{v}}_x\left[\int_0^{\breve{\tau}_r} (c_{\hat{v}}(X_t) - \hat{\varrho})\,\mathrm{d}t + \hat{V}(X_{\breve{\tau}_r})\right]. \tag{3.7.50}$$

Taking limits in (3.7.50) as $r \to 0$, yields $\hat{V} = \Psi^{\hat{v}}(x;\hat{\varrho})$. By (3.7.50) and Corollary 3.7.3, we have

$$\lim_{t\to\infty} \frac{1}{t}\ \mathbb{E}^{\hat{v}}_x[\hat{V}(X_t)] \to 0\,.$$

Therefore, applying Itô's formula to $\mathcal{L}^{\hat{v}}\hat{V} + c_{\hat{v}} = \hat{\varrho}$, dividing by t, and taking limits as $t \to \infty$, yields $\hat{\varrho} = \varrho_{\hat{v}}$. This completes the proof of (ii). □

Definition 3.7.9 For $v \in \tilde{\mathfrak{U}}_{\mathrm{ssm}}$, we refer to the function $V(x) = \Psi^v(x;\varrho_v) \in \mathscr{W}^{2,p}_{\mathrm{loc}}(\mathbb{R}^d)$ as the *canonical solution* of the Poisson equation $\mathcal{L}^v V + c_v = \varrho_v$ in $\mathbb{R}^d$.

Theorems 3.7.11 and 3.7.12 below assume (3.7.1). In other words, we assume that $1 + c$ is uniformly integrable with respect to $\{\pi_v : v \in \mathfrak{U}_{\mathrm{ssm}}\}$. Note that Assumption 3.7.1 is equivalent to the statement that $v \mapsto \varrho_v$ is a continuous map from $\mathfrak{U}_{\mathrm{ssm}}$ to $\mathbb{R}_+$. Theorem 3.7.12 asserts the uniqueness of the solution of the HJB equation in a certain class of functions which is defined as follows:

Definition 3.7.10 Let $\mathscr{V}$ be the class of nonnegative functions $\mathcal{V} \in C^2(\mathbb{R}^d)$ satisfying (3.7.3) for some nonnegative, inf-compact function $h \in \mathfrak{C}$, with $1 + c \in \mathfrak{o}(h)$. We denote by $\mathfrak{o}(\mathscr{V})$ the class of functions V satisfying $V \in \mathfrak{o}(\mathcal{V})$ for some $\mathcal{V} \in \mathscr{V}$.

Theorem 3.7.11 *Assume* (3.7.1) *holds. Then the HJB equation*

$$\min_{u\in\mathbb{U}} \left[\mathcal{L}^u V(x) + c(x,u)\right] = \varrho\,, \quad x \in \mathbb{R}^d\,, \tag{3.7.51}$$

admits a solution $V^* \in C^2(\mathbb{R}^d) \cap \mathfrak{o}(\mathscr{V})$ satisfying $V^*(0) = 0$, and with $\varrho = \varrho^*$. Moreover, if $v^* \in \mathfrak{U}_{\mathrm{sm}}$ satisfies

$$b^i_{v^*}(x)\,\partial_i V^*(x) + c_{v^*}(x) = \min_{u\in\mathbb{U}} \left[b^i(x,u)\,\partial_i V^*(x) + c(x,u)\right] \qquad a.e., \tag{3.7.52}$$

then

$$\varrho_{v^*} = \varrho^* = \inf_{U\in\mathfrak{U}}\ \liminf_{T\to\infty}\ \frac{1}{T}\ \mathbb{E}^U_x\left[\int_0^T c\,(X_t, U_t)\,\mathrm{d}t\right]. \tag{3.7.53}$$

Proof The existence of a solution to (3.7.51) in $C^2(\mathbb{R}^d)$ with $\varrho \le \varrho^*$ is asserted by Lemma 3.7.8. Combining (3.7.42) and (3.7.47) we obtain

$$|V(x)| \le \sup_{v\in\mathfrak{U}_{\mathrm{ssm}}} \mathbb{E}^v_x\left[\int_0^{\breve{\tau}_r} (c_v(X_t) + \varrho^*)\,\mathrm{d}t\right] + \sup_{B_r} V \qquad \forall x \in B^c_r\,.$$

Hence $V \in \mathfrak{o}(\mathcal{V})$ by (3.7.4). Suppose $v^* \in \mathfrak{U}_{\mathrm{sm}}$ satisfies (3.7.52). By Dynkin's formula,

$$\begin{aligned}\mathbb{E}_x^{v^*}\left[V(X_{t\wedge\tau_R})\right] - V(x) &= \mathbb{E}_x^{v^*}\left[\int_0^{t\wedge\tau_R} \mathcal{L}^{v^*} V(X_s)\,\mathrm{d}s\right]\\ &= \mathbb{E}_x^{v^*}\left[\int_0^{t\wedge\tau_R} [\varrho - c_{v^*}(X_s)]\,\mathrm{d}s\right]. \end{aligned}\tag{3.7.54}$$

Taking limits as $R \to \infty$ in (3.7.54), by applying (3.7.6) to the left-hand side, decomposing the right-hand side as

$$\varrho\,\mathbb{E}_x^{v^*}\left[t \wedge \tau_R\right] - \mathbb{E}_x^{v^*}\left[\int_0^{t\wedge\tau_R} c_{v^*}(X_s)\,\mathrm{d}s\right],$$

and employing monotone convergence, we obtain

$$\mathbb{E}_x^{v^*}\left[V(X_t)\right] - V(x) = \mathbb{E}_x^{v^*}\left[\int_0^{t} [\varrho - c_{v^*}(X_s)]\,\mathrm{d}s\right]. \tag{3.7.55}$$

Dividing (3.7.55) by t, letting $t \to \infty$, and applying (3.7.5), we obtain $\varrho_{v^*} = \varrho$, which implies $\varrho^* \le \varrho$. Since $\varrho \le \varrho^*$, we have equality. One more application of Itô's formula to (3.7.51), relative to $U \in \mathfrak{U}$, yields (3.7.53). □

Concerning uniqueness of solutions to the HJB equation, the following applies:

Theorem 3.7.12 *Let V^* denote the solution of* (3.7.51) *obtained via the vanishing discount limit in Theorem 3.7.11. The following hold:*

(i) $V^*(x) = \Psi^*(x;\varrho^*) = \Psi^{\hat{v}}(x;\varrho^*)$, *for any average-cost optimal* $\hat{v} \in \mathfrak{U}_{\mathrm{ssm}}$*;*
(ii) $\hat{v} \in \mathfrak{U}_{\mathrm{ssm}}$ *is average-cost optimal if and only if it is a measurable selector from the minimizer:* $\min_{u\in\mathbb{U}}\left[\mathcal{L}^u V^*(x) + c(x,u)\right]$*;*
(iii) *if a pair* $(\tilde{V},\tilde{\varrho}) \in (C^2(\mathbb{R}^d) \cap \mathfrak{o}(\mathcal{V})) \times \mathbb{R}$ *satisfies* (3.7.51) *and* $\tilde{V}(0) = 0$, *then* $(\tilde{V},\tilde{\varrho}) = (V^*,\varrho^*)$.

Proof By Lemma 3.7.8 (i), since V^* is obtained as a limit of $\bar{V}_{\alpha_n}$ as $\alpha_n \to 0$, we have $V^* \le \Psi^*(x;\varrho)$, and by Theorem 3.7.11, $\varrho = \varrho^*$. Suppose $\hat{v} \in \mathfrak{U}_{\mathrm{ssm}}$ is average-cost optimal. Let $\hat{V} \in \mathscr{W}^{2,p}_{\mathrm{loc}}(\mathbb{R}^d)$, $p > 1$, be the canonical solution to the Poisson equation $\mathcal{L}^{\hat{v}}\hat{V} + c_{\hat{v}} = \varrho_{\hat{v}}$ in $\mathbb{R}^d$. By the optimality of $\hat{v}$, we have $\varrho_{\hat{v}} = \varrho^*$. Hence,

$$\mathcal{L}^{\hat{v}}(V^* - \hat{V}) \ge \varrho^* - \varrho_{\hat{v}} = 0$$

and

$$V^*(x) - \hat{V}(x) \le \Psi^*(x;\varrho^*) - \Psi^{\hat{v}}(x;\varrho^*) \le 0\,.$$

Since $V^*(0) = \hat{V}(0)$, the strong maximum principle yields $V^* = \hat{V}$. This completes the proof of (i) – (ii).

Now suppose $(\tilde{V},\tilde{\varrho}) \in (C^2(\mathbb{R}^d)\cap\mathfrak{o}(\mathcal{V}))\times\mathbb{R}$ is any solution of (3.7.51), and $\tilde{v} \in \mathfrak{U}_{\mathrm{ssm}}$ is an associated measurable selector from the minimizer. We apply Dynkin's formula and (3.7.6), since $\tilde{V} \in \mathfrak{o}(\mathcal{V})$, to obtain (3.7.55) with $\tilde{V}$, $\tilde{v}$, and $\tilde{\varrho}$ replacing V, v^*, and

ϱ, respectively. Dividing by t, and taking limits as $t \to \infty$, using (3.7.5), we obtain $\varrho_{\tilde{v}} = \tilde{\varrho}$. Therefore $\varrho^* \le \tilde{\varrho}$. One more application of Itô's formula to (3.7.51) relative to an average-cost optimal control $v^* \in \mathfrak{U}_{\mathrm{sm}}$ yields

$$\mathbb{E}_x^{v^*}\big[\tilde{V}(X_t)\big] - \tilde{V}(x) \ge \mathbb{E}_x^{v^*}\left[\int_0^t [\tilde{\varrho} - c_{v^*}(X_s)]\,\mathrm{d}s\right]. \tag{3.7.56}$$

Once more, dividing (3.7.56) by t, letting $t \to \infty$, and applying (3.7.5), we obtain $\tilde{\varrho} \le \varrho^*$. Thus $\tilde{\varrho} = \varrho^*$. Next we show that $\tilde{V} \ge \Psi^*(x;\varrho^*)$. For $x \in \mathbb{R}^d$, choose $R > r > 0$ such that $r < |x| < R$. Using (3.7.51) and Dynkin's formula,

$$\tilde{V}(x) = \mathbb{E}_x^{\tilde{v}}\left[\int_0^{\breve{\tau}_r \wedge \tau_R} (c_{\tilde{v}}(X_t) - \varrho^*)\,\mathrm{d}t + \tilde{V}(X_{\breve{\tau}_r})\,\mathbb{I}\{\breve{\tau}_r < \tau_R\}\right.$$
$$\left. + \tilde{V}(X_{\tau_R})\,\mathbb{I}\{\breve{\tau}_r \ge \tau_R\}\right]. \tag{3.7.57}$$

By (3.7.8),

$$\mathbb{E}_x^{v}\big[\mathcal{V}(X_{\tau_R})\,\mathbb{I}\{\tau_R \le \breve{\tau}_r\}\big] \le k_0\,\mathbb{E}_x^{v}[\breve{\tau}_r] + \mathcal{V}(x) \qquad \forall v \in \mathfrak{U}_{\mathrm{ssm}}. \tag{3.7.58}$$

Since $\tilde{V} \in \mathfrak{o}(\mathcal{V})$, (3.7.58) implies that

$$\sup_{v \in \mathfrak{U}_{\mathrm{ssm}}} \mathbb{E}_x^{v}\Big[\tilde{V}(X_{\tau_R})\,\mathbb{I}\{\tau_R \le \breve{\tau}_r\}\Big] \xrightarrow[R\to\infty]{} 0.$$

Hence, letting $R \to \infty$ in (3.7.57), and using the monotone convergence theorem and Fatou's lemma, we obtain

$$\begin{aligned}\tilde{V}(x) &\ge \mathbb{E}_x^{\tilde{v}}\left[\int_0^{\breve{\tau}_r} (c_{\tilde{v}}(X_t) - \varrho^*)\,\mathrm{d}t + \tilde{V}(X_{\breve{\tau}_r})\right] \\ &\ge \inf_{v \in \mathfrak{U}_{\mathrm{ssm}}} \mathbb{E}_x^{v}\left[\int_0^{\breve{\tau}_r} (c_v(X_t) - \varrho^*)\,\mathrm{d}t\right] + \inf_{B_r} \tilde{V}.\end{aligned}$$

Next, letting $r \to 0$ and using the fact that $\tilde{V}(0) = 0$ yields $\tilde{V} \ge \Psi^*(x;\varrho^*)$. It follows that $V^* - \tilde{V} \le 0$ and $\mathcal{L}^{\tilde{v}}(V^* - \tilde{V}) \ge 0$. Therefore, by the strong maximum principle, $\tilde{V} = V^*$. This completes the proof of (iii). □

Let $v \in \tilde{\mathfrak{U}}_{\mathrm{ssm}}$ and let $(\mathcal{V}, h) \in C^2(\mathbb{R}^d) \times \mathfrak{C}$ be a pair of inf-compact functions satisfying the Lyapunov equation $\mathcal{L}^v\mathcal{V}(x) \le k_0 - h_v(x)$, with $k_0 \in \mathbb{R}$. Theorem 3.7.12 implies that if $c \in \mathfrak{o}(h)$, then there exists a unique solution $V \in \mathscr{W}^{2,p}_{\mathrm{loc}}(\mathbb{R}^d) \cap \mathfrak{o}(\mathcal{V})$ to the Poisson equation $\mathcal{L}^v V + c_v = \varrho_v$ in $\mathbb{R}^d$, and V agrees with the canonical solution $\Psi^v(x;\varrho_v)$ in Definition 3.7.9.

In what follows, we relax Assumption 3.7.1. Provided $\mathfrak{U}_{\mathrm{sm}}$ is uniformly stable and $\varrho_v < \infty$ for all $v \in \mathfrak{U}_{\mathrm{ssm}}$, we establish the existence of an average-cost optimal control in $\mathfrak{U}_{\mathrm{ssm}}$.

Corollary 3.7.13 *Suppose that $\mathfrak{U}_{\mathrm{sm}}$ is uniformly stable and that $\varrho_v < \infty$ for all $v \in \mathfrak{U}_{\mathrm{sm}}$. Then the HJB equation in* (3.7.51) *admits a solution $V^* \in C^2(\mathbb{R}^d)$ and $\varrho \in \mathbb{R}$. Moreover, $\varrho = \varrho^*$, and any $v^* \in \mathfrak{U}_{\mathrm{sm}}$ is average-cost optimal if and only if it satisfies* (3.7.52).

Proof By Lemma 3.7.8 (i) (which does not use Assumption 3.7.1), we obtain a solution (V,ϱ) to (3.7.51), via the vanishing discount limit, satisfying $\varrho \le \varrho^*$ and $V(0) = 0$. Decomposing V into its positive and negative parts as $V = V^+ - V^-$ in (3.7.54), we obtain

$$-\mathbb{E}_x^{v^*}[V^-(X_{t\wedge\tau_R})] - V(x) \le \varrho\,\mathbb{E}_x^{v^*}[t\wedge\tau_R] - \mathbb{E}_x^{v^*}\left[\int_0^{t\wedge\tau_R} c_{v^*}(X_s)\,\mathrm{d}s\right]. \qquad (3.7.59)$$

By uniform stability there exist nonnegative inf-compact functions $\mathcal{V} \in C^2(\mathbb{R}^d)$ and $h\colon \mathbb{R}^d \to \mathbb{R}_+$, which is locally Lipschitz, and a constant k_0 satisfying

$$\max_{u\in\mathbb{U}} \mathcal{L}^u\mathcal{V}(x) \le k_0 - h(x) \qquad \forall x \in \mathbb{R}^d .$$

Since for any fixed $r > 0$, $\sup_{v\in\mathfrak{U}_{\mathrm{ssm}}} \mathbb{E}_x^v[\breve{\tau}_r] \in \mathfrak{o}(\mathcal{V})$, it follows by (3.7.42) that $V^- \in \mathfrak{o}(\mathcal{V})$. Therefore, if v^* is a measurable selector from the minimizer in (3.7.51), then by Lemma 3.7.2, we have

$$\lim_{R\to\infty} \mathbb{E}_x^{v^*}[V^-(X_{t\wedge\tau_R})] = \mathbb{E}_x^{v^*}[V^-(X_t)]$$

and

$$\lim_{t\to\infty} \frac{1}{t}\,\mathbb{E}_x^{v^*}[V^-(X_t)] = 0 .$$

Thus, dividing (3.7.59) by t and taking limits, first as $R \to \infty$ and then as $t \to \infty$, we obtain

$$0 \le \varrho - \limsup_{t\to\infty} \frac{1}{t}\,\mathbb{E}_x^{v^*}\left[\int_0^t c_{v^*}(X_s)\,\mathrm{d}s\right].$$

Hence $\varrho_{v^*} \le \varrho$, and it follows that $\varrho = \varrho^*$.

To prove the second assertion, suppose that $\hat{v} \in \mathfrak{U}_{\mathrm{sm}}$ is average-cost optimal in $\mathfrak{U}_{\mathrm{sm}}$. Let $\hat{V} = \Psi^{\hat{v}}(x;\varrho_{\hat{v}})$ be the canonical solution to the Poisson equation $\mathcal{L}^{\hat{v}}\hat{V} + c_{\hat{v}} = \varrho_{\hat{v}}$. By optimality $\varrho_{\hat{v}} = \varrho^*$. Thus

$$\mathcal{L}^{\hat{v}}(V^* - \hat{V}) \ge \varrho^* - \varrho_{\hat{v}} = 0 .$$

Also, by Lemma 3.7.8,

$$V^*(x) \le \Psi^*(x;\varrho^*) .$$

Hence $V^* - \hat{V} \le 0$, and since $V^*(0) = \hat{V}(0)$, the strong maximum principle yields $V^* = \hat{V}$. Thus $\mathcal{L}^{\hat{v}}V^* + c_{\hat{v}} = \varrho^*$, and the proof is complete. □

We can improve the results in Corollary 3.7.13 by relaxing the hypothesis that $\mathfrak{U}_{\mathrm{ssm}} = \tilde{\mathfrak{U}}_{\mathrm{ssm}}$ and thus allowing for the existence of stationary Markov controls that yield infinite average cost. In the proof of the theorem below we use the fact that, provided $\mathscr{G}$ is tight, and therefore also compact, the map $v \mapsto \pi_v$ from $\mathfrak{U}_{\mathrm{ssm}}$ to $\mathscr{G}$ is continuous by Lemma 3.2.6 (b). It follows that $v \to \varrho_v$ is lower-semicontinuous and therefore also inf-compact.

Theorem 3.7.14 *Suppose $\mathfrak{U}_{\mathrm{sm}}$ is uniformly stable. Then the HJB equation* (3.7.51) *admits a solution (V^*, ϱ^*) with $V^* \in C^2(\mathbb{R}^d)$. Moreover, a stationary Markov control is average-cost optimal if and only if it is a measurable selector from the minimizer*

$$\min_{u\in\mathbb{U}} \left[b^i(x,u)\,\partial_i V^*(x) + c(x,u)\right]. \tag{3.7.60}$$

Proof The existence of an optimal $\hat{v} \in \mathfrak{U}_{\mathrm{ssm}}$ is clear from the remarks in the paragraph preceding the statement of the theorem. For $n \in \mathbb{N}$ define the truncated running cost $c^n := \min\{c, n\}$. Since c^n is bounded and $\mathfrak{U}_{\mathrm{sm}}$ is uniformly stable, Assumption 3.7.1 holds. Hence by Theorem 3.7.11, for each $n \in \mathbb{N}$, there exist $(V_n, \varrho_n) \in C^2(\mathbb{R}^d) \times \mathbb{R}_+$, with $V_n(0) = 0$ which solve

$$\min_{u\in\mathbb{U}} \left[\mathcal{L}^u V_n(x) + c^n(x,u)\right] = \varrho_n\,, \quad x \in \mathbb{R}^d\,.$$

The conclusions of Theorem 3.7.12 also hold. It is evident that $\varrho_n \le \varrho^*$. Therefore, since also $\inf_{v\in\mathfrak{U}_{\mathrm{ssm}}} \eta_v(B_R) > 0$ for any $R > 0$ by uniform stability, Theorem 3.7.4 implies that there exists a constant $\tilde{C} = \tilde{C}(R)$, such that

$$\left\|V_n\right\|_{\mathscr{W}^{2,p}(B_R)} \le \tilde{C}(R) \qquad \forall n \in \mathbb{N}\,, \quad \forall R > 0\,.$$

Applying Lemma 3.5.4, we deduce that $V_n \to V^* \in C^2(\mathbb{R}^d)$, uniformly over compact sets, $\varrho_n \to \varrho$, as $n \to \infty$ (over some subsequence), and that (V^*, ϱ) solve (3.7.51). Clearly, $\varrho \le \varrho^*$.

On the other hand, by (3.7.42) we have

$$V_n(x) \ge -\varrho_n \sup_{v\in\mathfrak{U}_{\mathrm{ssm}}} \mathbb{E}^v_x[\breve{\tau}_r] - \sup_{B_r} V_n \qquad \forall x \in B^c_r\,, \quad \forall n \in \mathbb{N}\,.$$

Taking limits as $n \to \infty$, it follows that V^* satisfies

$$V^*(x) \ge -\varrho^* \sup_{v\in\mathfrak{U}_{\mathrm{ssm}}} \mathbb{E}^v_x[\breve{\tau}_r] - \sup_{B_r} V^* \qquad \forall x \in B^c_r\,, \quad \forall n \in \mathbb{N}\,,$$

and as shown in the proof of Corollary 3.7.13 this implies that $\varrho = \varrho^*$, and that any measurable selector from the minimizer of (3.7.60) is average-cost optimal. It is also straightforward to show that $V^*(x) \le \Psi^*(x; \varrho^*)$. Therefore, any average-cost optimal stationary Markov control is necessarily a measurable selector from the minimizer in (3.7.60), as shown in the last paragraph of the proof of Corollary 3.7.13. □

By Lemma 3.7.8, provided $\varrho^* < \infty$, there always exists a solution $V \in C^2(\mathbb{R}^d)$ to the HJB equation (3.7.51) with $\varrho < \varrho^*$. It also follows from the proof of Corollary 3.7.13 that if

$$\liminf_{t\to\infty} \frac{1}{t}\, \mathbb{E}^{v^*}_x\left[V^-(X_t)\right] = 0\,,$$

where $v^* \in \mathfrak{U}_{\mathrm{sm}}$ satisfies (3.7.52), then $\varrho = \varrho^*$ and, provided v^* is stable, it is average-cost optimal. Conversely, as shown in the last paragraph of the proof of Corollary 3.7.13, if $v^* \in \mathfrak{U}_{\mathrm{ssm}}$ is average-cost optimal, then it necessarily satisfies (3.7.52). As a result, we can relax the hypothesis that $\mathfrak{U}_{\mathrm{sm}} = \mathfrak{U}_{\mathrm{ssm}}$ and assert the following.

Theorem 3.7.15 *Suppose that $\varrho^* < \infty$, and that over some sequence $\{\alpha_n\} \subset (0,1)$, converging to 0 as $n \to \infty$, the functions $\bar{V}_{\alpha_n}$, $n \in \mathbb{N}$, are all bounded below by some constant M_0. Then the HJB equation in* (3.7.51) *admits a solution $\varrho = \varrho^*$ and $V^* \in C^2(\mathbb{R}^d)$, and a stable stationary Markov control v^* is average-cost optimal if and only if it satisfies* (3.7.52).

Concerning the optimality of the control $v^* \in \mathfrak{U}_{\mathrm{ssm}}$ in Theorem 3.7.15 a standard application of a Tauberian theorem, which asserts that for all $U \in \mathfrak{U}$

$$\begin{aligned}\varrho_{v^*} = \varrho^* &= \limsup_{n\to\infty}\ \alpha_n V_{\alpha_n}(x)\\ &\le \limsup_{n\to\infty}\ \alpha_n \int_0^\infty \mathrm{e}^{-\alpha_n t}\,\mathbb{E}_x^U\,[\bar{c}(X_t,U_t)]\,\mathrm{d}t\\ &\le \limsup_{T\to\infty}\ \frac{1}{T}\int_0^T \mathbb{E}_x^U\,[\bar{c}(X_t,U_t)]\,\mathrm{d}t\,,\end{aligned}$$

shows that v^* is in fact average-cost optimal over all $\mathfrak{U}$. Compare this with the stronger form of optimality in (3.7.40) which holds under the hypothesis that $\mathfrak{U}_{\mathrm{sm}}$ is uniformly stable.

Remark 3.7.16 The hypothesis that the running cost is bounded below can be relaxed. If this is the case, we modify Assumption 3.7.1 by replacing c with $|c|$ in the integral of (3.7.1). Then the conclusions of Theorems 3.7.11 and 3.7.12 continue to hold. More general statements are also possible. For example, if we require that c^- satisfies Assumption 3.7.1, then the conclusions of Corollary 3.7.13 and Theorem 3.7.14 hold.

3.7.4 The HJB equation under a strong stability condition

We investigate the HJB equation (3.7.51) under the Lyapunov stability condition (L2.3) on p. 61. We need the following lemma.

Lemma 3.7.17 *Let $\mathcal{V}$ satisfy* (2.5.5). *Then for any $v \in \mathfrak{U}_{\mathrm{sm}}$, and any bounded domain D,*

$$\int_{\mathbb{R}^d} \mathcal{V}(x)\,\eta_v(\mathrm{d}x) \le \frac{k_0}{2k_1}\,, \tag{3.7.61a}$$

$$\lim_{t\to\infty} \frac{1}{t}\ \mathbb{E}_x^v[\mathcal{V}(X_t)] = 0 \qquad \forall x \in \mathbb{R}^d\,, \tag{3.7.61b}$$

and

$$\mathbb{E}_x^v\left[\int_0^{\tau(D^c)} \mathcal{V}(X_t)\,\mathrm{d}t\right] \in \mathscr{O}(\mathcal{V})\,. \tag{3.7.61c}$$

Proof Dividing (2.5.11) in Lemma 2.5.5 by t and taking limits as $t \to \infty$ yields (3.7.61b). Also, integrating (2.5.11) over $[0,t]$, dividing by t, and taking limits as

$t \to \infty$ yields (3.7.61a). With $R > 0$ large enough so that $x \in B_R$, applying Dynkin's formula, we obtain

$$\mathbb{E}_x^v[\mathcal{V}(X_{\tau(D^c)\wedge\tau_R})] - \mathcal{V}(x) \le \mathbb{E}_x^v\left[\int_0^{\tau(D^c)\wedge\tau_R} (k_0 - 2k_1\mathcal{V}(X_t))\,\mathrm{d}t\right].$$

Therefore

$$\mathbb{E}_x^v\left[\int_0^{\tau(D^c)\wedge\tau_R} \mathcal{V}(X_t)\,\mathrm{d}t\right] \le \frac{1}{2k_1}\mathcal{V}(x) + \frac{k_0}{2k_1}\,\mathbb{E}_x^v[\tau(D^c)\wedge\tau_R]. \tag{3.7.62}$$

It is evident that $\mathbb{E}_x^v[\tau(D^c)] \in \mathscr{O}(\mathcal{V})$. Therefore, taking limits as $R \to \infty$ in (3.7.62), (3.7.61c) follows. □

Suppose that

$$x \mapsto \max_{u\in\mathbb{U}} c(x,u) \in \mathscr{O}(\mathcal{V}).$$

It follows by Lemma 3.7.17 and Theorems 3.7.11–3.7.12 that the HJB equation (3.7.51) admits a unique solution $V^* \in C^2(\mathbb{R}^d)\cap\mathscr{O}(\mathcal{V})$ and $\varrho \in \mathbb{R}$, satisfying $V(0) = 0$. Moreover, V^* has the representation in Theorem 3.7.12 (i) and the necessary and sufficient condition for optimality in Theorem 3.7.12 (ii) holds.

3.7.5 Storage functions

In this section, we point out connections between the foregoing theory and the notion of storage functions (to be precise, a stochastic counterpart thereof) arising in the theory of dissipative dynamical systems [120].

Definition 3.7.18 A measurable function $V\colon \mathbb{R}^d \to \mathbb{R}$ is said to be a storage function associated with a supply rate function $g \in C(\mathbb{R}^d\times\mathbb{U})$ if it is bounded from below and $V(X_t)+\int_0^t \bar{g}(X_s,U_s)\,\mathrm{d}s$, $t \ge 0$, is an $(\mathfrak{F}_t^X)$-supermartingale for all (X,U) satisfying (2.2.1).

The storage function need not be unique. For example, we get another by adding a constant. Suppose that $g \in C(\mathbb{R}^d \times \mathbb{U})$ is such that the map

$$x \mapsto \inf_{t>0}\ \sup_{U\in\mathfrak{U}}\ \mathbb{E}_x^U\left[\int_0^t \bar{g}(X_s,U_s)\,\mathrm{d}s\right]$$

is bounded below in $\mathbb{R}^d$. We define

$$V_g(x) := \liminf_{t\to\infty}\ \sup_{U\in\mathfrak{U}}\ \mathbb{E}_x^U\left[\int_0^t \bar{g}(X_s,U_s)\,\mathrm{d}s\right], \quad x \in \mathbb{R}^d.$$

Lemma 3.7.19 *If V_g is finite, then it is a storage function.*

Proof For $t \geq 0$,

$$\begin{aligned} V_g(x) &= \liminf_{T\to\infty} \sup_{U\in\mathfrak{U}} \mathbb{E}_x^U\left[\int_0^T \bar{g}(X_s, U_s)\,\mathrm{d}s\right] \\ &= \liminf_{T\to\infty} \sup_{U_{[0,t]}} \mathbb{E}_x^U\left[\int_0^t \bar{g}(X_s, U_s)\,\mathrm{d}s + \sup_{U_{t+\cdot}} \mathbb{E}_{X_t}^U\left[\int_t^T \bar{g}(X_s, U_s)\,\mathrm{d}s\right]\right] \\ &\geq \sup_{U_{[0,t]}} \mathbb{E}_x^U\left[\int_0^t \bar{g}(X_s, U_s)\,\mathrm{d}s + \liminf_{T\to\infty} \sup_{U_{t+\cdot}} \mathbb{E}_{X_t}^U\left[\int_t^T \bar{g}(X_s, U_s)\,\mathrm{d}s\right]\right] \\ &= \sup_{U\in\mathfrak{U}} \mathbb{E}_x^U\left[\int_0^t \bar{g}(X_s, U_s)\,\mathrm{d}s + V_g(X_t)\right], \end{aligned} \tag{3.7.63}$$

where the second equality in (3.7.63) follows by a standard dynamic programming argument, and the inequality by Fatou's lemma. Let $0 \leq r \leq t$. Since by Lemma 2.3.7 the regular conditional law of $(X \circ \theta_r, U \circ \theta_r)$ given $\mathfrak{F}_r^X$ is again the law of a pair (X, U) satisfying (2.2.1) with initial condition X_r, a.s., using (3.7.63) we obtain

$$\begin{aligned} \mathbb{E}_x^U\left[\int_0^t \bar{g}(X_s, U_s)\,\mathrm{d}s + V_g(X_t)\,\middle|\,\mathfrak{F}_r^X\right] &= \int_0^r \bar{g}(X_s, U_s)\,\mathrm{d}s \\ &\quad + \mathbb{E}_{X_r}^U\left[\int_r^t \bar{g}(X_s, U_s)\,\mathrm{d}s + V_g(X_t)\right] \\ &\leq \int_0^r \bar{g}(X_s, U_s)\,\mathrm{d}s + V_g(X_r)\,. \end{aligned}$$

It follows that

$$\int_0^t \bar{g}(X_s, U_s)\,\mathrm{d}s + V_g(X_t)\,, \quad t \geq 0\,,$$

is an $(\mathfrak{F}_t^X)$-supermartingale. Since also by definition V_g is bounded below, it is a storage function. □

We now discuss an ergodic control problem which involves maximization of the average reward. Let $h\colon \mathbb{R}^d \times \mathbb{U}$ be a nonnegative, continuous function, which is locally Lipschitz in its first argument uniformly in the second. We wish to maximize over all admissible $U \in \mathfrak{U}$ the reward

$$\liminf_{t\to\infty} \frac{1}{t} \int_0^t \mathbb{E}^U[\bar{h}(X_s, U_s)]\,\mathrm{d}s\,. \tag{3.7.64}$$

Let k^* denote the supremum of (3.7.64), and suppose k^* is finite. Under suitable conditions, this problem has an optimal stationary solution.

Definition 3.7.20 A measurable map $\psi\colon \mathbb{R}^d \to \mathbb{R}$ is said to be a value function for the optimal average reward problem in (3.7.64) if for all (X, U) satisfying (2.2.1), the process

$$\psi(X_t) + \int_0^t [\bar{h}(X_s, U_s) - k^*]\,\mathrm{d}s\,, \quad t \geq 0\,,$$

is an $(\mathfrak{F}_t^X)$-supermartingale and is a martingale if and only if (X, U) is an optimal pair.

Proving a martingale dynamic programming principle amounts to exhibiting a value function ψ. The following lemmas establish the link with storage functions.

Lemma 3.7.21 *If $\psi \geq 0$ is as in Definition 3.7.20, then ψ is a storage function for $g = h - k^*$.*

This is immediate from the definition. Note that if ψ is a value function, so is $\psi + C$ for any scalar C. In particular, there exists a nonnegative value function whenever there exists one that is bounded from below.

Going in the other direction, we have

Lemma 3.7.22 *If V_g is finite a.e. for $g = h - k^*$,*

$$V_g(X_t) + \int_0^t [\bar{h}(X_s, U_s) - k^*]\, \mathrm{d}s\,, \quad t \geq 0\,, \tag{3.7.65}$$

is an $(\mathfrak{F}_t^X)$-supermartingale for all (X, U) as in (2.2.1). Moreover, provided (X, U) is a stationary solution and $V_g(X_t)$ is integrable under this stationary law, then (X, U) is optimal if and only if the process in (3.7.65) is a martingale.

Proof The first claim is immediate. For stationary optimal (X, U),

$$\mathbb{E}\left[V_g(X_0)\right] \geq \int_0^t \mathbb{E}[\bar{h}(X_s, U_s) - k^*]\, \mathrm{d}s + \mathbb{E}\left[V_g(X_t)\right].$$

Hence

$$0 \geq \mathbb{E}\,[\bar{h}(X_s, U_s)] - k^*\,.$$

But since (X, U) is stationary, the corresponding reward

$$\limsup_{T\to\infty} \frac{1}{T} \int_0^T \mathbb{E}\,[\bar{h}(X_s, U_s)]\, \mathrm{d}s$$

in fact equals $\mathbb{E}\,[\bar{h}(X_s, U_s)]$. Since it is optimal, this equals k^*, so equality must hold throughout, which is possible only if

$$V_g(X_t) + \int_0^t [\bar{h}(X_s, U_s) - k^*]\, \mathrm{d}s\,, \quad t \geq 0\,,$$

is in fact an $(\mathfrak{F}_t^X)$-martingale. The converse easily follows by the same arguments. □

Now suppose that the Lyapunov stability condition (L2.3) on p. 61 holds, and that $h \in \mathfrak{C}$ (see Assumption A3.5.1) is an inf-compact function which satisfies

$$x \mapsto \max_{u \in \mathbb{U}} h(x, u) \in \mathscr{O}(\mathcal{V})\,. \tag{3.7.66}$$

By Corollary 3.3.2,

$$k^* := \sup_{v \in \mathfrak{U}_{\mathrm{sm}}} \int_{\mathbb{R}^d \times \mathbb{U}} h(x, u)\, \pi_v(\mathrm{d}x, \mathrm{d}u) < \infty\,. \tag{3.7.67}$$

Then by Theorem 3.7.4, for each fixed $\alpha > 0$, J^v_α is bounded in $\mathscr{W}^{2,p}(B_R)$, $p > 1$, for each $R > 0$, uniformly over $v \in \mathfrak{U}_{\mathrm{ssm}}$. Thus we can use Lemma 3.5.4 in the same manner that we did for the minimization problem to show that the function $F_\alpha(x) := \sup_{v\in\mathfrak{U}_{\mathrm{ssm}}} J^v_\alpha(x)$ satisfies

$$\max_{u\in\mathbb{U}} \left[\mathcal{L}^u F_\alpha(x) + h(x,u)\right] = \alpha F_\alpha(x)\,. \tag{3.7.68}$$

Moreover, F_α is the minimal nonnegative solution of (3.7.68) and $v \in \mathfrak{U}_{\mathrm{sm}}$ is optimal for the maximization problem if and only if it is a measurable selector a.e. from the maximizer in (3.7.68). On the other hand, again by Theorem 3.7.4, the function $\bar F_\alpha := F_\alpha - F_\alpha(0)$ is bounded in $\mathscr{W}^{2,p}(B_R)$, $p > 1$, and $x \mapsto \alpha F_\alpha(x)$ is bounded in B_R, uniformly in $\alpha \in (0,1)$ for any $R > 0$. Passing to the limit in (3.7.68) as $\alpha \downarrow 0$ along a subsequence we obtain a solution $V \in C^2(\mathbb{R}^d)$ to the equation

$$\max_{u\in\mathbb{U}} \left[\mathcal{L}^u V(x) + h(x,u)\right] = k \tag{3.7.69}$$

by Lemma 3.5.4. In this case $k \geq k^*$. As in the proof of Lemma 3.7.8, for any $r > 0$, we have

$$\begin{aligned}\bar F_\alpha(x) = \mathbb{E}^{v_\alpha}_x\left[\int_0^{\breve\tau_r} \mathrm{e}^{-\alpha t}\left(h_{v_\alpha}(X_t) - k\right)\mathrm{d}t\right] + \mathbb{E}^{v_\alpha}_x\left[\bar F_\alpha(X_{\breve\tau_r})\right]\\ + \mathbb{E}^{v_\alpha}_x\left[\alpha^{-1}(1-\mathrm{e}^{-\alpha\breve\tau_r})\left[k - \alpha F_\alpha(X_{\breve\tau_r})\right]\right], \quad x \in B^c_r\,. \end{aligned}\tag{3.7.70}$$

Taking absolute values in (3.7.70), then strengthening the equality to an inequality and passing to the limit as $\alpha \to 0$, we obtain

$$\left|V(x)\right| \leq \sup_{v\in\mathfrak{U}_{\mathrm{sm}}} \mathbb{E}^v_x\left[\int_0^{\breve\tau_r} \left(h_{v_\alpha}(X_t) + k\right)\mathrm{d}t\right] + \sup_{B_r} V\,, \quad x \in B^c_r\,. \tag{3.7.71}$$

Since $h \in \mathscr{O}(\mathcal{V})$, it follows by (3.7.61c) and (3.7.71) that $V \in \mathscr{O}(\mathcal{V})$. Thus, following the method in the proof of Theorems 3.7.11–3.7.12, using (3.7.61b) and (3.7.68), we verify that completely analogous optimality results hold for the maximization problem. It is also clear that V is bounded below in $\mathbb{R}^d$. Indeed, one can either repeat the argument in the proof of Lemma 3.6.1 to show that

$$\operatorname*{Arg\,min}_{\mathbb{R}^d}\ F_\alpha \subset \left\{x \in \mathbb{R}^d : \min_{u\in\mathbb{U}}\ h(x,u) \leq k^*\right\},$$

or choose to argue that if $r > 0$ is large enough so that $\min_{u\in\mathbb{U}}\ h(x,u) > k^*$ for $x \in B^c_r$, then the first term on the right-hand side of (3.7.70) is bounded below by the nonnegative inf-compact function

$$x \mapsto \inf_{v\in\mathfrak{U}_{\mathrm{sm}}} \mathbb{E}^v_x\left[\int_0^{\breve\tau_r} \mathrm{e}^{-t}\left(h_v(X_t) - k^*\right)\mathrm{d}t\right]$$

for all $\alpha \in (0,1)$, while the other two are bounded uniformly in $\alpha \in (0,1)$. As a result, V is inf-compact. It follows that V is a storage function with supply rate

$g = h - k^*$. Let v^* be a measurable selector from the maximizer in (3.7.68). Then since $V \in \mathcal{O}(\mathcal{V})$, it holds that $\int V \, \mathrm{d}\eta_{v^*} < \infty$ by (3.7.68). Let

$$K_0 := \limsup_{t\to\infty} \inf_{U\in\mathfrak{U}} \mathbb{E}_x^U \left[V(X_t)\right] .$$

The constant K_0 is finite by (2.5.11). Using these properties and (3.7.69) it easily follows that V_g is well defined and satisfies

$$V(x) - \int V \, \mathrm{d}\eta_{v*} \le V_g(x) \le V(x) - K_0 , \quad x \in \mathbb{R}^d . \tag{3.7.72}$$

Hence $V_g \in \mathcal{O}(\mathcal{V})$. Let $v \in \mathfrak{U}_{\mathrm{sm}}$ be a measurable selector from the minimizer in (3.7.69). Since for the controlled diffusion under v^* with $X_0 = x \in \mathbb{R}^d$ the process $V(X_t)$ is an $(\mathfrak{F}_t^X)$-martingale, while $V_g(X_t)$ is an $(\mathfrak{F}_t^X)$-supermartingale, it follows that $M_t := V(X_t) - V_g(X_t)$ is an $(\mathfrak{F}_t^X)$-submartingale. Consequently, since it is bounded by (3.7.72), M_t converges a.s. as $t \to \infty$. Since $V(x) - V_g(x)$ is integrable under the invariant probability distribution η_{v^*}, the limit must be a constant a.e. Therefore V and V_g differ by a constant a.e. We have proved the following.

Theorem 3.7.23 *Let the Lyapunov stability condition (L2.3) hold, and let $h \in \mathfrak{C}$ be an inf-compact function satisfying* (3.7.66). *Define k^* by* (3.7.67) *and set $g = h - k^*$. Then $V_g - \min_{\mathbb{R}^d} V_g$ is a.e. equal to the minimal nonnegative solution in $C^2(\mathbb{R}^d)$ of* (3.7.69).

3.8 One-dimensional controlled diffusions

In this section we specialize the results of the previous sections to one-dimensional diffusions. We exhibit explicit solutions for the optimal control when $d = 1$.

For $v \in \mathfrak{U}_{\mathrm{ssd}}$, let φ_v denote the density of the corresponding invariant probability measure η_v. Then φ_v is the unique solution of

$$\tfrac{1}{2}(\sigma^2(x)\varphi_v(x))'' - (b_v(x)\varphi_v(x))' = 0 ,$$
$$\varphi_v \ge 0 , \qquad \int_{-\infty}^{\infty} \varphi_v(x) \, \mathrm{d}x = 1 .$$

If we define

$$\begin{aligned} \beta_v(x) &:= \exp\left(\int_0^x \frac{2b_v(y)}{\sigma^2(y)} \, \mathrm{d}y\right) \\ K_v &:= \int_{-\infty}^{\infty} \frac{\beta_v(x)}{\sigma^2(x)} \, \mathrm{d}x , \end{aligned} \tag{3.8.1}$$

φ_v takes the form

$$\varphi_v(x) = \frac{\beta_v(x)}{K_v \sigma^2(x)} .$$

Also, a stationary Markov control v is stable if and only if $K_v < \infty$.

Let the cost function $c\colon \mathbb{R}\times\mathbb{U}\to\mathbb{R}$, satisfy the near-monotone Assumption 3.4.2. Recall that for $v\in\mathfrak{U}_{\mathrm{ssd}}$,

$$\varrho_v = \int_{\mathbb{R}} c_v(x)\,\eta_v(\mathrm{d}x)$$

and

$$\varrho^* = \inf_{v\in\mathfrak{U}_{\mathrm{ssd}}} \varrho_v\,.$$

Let $v\in\mathfrak{U}_{\mathrm{ssd}}$ and X the corresponding state process. For $x_0\in\mathbb{R}$, let

$$\breve{\tau}_{x_0} = \inf\,\{t\ge 0 : X_t = x_0\}\,. \tag{3.8.2}$$

Lemma 3.8.1 *For $x < x_0$,*

$$\mathbb{E}^v_x\,[\breve{\tau}_{x_0}] = 2\int_x^{x_0} \frac{\mathrm{d}y}{\beta_v(y)}\left[\int_{-\infty}^{y} \frac{\beta_v(z)}{\sigma^2(z)}\,\mathrm{d}z\right], \tag{3.8.3}$$

while for $x > x_0$,

$$\mathbb{E}^v_x\,[\breve{\tau}_{x_0}] = 2\int_{x_0}^{x} \frac{\mathrm{d}y}{\beta_v(y)}\left[\int_{y}^{\infty} \frac{\beta_v(z)}{\sigma^2(z)}\,\mathrm{d}z\right]. \tag{3.8.4}$$

Proof For $y_0 < x \le x_0$ or $x_0 \le x < y_0$ let

$$g_{y_0}(x) := \mathbb{E}^v_x\left[\breve{\tau}_{x_0}\wedge\breve{\tau}_{y_0}\right].$$

Then $g_{y_0}(x)$ is the unique solution to

$$\begin{aligned}\tfrac{1}{2}\sigma^2(x)g''_{y_0}(x) + b_v(x)g'_{y_0}(x) &= -1\,,\\ g_{y_0}(x_0) = g_{y_0}(y_0) &= 0\,.\end{aligned} \tag{3.8.5}$$

Solving (3.8.5) and letting $y_0\to-\infty$ or $y_0\to\infty$, we obtain (3.8.3) or (3.8.4), respectively. □

The HJB equation when $d = 1$ simplifies to the following second order nonlinear ordinary differential equation:

$$\tfrac{1}{2}\sigma^2(x)V''(x) + \min_{u\in\mathbb{U}}\,[b(x,u)V'(x) + c(x,u)] = \varrho\,. \tag{3.8.6}$$

Let

$$G := \left\{V\in C^2(\mathbb{R}) : \inf_{\mathbb{R}} V > -\infty\,,\; V(0) = 0\right\}$$

and

$$H := G\times(-\infty,\varrho^*]\,.$$

Theorem 3.8.2 *Under the near-monotone Assumption 3.4.2,* (3.8.6) *has a unique solution (V,ϱ) in the class H. This solution satisfies $\varrho = \varrho^*$. Moreover, a control $v\in\mathfrak{U}_{\mathrm{sm}}$ is optimal if and only if*

$$b_v(x)V'(x) + c_v(x) = \min_{u\in\mathbb{U}}\,[b(x,u)V'(x) + c(x,u)]\,.$$

Proof Let v^* be a stable stationary Markov control which is optimal. With $\beta \equiv \beta_{v^*}$, define

$$V(x) := \begin{cases} 2\int_x^0 \frac{\mathrm{d}y}{\beta(y)} \left[\int_{-\infty}^y \frac{\beta(z)}{\sigma^2(z)}(c_{v^*}(z) - \varrho^*)\,\mathrm{d}z\right] & \text{if } x \le 0\,, \\ 2\int_0^x \frac{\mathrm{d}y}{\beta(y)} \left[\int_y^{\infty} \frac{\beta(z)}{\sigma^2(z)}(c_{v^*}(z) - \varrho^*)\,\mathrm{d}z\right] & \text{if } x \ge 0\,. \end{cases}$$

By direct verification we can show that V'' is continuous and satisfies

$$\tfrac{1}{2}\sigma^*(x)V''(x) + b_{v^*}(x)V'(x) + c_{v^*}(x) = \varrho^*\,.$$

Then by the Dynkin formula, with $\check{\tau}_0$ as defined in (3.8.2), V has the stochastic representation

$$V(x) = \mathbb{E}_x^{v^*}\left[\int_0^{\check{\tau}_0} [c_{v^*}(X_t) - \varrho^*]\,\mathrm{d}t\right]. \tag{3.8.7}$$

Clearly, $V(0) = 0$. In order to show that $V \in G$ it suffices to prove that $\inf_{\mathbb{R}} V > -\infty$. Choose $\varepsilon > 0$ and $R_\varepsilon > 0$ such that

$$\varepsilon < \liminf_{|x|\to\infty} \min_{u\in\mathbb{U}} c(x,u) - \varrho^*\,,$$

and

$$\varepsilon < \min_{u\in\mathbb{U}} c(x,u) - \varrho^*\,, \quad \text{if } |x| > R_\varepsilon\,.$$

Let

$$\hat{\tau} := \inf\{t \ge 0 : |X_t| \le R_\varepsilon\}\,.$$

Then, for $|x| > R_\varepsilon$, by decomposing the integral in (3.8.7) and applying the strong Markov property, we obtain

$$V(x) \ge \mathbb{E}_x^{v^*}\left[\int_0^{\hat{\tau}} [c_{v^*}(X_t) - \varrho^*]\,\mathrm{d}t\right] + \inf_{|x|=R_\varepsilon} \mathbb{E}_x^{v^*}\left[\int_0^{\check{\tau}_0} [c_{v^*}(X_t) - \varrho^*]\,\mathrm{d}t\right]. \tag{3.8.8}$$

Since v^* is stable, the second term in the right-hand side of (3.8.8) is bounded uniformly in $x \in \mathbb{R}$. Also, the first term is bounded below by $\varepsilon\,\mathbb{E}_x^{v^*}[\hat{\tau}]$. Also, for any $T > 0$,

$$\begin{aligned} \mathbb{P}_x^{v^*}(\hat{\tau} \le T) &\le \mathbb{P}_x^{v^*}\left(\sup_{0\le t\le T} |X_t - x|^2 \ge (|x| - R_\varepsilon)^2\right) \\ &\le \frac{C_T}{(|x| - R_\varepsilon)^2} \xrightarrow[|x|\to\infty]{} 0 \end{aligned}$$

for some constant C_T, thus implying, since T is arbitrary,

$$\liminf_{|x|\to\infty} \mathbb{E}_x^{v^*}[\hat{\tau}] = \infty\,.$$

Therefore $V(x) \to \infty$ as $|x| \to \infty$, and in particular $\inf_{\mathbb{R}} V > -\infty$. □

3.8.1 Non-uniqueness of solutions of the HJB equation

We show via an example that under the near-monotone hypothesis the HJB equation can in general admit multiple solutions (V, ϱ) with V bounded below and $\varrho > \varrho^*$, even though the Markov controls that are selectors from the corresponding minimizers are all stable.

Example 3.8.3 Consider the controlled stochastic differential equation

$$\mathrm{d}X_t = U_t\,\mathrm{d}t + \mathrm{d}W_t\,, \qquad X_0 = x\,,$$

where $U_t \in [-1, 1]$ is the control variable. Let $c(x) = 1 - \mathrm{e}^{-|x|}$ be the running cost function. Clearly, the near-monotone property holds. The HJB equation (3.8.6), reduces to

$$\tfrac{1}{2}V''(x) + \min_u \{uV'(x)\} + c(x) = \varrho\,,$$

which is equivalent to

$$\tfrac{1}{2}V''(x) - |V'(x)| + c(x) = \varrho\,. \tag{3.8.9}$$

Let $\hat{c}$ be defined by

$$\hat{c}(x) = 2\int_{-\infty}^{x} \mathrm{e}^{2(y-x)} c(y)\,\mathrm{d}y\,, \qquad x \in \mathbb{R}\,.$$

The following properties can be easily verified:

(i) $\hat{c}(-x) > c(x) > \hat{c}(x)$ for all sufficiently large positive x.

(ii) $\hat{c}'(x) = 2[c(x) - \hat{c}(x)]$.

(iii) $\hat{c}(x) = \begin{cases} 1 - \frac{2}{3}\mathrm{e}^{x} & \text{if } x \le 0\,, \\ \frac{1}{3} & \text{if } x = 0\,, \\ 1 - 2\mathrm{e}^{-x} + \frac{4}{3}\mathrm{e}^{-2x} & \text{if } x > 0\,. \end{cases}$

(iv) $c\left(\log \frac{4}{3}\right) = \hat{c}\left(\log \frac{4}{3}\right)$. Also, $\hat{c}$ is strictly decreasing in $\left(-\infty, \log \frac{4}{3}\right)$ and satisfies $\hat{c} > c$, and it is strictly increasing in $\left(\log \frac{4}{3}, \infty\right)$, satisfying $\hat{c} < c$.

(v) $\lim_{|x|\to\infty} c(x) = \lim_{|x|\to\infty} \hat{c}(x) = 1$.

(vi) The function $\hat{c}(x) + \hat{c}(-x)$ has a unique minimum, which is attained at zero, and this minimum value is $\frac{2}{3}$.

Consider the Markov control $v^*(x) = -\operatorname{sign}(x)$. This is a stable control and the corresponding invariant probability measure is given by

$$\eta_{v^*}(\mathrm{d}x) = \mathrm{e}^{-2|x|}\mathrm{d}x\,.$$

Also,

$$\varrho_{v^*}(\mathrm{d}x) = \int_{-\infty}^{\infty} c(x)\,\eta_{v^*}(\mathrm{d}x) = \tfrac{1}{3}\,,$$

implying that $\varrho^* \le \frac{1}{3}$. Let $\varrho \mapsto \xi_\varrho$ be the map from $[1/3, 1)$ to $(-\infty, 0]$, defined by $\hat{c}(\xi_\varrho) = \varrho$, i.e.,

$$\xi_\varrho = \log \frac{3}{2} + \log(1-\varrho)\,.$$

Define

$$V_\varrho(x) := 2 \int_{-\infty}^{x} \mathrm{e}^{2|y-\xi_\varrho|} \mathrm{d}y \int_{-\infty}^{y} \mathrm{e}^{-2|z-\xi_\varrho|} (\varrho - c(z))\,\mathrm{d}z\,, \quad x \in \mathbb{R}\,.$$

Direct verification shows that V_ϱ is C^2 and

$$\tfrac{1}{2} V_\varrho''(x) - \frac{x-\xi_\varrho}{|x-\xi_\varrho|} V_\varrho'(x) + c(x) = \varrho\,, \quad x \in \mathbb{R}\,. \tag{3.8.10}$$

Also, the derivative of V_ϱ evaluates to

$$\begin{aligned} V_\varrho'(x) &= 2\mathrm{e}^{2|x-\xi_\varrho|} \int_{-\infty}^{x} \mathrm{e}^{-2|y-\xi_\varrho|} (\varrho - c(y))\,\mathrm{d}y \\ &= \begin{cases} \varrho - \hat{c}(x) & \text{if } x \le \xi_\varrho\,, \\ \mathrm{e}^{2(x-\xi_\varrho)}(\varrho - \hat{c}(-\xi_\varrho)) + (\hat{c}(-x) - \varrho) & \text{if } x > \xi_\varrho \end{cases} \\ &= \begin{cases} \frac{2}{3}(\mathrm{e}^{x} - \mathrm{e}^{\xi_\varrho}) & \text{if } x \le \xi_\varrho\,, \\ \frac{4}{3}\mathrm{e}^{2x}\mathrm{e}^{-\xi_\varrho}(1 - \mathrm{e}^{\xi_\varrho}) + (\hat{c}(-x) - \varrho) & \text{if } x > \xi_\varrho\,. \end{cases} \end{aligned} \tag{3.8.11}$$

We claim that $V_\varrho' < 0$ in $(-\infty, \xi_\varrho)$, and $V_\varrho' > 0$ in (ξ_ϱ, ∞). It follows by (3.8.11) that $V_\varrho' < 0$ in $(-\infty, \xi_\varrho)$, $V_\varrho'(\xi_\varrho) = 0$ and $V_\varrho' > 0$ in $[-\log(1-\varrho), \infty)$. On the other hand, since $V_\varrho'(\xi_\varrho) = 0$, decomposing the integral in (3.8.11) we obtain

$$V_\varrho'(x) = 2\mathrm{e}^{2(x-\xi_\varrho)} \int_{\xi_\varrho}^{x} \mathrm{e}^{-2(y-\xi_\varrho)} (\varrho - c(y))\,\mathrm{d}y\,, \quad \xi_\varrho < x < -\log(1-\varrho)\,.$$

Since $\varrho > c(x)$ for $x \in (\xi_\varrho, -\log(1-\varrho))$, it follows that $V_\varrho'(x) > 0$ on this interval. This settles the claim. Therefore, by (3.8.10), we obtain

$$\tfrac{1}{2} V_\varrho''(x) - |V_\varrho'(x)| + c(x) = \varrho\,, \qquad \forall \varrho \in [1/3, 1)\,, \tag{3.8.12}$$

and thus the HJB equation (3.8.9) admits a continuum of solution pairs (V_ϱ, ϱ), which satisfy $\inf_{\mathbb{R}} V_\varrho > -\infty$, for $1/3 \le \varrho < 1$. The stationary Markov control corresponding to the solution pair (V_ϱ, ϱ) of the HJB in (3.8.12) is given by $v_\varrho(x) = -\operatorname{sign}(x - \xi_\varrho)$. The controlled process under v_ϱ has invariant probability density $\varphi_\varrho(x) = \mathrm{e}^{-2|x-\xi_\varrho|}$. A simple computation shows that

$$\int_{-\infty}^{\infty} c(x)\varphi_\varrho(x)\,\mathrm{d}x = \varrho - \tfrac{9}{8}(1-\varrho)(3\varrho - 1) < \varrho\,, \quad \forall \varrho \in (1/3, 1)\,.$$

Therefore, if $\varrho > 1/3$, then V_ϱ is not a canonical solution of the Poisson equation corresponding to the stable control v_ϱ, and it holds that

$$\lim_{t\to\infty} \frac{\mathbb{E}_x^{v_\varrho}[V_\varrho(X_t)]}{t} = \tfrac{9}{8}(1-\varrho)(3\varrho - 1)\,, \quad \forall x \in \mathbb{R}\,.$$

Now, consider the solution of (3.8.12) corresponding to $\varrho = \frac{1}{3}$. We obtain

$$V_{1/3}(x) = \tfrac{2}{3}(\mathrm{e}^{-|x|} + |x| - 1)\,.$$

Therefore $V_{1/3}$ is integrable with respect to η_{v^*} and it follows that

$$\lim_{t\to\infty} \frac{\mathbb{E}_x^{v^*}[V_{1/3}(X_t)]}{t} = 0 \qquad \forall x \in \mathbb{R}\,.$$

An application of Itô's formula yields

$$\lim_{t\to\infty} \frac{1}{t}\ \mathbb{E}_x^{v^*}\left[\int_0^t c(X_s)\,\mathrm{d}s\right] = \frac{1}{3}\,.$$

We need to show that v^* is optimal. Let v be any stable Markov control. For an arbitrary $x' > 0$ define

$$v'(x) = \begin{cases} v(x) & \text{if } x \in (-x', x')\,,\\ -\operatorname{sign}(x) & \text{otherwise.}\end{cases}$$

It is evident by (3.8.1) that $K_{v'} < K_v$, and hence $\varphi_{v'} > \varphi_v$ on $(-x', x')$. As a result $\varrho_{v'} < \varrho_v$, which implies that any optimal control equals v^* a.e. Therefore v^* is optimal and $\varrho^* = \frac{1}{3}$.

3.8.2 The upper envelope of invariant probability measures

Suppose $\mathfrak{U}_{\mathrm{sm}} = \mathfrak{U}_{\mathrm{ssm}}$. Let $\overline{v}$ and $\underline{v}$ be measurable selectors satisfying

$$\overline{v}(x) \in \operatorname*{Arg\,max}_{u\in\mathbb{U}} \{\operatorname{sign}(x) b(x,u)\}$$

$$\underline{v}(x) = \operatorname*{Arg\,min}_{u\in\mathbb{U}} \{\operatorname{sign}(x) b(x,u)\}\,.$$

It follows by (3.8.1), that

$$\beta_{\underline{v}}(x) \le \beta_v(x) \le \beta_{\overline{v}}(x) \qquad \forall x \in \mathbb{R}\,, \quad \forall v \in \mathfrak{U}_{\mathrm{sm}}\,,$$

and as a result

$$0 < K_{\underline{v}} \le K_v \le K_{\overline{v}} < \infty \qquad \forall v \in \mathfrak{U}_{\mathrm{sm}}\,.$$

Therefore

$$\varphi_v(x) \le \frac{K_{\overline{v}}}{K_{\underline{v}}}\varphi_{\overline{v}}(x) \qquad \forall x \in \mathbb{R}\,, \quad \forall v \in \mathfrak{U}_{\mathrm{sm}}\,. \tag{3.8.13}$$

Hence, (3.8.13) implies that the upper envelope of $\mathscr{H}$ is a finite measure, in particular that $\mathscr{H}$ is tight.

3.9 Bibliographical note

Sections 3.2, 3.4 and 3.6. We follow primarily [28] which essentially builds upon [37, 39]. We extend the results to unbounded data for the controlled diffusion, and unbounded costs. See also [17] for an example of an unbounded control space.

Sections 3.3 and 3.7. These are based on [4] and [41] with corrections.

Section 3.8. See [15, 16].

4

Various Topics in Nondegenerate Diffusions

4.1 Introduction

This chapter consists of several topics in nondegenerate controlled diffusions. In Section 4.2 we study constrained and multi-objective problems. Section 4.3 deals with a singularly perturbed ergodic control problem involving two time-scales. Section 4.4 gives a brief account of ergodic control in bounded domains and lastly Section 4.5 presents a detailed solution of a problem arising in practice.

4.2 Multi-objective problems

This section is devoted to the study of multi-objective optimal control. We first study a class of constrained optimal control problems and then continue the analysis for models endowed with multiple cost criteria.

4.2.1 Optimal control under constraints

The ergodic control problem with constraints can be described as follows. Let

$$c_i \colon \mathbb{R}^d \times \mathbb{U} \to \mathbb{R}_+ , \quad 0 \le i \le \ell ,$$

be continuous functions and $\underline{m}_i, \overline{m}_i \in \mathbb{R}_+$, $1 \le i \le \ell$, given constants with $\underline{m}_i \le \overline{m}_i$. The objective here is to minimize

$$\int_{\mathbb{R}^d \times \mathbb{U}} c_0 \, \mathrm{d}\pi \tag{4.2.1}$$

over $\pi \in \mathscr{G}$, subject to

$$\underline{m}_i \le \int_{\mathbb{R}^d \times \mathbb{U}} c_i \, \mathrm{d}\pi \le \overline{m}_i , \quad 1 \le i \le \ell .$$

Let

$$H := \left\{ \pi \in \mathscr{G} : \underline{m}_i \le \int_{\mathbb{R}^d \times \mathbb{U}} c_i \, \mathrm{d}\pi \le \overline{m}_i , \; 1 \le i \le \ell \right\} .$$

We assume that H is non-empty. Also, assume that

$$\varrho_0 := \inf_{\pi \in H} \int_{\mathbb{R}^d \times \mathbb{U}} c_0 \, \mathrm{d}\pi < \infty . \tag{4.2.2}$$

As in the unconstrained problem, we consider two cases.

Assumption 4.2.1 (near-monotone) There are no lower constraints, i.e., $\underline{m}_i = 0$, $i = 1, \dots, \ell$. Also

$$\liminf_{|x| \to \infty} \min_{u \in \mathbb{U}} c_0(x, u) > \varrho_0$$

$$\liminf_{|x| \to \infty} \min_{u \in \mathbb{U}} c_i(x, u) > \overline{m}_i , \quad i = 1, \dots, \ell .$$

Assumption 4.2.2 (stable) $\mathscr{G}$ is compact, and H closed, hence also compact.

Lemma 4.2.3 *Under either Assumption 4.2.1 or 4.2.2, the minimization problem* (4.2.1) *subject to* (4.2.2) *has a solution in H_e, the set of extreme points of H.*

Proof First note that under either assumption, H is closed and convex. Under Assumption 4.2.2 the result follows by compactness of H. Next, consider Assumption 4.2.1. Let $\{\pi_n\}$ be a sequence in H such that

$$\int_{\mathbb{R}^d \times \mathbb{U}} c_0 \, \mathrm{d}\pi_n \downarrow \varrho_0 \quad \text{as } n \to \infty .$$

Let $\bar{H}$ be the closure of H in $\mathscr{P}(\bar{\mathbb{R}}^d \times \mathbb{U})$. Since $\bar{H}$ is compact, dropping to a subsequence if necessary, we may assume that $\pi_n \to \hat{\pi} \in \bar{H}$. We write $\hat{\pi}$ as

$$\hat{\pi}(A) = \delta \hat{\pi}_1(A \cap (\mathbb{R}^d \times \mathbb{U})) + (1 - \delta) \hat{\pi}_2(A \cap (\{\infty\} \times \mathbb{U})) \tag{4.2.3}$$

for $A \in \mathscr{B}(\bar{\mathbb{R}}^d \times \mathbb{U})$, where $\hat{\pi}_1 \in \mathscr{P}(\mathbb{R}^d \times \mathbb{U})$, $\hat{\pi}_2 \in \mathscr{P}(\{\infty\} \times \mathbb{U})$ and $\delta \in [0, 1]$. We first show that $\delta > 0$. Set $\overline{m}_0 \equiv \varrho_0$, and choose $\varepsilon > 0$ and $r > 0$ such that

$$\min_{u \in \mathbb{U}} c_j(x, u) \geq \overline{m}_j + \varepsilon , \quad \text{for } |x| > r , \quad 0 \leq j \leq \ell .$$

For $n \in \mathbb{N}$, let

$$c_j^n(x, u) := \frac{\mathfrak{d}(x, B_{n+1}^c) c_j(x, u) + \mathfrak{d}(x, B_n)(\overline{m}_j + \varepsilon)}{\mathfrak{d}(x, B_n) + \mathfrak{d}(x, B_{n+1}^c)} , \quad 0 \leq j \leq \ell ,$$

where, as defined earlier, $\mathfrak{d}(x, A)$ denotes the Euclidean distance of the point x from the set A. Observe that, by defining $c_j^n(\infty, u) = \overline{m}_j + \varepsilon$, it follows that $c_j^n(x, u)$ has a continuous extension on $\bar{\mathbb{R}}^d \times \mathbb{U}$. Thus, for $n > r$,

$$\begin{aligned}
\overline{m}_j &\geq \liminf_{k \to \infty} \int_{\mathbb{R}^d \times \mathbb{U}} c_j \, \mathrm{d}\pi_k \\
&\geq \lim_{k \to \infty} \int_{\bar{\mathbb{R}}^d \times \mathbb{U}} c_j^n \, \mathrm{d}\pi_k \\
&= \delta \int_{\mathbb{R}^d \times \mathbb{U}} c_j^n \, \mathrm{d}\hat{\pi}_1 + (1 - \delta)(\overline{m}_j + \varepsilon) .
\end{aligned} \tag{4.2.4}$$

Taking limits in (4.2.4) as $n \to \infty$, using monotone convergence, we obtain

$$\overline{m}_j \geq \delta \int_{\mathbb{R}^d \times \mathbb{U}} c_j \, \mathrm{d}\hat{\pi}_1 + (1-\delta)(\overline{m}_j + \varepsilon_j), \quad 0 \leq j \leq \ell. \tag{4.2.5}$$

By (4.2.5), $\delta > 0$, and

$$\int c_j \, \mathrm{d}\hat{\pi}_1 \leq \overline{m}_j, \quad 0 \leq j \leq \ell. \tag{4.2.6}$$

Since for each compactly supported $f \in C_c^2(\mathbb{R}^d)$,

$$\int_{\mathbb{R}^d \times \mathbb{U}} \mathcal{L}^u f(x) \, \pi_n(\mathrm{d}x, \mathrm{d}u) = 0 \qquad \forall n \in \mathbb{N},$$

passing to the limit as $n \to \infty$, we obtain

$$\delta \int_{\mathbb{R}^d \times \mathbb{U}} \mathcal{L}^u f(x) \, \hat{\pi}_1(\mathrm{d}x, \mathrm{d}u) = 0 \qquad \forall f \in C_c^2(\mathbb{R}^d).$$

Since $\delta > 0$, using Lemma 3.2.2 we obtain $\hat{\pi}_1 \in \mathscr{G}$. In turn, by (4.2.6), $\hat{\pi}_1 \in H$. This of course implies that $\int c_0 \, \mathrm{d}\hat{\pi}_1 \geq \varrho_0$. Using (4.2.5) with $j = 0$, we deduce that $\delta = 1$ and

$$\varrho_0 = \overline{m}_0 = \int_{\mathbb{R}^d \times \mathbb{U}} c_0 \, \mathrm{d}\hat{\pi}_1 ,$$

to conclude that (4.2.1) attains its minimum in H at $\hat{\pi} = \hat{\pi}_1$. To complete the proof, we need to show that $\hat{\pi}$ is the barycenter of a probability measure supported on H_e. By Choquet's theorem, $\hat{\pi} \in H$ is the barycenter of a probability measure Ψ on $\bar{H}_e$. By (4.2.3), since $\delta = 1$, $\Psi(\bar{H}_e \setminus H_e) = 0$, or equivalently, $\Psi(H_e) = 1$. Hence,

$$\varrho_0 = \int_{\mathbb{R}^d \times \mathbb{U}} c_0 \, \mathrm{d}\hat{\pi} = \int_{H_e} \Psi(\mathrm{d}\xi) \int_{\mathbb{R}^d \times \mathbb{U}} c_0(x,u) \, \xi(\mathrm{d}x, \mathrm{d}u). \tag{4.2.7}$$

Therefore, since

$$\int_{\mathbb{R}^d \times \mathbb{U}} c_0 \, \mathrm{d}\xi \geq \varrho_0 \qquad \forall \xi \in H_e ,$$

we conclude from (4.2.7) that $\int c_0 \, \mathrm{d}\hat{\xi} = \varrho_0$ for some $\hat{\xi} \in H_e$. □

Next we show that $H_e \subset \mathscr{G}_e$. The intuition behind this somewhat surprising result is that if $\pi \in \mathscr{G}$ is not an extreme point, then, viewed as a subset of the space of finite signed measures on $\mathbb{R}^d \times \mathbb{U}$, $\mathscr{G}$ is locally of infinite dimension at π. In other words, $\mathscr{G}$ does not have any finite dimensional faces other than its extreme points. Since the intersection of a finite number of half-spaces has finite co-dimension, there are no extreme points in H, other than the ones in $\mathscr{G}_e$.

Lemma 4.2.4 *Let π_0 in $\mathscr{G}$ take the form*

$$\pi_0 = \tfrac{1}{2}(\pi_1 + \pi_2)\,,$$

with π_1, $\pi_2 \in \mathscr{G}$, $\pi_1 \neq \pi_2$. Then, for any $n \in \mathbb{N}$, there exists a one-to-one homomorphism h mapping $\mathbb{R}^n$ into the space of finite signed measures on $\mathbb{R}^d \times \mathbb{U}$ such that $\pi_0 + h(\boldsymbol{\lambda}) \in \mathscr{G}$ for all

$$\boldsymbol{\lambda} \in \Lambda_n := \left\{(\lambda_1, \dots, \lambda_n) \in [-1,1]^n : \sum_{i=1}^{n} |\lambda_i| \le 1\right\}.$$

Proof Using the notation introduced in Definition 3.2.1, p. 87, we write $\pi_i = \eta_{v_i} \circledast v_i$, $i = 0, 1, 2$. According to the hypothesis of the lemma, $v_1 \neq v_2$. Therefore there exists a bounded measurable set $A \subset \mathbb{R}^d$ such that

$$\left\| v_1(\cdot \mid x) - v_2(\cdot \mid x) \right\|_{\mathrm{TV}} > 0 \qquad \forall x \in A\,.$$

Let $\{A_1, \dots, A_n\}$ be any partition of A consisting of sets of positive Lebesgue measure. Define

$$\tilde{v}_i^k(\cdot \mid x) = v_i(\cdot \mid x)\,\mathbb{I}_{A_k}(x) + v_0(\cdot \mid x)\,\mathbb{I}_{A_k^c}(x)\,, \quad i = 1, 2\,, \quad k = 1, \dots, n\,.$$

Since

$$v_0(\cdot \mid x) = \frac{1}{2} \sum_{i=1,2} \frac{\mathrm{d}\eta_{v_i}}{\mathrm{d}\eta_{v_0}}(x) v_i(\cdot \mid x)\,,$$

it follows that

$$v_0(\cdot \mid x) = \frac{1}{2} \sum_{i=1,2} \left(\mathbb{I}_{A_k}(x) \frac{\mathrm{d}\eta_{v_i}}{\mathrm{d}\eta_{v_0}}(x) + \mathbb{I}_{A_k^c}(x) \right) \tilde{v}_i^k(\cdot \mid x)\,, \quad 1 \le k \le n\,.$$

Since, by Lemma 3.2.4 (b), $\frac{\mathrm{d}\eta_{v_i}}{\mathrm{d}\eta_{v_0}}$, $i = 0, 1$, is bounded away from zero on bounded sets in $\mathbb{R}^d$, using Lemma 3.2.7, we conclude that for each $k = 1, \dots, n$, there exists a pair (v_1^k, v_2^k) of points in $\mathfrak{U}_{\mathrm{ssm}}$, which agree on A_k^c and differ a.e. on A_k, and such that

$$\pi_{v_0} = \pi_0 = \tfrac{1}{2}(\pi_{v_1^k} + \pi_{v_2^k})\,. \tag{4.2.8}$$

Define

$$\begin{aligned} \xi_k &:= \tfrac{1}{2}(\pi_{v_1^k} - \pi_{v_2^k})\,, \quad k = 1, \dots, n\,, \\ h(\boldsymbol{\lambda}) &:= \sum_{k=1}^{n} \lambda_k \xi_k\,, \qquad \boldsymbol{\lambda} = (\lambda_1, \dots, \lambda_n) \in \Lambda_n\,. \end{aligned} \tag{4.2.9}$$

Note that $h(\Lambda_n)$ is a rectangle in the space of signed measures with corner points $(\pm\xi_1, \dots, \pm\xi_n)$. With ‘co’ denoting the convex hull, it follows that

$$\pi_0 + h(\Lambda_n) = \Delta_n := \mathrm{co}\{\pi_{v_1^k}, \pi_{v_2^k} : 1 \le k \le n\}\,,$$

and π_0 is in its relative interior. It remains to show that $\{\xi_k : 1 \le k \le n\}$ are linearly independent. If not, assume without loss of generality that $\xi_1 = \sum_{k=2}^n r_k \xi_k$ for some constants $r_2, \dots, r_n$. Let $\hat{r} \in \mathbb{R}$ satisfy $\hat{r}^{-1} > \max\left\{1, \sum_{k=2}^n |r_k|\right\}$. It follows that

$$\pi_0 + \hat{r}\xi_1 \in \mathrm{co}\{\pi_{v_1^k}, \pi_{v_2^k} : 2 \le k \le n\}.$$

Let $\hat{v} \in \mathfrak{U}_{\mathrm{sm}}$ such that $\pi_{\hat{v}} = \pi_0 + \hat{r}\xi_1$. Note that since

$$\pi_{\hat{v}} \in \mathrm{co}\{\pi_{v_1^k}, \pi_{v_2^k} : 2 \le k \le n\},$$

then $\hat{v}$ must agree with v_0 on A_1. Since $\hat{r} \in (0, 1)$, we obtain $\pi_{\hat{v}} = \hat{r}\pi_{v_1^1} + (1 - \hat{r})\pi_0$, and by the remark in the paragraph preceding Lemma 3.2.7 in p. 91, since $\hat{v}$ agrees with v_0 on A_1, then v_1^1 should also agree with v_0 on A_1. In turn, by (4.2.8), v_1^1 and v_2^1 should agree on A_1 which is a contradiction. This completes the proof. □

The next lemma follows from Lemma 4.2.4 and the more general results of Dubins [48]. We present a simple algebraic proof which is tailored for the particular problem.

Lemma 4.2.5 *Let H_i, $i = 1, \dots, \ell$, be a collection of half-spaces in the space of finite signed measures on $\mathbb{R}^d \times \mathbb{U}$ of the form*

$$H_i = \{\xi : \textstyle\int h_i \,\mathrm{d}\xi \le 0\},$$

and let L

$$H := \mathscr{G} \cap H_1 \cap \cdots \cap H_\ell.$$

Then $H_e \subset \mathscr{G}_e$.

Proof Suppose $\pi_0 \in H_e \setminus \mathscr{G}_e$. Choose $n > \ell$, and let $\{\xi_1, \dots, \xi_n\}$ and $h(\boldsymbol{\lambda})$ be as in (4.2.9). We use the convenient notation $\langle h_i, \xi_j \rangle := \int h_i \,\mathrm{d}\xi_j$. Since $n > \ell$, we can choose $\boldsymbol{\lambda} \in \Lambda_n$ such that $\sum_{k=1}^n |\lambda_k| = 1$ and

$$\sum_{k=1}^n \lambda_k \langle h_j, \xi_k \rangle = 0 \qquad \forall j \in \{1, \dots, \ell\}.$$

Then, for $\delta \in [-1, 1]$, $\pi(\delta) \in \mathscr{G}$ defined by

$$\pi(\delta) := \pi_0 + \delta \sum_{k=1}^n \lambda_k \xi_k = \pi_0 + h(\delta\boldsymbol{\lambda})$$

satisfies $\langle h_j, \pi(\delta) \rangle \le 0$. Therefore $\pi(\delta) \in H$ for $\delta \in [-1, 1]$, and since $\pi(0) = \pi_0$ is in the relative interior of $\pi([-1, 1])$, it follows that $\pi_0 \notin H_e$ contradicting the original hypothesis. □

The existence of a stable stationary Markov optimal control follows directly from Lemmas 4.2.3–4.2.5.

Theorem 4.2.6 *Under either Assumption 4.2.1 or 4.2.2, the constrained problem (4.2.1) – (4.2.2) has an optimal solution $v \in \mathfrak{U}_{\mathrm{ssd}}$.*

Suppose there is a feasible π for which the constraints are satisfied with strict inequality. Then standard Lagrange multiplier theory [88, pp. 216–219] implies that the constrained problem is equivalent to the unconstrained minimization of the functional

$$J(\pi,\gamma,\kappa) = \int c_j \,\mathrm{d}\pi + \sum_1^{\ell} \gamma_i \left(\int c_i \,\mathrm{d}\pi - \overline{m}_i \right) + \sum_1^{\ell} \kappa_i \left(\underline{m}_i - \int c_i \,\mathrm{d}\pi \right), \qquad (4.2.10)$$

when the weights $(\gamma,\kappa) = \{\gamma_i, \kappa_i, 1 \le i \le \ell\}$ are equal to the Lagrange multipliers $(\gamma^*,\kappa^*) \in \mathbb{R}_+^{2\ell}$. Moreover, if we use the optimal value π^* of π in (4.2.10), then as a function of (γ,κ), (4.2.10) is maximized over the nonnegative quadrant at the Lagrange multipliers. In other words, the saddle point property holds:

$$J(\pi^*,\gamma,\kappa) \le J(\pi^*,\gamma^*,\kappa^*) \le J(\pi,\gamma^*,\kappa^*), \quad \pi \in \mathscr{G}, \ (\gamma,\kappa) \in \mathbb{R}_+^{2\ell}.$$

4.2.2 A general multi-objective problem

In Section 4.2.1, we considered the problem of minimizing the ergodic cost of a functional c_0, subject to constraints on the ergodic cost of functionals $c_1,\dots,c_\ell$. In the traditional multi-objective framework optimality is relative to a chosen partial ordering in $\mathbb{R}^{\ell+1}$. Let

$$\hat{c}(\pi) := \left(\int c_0 \,\mathrm{d}\pi, \ \int c_1 \,\mathrm{d}\pi, \dots, \int c_\ell \,\mathrm{d}\pi \right)$$

denote the ergodic cost vector corresponding to $\pi \in \mathscr{G}$, and

$$W := \{\hat{c}(\pi) : \pi \in \mathscr{G}\}.$$

Note that W is a closed, convex subset of $\mathbb{R}^{\ell+1}$. Define the partial ordering $\prec$ on W by

$$(x_0, x_1, \dots, x_\ell) \prec (y_0, y_1, \dots, y_\ell)$$

if $x_i \le y_i$ for all $i = 0, 1, \dots, \ell$, with strict inequality for at least one i. We call $x^* \in W$ a *Pareto point* if there is no $z \in W$ for which $z \prec x^*$. Note that a minimizer of $\langle \lambda, x \rangle = \sum_{i=0}^{\ell} \lambda_i x_i$ over W for any choice of $\lambda_i > 0$, $0 \le i \le \ell$, is a Pareto point. In fact, if W is a polytope, then the minimizers

$$\left\{ \hat{x} \in W : \exists \lambda_i > 0 \,,\ 0 \le i \le \ell \,,\ \min_{x \in W} \langle \lambda, x \rangle = \langle \lambda, \hat{x} \rangle \right\} \qquad (4.2.11)$$

include all Pareto points. But, in general, W is not a polytope. In such a case we have to settle for a weaker result, namely, that the set in (4.2.11) of minimizers is dense in the set of all Pareto points. This is a consequence of the celebrated Arrow–Barankin–Blackwell theorem [9].

More generally, we can obtain Pareto points by minimizing a continuous functional $U_0 \colon W \to \mathbb{R}$ having the property that $U_0(x) < U_0(y)$ whenever $x \prec y$. Such a functional is called a *utility function*. The functional $x \mapsto \langle \lambda, x \rangle$, with $\lambda_i > 0$,

$0 \le i \le \ell$, is an example of a utility function. Another important example is constructed as follows: Let

$$c^* = \left(\min_{\pi \in \mathscr{G}} \int c_0 \, \mathrm{d}\pi, \ \min_{\pi \in \mathscr{G}} \int c_1 \, \mathrm{d}\pi, \dots, \ \min_{\pi \in \mathscr{G}} \int c_\ell \, \mathrm{d}\pi \right).$$

The vector $c^* \in \mathbb{R}^{\ell+1}$ is called an *ideal point*. In general the ideal point c^* does not belong to the feasible domain W. Let $\tilde{c}$ be the unique point in W such that

$$|c^* - \tilde{c}| = \min_{c \in W} |c^* - c|.$$

The point $\tilde{c}$ corresponds to minimization with respect to the utility function $U(x) = |x - c^*|$ and is a Pareto point, often referred to as the *shadow minimum.*

Since $\tilde{c} \in W$, there exists a $\tilde{\pi} \in \mathscr{G}$ such that

$$\tilde{c} = \left(\int c_0 \, \mathrm{d}\tilde{\pi}, \int c_1 \, \mathrm{d}\tilde{\pi}, \dots, \int c_\ell \, \mathrm{d}\tilde{\pi} \right). \tag{4.2.12}$$

But $\tilde{\pi}$ may or may not belong to $\mathscr{G}_e$. For simplicity, we assume that c_i, $i = 0, 1, \dots, \ell$, are bounded. By Caratheodory's theorem [101], $\tilde{c}$ can always be expressed as a convex combination of at most $\ell + 2$ points of W. On the other hand, it is evident that any extreme point of W is of the form $\hat{c}(\pi)$ for some $\pi \in \mathscr{G}_e$. Consequently $\tilde{c} = \hat{c}(\pi')$ for some $\pi' \in \mathscr{G}$ that is a convex combination of at most $\ell + 2$ points of $\mathscr{G}_e$. In fact the extreme points of the subset of $\mathscr{G}$ satisfying (4.2.12) are in $\mathscr{G}_e$, as shown in Lemma 4.2.5.

4.3 Singular perturbations in ergodic control

In this section, we consider ergodic control of singularly perturbed diffusions, in which the singular perturbation parameter $\varepsilon > 0$ is introduced in such a way that the state variables are decomposed into a group of slow variables that change their values with rates of order $\mathscr{O}(1)$, and a group of fast ones that change their values with rates of order $\mathscr{O}(\frac{1}{\varepsilon})$. Our objective is to relate this problem to the ergodic control problem for the *averaged* system obtained in the $\varepsilon \to 0$ limit, i.e., to prove that the latter (lower dimensional) problem is a valid approximation to the original problem for small ε.

Our approach relies on the intuition that for small ε, the time-scales are sufficiently separated so that the fast component sees the slow component as quasistatic and can be analyzed, within an approximation error that becomes asymptotically negligible as $\varepsilon \downarrow 0$, by freezing the slow component at a constant value. In turn, the slow component sees the fast component as quasi-equilibrated and can be analyzed by averaging out its dynamics with respect to the equilibrium behavior of the latter. We make this precise in what follows.

The model is a coupled pair of stochastic differential equations in $\mathbb{R}^d \times \mathbb{R}^s$ given by

$$dZ_t^\varepsilon = \bar{h}(Z_t^\varepsilon, X_t^\varepsilon, U_t)\,dt + \gamma(Z_t^\varepsilon)\,dB_t\,, \tag{4.3.1a}$$

$$dX_t^\varepsilon = \frac{1}{\varepsilon}\bar{b}(Z_t^\varepsilon, X_t^\varepsilon, U_t)\,dt + \frac{1}{\sqrt{\varepsilon}}\sigma(Z_t^\varepsilon, X_t^\varepsilon)\,dW_t\,. \tag{4.3.1b}$$

Here $\bar{h}$ and $\bar{b}$ are the relaxed versions of drift vector fields $h\colon \mathbb{R}^d \times \mathbb{R}^s \times \mathbb{U} \to \mathbb{R}^d$ and $b\colon \mathbb{R}^d \times \mathbb{R}^s \times \mathbb{U} \to \mathbb{R}^s$, i.e., $\bar{h}(z,x,v) = \int_{\mathbb{U}} h(z,x,u)v(du)$, $v \in \mathscr{P}(\mathbb{U})$, and similarly for $\bar{b}$. The action set $\mathbb{U}$ is a compact metric space. The standard assumptions in Section 2.2 of Lipschitz continuity and affine growth for the drift terms (h, b) and diffusion matrix $\operatorname{diag}(\gamma, \sigma)$ are in effect here. Also, B and W are d- and s-dimensional independent standard Brownian motions, respectively, and the least eigenvalues of $\gamma(z)\gamma^{\mathsf{T}}(z)$ and $a(z,x) := \frac{1}{2}\sigma(z,x)\sigma^{\mathsf{T}}(z,x)$ are uniformly bounded away from zero (nondegeneracy assumption).

As usual, U is a $\mathscr{P}(\mathbb{U})$-valued control process with measurable paths satisfying the *non-anticipativity* condition: for $t \ge s$, $(B_t - B_s, W_t - W_s)$ is independent of $\mathfrak{F}_s$, the completion of

$$\bigcap_{s'>s} \sigma(Z_r^\varepsilon, X_r^\varepsilon, U_r, B_r, W_r : r \le s')\,.$$

As before, we call such U an admissible control, and denote this set as $\mathfrak{U}$.

Given a running cost in the form of a continuous map $c\colon \mathbb{R}^d \times \mathbb{R}^s \times \mathbb{U} \to \mathbb{R}_+$, we set $\bar{c}(z,x,v) = \int_{\mathbb{U}} c(z,x,u)v(du)$. We seek to minimize over all admissible U the ergodic cost in its average formulation

$$\limsup_{t\uparrow\infty} \frac{1}{t}\int_0^t \mathbb{E}\left[\bar{c}(Z_s^\varepsilon, X_s^\varepsilon, U_s)\right] ds\,.$$

For $f \in C^2(\mathbb{R}^s)$, define $\mathcal{L}_z^u\colon C^2(\mathbb{R}^s) \to C(\mathbb{R}^d \times \mathbb{R}^s \times \mathbb{U})$ by

$$\mathcal{L}_z^u f(x) := \operatorname{tr}\left(a(z,x)\nabla^2 f(x)\right) + \langle \nabla f(x), b(z,x,u)\rangle\,, \tag{4.3.2}$$

and $\hat{\mathcal{L}}_\varepsilon^u\colon C^2(\mathbb{R}^d \times \mathbb{R}^s) \to C(\mathbb{R}^d \times \mathbb{R}^s \times \mathbb{U})$ by

$$\hat{\mathcal{L}}_\varepsilon^u f(z,x) := \frac{1}{2}\operatorname{tr}\left(\gamma(z)\gamma^{\mathsf{T}}(z)\nabla_z^2 f(z,x)\right) + \langle \nabla_z f(z,x), h(z,x,u)\rangle + \frac{1}{\varepsilon}\mathcal{L}_z^u f(z,x)\,,$$

where ∇_y and ∇_y^2 denote the gradient and the Hessian in the variable y, respectively.

We work with the weak formulation of the above control problem. We also impose the following conditions, which are in effect throughout unless otherwise stated.

Assumption 4.3.1 We assume the following:

(a) c is locally Lipschitz continuous in (x,z) uniformly in $u \in \mathbb{U}$.
(b) c is inf-compact, i.e.,

$$\lim_{|(z,x)|\to\infty} \min_{u\in\mathbb{U}} c(z,x,u) = \infty\,. \tag{4.3.3}$$

(c) There exists an inf-compact function $\mathcal{V} \in C^2(\mathbb{R}^s)$ and a function $g \in C(\mathbb{R}^d \times \mathbb{R}^s)$ such that

$$\lim_{|x|\to\infty} g(z, x) = \infty\,,$$

uniformly in z in compact subsets of $\mathbb{R}^d$, satisfying

$$\mathcal{L}_z^u \mathcal{V}(x) \le -g(z, x)\,. \tag{4.3.4}$$

(d) There exists a constant $M^* > 0$ such that for each $\varepsilon \in (0, 1)$, the average cost for at least one admissible U does not exceed M^*.

For $U \in \mathfrak{U}$ and $\varepsilon > 0$, define the empirical measures $\{\zeta_t^{U,\varepsilon} : t > 0\}$, and the mean empirical measures $\{\bar{\zeta}_t^{U,\varepsilon} : t > 0\}$ by

$$\int_{\mathbb{R}^d\times\mathbb{R}^s\times\mathbb{U}} f\, \mathrm{d}\zeta_t^{U,\varepsilon} := \frac{1}{t}\int_0^t f(Z_s^\varepsilon, X_s^\varepsilon, U_s)\, \mathrm{d}s\,,$$

$$\int_{\mathbb{R}^d\times\mathbb{R}^s\times\mathbb{U}} f\, \mathrm{d}\bar{\zeta}_t^{U,\varepsilon} := \frac{1}{t}\int_0^t \mathbb{E}^U\left[f(Z_s^\varepsilon, X_s^\varepsilon, U_s)\right] \mathrm{d}s$$

for $f \in C_b(\mathbb{R}^d \times \mathbb{R}^s \times \mathbb{U})$.

Let

$$\Phi_v^\varepsilon(\mathrm{d}z, \mathrm{d}x, \mathrm{d}u) := \eta_v^\varepsilon(\mathrm{d}z, \mathrm{d}x)\, v(\mathrm{d}u \mid z, x) \tag{4.3.5}$$

denote the ergodic occupation measure associated with a Markov control $v \in \mathfrak{U}_{\mathrm{ssm}}$ and denote by $\mathscr{G}^\varepsilon$ the set of all ergodic occupation measures Φ_v^ε as v varies over all stable Markov controls.

The following is immediate from Theorem 3.4.7.

Theorem 4.3.2 *Under Assumption 4.3.1 (a), (b) and (d), there exists $v_\varepsilon^* \in \mathfrak{U}_{\mathrm{ssd}}$ such that if $\Phi_*^\varepsilon := \Phi_{v_\varepsilon^*}^\varepsilon$ is the corresponding ergodic occupation measure, then, under any admissible U,*

$$\liminf_{t\uparrow\infty} \int c\, \mathrm{d}\zeta_t^{U,\varepsilon} \ge \int c\, \mathrm{d}\Phi_*^\varepsilon \quad \textit{a.s.},$$

$$\liminf_{t\uparrow\infty} \int c\, \mathrm{d}\bar{\zeta}_t^{U,\varepsilon} \ge \int c\, \mathrm{d}\Phi_*^\varepsilon\,.$$

4.3.1 The averaged system

We introduce the *averaged system* and show that the optimal cost for the associated ergodic control problem serves in general as an asymptotic lower bound for the optimal cost for the original problem in the $\varepsilon \downarrow 0$ limit. The argument is based on the tightness of the optimal ergodic occupation measures as ε is reduced to zero, implying their relative compactness in the Prohorov topology.

Setting $\tau = \frac{t}{\varepsilon}$, $\bar{X}_\tau = X^\varepsilon_{\varepsilon\tau}$, $\bar{Z}_\tau = Z^\varepsilon_{\varepsilon\tau}$, $\bar{U}_\tau = U_{\varepsilon\tau}$, and $\bar{W}_\tau = \frac{1}{\sqrt{\varepsilon}} W_{\varepsilon\tau}$, (4.3.1b) becomes

$$\mathrm{d}\bar{X}_\tau = \bar{b}(\bar{Z}_\tau, \bar{X}_\tau, \bar{U}_\tau)\,\mathrm{d}\tau + \sigma(\bar{Z}_\tau, \bar{X}_\tau)\,\mathrm{d}\bar{W}_\tau\,,$$

which does not depend on ε explicitly. We define the *associated system*

$$\mathrm{d}X^z_\tau = \bar{b}(z, X^z_\tau, U_\tau)\,\mathrm{d}\tau + \sigma(z, X^z_\tau)\,\mathrm{d}W_\tau\,, \tag{4.3.6}$$

where $z \in \mathbb{R}^s$ is fixed, W a standard Brownian motion independent of X_0, and the admissibility of $\bar{U}$ is defined by: for $t > s$, $W_t - W_s$ is independent of $\mathfrak{G}_s$, the completion of $\bigcap_{s'>s} \sigma(X^z_r, U_r, W_r : r \le s')$. The definition of the associated system is motivated by the intuition that the fast system may be analyzed by freezing the slow dynamics as the separation of the time-scales increases with $\varepsilon \downarrow 0$.

Remark 4.3.3 Equation (4.3.6) is the relaxed control form of the associated system. One also has the pre-relaxation form

$$\mathrm{d}X_\tau = b(z, X_\tau, U_\tau)\,\mathrm{d}\tau + \sigma(z, X_\tau)\,\mathrm{d}W_\tau\,, \tag{4.3.7}$$

which we shall have an occasion to use later.

Let

$$\mathscr{G}_z := \left\{\pi_z \in \mathscr{P}(\mathbb{R}^s \times \mathbb{U}) : \int_{\mathbb{R}^s \times \mathbb{U}} \mathcal{L}^u_z f(x)\,\pi_z(\mathrm{d}x, \mathrm{d}u) = 0 \quad \forall f \in C^2_0(\mathbb{R}^s)\right\},$$

where $\mathcal{L}^u_z$ is defined in (4.3.2). The next lemma in particular characterizes this as the set of ergodic occupation measures for the associated system.

Lemma 4.3.4 *The class $\mathscr{G}_z$ is the set of $\pi_z \in \mathscr{P}(\mathbb{R}^s \times \mathbb{U})$ taking the form*

$$\pi_z(\mathrm{d}x, \mathrm{d}u) = \eta_{z,v}(\mathrm{d}x)\,v(\mathrm{d}u \mid x)\,,$$

where $\eta_{z,v}$ is the unique invariant distribution of the time-homogeneous diffusion X given by (4.3.6) *when $U_t = v(X_t) := v(\mathrm{d}u \mid X_t)$. The set-valued map $z \mapsto \mathscr{G}_z$ is convex compact-valued and continuous. Moreover, for any compact set $K \subset \mathbb{R}^d$, $\cup_{z\in K}\mathscr{G}_z$ is compact.*

Proof The first claim follows from Lemma 2.6.14. That $\mathscr{G}_z$ is convex and closed for each $z \in \mathbb{R}^d$ is easily verified from the definition. It is also straightforward to verify that it is compact-valued and continuous. Let K be a compact set in $\mathbb{R}^d$, and write π_z as $\pi_{z,v}$ to explicitly denote its dependence on v. Viewing (4.3.7) as a controlled diffusion with action space $K \times \mathbb{U}$, it follows by (4.3.4) and (iv) of Lemma 3.3.4, that the diffusion is uniformly stable. Consider the set of controls $\mathcal{U}$ of the form $\tilde{v} = \delta_z(\mathrm{d}z')v(\mathrm{d}u \mid x)$, $z \in K$. It follows that $\mathscr{H}_{\mathcal{U}}$ (defined as in Section 3.2 on p. 89) is tight uniformly in $z \in K$, and since $\mathcal{U}$ is closed, Lemma 3.2.6 implies that the map $\tilde{v} \mapsto \pi_{z,v} \in \cup_{z'\in K}\mathscr{G}_{z'}$ is continuous. Since $\mathcal{U}$ is compact and the map is onto, the proof is complete. □

It follows by Lemma 4.3.4 that the graph $\{(z, \pi_z) : z \in \mathbb{R}^s, \pi_z \in \mathcal{G}_z\}$ is closed and the set $\{(z, \pi_z) : z \in K, \pi_z \in \mathcal{G}_z\}$ is compact for any compact set $K \subset \mathbb{R}^s$. For $\mu \in \mathcal{G}_z$, define

$$\tilde{h}(z, \mu) := \int_{\mathbb{R}^s \times \mathbb{U}} h(z, x, u)\, \mu(\mathrm{d}x, \mathrm{d}u),$$

$$\tilde{c}(z, \mu) := \int_{\mathbb{R}^s \times \mathbb{U}} c(z, x, u)\, \mu(\mathrm{d}x, \mathrm{d}u).$$

The averaged system is defined by

$$\mathrm{d}Z_t = \tilde{h}(Z_t, \mu_t)\,\mathrm{d}t + \gamma(Z_t)\,\mathrm{d}\tilde{B}_t, \quad \mu_t \in \mathcal{G}_{Z_t}, \quad \forall t \geq 0. \tag{4.3.8}$$

The averaged system corresponds to the slow variables with their dynamics averaged over the equilibrium behavior of the fast variables. Here $Z_0 = Z_0^{\varepsilon} = z_0$, $\varepsilon > 0$, $\tilde{B}$ is a standard Brownian motion in $\mathbb{R}^d$, while μ satisfies $\mu_t \in \mathcal{G}_{Z_t}$ and the non-anticipativity condition: for $t \geq s \geq 0$, $\tilde{B}_t - \tilde{B}_s$ is independent of the completion of

$$\bigcap_{s' > s} \sigma(Z_r, \tilde{B}_r, \mu_r : r \leq s').$$

We may view μ as the *effective control process* for the averaged system. By analogy, we call μ a Markov control if $\mu_t = \mu(\mathrm{d}x, \mathrm{d}u \mid Z_t)$ for all t, identified with the measurable map $\mu\colon \mathbb{R}^d \to \mathcal{P}(\mathbb{R}^s \times \mathbb{U})$. Call it a stable Markov control if in addition the resulting time-homogeneous Markov process Z is positive recurrent. In the latter case, Z has a unique invariant probability distribution $\varphi_\mu(\mathrm{d}z)$. Let $\mathcal{G}$ denote the set of the corresponding ergodic occupation measures for the averaged system in (4.3.8). For $\xi \in \mathcal{G}$, we have

$$\xi(\mathrm{d}z, \mathrm{d}x, \mathrm{d}u) := \varphi_\mu(\mathrm{d}z)\, \mu(\mathrm{d}x, \mathrm{d}u \mid z),$$

or equivalently, $\xi = \varphi_\mu \circledast \mu$. Then as before, one has the following characterization. Define $\tilde{\mathcal{L}}\colon C^2(\mathbb{R}^d) \to C(\mathbb{R}^d \times \mathcal{P}(\mathbb{R}^s \times \mathbb{U}))$ by

$$\tilde{\mathcal{L}}^\mu f(z) = \frac{1}{2} \operatorname{tr}\left(\gamma(z)\gamma^{\mathsf{T}}(z)\nabla^2 f(z)\right) + \langle \nabla f(z), \tilde{h}(z, \mu)\rangle.$$

Lemma 4.3.5 *The set $\mathcal{G}$ is the subset of $\mathcal{P}(\mathbb{R}^d \times \mathbb{R}^s \times \mathbb{U})$ consisting of all $\xi = \varphi_\mu \circledast \mu$ such that $\mu(\cdot \mid z) \in \mathcal{G}_z$ for all $z \in \mathbb{R}^s$ and*

$$\int_{\mathbb{R}^s} \tilde{\mathcal{L}}^\mu f(z)\, \varphi_\mu(\mathrm{d}z) = 0 \qquad \forall f \in C_0^2(\mathbb{R}^d).$$

This again follows from Lemma 2.6.14. The ergodic control problem for the averaged system takes the form: minimize over all admissible μ

$$\limsup_{t \uparrow \infty} \frac{1}{t}\, \mathbb{E}\left[\int_0^t \tilde{c}(Z_s, \mu_s)\,\mathrm{d}s\right].$$

We then obtain the following, in analogy to Theorem 3.4.7.

Theorem 4.3.6 *There exists a stable optimal Markov control μ^* for the averaged system such that if $\xi^* = \varphi_{\mu^*} \circledast \mu^*$ is the corresponding ergodic occupation measure, then for any admissible μ as above*

$$\liminf_{t\uparrow\infty} \frac{1}{t}\int_0^t \tilde{c}(Z_s, \mu_s)\,\mathrm{d}s \geq \int c\,\mathrm{d}\xi^* \quad a.s.,$$

$$\liminf_{t\uparrow\infty} \frac{1}{t}\int_0^t \mathbb{E}^{\mu^*}[\tilde{c}(Z_s, \mu_s)]\,\mathrm{d}s \geq \int c\,\mathrm{d}\xi^*.$$

Let $\mathscr{G}^*$ denote the set of optimal ergodic occupation measures, i.e.,

$$\mathscr{G}^* = \operatorname*{Arg\,min}_{\xi\in\mathscr{G}} \int c\,\mathrm{d}\xi\,.$$

If μ is an ergodic occupation measure for the associated system in (4.3.6), i.e., $\mu \in \mathscr{G}_z$, we decompose it as $\mu = \mu_v \circledast v$ where μ_v is the unique invariant distribution. In particular, for an optimal μ^*, we obtain

$$\mu^*(\mathrm{d}x, \mathrm{d}u \mid z) = \mu_{v^*}(\mathrm{d}x \mid z)\, v^*(\mathrm{d}u \mid z, x)\,. \tag{4.3.9}$$

Also define η_v via the decomposition

$$\xi(\mathrm{d}z, \mathrm{d}x, \mathrm{d}u) = \eta_v(\mathrm{d}z, \mathrm{d}x)\, v(\mathrm{d}u \mid z, x)\,.$$

We now consider the limit as $\varepsilon \downarrow 0$. Let Φ_*^ε be as in Theorem 4.3.2. Then by Assumption 4.3.1 (d), it follows that $\sup_{0<\varepsilon<1} \int c\,\mathrm{d}\Phi_*^\varepsilon \leq M^*$. In turn, by (4.3.3) and the Chebyshev inequality, it then follows that $\{\Phi_*^\varepsilon : \varepsilon \in (0,1)\}$ is tight. The following theorem characterizes the limit points of this collection.

Theorem 4.3.7 *Any limit point Φ_*^0 of Φ_*^ε as $\varepsilon \downarrow 0$ lies in $\mathscr{G}$.*

Proof Disintegrate Φ_*^0 as

$$\Phi_*^0(\mathrm{d}z, \mathrm{d}x, \mathrm{d}u) = \breve{\varphi}(\mathrm{d}z)\, \breve{\mu}(\mathrm{d}x, \mathrm{d}u \mid z)\,.$$

With $f_1 \in C_0^2(\mathbb{R}^d)$ and $f_2 \in C_0^2(\mathbb{R}^s)$, let $\varepsilon \downarrow 0$ in the equation $\varepsilon \int \hat{\mathcal{L}}_\varepsilon^u(f_1 f_2)\,\mathrm{d}\Phi_*^\varepsilon = 0$ to obtain

$$\int_{\mathbb{R}^d} f_1(z) \left[\int_{\mathbb{R}^s\times\mathbb{U}} \mathcal{L}_z^u f_2(x)\, \breve{\mu}(\mathrm{d}x, \mathrm{d}u \mid z)\right] \breve{\varphi}(\mathrm{d}z) = 0\,. \tag{4.3.10}$$

Then as (4.3.10) holds for all $f_1 \in C_0^2(\mathbb{R}^d)$, we conclude that for $\breve{\varphi}$-a.s. z,

$$\int_{\mathbb{R}^s\times\mathbb{U}} \mathcal{L}_z^u f_2(x)\, \breve{\mu}(\mathrm{d}x, \mathrm{d}u \mid z) = 0\,,$$

implying that $\breve{\mu}(\cdot \mid z) \in \mathscr{G}_z$. The qualification "$\breve{\varphi}$-a.s." may be dropped by choosing a suitable version. Now for $f \in C_0^2(\mathbb{R}^d)$, let $\varepsilon \downarrow 0$ in $\int \hat{\mathcal{L}}_\varepsilon^u f\,\mathrm{d}\Phi_*^\varepsilon = 0$ to obtain

$$\int \hat{\mathcal{L}}_0^u f\,\mathrm{d}\Phi_*^0 = \int_{\mathbb{R}^d} \tilde{\mathcal{L}}^{\breve{\mu}} f(z)\, \breve{\varphi}(\mathrm{d}z) = 0\,. \tag{4.3.11}$$

By Lemma 4.3.5, (4.3.11) implies that $\breve{\varphi}$ is the unique stationary distribution $\varphi_{\breve{\mu}}$ under $\breve{\mu}$ for the averaged system. It follows that $\Phi_*^0 \in \mathscr{G}$. □

Corollary 4.3.8 $\liminf_{\varepsilon\downarrow 0} \int c\, d\Phi_*^\varepsilon \geq \int c\, d\xi^*$.

This shows that the optimal ergodic cost for the averaged system provides an asymptotic lower bound (as $\varepsilon \downarrow 0$) for the optimal ergodic cost of the original problem. To show that it is in fact a valid approximation, we must replace the "lim inf" by "lim" in Corollary 4.3.8 and the inequality by an equality. We do so under additional assumptions in the following sections.

4.3.2 The affine case

We show that in the special case of the control entering the drift in an affine manner and the running cost being strict convex in the control, we have

$$\lim_{\varepsilon\downarrow 0} \int c\, d\Phi_*^\varepsilon = \int c\, d\xi^*.$$

This crucially depends on the fact that under these conditions, the expression being minimized over the control parameter in the associated Hamilton–Jacobi–Bellman equation is strictly convex and therefore the minimizer is unique and continuously varying with ε.

Assume the following.

Assumption 4.3.9 (i) $\mathbb{U}$ is a compact subset of $\mathbb{R}^m$ for some $m \geq 1$ and for each $(z, x) \in \mathbb{R}^{d+s}$, $h(z, x, \cdot)$, $b(z, x, \cdot)$ are componentwise affine and $c(z, x, \cdot)$ is strictly convex.

(ii) $|h(z, x, u)| \in \mathfrak{o}(c(z, x, u))$ and

$$\max_{u\in\mathbb{U}} c(z, x, u) \in \mathfrak{o}(|g(z, x)|)$$

with g as in (4.3.4).

(iii) $v^* = v^*(du \mid z, x)$ in (4.3.9) is a stable Markov control for (4.3.1) for sufficiently small $\varepsilon > 0$ (say, $\varepsilon < \varepsilon_0$), and the corresponding stationary distributions $\eta_{v^*}^\varepsilon(dz, dx)$, $\varepsilon \in (0, \varepsilon_0)$, are tight.

The next lemma, which uses only Assumption 4.3.9 (i) and (ii), shows in particular that v^* in (4.3.9) is unique. Thus part (iii) of Assumption 4.3.9 is unambiguous.

Lemma 4.3.10 *Under Assumption 4.3.9 (i) and (ii), $v^*(du \mid z, x)$ in (4.3.9) is unique and continuous in $(z, x) \in \mathbb{R}^{d+s}$. In addition, the corresponding μ^* in (4.3.9) and the associated $\xi^* \in \mathscr{G}^*$ are both unique, or equivalently, $\mathscr{G}^*$ is a singleton.*

Proof By Theorem 3.6.10, a necessary and sufficient condition for the optimality of μ^* is that $\mu^*(z)$ minimizes the function

$$\mu \mapsto \tilde{c}(z, \mu) + \langle \nabla\Psi(z), \tilde{h}(z, \mu)\rangle,$$

over $\mathscr{G}_z$ for a.e. z, where $\Psi \in C^2(\mathbb{R}^d)$ is the value function for the ergodic control problem for the averaged system. Note that Theorem 3.6.10 proves the existence of a C^2 value function and the associated "verification theorem" for the case when $\mathscr{G}_z$ is independent of z. Condition (4.3.3) is a special case of near-monotonicity. The modifications required to handle the more general state-dependent control space needed here are minor in view of the upper semicontinuity of the set-valued map $z \mapsto \mathscr{G}_z$ already established. The details are omitted.

We may drop the qualification "for a.e. z" by taking an appropriate version. Now for fixed $z \in \mathbb{R}^d$, consider the ergodic control problem for the associated system (4.3.7) with cost

$$\limsup_{t\uparrow\infty} \frac{1}{t} \int_0^t \mathbb{E}\left[\ell_z(\bar{X}_s, U_s)\right] \mathrm{d}s\,,$$

where $\ell_z \in C(\mathbb{R}^s \times \mathbb{U})$ is defined by

$$\ell_z(x,u) := c(z,x,u) + \langle \nabla\Psi(z), h(z,x,u)\rangle\,.$$

Since $\mathscr{G}_z$ is precisely the set of ergodic occupation measures for the associated system, μ^* is the optimal ergodic occupation measure for the above problem. Note that for each z, the cost function ℓ_z is inf-compact, because of the first half of Assumption 4.3.9 (ii). By Assumption 4.3.9 (ii), $\langle \nabla\Psi(z), h(z,x,u)\rangle \in \mathfrak{o}(g)$ for each $z \in \mathbb{R}^d$. Thus Theorem 3.6.10 can be applied again to this new control problem, in order to conclude as above that $v^*(\mathrm{d}u \mid z, x)$ minimizes

$$\mu \mapsto \int_{\mathbb{R}^s} \bigl(\ell_z(x,\cdot) + \langle \nabla\tilde{\Psi}_z(x), b(z,x,\cdot)\rangle\bigr)\,\mu(\mathrm{d}x \mid z)\,,$$

where $\tilde{\Psi}_z \in C^2(\mathbb{R}^s)$ is the value function for this new ergodic control problem for the model in (4.3.7). Let

$$G(z,x,u) := \ell_z(x,u) + \langle \nabla\tilde{\Psi}_z(x), b(z,x,u)\rangle\,.$$

By Theorem 3.6.10, the map $(z,x) \mapsto \inf_{u\in\mathbb{U}} G(z,x,u)$ is locally Lipschitz continuous, while Assumption 4.3.9 (i) implies that $u \mapsto G(z,x,u)$ is strictly convex. It follows that $\operatorname{Arg\,min}_{u\in\mathbb{U}} G(z,x,u)$ is a singleton, and is continuous in $(z,x) \in \mathbb{R}^{d+s}$. In other words, $(z,x) \mapsto v^*$ is continuous.

Recall (4.3.9), where μ_{v^*} is the unique stationary distribution for the associated system under v^*. Uniqueness of μ^* follows. In turn, $\xi^* = \varphi_{\mu^*} \circledast \mu^*$, where φ_{μ^*} is the unique stationary distribution of the averaged system under μ^*. Thus ξ^* is unique. □

Remark 4.3.11 Note that $v^*(\mathrm{d}u \mid z, x)$ is in fact Dirac for all $(z,x) \in \mathbb{R}^{d+s}$. Also note that the assumption of affine dependence of h and b on u is crucial in preserving the strict convexity of the running cost.

Theorem 4.3.12 *With v^* as in (4.3.9), $\Phi^\varepsilon_{v^*}$ defined in (4.3.5) converges to ξ^* in $\mathscr{P}(\mathbb{R}^d \times \mathbb{R}^s \times \mathbb{U})$ as $\varepsilon \downarrow 0$. Moreover, if the function g in Assumption 4.3.9 (ii) satisfies*

$$\limsup_{\varepsilon\downarrow 0} \int_{\mathbb{R}^{d+s}} g(z,x)\,\eta^\varepsilon_{v^*}(\mathrm{d}z, \mathrm{d}x) < \infty\,, \tag{4.3.12}$$

then

$$\lim_{\varepsilon\downarrow 0} \int c\,\mathrm{d}\Phi^\varepsilon_{v^*} = \int c\,\mathrm{d}\xi^* \tag{4.3.13}$$

and

$$\lim_{\varepsilon\downarrow 0}\int c\,\mathrm{d}\Phi^{\varepsilon}_{*} = \int c\,\mathrm{d}\xi^{*}\,. \tag{4.3.14}$$

Proof In view of Theorem 4.3.7, (4.3.13) implies (4.3.14). Therefore it suffices to prove the convergence of $\Phi^{\varepsilon}_{v^*}$ to ξ^* and show the validity of (4.3.13).

Suppose that

$$\eta^{\varepsilon}_{v^*}(\mathrm{d}z,\mathrm{d}x) \to \breve{\eta}(\mathrm{d}z,\mathrm{d}x) = \breve{\varphi}(\mathrm{d}z)\,\breve{\mu}(\mathrm{d}x \mid z)$$

along a subsequence as $\varepsilon \downarrow 0$. In view of the continuity of $v^*(\mathrm{d}u \mid \cdot,\cdot)$, we may pass to the limit along this subsequence in

$$\varepsilon\int_{\mathbb{R}^d\times\mathbb{R}^s}\left[\int_{\mathrm{U}}\hat{\mathcal{L}}^{u}_{\varepsilon}f(z,x)\,v^*(\mathrm{d}u \mid z,x)\right]\eta^{\varepsilon}_{v^*}(\mathrm{d}z,\mathrm{d}x) = 0\,,\quad f\in C^2_0(\mathbb{R}^{d+s})\,,$$

with $f = f_1 f_2$, $f_1 \in C^2_0(\mathbb{R}^d)$ and $f_2 \in C^2_0(\mathbb{R}^s)$, and argue as in Theorem 4.3.7 to obtain

$$\int_{\mathbb{R}^d\times\mathbb{R}^s}\left[\int_{\mathrm{U}}\mathcal{L}^{u}_{z}f(z,x)\,v^*(\mathrm{d}u \mid z,x)\right]\breve{\eta}(\mathrm{d}z,\mathrm{d}x) = 0\,,\quad f\in C^2_0(\mathbb{R}^{d+s})\,.$$

Hence $\breve{\mu}(\mathrm{d}x \mid z)$ is in fact the unique stationary distribution for the associated system controlled by $v^*(\mathrm{d}u \mid z,x)$ (i.e., $\breve{\mu}(\mathrm{d}x \mid z) = \mu_{v^*}(\mathrm{d}x \mid z)$) for $\breve{\varphi}$-a.s. z. The latter qualification may be dropped by choosing an appropriate version. Recall (4.3.9). Let $\varepsilon \downarrow 0$ in

$$\int_{\mathbb{R}^d\times\mathbb{R}^s}\left[\int_{\mathrm{U}}\hat{\mathcal{L}}^{u}_{\varepsilon}f(z,x)\,v^*(\mathrm{d}u \mid z,x)\right]\eta^{\varepsilon}_{v^*}(\mathrm{d}z,\mathrm{d}x) = 0$$

for $f = f_1 \in C^2_0(\mathbb{R}^d)$ (i.e., f is a C^2 function of the z variable alone). A similar argument then yields

$$\int_{\mathbb{R}^d}\tilde{\mathcal{L}}^{\mu^*}f(z)\,\breve{\varphi}(\mathrm{d}z) = 0\,,\quad f\in C^2_0(\mathbb{R}^d)\,.$$

Thus $\breve{\varphi}(\mathrm{d}z)$ is the unique stationary distribution for the averaged system controlled by the stable Markov control μ^*, i.e., $\breve{\varphi} = \varphi_{\mu^*}$. Thus

$$\breve{\eta}(\mathrm{d}z,\mathrm{d}x)\,v^*(\mathrm{d}u \mid z,x) = \xi^*(\mathrm{d}z,\mathrm{d}x,\mathrm{d}u)\,,$$

implying $\Phi^{\varepsilon}_{v^*} \to \xi^*$. By (4.3.12) and the second half of Assumption 4.3.9 (ii) there exists $\varepsilon_0 > 0$ such that c is uniformly integrable over $\{\Phi^{\varepsilon}_{v^*}\}$ for $\varepsilon \in [0,\varepsilon_0]$. Hence (4.3.13) holds. □

4.3.3 The general case

We drop part (i) of Assumption 4.3.9, retain part (ii), and modify part (iii) in order to extend the results of Section 4.3.2 under some technical assumptions. These assumptions basically allow us, given an optimal control for the averaged system, to approximate the optimal process in law by the ε-indexed processes we start with.

Define $v^*_\delta(\mathrm{d}u \mid z, x)$, for δ in some interval $(0, \delta_0]$, by

$$v^*_\delta(\mathrm{d}u \mid z, x) := \int_{\mathbb{R}^d \times \mathbb{R}^s} v^*(\mathrm{d}u \mid z', x')\,\rho_\delta(z - z', x - x')\,\mathrm{d}z'\mathrm{d}x',$$

where $\rho_\delta \colon \mathbb{R}^{d+s} \to \mathbb{R}_+$, $\delta \in (0, \delta_0]$, are smooth mollifiers supported on a ball of radius δ, centered at the origin. In what follows,

$$v^*_0(\mathrm{d}u \mid z, x) := v^*(\mathrm{d}u \mid z, x),$$

and the variables $\hat{\eta}^\varepsilon_\delta$, $\hat{\mu}_\delta$, $\hat{\hat{\mu}}_\delta$, $\hat{\varphi}_\delta$, and $\hat{\Phi}^\varepsilon_\delta$ introduced below, correspond to v^* when $\delta = 0$. Replace Assumption 4.3.9 (iii) by Assumptions 4.3.13–4.3.14 below:

Assumption 4.3.13 The kernel $v^*_\delta(\mathrm{d}u \mid z, x)$ is a stable Markov control for (4.3.1) for all $\delta \in [0, \delta_0]$, and $\varepsilon \in (0, \varepsilon_0)$. Moreover, there exists an inf-compact $\hat{g} \in C(\mathbb{R}^{d+s})$ satisfying

$$\sup_{u \in \mathbb{U}} c(z, x, u) \in \mathfrak{o}(|\hat{g}(z, x)|), \tag{4.3.15}$$

such that the stationary distributions $\eta^\varepsilon_{v^*_\delta}$ of (4.3.1) corresponding to $\{v^*_\delta\}$, denoted by $\hat{\eta}^\varepsilon_\delta(\mathrm{d}z, \mathrm{d}x)$, $0 < \varepsilon < \varepsilon_0$, satisfy

$$\sup_{0<\varepsilon<\varepsilon_0} \int_{\mathbb{R}^d \times \mathbb{R}^s} \hat{g}(z, x)\,\hat{\eta}^\varepsilon_\delta(\mathrm{d}z, \mathrm{d}x) < \infty \qquad \forall \delta \in [0, \delta_0]. \tag{4.3.16}$$

Once again, in view of the nondegeneracy assumption, the transition probabilities for $t > 0$ of the time-homogeneous Markov process described by (4.3.6) under the Markov control v^*_δ, $\delta \in [0, \delta_0]$, have densities w.r.t. the Lebesgue measure. Therefore the same applies for the corresponding invariant probability measures $\hat{\mu}_\delta := \mu_{v^*_\delta}$ for the associated system. Let

$$\hat{\hat{\mu}}_\delta(\mathrm{d}x, \mathrm{d}u \mid z) := \hat{\mu}_\delta(\mathrm{d}x \mid z)\, v^*_\delta(\mathrm{d}u \mid z, x)$$

and $\hat{\varphi}_\delta$ be the unique stationary distribution for (4.3.8) under the Markov control $\hat{\hat{\mu}}_\delta$. Also, for $\delta \in [0, \delta_0]$, let

$$\hat{\eta}^0_\delta(\mathrm{d}z, \mathrm{d}x) := \hat{\varphi}_\delta(\mathrm{d}z)\,\hat{\mu}_\delta(\mathrm{d}x \mid z)$$
$$\hat{\Phi}^0_\delta(\mathrm{d}z, \mathrm{d}x, \mathrm{d}u) := \hat{\eta}^0_\delta(\mathrm{d}z, \mathrm{d}x)\, v^*_\delta(\mathrm{d}u \mid z, x).$$

Note that $\hat{\Phi}^0_0 \in \mathscr{G}^*$. We also assume:

Assumption 4.3.14 The kernel $\hat{\hat{\mu}}_\delta(\mathrm{d}x, \mathrm{d}u \mid z)$ is a stable Markov control for (4.3.8) for $\delta \in [0, \delta_0]$, and for $\hat{g}$ as in Assumption 4.3.13,

$$\sup_{\delta \in [0,\delta_0]} \int_{\mathbb{R}^d \times \mathbb{R}^s} \hat{g}(z, x)\,\hat{\eta}^0_\delta(\mathrm{d}z, \mathrm{d}x) < \infty. \tag{4.3.17}$$

Lemma 4.3.15 *As $(\delta_n, z_n) \to (\delta, z)$ in $[0, \delta^*] \times \mathbb{R}^d$, $\hat{\mu}_{\delta_n}(\mathrm{d}x \mid z_n) \to \hat{\mu}_\delta(\mathrm{d}x \mid z)$ in total variation.*

Proof By Assumption 4.3.1 (d), $\{v^*_\delta : \delta \in [0,\delta^*]\,,\ z \in K\}$, with K a compact subset of $\mathbb{R}^d$, are uniformly stable controls for the associated system. The continuity of the map $(\delta, z) \mapsto v^*_\delta(\,\cdot \mid z, x)$, along with Lemma 3.2.6 (a), yield the desired result. □

We also let $\hat{\Phi}^\varepsilon_\delta \in \mathscr{P}(\mathbb{R}^d \times \mathbb{R}^s \times \mathbb{U})$, $\delta > 0$, denote the ergodic occupation measure for (4.3.1) under the control v^*_δ.

Lemma 4.3.16 *$\int c\,\mathrm{d}\hat{\Phi}^0_\delta \to \int c\,\mathrm{d}\hat{\Phi}^0_0$ as $\delta \downarrow 0$, and $\int c\,\mathrm{d}\hat{\Phi}^\varepsilon_\delta \to \int c\,\mathrm{d}\hat{\Phi}^0_\delta$ as $\varepsilon \downarrow 0$.*

Proof By (4.3.3), (4.3.15), (4.3.16), and the Chebyshev inequality,

$$\{\hat{\eta}^\varepsilon_\delta : \delta \in [0,\delta_0]\,,\ \varepsilon \in [0,\varepsilon_0]\}\,,$$

and therefore, also $\{\hat{\varphi}_\delta : \delta \in [0,\delta_0]\}$ are tight. Let $\hat{\varphi}$ be any limit point of $\hat{\varphi}_\delta$ as $\delta \downarrow 0$. Since $\hat{\varphi}_\delta$ is characterized by

$$\int_{\mathbb{R}^d} \tilde{\mathcal{L}}^{\hat{\mu}_\delta} f(z)\,\hat{\varphi}_\delta(\mathrm{d}z) = 0\,, \quad f \in C^2(\mathbb{R}^s)\,, \tag{4.3.18}$$

then the standard argument based on Harnack's inequality used in the proofs of Lemmas 3.2.4 and 3.2.5 implies that the convergence of $\hat{\varphi}_\delta \to \hat{\varphi}$ is in fact in total variation. Now for $f \in C_b(\mathbb{R}^d \times \mathbb{R}^s \times \mathbb{U})$,

$$\int_{\mathbb{U}} f(z,x,u)\,v^*_\delta(\mathrm{d}u \mid z,x) \xrightarrow[\delta\to 0]{} \int_{\mathbb{U}} f(z,x,u)\,v^*_0(\mathrm{d}u \mid z,x)$$

a.e. along a subsequence $\delta_m \downarrow 0$. Hence, if we define

$$\bar{f}(z,x,v) := \int_{\mathbb{U}} f(z,x,u)\,v(\mathrm{d}u \mid z,x)\,,$$

then by Lemma 4.3.15, along this subsequence,

$$\int_{\mathbb{R}^s} \bar{f}(z,x,v^*_\delta)\,\hat{\mu}_\delta(\mathrm{d}x \mid z) \xrightarrow[\delta\downarrow 0]{} \int_{\mathbb{R}^s} \bar{f}(z,x,v^*_0)\,\hat{\mu}_0(\mathrm{d}x \mid z) \quad \text{a.e.},$$

which in turn leads to

$$\int_{\mathbb{R}^s\times\mathbb{R}^d} \bar{f}(z,x,v^*_\delta)\,\hat{\mu}_\delta(\mathrm{d}x \mid z)\,\hat{\varphi}_\delta(\mathrm{d}z) \xrightarrow[\delta\downarrow 0]{} \int_{\mathbb{R}^s\times\mathbb{R}^d} \bar{f}(z,x,v^*_0)\,\hat{\mu}_0(\mathrm{d}x \mid z)\,\hat{\varphi}(\mathrm{d}z)\,.$$

In particular, letting $\delta \downarrow 0$ along an appropriate subsequence in (4.3.18), we obtain

$$\int_{\mathbb{R}^d} \tilde{\mathcal{L}}^{\hat{\mu}_0} f(z)\,\hat{\varphi}(\mathrm{d}z) = 0\,, \quad f \in C^2(\mathbb{R}^s)\,,$$

i.e., $\hat{\varphi} = \hat{\varphi}_0$. Thus

$$\hat{\Phi}^0_\delta = \hat{\varphi}_\delta(\mathrm{d}z)\,\hat{\mu}_\delta(\mathrm{d}x,\mathrm{d}u \mid z) \xrightarrow[\delta\downarrow 0]{} \hat{\Phi}^0_0 = \hat{\varphi}_0(\mathrm{d}z)\,\hat{\mu}_0(\mathrm{d}x,\mathrm{d}u \mid z)\,.$$

The convergence $\hat{\Phi}^\varepsilon_\delta \to \hat{\Phi}^0_\delta$ follows along the same lines. Equations (4.3.15) – (4.3.17) ensure uniform integrability of c under these measures, which in turn implies the convergence claimed in the lemma. □

Theorem 4.3.17 $\lim_{\varepsilon\downarrow 0} \int c\,\mathrm{d}\Phi^\varepsilon_* = \int c\,\mathrm{d}\Phi^0_*$.

Proof Fix $\alpha > 0$ and take $\delta > 0$ small enough such that

$$\left|\int c\,\mathrm{d}\hat{\Phi}^0_\delta - \int c\,\mathrm{d}\hat{\Phi}^0_0\right| < \frac{\alpha}{2}\,.$$

Then pick $\varepsilon > 0$ small enough so that

$$\left|\int c\,\mathrm{d}\hat{\Phi}^\varepsilon_\delta - \int c\,\mathrm{d}\hat{\Phi}^0_\delta\right| < \frac{\alpha}{2}\,.$$

Thus

$$\begin{aligned}\limsup_{\varepsilon\downarrow 0} \int c\,\mathrm{d}\Phi^\varepsilon_* &\le \limsup_{\varepsilon\downarrow 0} \int c\,\mathrm{d}\hat{\Phi}^\varepsilon_\delta \\ &\le \int c\,\mathrm{d}\hat{\Phi}^0_0 + \alpha\,.\end{aligned}$$

Since $\alpha > 0$ is arbitrary, the claim follows in view of Corollary 4.3.8. □

Remark 4.3.18 Condition (4.3.3) can be replaced by the weaker requirement

$$\lim_{\|z\|\to\infty} \inf_{x,u} c(z, x, u) > \beta^* := \sup_{0\le\varepsilon<\varepsilon_0} \beta^\varepsilon \tag{4.3.19}$$

for some $\varepsilon_0 > 0$, where β^ε, $\varepsilon > 0$, is the optimal cost for the ergodic control problem ($\varepsilon = 0$ yields the corresponding condition for the averaged system). This goes exactly along the lines of Sections 3.4–3.5. Since in particular this presupposes that β^ε are uniformly bounded for $\varepsilon \in (0, \varepsilon_0)$, in view of Theorem 4.3.7, we may replace the "$\sup_{0\le\varepsilon<\varepsilon_0}$" in (4.3.19) by "$\sup_{0<\varepsilon<\varepsilon_0}$."

We briefly indicate the corresponding developments when a blanket stability condition is available. We do not assume (4.3.3) or its generalization (4.3.19), but require that c be bounded from below. Suppose for $\varepsilon \in (0, \varepsilon_0)$ there exist a bounded ball $B \subset \mathbb{R}^{d+s}$ and a pair of nonnegative, inf-compact functions $\mathcal{V}^{(i)}_\varepsilon \in C^2(\mathbb{R}^{d+s})$, $i = 1, 2$, satisfying

$$\begin{aligned}\hat{\mathcal{L}}^u_\varepsilon \mathcal{V}^{(1)}_\varepsilon(z, x) &\le -1\,,\\ \hat{\mathcal{L}}^u_\varepsilon \mathcal{V}^{(2)}_\varepsilon(z, x) &\le -\mathcal{V}^{(1)}_\varepsilon(z, x)\end{aligned} \tag{4.3.20}$$

for all $(z, x) \in D^c$. The Lyapunov condition (4.3.20) is the same as Assumption 3.4.9, and implies as in Corollary 3.4.10 that the second moments of hitting times of bounded domains are uniformly bounded over all admissible controls. One can then argue as in Theorem 3.4.7 to derive Theorem 4.3.2. By Theorem 3.4.11, (4.3.20) ensures that $\{\mu_t\}$ remain a.s. tight (so do $\{\bar{\mu}_t\}$) and $\mathscr{G}$ is compact. This allows one to derive Theorem 4.3.2 without the need to use (4.3.3). Conditions similar to (4.3.20) imposed on (4.3.8) ensure Theorem 4.3.6. Next, for obtaining the counterparts of the results of Section 4.3.3 for the affine case, assume the additional conditions stipulated in Theorem 3.7.11 to ensure the existence of C^2 value functions for the two ergodic control problems that feature in the proof of Lemma 4.3.10. The rest remains as before. We omit the details as they are straightforward adaptations of the foregoing.

4.4 Control over a bounded domain

4.4.1 Controlled diffusions with periodic coefficients

Consider a controlled diffusion governed by (2.2.1), where for each $i = 1, \dots, d$, $b(x,u)$, $\sigma(x)$, and the running cost $c(x,u)$ are periodic in x_i with period T_i. The state space can be viewed as the d-dimensional torus

$$\mathbb{T} := \mathbb{R}^d \big/ \textstyle\prod_{i=1}^d T_i\mathbb{Z} \,.$$

Since the state space is compact, it follows by Theorem 1.5.15 that under any control $v \in \mathfrak{U}_{\mathrm{sm}}$ the corresponding process has an invariant probability measure η_v. Since the diffusion is nondegenerate, this invariant probability measure is unique. Without loss of generality, we assume that for each i, $T_i = 1$.

We follow the vanishing discount approach. Let α be the discount factor. As usual, for $U \in \mathfrak{U}$, define

$$J_\alpha^U(x) := \mathbb{E}_x^U\left[\int_0^\infty \mathrm{e}^{-\alpha t}\bar{c}(X_t, U_t)\,\mathrm{d}t\right],$$

$$V_\alpha(x) := \inf_{U\in\mathfrak{U}} J_\alpha^U(x)\,.$$

It is evident that Theorem 3.5.6 has the following version in the periodic case:

Theorem 4.4.1 *The optimal α-discounted value function V_α is periodic in each coordinate with period* 1 *and is the unique solution in $C^2(\mathbb{R}^d) \cap C_b(\mathbb{R}^d)$ of*

$$\min_{u\in\mathbb{U}} \left[\mathcal{L}^u V_\alpha(x) + c(x,u)\right] = \alpha V_\alpha(x)\,.$$

Also, a stationary Markov control v is α-discounted optimal if and only if

$$\sum_{i=1}^d b_v^i(x)\frac{\partial V_\alpha}{\partial x_i}(x) + c_v(x) = \min_{u\in\mathbb{U}}\left[\sum_{i=1}^d b^i(x,u)\frac{\partial V_\alpha}{\partial x_i}(x) + c(x,u)\right] \quad a.e. \qquad (4.4.1)$$

Clearly V_α assumes its global minimum at some $x_\alpha \in \mathbb{T}$. By Lemma 3.6.3, V_α has bounded oscillation on every ball $B_R \subset \mathbb{R}^d$ uniformly in $\alpha > 0$. Since it is periodic, the same applies over $\mathbb{R}^d$. Moreover, by the same lemma,

$$\sup_{\alpha>0} \left\|\bar{V}_\alpha - \bar{V}_\alpha(0)\right\|_{\mathscr{W}^{2,p}(\mathbb{T})} < \infty\,.$$

Thus, letting $\alpha \to 0$ and proceeding as in Section 3.6, we obtain the following.

Theorem 4.4.2 *There exists a unique pair $(V, \varrho) \in C^2(\mathbb{R}^d) \times \mathbb{R}$, such that $V(0) = 0$ and V is periodic in each coordinate with period* 1, *satisfying*

$$\min_{u\in\mathbb{U}} \left[\mathcal{L}^u V(x) + c(x,u)\right] = \varrho\,, \quad x \in \mathbb{R}^d\,.$$

Moreover, a stationary Markov control is optimal if and only if

$$\sum_{i=1}^d b_v^i(x)\frac{\partial V}{\partial x_i}(x) + c_v(x) = \min_{u\in\mathbb{U}}\left[\sum_{i=1}^d b^i(x,u)\frac{\partial V}{\partial x_i}(x) + c(x,u)\right] \quad a.e. \qquad (4.4.2)$$

and

$$\varrho = \varrho^* := \min_{v \in \mathfrak{U}_{\mathrm{sd}}} \int_{\mathbb{T}} c_v(x)\, \eta_v(\mathrm{d}x)\,.$$

4.4.2 Controlled reflected diffusions

Let $D \subset \mathbb{R}^d$ be a bounded domain with C^3 boundary ∂D. Let X_t be a $\bar{D}$-valued controlled reflected diffusion governed by

$$\mathrm{d}X_t = b(X_t, U_t)\,\mathrm{d}t + \sigma(X_t)\,\mathrm{d}W_t - r(X_t)\,\mathrm{d}\xi_t\,, \qquad (4.4.3)$$

and satisfying the following:

(i) $b \in C(\bar{D} \times \mathbb{U}; \mathbb{R}^d)$, where $\mathbb{U}$ is a compact metric space, and is Lipschitz continuous in its first argument.
(ii) $\sigma\colon \bar{D} \to \mathbb{R}^{d\times d}$ is Lipschitz continuous.
(iii) X_0 is prescribed in law, and W is a d-dimensional standard Brownian motion independent of X_0.
(iv) U is a $\mathbb{U}$-valued control process with measurable sample paths satisfying the following non-anticipativity condition: For $t \geq s$, $W_t - W_s$ is independent of $\{X_0, W_{s'}, U_{s'} : s' \leq s\}$.
(v) ξ is an $\mathbb{R}$-valued continuous, nondecreasing process ("local time on the boundary") satisfying

$$\xi_t = \int_0^t \mathbb{I}_{\partial D}(X_s)\,\mathrm{d}\xi_s\,. \qquad (4.4.4)$$

(vi) There exists a C^2 function $\eta\colon \mathbb{R}^d \to \mathbb{R}$ satisfying

$$D = \{x \in \mathbb{R}^d : \eta(x) < 0\}\,, \qquad \partial D = \{x \in \mathbb{R}^d : \eta(x) = 0\}\,,$$

and

$$|\nabla \eta| \geq 1 \quad \text{on } \partial D\,.$$

(vii) $r\colon \bar{D} \to \mathbb{R}^d$ is a C^2 vector field such that, for some $\delta > 0$, $r(x)\cdot\nabla\eta(x) \geq \delta$ for all $x \in \partial D$.

Under assumptions (i) – (vii), given the joint law of (X_0, W, U) there exists a unique weak solution of (4.4.3) – (4.4.4) [86, 100].

The following theorem characterizing the α-discounted control problem is standard [61, 77].

Theorem 4.4.3 *For any $\alpha > 0$, the α-discounted optimal cost V_α is the unique solution in $C^2(D) \cap C^1(\bar{D})$ of*

$$\begin{aligned} \min_{u \in \mathbb{U}} \left[\mathcal{L}^u V_\alpha(x) + c(x,u)\right] &= \alpha V_\alpha(x) \qquad \forall x \in D\,,\\ r(x)\cdot\nabla V_\alpha(x) &= 0 \qquad \forall x \in \partial D\,. \end{aligned} \qquad (4.4.5)$$

Moreover, a stationary Markov control v is α-discounted optimal if and only if it satisfies (4.4.1).

Note that since the state space is compact, the controlled process has an invariant probability measure η_v for every stationary Markov control v. As in Section 4.4.1, we have the following:

Theorem 4.4.4 *Let $x_0 \in D$ be fixed. Then there exists a unique pair (V, ϱ^*), with $V \in C^2(D) \cap C^1(\bar{D})$, $V(x_0) = 0$, and $\varrho^* \in \mathbb{R}$, which satisfies the HJB equation:*

$$\begin{aligned} \min_{u \in \mathbb{U}} \, [\mathcal{L}^u V(x) + c(x,u)] &= \varrho^* \qquad \forall x \in D\,, \\ r(x) \cdot \nabla V(x) = 0 & \qquad \forall x \in \partial D\,. \end{aligned} \tag{4.4.6}$$

Moreover, ϱ^* is the optimal cost and a stationary Markov control v is optimal if and only if it satisfies (4.4.2).

Proof For $\alpha > 0$, let v_α be a measurable selector from the minimizer in (4.4.5). Since under any stationary Markov control the process is ergodic and the diffusion is nondegenerate, it is straightforward to show that $\operatorname{osc}_D V_\alpha$ is uniformly bounded over $\alpha \in (0,1)$. This can be accomplished for example by the technique of the proof of Theorem 3.7.6. Therefore $\bar{V}_\alpha := V_\alpha - V_\alpha(0)$ is bounded, uniformly in $\alpha \in (0,1)$, and satisfies

$$\mathcal{L}^{v_\alpha} \bar{V}_\alpha - \alpha \bar{V}_\alpha = -c_{v_\alpha} + \alpha V_\alpha(0) \quad \text{in } D\,.$$

Since $\alpha V_\alpha(0) \le \max_{\bar{D}\times\mathbb{U}} c$ for all $\alpha > 0$, by [1, theorem 15.2] we obtain the estimate

$$\|\bar{V}_\alpha\|_{\mathscr{W}^{2,p}(D)} \le C_0\Big(\|\bar{V}_\alpha\|_{L^p(D)} + 2\sqrt[p]{|D|}\max_{\bar{D}\times\mathbb{U}} c\Big) \quad \forall \alpha \in (0,1)\,,$$

for some constant C_0. Therefore we can use Lemma 3.5.4, together with the fact that by Theorem A.2.15 the embedding $\mathscr{W}^{2,p}(D) \hookrightarrow C^{1,r}(\bar{D})$ is compact for $r < 1 - \frac{d}{p}$, in order to take limits along some subsequence $\alpha_n \downarrow 0$ in (4.4.5) and derive (4.4.6). □

For related work, see [18, 77, 100]. If the domain D is not smooth, the problem is quite challenging since the reflection field is oblique, discontinuous, and/or multi-valued at the boundary points which lack local smoothness. The theory of existence and uniqueness of strong solutions for PDEs in such domains is not available. For a treatment of this problem in polyhedral cones using viscosity solutions, see Borkar and Budhiraja [34].

4.5 An application

We apply the theory developed in this section to the problem of energy-efficient scheduling over a fading wireless channel. For a given data-arrival rate, the minimum power required to stabilize the queue can be computed directly from the capacity of the channel. However, with this minimum power, it is well known from queueing theory that the associated queueing delay is unbounded. Excess power needs to be allocated to stabilize the queue and in this section we deal with the problem of minimizing the queueing delay for a time-varying channel with a single queue, subject to constraints on both the average and peak excess power allocated.

Under the assumption of fast channel variation, i.e., if the channel state changes much faster than the queueing dynamics, we consider the heavy traffic limit and associate a monotone cost function with the limiting queue-length process. By separating the time-scales of the arrival process, the channel process and the queueing dynamics a heavy traffic limit for the queue length can be obtained in the form of a reflected diffusion process on $\mathbb{R}_+$, given by

$$\mathrm{d}X_t = -b(U_t)\,\mathrm{d}t + \sigma\,\mathrm{d}W_t + \mathrm{d}Z_t\,, \tag{4.5.1}$$

where

$$b(u) = \sum_{j=1}^{N} \gamma_j \pi_j u_j\,.$$

Here X_t is the queue-length process, and satisfies $X_t \geq 0$ for all $t \geq 0$, W_t is the standard Wiener process, σ is a positive constant, and Z_t is a nondecreasing process and grows only at those points t for which $X_t = 0$. The control variable u_j is the power allocated when the channel is in state j. The process Z_t, which ensures that the queue-length X_t remains nonnegative, is uniquely defined. There are N channel states and $\pi = (\pi_1, \ldots, \pi_N)$ is their stationary distribution, which is constant at the diffusion time-scale. The constants γ_j can be interpreted as the efficiency of the channel when in state j.

4.5.1 The optimal control problem for the heavy traffic model

The optimization problem of interest for the non-scaled queueing system is to minimize (pathwise, a.s.) the long-term average queueing length (and thus, from Little's law, the mean delay) or more generally, to minimize the long-term average value of some penalty function $c\colon \mathbb{R}_+ \to \mathbb{R}$, i.e., subject to a constraint on the average available power.

It is well known from queueing theory that if only the basic power is allocated, which matches the service rate to the arrival rate, then the resulting traffic intensity is equal to 1, and the queueing delay diverges. However, choosing the control U appropriately can result in a bounded average queue length. Thus the original optimization problem transforms to an analogous problem in the limiting system, namely,

$$\begin{aligned} &\text{minimize} && \limsup_{T\to\infty} \frac{1}{T}\int_0^T c(X_t)\,\mathrm{d}t \quad \text{a.s.} \\ &\text{subject to} && \limsup_{T\to\infty} \frac{1}{T}\int_0^T h(U_t)\,\mathrm{d}t \leq \bar{p} \quad \text{a.s.} \end{aligned} \tag{4.5.2}$$

where

$$h(u) = h(u_1, \dots, u_N) = \sum_{j=1}^{N} \pi_j u_j \,.$$

The control variable u takes values in $\mathbb{U} := [0, p_{\max}]^N$, with $p_{\max}$ denoting the (excess) peak power, and $\bar{p}$ denoting the (excess) average power. Naturally, for the constraint in (4.5.2) to be nontrivial, $\bar{p} \le p_{\max}$.

Remark 4.5.1 It is important to note that although the running cost c satisfies the near-monotone condition in a natural manner (e.g., for the minimization of the queue length $c(x) = x$), the constraint in (4.5.2) fails to satisfy Assumption 4.2.1 on p. 155. Therefore the results in Section 4.2.1 cannot be applied directly to even assert the existence of a Markov optimal control. In studying this application we intermarry the rich body of results for nondegenerate diffusions presented in Chapter 3 with the standard theory of Lagrange multipliers to transform to an equivalent unconstrained problem that satisfies the near-monotone hypothesis in Section 3.4 and then use the dynamic programming formulation in Section 3.6 to obtain an explicit solution for a Markov optimal control.

We first prove the existence of a stationary Markov optimal control, and then show that this is a channel-state based threshold control. In other words, for each channel state j, there is a queue-length threshold. The optimal control transmits at peak power over channel state j only if the queue length exceeds the threshold, and does not transmit otherwise.

The boundary at 0 imposes restrictions on the domain of the infinitesimal generator $\mathcal{L}^u$ given by

$$\mathcal{L}^u := \frac{\sigma^2}{2}\frac{\mathrm{d}^2}{\mathrm{d}x^2} - b(u)\frac{d}{\mathrm{d}x}\,, \qquad u \in \mathbb{U}\,.$$

If f is a bounded measurable function on $\mathbb{R}_+$, then $\varphi(x,t) = \mathbb{E}^v_x[f(X_t)]$ is a generalized (mild) solution of the problem

$$\frac{\partial\varphi}{\partial t}(x,t) = \mathcal{L}^v\varphi(x,t)\,, \qquad x \in (0,\infty)\,,\ t > 0\,,$$

$$\varphi(x,0) = f(x)\,, \qquad \frac{\partial\varphi}{\partial x}(0,t) = 0\,.$$

Itô's formula can be applied as follows [77, p. 500, lemma 4],[78]: If $\varphi \in \mathscr{W}^{2,p}_{\mathrm{loc}}(\mathbb{R}_+)$ is a bounded function satisfying $\frac{d\varphi}{\mathrm{d}x}(0) = 0$, then, for $t \ge 0$,

$$\mathbb{E}^v_x[\varphi(X_t)] - \varphi(x) = \mathbb{E}^v_x\left[\int_0^t \mathcal{L}^v\varphi(X_t)\,\mathrm{d}t\right].$$

Moreover, f_v is the density of an invariant probability measure μ_v if and only if it is a solution of the equation

$$(\mathcal{L}^v)^* f_v(x) = \frac{d}{\mathrm{d}x}\left(\frac{\sigma^2}{2}\frac{\mathrm{d}f_v}{\mathrm{d}x}(x) + b(v(x))f_v(x)\right) = 0\,. \tag{4.5.3}$$

Solving (4.5.3), we deduce that $v \in \mathfrak{U}_{\mathrm{sm}}$ is stable if and only if

$$A_v := \int_0^\infty \exp\left(-\frac{2}{\sigma^2}\int_0^x b(v(y))\,\mathrm{d}y\right)\mathrm{d}x < \infty\,,$$

in which case the solution of (4.5.3) takes the form

$$f_v(x) = A_v^{-1}\exp\left(-\frac{2}{\sigma^2}\int_0^x b(v(y))\,\mathrm{d}y\right). \tag{4.5.4}$$

We work under the assumption that c has the following monotone property:

Assumption 4.5.2 The function c is continuous and is either inf-compact and, if c is bounded, then it is strictly increasing. In the latter case we define

$$c_\infty := \lim_{x\to\infty} c(x)\,.$$

The analysis and solution of the optimization problem proceeds as follows: We first show that optimality is achieved for (4.5.2) relative to the class of stationary controls. Next, in Section 4.5.3 using the theory of Lagrange multipliers we formulate an equivalent unconstrained optimization problem. We show that an optimal control for the unconstrained problem can be characterized via the HJB equation. This accomplishes two tasks. First, it enables us to study the structure of the optimal controls. Second, we show that this control is optimal among all controls in $\mathfrak{U}$. An analytical solution of the HJB equation is presented in Section 4.5.5.

4.5.2 Existence of optimal stationary controls

A control $v \in \mathfrak{U}_{\mathrm{ssm}}$ is called *bang-bang*, or *extreme*, if $v(x) \in \{0, p_{\max}\}^N$, for almost all $x \in \mathbb{R}_+$. We refer to the class of extreme controls in $\mathfrak{U}_{\mathrm{ssm}}$ as *stable extreme* controls and denote it by $\mathfrak{U}_{\mathrm{se}}$. In this subsection, we show that if the optimization problem in (4.5.2) is restricted to stationary controls, then there exists $v \in \mathfrak{U}_{\mathrm{se}}$ which is optimal.

We take advantage of the fact that the set of power levels $\mathbb{U}$ is convex and avoid transforming the problem to the relaxed control framework. Instead, we view $\mathbb{U}$ as the space of product probability measures on $\{0, p_{\max}\}^N$. This is simply stating that for each j, u_j may be represented as a convex combination of the "0" power level and the peak power $p_{\max}$. In other words, $\mathbb{U}$ is viewed as a space of relaxed controls relative to the discrete control input space $\{0, p_{\max}\}^N$. This has the following advantage: by showing that optimality is attained in the set of precise controls, we assert the existence of a control in $\mathfrak{U}_{\mathrm{se}}$ which is optimal.

Let $\mathscr{H} \subset \mathscr{P}(\mathbb{R}_+)$ denote the set of all invariant probability measures μ_v of the process X_t under the controls $v \in \mathfrak{U}_{\mathrm{ssm}}$. Let $\widetilde{\mathbb{U}} := \{0, p_{\max}\}^N$. The generic element of $\widetilde{\mathbb{U}}$ takes the form $\tilde{u} = (\tilde{u}_1, \dots, \tilde{u}_N)$, with $\tilde{u}_i \in \{0, p_{\max}\}$, $i = 1, \dots, N$. There exists a natural isomorphism between $\mathbb{U}$ and the space of product probability measures on $\widetilde{\mathbb{U}}$ which we denote by $\mathcal{P}_\otimes(\widetilde{\mathbb{U}})$. This is viewed as follows. Let δ_p denote the Dirac probability measure concentrated at $p \in \mathbb{R}_+$. For $u \in \mathbb{U}$, we associate the probability measure $\tilde{\eta}_u \in \mathcal{P}_{\otimes}(\widetilde{\mathbb{U}})$ defined by

$$\tilde{\eta}_u(\tilde{u}) := \bigotimes_{i=1}^{N} \left[\left(1 - \frac{u_i}{p_{\max}}\right) \delta_0(\tilde{u}_i) + \frac{u_i}{p_{\max}} \delta_{p_{\max}}(\tilde{u}_i) \right]$$

for $\tilde{u} \in \widetilde{\mathbb{U}}$. Similarly, given $v \in \mathfrak{U}_{\mathrm{ssm}}$ we define $\eta_v \colon \mathbb{R}_+ \to \mathcal{P}_{\otimes}(\widetilde{\mathbb{U}})$ and $\pi_v \in \mathscr{P}(\mathbb{R}_+ \times \widetilde{\mathbb{U}})$ by

$$\eta_v(x, \mathrm{d}\tilde{u}) := \tilde{\eta}_{v(x)}(\mathrm{d}\tilde{u}) ,$$

$$\pi_v(\mathrm{d}x, \mathrm{d}\tilde{u}) := \mu_v(\mathrm{d}x)\, \eta_v(x, \mathrm{d}\tilde{u}) ,$$

where $\mu_v \in \mathscr{H}$ is the invariant probability measure of the process under the control $v \in \mathfrak{U}_{\mathrm{ssm}}$.

Due to the linearity of $u \mapsto h(u)$, we have the following identity (which we choose to express as an integral rather than a sum, despite the fact that $\widetilde{\mathbb{U}}$ is a finite space):

$$h(v(x)) = \int_{\widetilde{\mathbb{U}}} h(\tilde{u})\, \eta_v(x, \mathrm{d}\tilde{u}) , \qquad v \in \mathfrak{U}_{\mathrm{ssm}} .$$

As a point of clarification, the function h inside this integral is interpreted as the restriction of h on $\widetilde{\mathbb{U}}$. The analogous identity holds for $b(u)$.

In this manner we have defined a model whose input space $\widetilde{\mathbb{U}}$ is discrete, and for which the original input space $\mathbb{U}$ provides an appropriate convexification. Note however that $\mathbb{U} \sim \mathcal{P}_{\otimes}(\widetilde{\mathbb{U}})$ is not the input space corresponding to the relaxed controls based on $\widetilde{\mathbb{U}}$. The latter is $\mathscr{P}(\widetilde{\mathbb{U}})$, which is isomorphic to a 2^N-simplex in $\mathbb{R}^{2^N-1}$, whereas $\mathcal{P}_{\otimes}(\widetilde{\mathbb{U}})$ is isomorphic to a cube in $\mathbb{R}^N$. We select $\mathcal{P}_{\otimes}(\widetilde{\mathbb{U}})$ as the input space mainly because it is isomorphic to $\mathbb{U}$. Since there is a one-to-one correspondence between the extreme points of $\mathcal{P}_{\otimes}(\widetilde{\mathbb{U}})$ and $\mathscr{P}(\widetilde{\mathbb{U}})$, had we chosen to use the latter, the analysis and results would have remained unchanged. Even though we are not using the standard relaxed control setting, since $\mathcal{P}_{\otimes}(\widetilde{\mathbb{U}})$ is closed under convex combinations and limits, the theory goes through without any essential modifications.

For $\bar{p} \in (0, p_{\max}]$, let

$$H(\bar{p}) := \left\{ \pi \in \mathscr{G} : \int_{\mathbb{R}_+ \times \widetilde{\mathbb{U}}} h(\tilde{u})\, \pi(\mathrm{d}x, \mathrm{d}\tilde{u}) \le \bar{p} \right\} .$$

Then $H(\bar{p})$ is a closed, convex subset of $\mathscr{G}$. It is easy to see that it is also non-empty, provided $\bar{p} > 0$. Indeed, let $x' \in \mathbb{R}_+$ and consider the control $v_{x'}$ defined by

$$(v_{x'})_i = \begin{cases} 0 & \text{if } x \le x' , \\ p_{\max} & \text{if } x > x' , \end{cases} \qquad i = 1, \dots, N .$$

Under this control, the diffusion process in (4.5.1) is positive recurrent and its invariant probability measure has a density $f_{x'}$ which is a solution of (4.5.3). Let

$$\alpha_k := \frac{2 p_{\max}}{\sigma^2} \sum_{i=1}^{k} \gamma_i \pi_i , \quad k = 1, \dots, N . \tag{4.5.5}$$

The solution of (4.5.3) takes the form

$$f_{x'}(x) = \frac{\alpha_N e^{-\alpha_N (x-x')^+}}{1+\alpha_N x'}$$

where $(y)^+ := \max(y,0)$. Then

$$\int_{\mathbb{R}_+} h(v(x)) f_{x'}(x)\,\mathrm{d}x = \frac{p_{\max}}{1+\alpha_N x'}\,,$$

and it follows that $\pi_{v_{x'}} \in H(\bar{p})$, provided

$$x' \geq \frac{1}{\alpha_N}\Big(\frac{p_{\max}}{\bar{p}} - 1\Big).$$

Thus the optimization problem in (4.5.2) when restricted to stationary, stable controls is equivalent to

$$\text{minimize over } \pi \in H(\bar{p}) \quad \int_{\mathbb{R}_+\times\widetilde{\mathbb{U}}} c(x)\,\pi(\mathrm{d}x,\mathrm{d}\tilde{u})\,. \tag{4.5.6}$$

We also define

$$J^*(\bar{p}) := \inf_{\pi\in H(\bar{p})} \int_{\mathbb{R}_+\times\widetilde{\mathbb{U}}} c\,\mathrm{d}\pi\,. \tag{4.5.7}$$

As mentioned earlier, Assumption 4.2.1 does not hold, and hence the results in Section 4.2.1 cannot be quoted to assert existence. So we show directly in Theorem 4.5.3 that (4.5.6) attains a minimum in $H(\bar{p})$, and more specifically that this minimum is attained in $\mathfrak{U}_{\mathrm{se}}$.

Theorem 4.5.3 *Under Assumption 4.5.2, for any $\bar{p} \in (0, p_{\max}]$, there exists $v^* \in \mathfrak{U}_{\mathrm{se}}$ such that π_{v^*} attains the minimum in* (4.5.6).

Proof First suppose c is unbounded. Fix $\bar{p} \in (0, p_{\max}]$ and let $\{\pi_k\}$ be a sequence in $H(\bar{p})$ such that

$$\lim_{k\to\infty} \int_{\mathbb{R}_+\times\widetilde{\mathbb{U}}} c\,\mathrm{d}\pi_k \to J^*(\bar{p})\,. \tag{4.5.8}$$

Since the running cost c was assumed inf-compact, it follows that the sequence $\{\pi_k\}$ is tight in $\mathscr{P}(\mathbb{R}_+\times\widetilde{\mathbb{U}})$ and hence some subsequence converges weakly to some π^* in $\mathscr{P}(\mathbb{R}_+\times\widetilde{\mathbb{U}})$. Clearly, $\pi^* \in \mathscr{G}$. On the other hand, since h is continuous and bounded, and $\pi_k \to \pi^*$, weakly, we obtain

$$\int_{\mathbb{R}_+\times\widetilde{\mathbb{U}}} h\,\mathrm{d}\pi^* = \lim_{k\to\infty} \int_{\mathbb{R}_+\times\widetilde{\mathbb{U}}} h\,\mathrm{d}\pi_k \leq \bar{p}\,.$$

Hence, $\pi^* \in H(\bar{p})$. Since the map $\pi \mapsto \int c\,\mathrm{d}\pi$ is lower semicontinuous on $\mathscr{G}$, we obtain

$$\int_{\mathbb{R}_+\times\widetilde{\mathbb{U}}} c\,\mathrm{d}\pi^* \leq \liminf_{k\to\infty} \int_{\mathbb{R}_+\times\widetilde{\mathbb{U}}} c\,\mathrm{d}\pi_k = J^*(\bar{p})\,,$$

and thus π^* attains the infimum in (4.5.6).

Now suppose c is bounded. As before, let $\{\pi_k\}$ be a sequence in $\mathscr{G}$ satisfying (4.5.8) and let $\tilde{\pi}$ be a limit point of $\{\pi_k\}$ in $\bar{\mathscr{G}}$. Dropping to a subsequence if necessary, we suppose without changing the notation that $\pi_k \to \tilde{\pi}$ in $\bar{\mathscr{G}}$, and we decompose $\tilde{\pi}$ as

$$\tilde{\pi} = \delta\,\tilde{\pi}' + (1-\delta)\,\tilde{\pi}'' ,$$

with $\tilde{\pi}' \in \mathscr{G}$, $\tilde{\pi}'' \in \mathscr{P}(\{\infty\} \times \widetilde{\mathbb{U}})$, and $\delta \in [0,1]$. Then on the one hand

$$\delta \int_{\mathbb{R}_+ \times \widetilde{\mathbb{U}}} h\,\mathrm{d}\tilde{\pi}' \le \liminf_{k\to\infty} \int_{\mathbb{R}_+ \times \widetilde{\mathbb{U}}} h\,\mathrm{d}\pi_k \le \bar{p}\,, \tag{4.5.9}$$

while on the other, since c has a continuous extension on $\bar{\mathbb{R}}_+$,

$$\begin{aligned} J^*(\bar{p}) &= \lim_{k\to\infty} \int_{\bar{\mathbb{R}}_+ \times \widetilde{\mathbb{U}}} c\,\mathrm{d}\pi_k \\ &= \delta \int_{\mathbb{R}_+ \times \widetilde{\mathbb{U}}} c\,\mathrm{d}\tilde{\pi}' + (1-\delta)c_\infty\,. \end{aligned} \tag{4.5.10}$$

Note that since c is not a constant by Assumption 4.5.2, $J^*(\bar{p}) < c_\infty$, and hence, by (4.5.10), $\delta > 0$. Let $\tilde{v} \in \mathfrak{U}_{\mathrm{ssm}}$ be the control associated with $\tilde{\pi}'$ and $f_{\tilde{v}}$ the corresponding density of the invariant probability measure. Let $\hat{x} \in \mathbb{R}_+$ have the value

$$\hat{x} = \frac{1-\delta}{\delta f_{\tilde{v}}(0)}\,,$$

and $v^* \in \mathfrak{U}_{\mathrm{ssm}}$ defined by

$$v^*(x) = \begin{cases} 0 & \text{if } x \le \hat{x}\,, \\ \tilde{v}(x-\hat{x}) & \text{otherwise.} \end{cases}$$

The corresponding density is

$$f_{v^*}(x) = \begin{cases} \delta f_{\tilde{v}}(0) & \text{if } x \le \hat{x}\,, \\ \delta f_{\tilde{v}}(x-\hat{x}) & \text{otherwise.} \end{cases}$$

By (4.5.9),

$$\begin{aligned} \int_{\mathbb{R}_+} h(v^*(x)) f_{v^*}(x)\,\mathrm{d}x &= \delta \int_{\hat{x}}^{\infty} h(v^*(x)) f_{\tilde{v}}(x-\hat{x})\,\mathrm{d}x \\ &= \delta \int_{\mathbb{R}_+ \times \widetilde{\mathbb{U}}} h\,\mathrm{d}\tilde{\pi}' \\ &\le \bar{p}\,. \end{aligned}$$

By construction $f_{v^*}(x) \ge \delta f_{\tilde{v}}(x)$ for all $x \in \mathbb{R}_+$. Hence,

$$\int_{\mathbb{R}_+} c(x)[f_{v^*}(x) - \delta f_{\tilde{v}}(x)]\,\mathrm{d}x \le (1-\delta)c_\infty\,. \tag{4.5.11}$$

By (4.5.10) – (4.5.11),

$$\begin{aligned}\int_{\mathbb{R}_+} c(x) f_{v^*}(x)\,\mathrm{d}x &\le \delta \int_{\mathbb{R}_+} c(x) f_{\tilde{v}}(x)\,\mathrm{d}x + (1-\delta)c_\infty \\ &= J^*(\bar{p})\,.\end{aligned}$$

Therefore $v^* \in \mathfrak{U}_{\mathrm{ssm}}$ is optimal for (4.5.6). □

4.5.3 Lagrange multipliers and the HJB equation

In order to study the stationary Markov optimal controls for (4.5.6), we introduce a parameterized family of unconstrained optimization problems that is equivalent to the problem in (4.5.2) in the sense that stationary optimal controls for the former are also optimal for the latter and vice versa. We show that optimal controls for the unconstrained problem can be derived from the associated HJB equation. Hence, by studying the HJB equation we characterize the stationary Markov optimal controls in (4.5.6). We show that these are of a multi-threshold type and this enables us to reduce the optimal control problem to that of solving a system of $N+1$ algebraic equations. In addition, we show that optimality is achieved over the class of all admissible controls $\mathfrak{U}$, and not only over $\mathfrak{U}_{\mathrm{sm}}$.

With $\lambda \in \mathbb{R}_+$ playing the role of a Lagrange multiplier, we define

$$\begin{aligned}L(x,u,\bar{p},\lambda) &:= c(x) + \lambda(h(u) - \bar{p}) \\ \tilde{J}(v,\bar{p},\lambda) &:= \limsup_{T\to\infty} \frac{1}{T}\int_0^T L(X_t, v(t), \bar{p}, \lambda)\,\mathrm{d}t \\ \tilde{J}^*(\bar{p},\lambda) &:= \inf_{v\in\mathfrak{U}_{\mathrm{ssm}}} \tilde{J}(v,\bar{p},\lambda)\,.\end{aligned} \tag{4.5.12}$$

The choice of the optimization problem in (4.5.12) is motivated by the fact that $J^*(\bar{p})$, defined in (4.5.7), is a convex, decreasing function of $\bar{p}$. This is rather simple to establish. Let $\bar{p}', \bar{p}'' \in (0, p_{\max}]$ and denote by π', π'' the corresponding ergodic occupation measures that achieve the minimum in (4.5.6). Then, for any $\delta \in [0,1]$, $\pi_0 := \delta\pi' + (1-\delta)\pi''$ satisfies $\int h\,\mathrm{d}\pi_0 = \delta\bar{p}' + (1-\delta)\bar{p}''$, and since π_0 is sub-optimal for the optimization problem in (4.5.6) with power constraint $\delta\bar{p}' + (1-\delta)\bar{p}''$, we have

$$J^*(\delta\bar{p}' + (1-\delta)\bar{p}'') \le \int c\,\mathrm{d}\pi_0 = \delta J^*(\bar{p}') + (1-\delta)J^*(\bar{p}'')\,.$$

A separating hyperplane which is tangent to the graph of the function J^* at a point $(\bar{p}_0, J^*(\bar{p}_0))$, with $\bar{p}_0 \in (0, p_{\max}]$, takes the form

$$\{(\bar{p}, J) : J + \lambda_{\bar{p}_0}(\bar{p} - \bar{p}_0) = J^*(\bar{p}_0)\}$$

for some $\lambda_{\bar{p}_0} \in \mathbb{R}_+$ (see Figure 4.1).

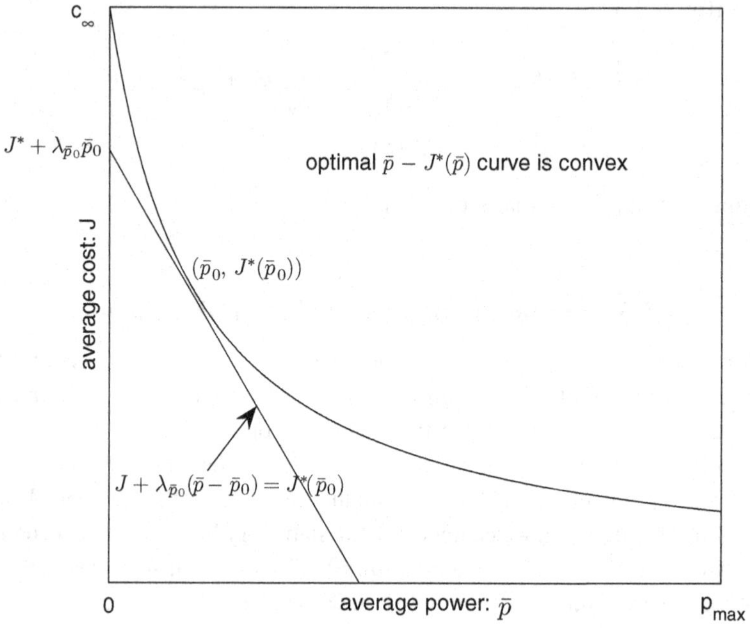

Figure 4.1 The separating hyperplane through $(\bar{p}_0, J^*(\bar{p}_0))$

Standard Lagrange multiplier theory yields the following [88, p. 217, theorem 1]:

Theorem 4.5.4 *Let $\bar{p}_0 \in (0, p_{\max}]$. There exists $\lambda_{\bar{p}_0} \in \mathbb{R}_+$ such that the minimization problem in* (4.5.6) *over $H(\bar{p}_0)$ and the problem*

$$\textit{minimize} \quad \int_{\mathbb{R}_+\times\tilde{\mathbb{U}}} L(x, \tilde{u}, \bar{p}_0, \lambda_{\bar{p}_0})\, \pi(\mathrm{d}x, \mathrm{d}\tilde{u}) \qquad \textit{over } \pi \in \mathscr{G} \tag{4.5.13}$$

attain the same minimum value $J^*(\bar{p}_0) = \tilde{J}^*(\bar{p}_0, \lambda_{\bar{p}_0})$ at some $\pi_0 \in H(\bar{p}_0)$. In particular,

$$\int_{\mathbb{R}_+\times\tilde{\mathbb{U}}} h(\tilde{u})\, \pi_0(\mathrm{d}x, \mathrm{d}\tilde{u}) = \bar{p}_0\,.$$

Characterizing the optimal control via the HJB equation associated with the unconstrained problem in (4.5.13), is made possible by first showing that under Assumption 4.5.2 the cost $L(x, u, \bar{p}, \lambda)$ is near-monotone and then employing the results in Sections 3.5–3.6. It is not difficult to show that under Assumption 4.5.2

$$\lim_{\bar{p}\to 0} J^*(\bar{p}) = \lim_{x\to\infty} c(x)\,. \tag{4.5.14}$$

Indeed, for $\bar{p} \in (0, p_{\max}]$, suppose $v \in \mathfrak{U}_{\mathrm{ssm}}$ is such that $\pi_v \in H(\bar{p})$. Letting

$$\gamma_{\max} := \max_i \{\gamma_i\}\,,$$

and using (4.5.4) we obtain

$$\begin{aligned}\bar{p} &\geq \int_0^\infty h(v(x)) f_v(x)\,\mathrm{d}x \\ &\geq \gamma_{\max}^{-1} \int_0^\infty b(v(x)) f_v(x)\,\mathrm{d}x \\ &= \frac{\sigma^2}{2\gamma_{\max} A_v}\,.\end{aligned}$$

Therefore

$$\begin{aligned}\int_0^\infty c(x) f_v(x)\,\mathrm{d}x &\geq \min_{x \geq \frac{1}{\sqrt{\bar{p}}}} \{c(x)\} \int_{\frac{1}{\sqrt{\bar{p}}}}^\infty f_v(x)\,\mathrm{d}x \\ &= \min_{x \geq \frac{1}{\sqrt{\bar{p}}}} \{c(x)\} \left(1 - \left(A_v \sqrt{\bar{p}}\right)^{-1}\right) \\ &\geq \min_{x \geq \frac{1}{\sqrt{\bar{p}}}} \{c(x)\} \left(1 - \frac{2\gamma_{\max}}{\sigma^2} \sqrt{\bar{p}}\right).\end{aligned}$$

Hence,

$$J^*(\bar{p}) \geq \min_{x \geq \frac{1}{\sqrt{\bar{p}}}} \{c(x)\} \left(1 - \frac{2\gamma_{\max}}{\sigma^2} \sqrt{\bar{p}}\right)$$

and (4.5.14) follows. We need the following lemma.

Lemma 4.5.5 *Let Assumption 4.5.2 hold and suppose c is bounded. Then for any $\bar{p} \in (0, p_{\max}]$, we have*

$$J^*(\bar{p}/2) < \tfrac{1}{2}\left(J^*(\bar{p}) + c_\infty\right).$$

Proof For $\bar{p} \in (0, p_{\max}]$, let $\pi^{(\bar{p})} \in H(\bar{p})$ be an optimal ergodic occupation measure, i.e.,

$$\int_{\mathbb{R}_+ \times \tilde{\mathbb{U}}} c\,\mathrm{d}\pi^{(\bar{p})} = J^*(\bar{p})\,.$$

Denote by $v^{(\bar{p})} \in \mathfrak{U}_{\mathrm{ssm}}$ the associated optimal control, and let $f_{v^{(\bar{p})}}$ stand for the density of the invariant probability measure. Set $\hat{x} = [f_{v^{(\bar{p})}}(0)]^{-1}$, and define $v^* \in \mathfrak{U}_{\mathrm{ssm}}$ by

$$v^*(x) := \begin{cases} 0 & \text{if } x \leq \hat{x}\,, \\ v^{(\bar{p})}(x - \hat{x}) & \text{otherwise.} \end{cases}$$

We compute the density of the invariant probability measure as

$$f_{v^*}(x) = \begin{cases} \frac{f_{v^{(\bar{p})}}(0)}{2} & \text{if } x \leq \hat{x}\,, \\ \frac{f_{v^{(\bar{p})}}(x - \hat{x})}{2} & \text{otherwise.} \end{cases}$$

Then

$$\int_{\mathbb{R}_+} h(v^*(x)) f_{v^*}(x)\,\mathrm{d}x = \frac{\bar{p}}{2}\,.$$

Observe that $f_{v^*}(x) \geq \frac{1}{2} f_{v(\bar{p})}(x)$ for all $x \in \mathbb{R}_+$. Hence, since $c(x) < c_\infty$ for all $x \in \mathbb{R}_+$, we obtain

$$\begin{aligned} J^*(\bar{p}/2) - \tfrac{1}{2} J^*(\bar{p}) &\leq \int_{\mathbb{R}_+} c(x)[f_{v^*}(x) - \tfrac{1}{2} f_{v(\bar{p})}(x)]\, \mathrm{d}x \\ &< \tfrac{1}{2} c_\infty , \end{aligned}$$

which yields the desired result. □

We are now ready to establish the near-monotone property of L. First, we introduce some new notation. For $\bar{p} \in (0, p_{\max}]$, let

$$\Lambda(\bar{p}) := \{\lambda \in \mathbb{R}_+ : J^*(\bar{p}') \geq J^*(\bar{p}) + \lambda(\bar{p} - \bar{p}') \quad \forall \bar{p}' \in (0, p_{\max}]\} ,$$

and

$$\Lambda := \bigcup_{\bar{p} \in (0, p_{\max}]} \Lambda(\bar{p}) .$$

Remark 4.5.6 It follows from the definition of $\Lambda(\bar{p})$ that

$$\inf_{\pi \in \mathscr{G}} \int_{\mathbb{R}_+ \times \widetilde{\mathbb{U}}} [c(x) + \lambda h(\tilde{u})]\, \pi(\mathrm{d}x, \mathrm{d}\tilde{u}) = J^*(\bar{p}) + \lambda \bar{p}$$

for all $\lambda \in \Lambda(\bar{p})$. To show this, let $\pi' \in \mathscr{G}$ and set $\bar{p}' := \int_{\mathbb{R}_+ \times \widetilde{\mathbb{U}}} h(\tilde{u})\, \pi'(\mathrm{d}x, \mathrm{d}\tilde{u})$. Then for $\lambda \in \Lambda(\bar{p})$, we have

$$\int_{\mathbb{R}_+ \times \widetilde{\mathbb{U}}} [c(x) + \lambda h(\tilde{u})]\, \pi'(\mathrm{d}x, \mathrm{d}\tilde{u}) \geq J^*(\bar{p}') + \lambda \bar{p}' \geq J^*(\bar{p}) + \lambda \bar{p} ,$$

where the second inequality follows by the definition of $\Lambda(\bar{p})$. Conversely,

$$\begin{aligned} \inf_{\pi \in \mathscr{G}} \int_{\mathbb{R}_+ \times \widetilde{\mathbb{U}}} [c(x) + \lambda h(\tilde{u})]\, \pi(\mathrm{d}x, \mathrm{d}\tilde{u}) &\leq \inf_{\pi \in H(\bar{p})} \int_{\mathbb{R}_+ \times \widetilde{\mathbb{U}}} [c(x) + \lambda h(\tilde{u})]\, \pi(\mathrm{d}x, \mathrm{d}\tilde{u}) \\ &\leq J^*(\bar{p}) + \lambda \bar{p} . \end{aligned}$$

Also, it is rather straightforward to show that $\Lambda = [0, \bar{\lambda})$ for some $\bar{\lambda} \in \mathbb{R}_+ \cup \{\infty\}$.

Lemma 4.5.7 *Let Assumption 4.5.2 hold. Then, for all $\bar{p} \in (0, p_{\max}]$ and $\lambda \in \Lambda$,*

$$\liminf_{x \to \infty} \inf_{\tilde{u} \in \widetilde{\mathbb{U}}} L(x, \tilde{u}, \bar{p}, \lambda) > \tilde{J}^*(\bar{p}, \lambda) . \tag{4.5.15}$$

Proof If c is inf-compact, (4.5.15) always follows. Otherwise, fix $\bar{p} \in (0, p_{\max}]$ and $\lambda \in \Lambda$. Let $\bar{p}' \in (0, p_{\max}]$ be such that $\lambda \in \Lambda(\bar{p}')$. By the definition of $\Lambda(\bar{p}')$, we have

$$J^*(\bar{p}'/2) \geq J^*(\bar{p}') + \lambda \tfrac{\bar{p}'}{2} .$$

Thus, using Lemma 4.5.5, we obtain

$$J^*(\bar{p}') + \lambda \bar{p}' < c_\infty . \tag{4.5.16}$$

The definition of $\tilde{J}^*(\bar{p},\lambda)$ in (4.5.12) yields

$$\begin{aligned}\tilde{J}^*(\bar{p},\lambda) &:= \inf_{\pi\in\mathscr{G}} \int_{\mathbb{R}_+\times\tilde{\mathbb{U}}} L(x,\tilde{u},\bar{p},\lambda)\,\pi(\mathrm{d}x,\mathrm{d}\tilde{u}) \\ &= -\lambda\bar{p} + \inf_{\pi\in\mathscr{G}} \int_{\mathbb{R}_+\times\tilde{\mathbb{U}}} (c(x)+\lambda h(\tilde{u}))\,\pi(\mathrm{d}x,\mathrm{d}\tilde{u})\,.\end{aligned}$$

Using this along with (4.5.16) and Remark 4.5.6 (since $\lambda\in\Lambda(\bar{p}')$), we obtain

$$\begin{aligned}\liminf_{x\to\infty}\ \inf_{\tilde{u}\in\tilde{\mathbb{U}}} L(x,\tilde{u},\bar{p},\lambda)+\lambda\bar{p} &= \liminf_{x\to\infty}\ c(x) \\ &> J^*(\bar{p}')+\lambda\bar{p}' \\ &= \inf_{\pi\in\mathscr{G}} \int_{\mathbb{R}_+\times\tilde{\mathbb{U}}} [c(x)+\lambda h(\tilde{u})]\,\pi(\mathrm{d}x,\mathrm{d}\tilde{u}) \\ &= \tilde{J}^*(\bar{p},\lambda)+\lambda\bar{p}\,,\end{aligned}$$

and the proof is complete. □

4.5.4 The structure of the optimal control

We can characterize optimality via the HJB equation. This is summarized as follows:

Theorem 4.5.8 *Let Assumption 4.5.2 hold. Fix $\bar{p}\in(0,p_{\max}]$ and $\lambda_{\bar{p}}\in\Lambda(\bar{p})$. Then there exists a unique solution pair (V,β), with $V\in C^2(\mathbb{R}_+)$ and $\beta\in\mathbb{R}$, to the HJB*

$$\min_{\tilde{u}\in\tilde{\mathbb{U}}} \left[\mathcal{L}^{\tilde{u}}V(x)+L(x,\tilde{u},\bar{p},\lambda_{\bar{p}})\right]=\beta\,, \tag{4.5.17}$$

subject to the boundary condition $\frac{\mathrm{d}V}{\mathrm{d}x}(0)=0$, and also satisfying

(a) $V(0)=0$;

(b) $\inf_{x\in\mathbb{R}_+}\ V(x)>-\infty$;

(c) $\beta\le\tilde{J}^*(\bar{p},\lambda_{\bar{p}})$.

Moreover, if the function v^* is a measurable selector from the minimizer in (4.5.17)*, then $v^*\in\mathfrak{U}_{\mathrm{se}}\subset\mathfrak{U}_{\mathrm{ssm}}$, and v^* is an optimal control for* (4.5.13)*, or equivalently, for* (4.5.6)*. Also, $\beta=\tilde{J}^*(\bar{p},\lambda_{\bar{p}})=J^*(\bar{p})$ (the second equality follows by Theorem 4.5.4).*

From the results of Section 3.4 it follows that the stationary Markov control v^* in Theorem 4.5.8 is optimal among all admissible controls $\mathfrak{U}$, and hence is a minimizer for (4.5.2).

Theorem 4.5.9 *Under Assumption 4.5.2, for any $\bar{p}\in(0,p_{\max}]$, there exists $v^*\in\mathfrak{U}_{\mathrm{se}}$ which attains the minimum in* (4.5.2) *over all controls in $\mathfrak{U}$.*

If $\Lambda(\bar{p})$ and $J^*(\bar{p})$ were known, then one could solve (4.5.17) and derive the optimal control. Since this is not the case, we embark on a different approach. We write (4.5.17) as

$$\min_{\tilde{u}\in\tilde{\mathbb{U}}} \left[\mathcal{L}^{\tilde{u}}V(x)+c(x)+\lambda_{\bar{p}}h(\tilde{u})\right]=\beta+\lambda_{\bar{p}}\,\bar{p}\,. \tag{4.5.18}$$

By Theorem 4.5.8, $J^*(\bar{p})$ is the smallest value of β for which there exists a solution pair (V,β) to (4.5.17), satisfying (b). This yields the following corollary:

Corollary 4.5.10 *Let Assumption 4.5.2 hold. For $\lambda \in \Lambda$, consider the HJB equation*

$$\min_{\tilde{u}\in\tilde{\mathbb{U}}} \left[\mathcal{L}^{\tilde{u}} V(x) + c(x) + \lambda h(\tilde{u})\right] = \varrho\,, \tag{4.5.19a}$$

subject to the boundary condition

$$\frac{\mathrm{d}V}{\mathrm{d}x}(0) = 0\,, \tag{4.5.19b}$$

and define

$$\mathcal{Q}_\lambda := \left\{(V,\varrho) \text{ solves } (4.5.19) : \inf_{x\in\mathbb{R}_+} V(x) > -\infty\right\} \tag{4.5.20a}$$

$$\varrho_\lambda := \min\left\{\varrho : (V,\varrho) \in \mathcal{Q}_\lambda\right\}. \tag{4.5.20b}$$

Then

$$\varrho_\lambda = \min_{v\in\mathfrak{U}_{\mathrm{ssm}}} \int_{\mathbb{R}_+\times\tilde{\mathbb{U}}} \left[c(x) + \lambda h(\tilde{u})\right] \pi_v(\mathrm{d}x, \mathrm{d}\tilde{u})\,. \tag{4.5.21}$$

Moreover, if $\bar{p}$ is a point in $(0, p_{\max}]$ such that $\lambda \in \Lambda(\bar{p})$, then

$$\varrho_\lambda = J^*(\bar{p}) + \lambda\bar{p}\,,$$

and if v^*_λ is a measurable selector of the minimizer in (4.5.19a) with $\varrho = \varrho_\lambda$, then v^*_λ is a stationary Markov optimal control for (4.5.13).

The minimizer in (4.5.19a) satisfies

$$\min_{\tilde{u}\in\tilde{\mathbb{U}}} \left[-b(\tilde{u})\frac{\mathrm{d}V}{\mathrm{d}x} + \lambda h(\tilde{u})\right] = \min_{\tilde{u}\in\tilde{\mathbb{U}}} \sum_j \left(\lambda - \gamma_j \frac{\mathrm{d}V}{\mathrm{d}x}\right) \pi_j \tilde{u}_j\,.$$

Thus the optimal control v^*_λ takes the following simple form: for $i = 1,\dots,N$ and $x \in \mathbb{R}_+$,

$$(v^*_\lambda)_i(x) = \begin{cases} 0 & \text{if } \gamma_i \frac{\mathrm{d}V}{\mathrm{d}x}(x) < \lambda\,, \\ p_{\max} & \text{if } \gamma_i \frac{\mathrm{d}V}{\mathrm{d}x}(x) \geq \lambda\,. \end{cases} \tag{4.5.22}$$

Thus, provided $\frac{\mathrm{d}V}{\mathrm{d}x}$ is monotone, the optimal control v^*_λ is of multi-threshold type, i.e., for each channel state j, there exists a queue-threshold $\hat{x}_j$ such that, at any time t, the optimal control transmits at peak power $p_{\max}$ over channel state j, if the queue length $X_t > \hat{x}_j$, and does not transmit otherwise.

The following lemma asserts the monotonicity of $\frac{\mathrm{d}V}{\mathrm{d}x}$, under the additional assumption that c is nondecreasing.

Lemma 4.5.11 *Suppose c is nondecreasing on $[0,\infty)$, and Assumption 4.5.2 holds. Then every $(V,\varrho) \in \mathcal{Q}_\lambda$ satisfies*

(a) $\frac{dV}{dx}$ *is nondecreasing;*

(b) *if c is unbounded, then $\frac{dV}{dx}$ is unbounded.*

Proof Equation (4.5.19a) takes the form

$$\frac{\sigma^2}{2}\frac{d^2V}{dx^2}(x) = \sum_j \pi_j p_{\max}\left[\gamma_j \frac{dV}{dx}(x) - \lambda\right]^+ + \varrho - c(x)\,, \qquad (4.5.23)$$

where the initial condition is given by (4.5.19b). Since c is nondecreasing, then by (4.5.21), $\varrho > c(0)$. We argue by contradiction. Suppose that for some $x' \in \mathbb{R}_+$, $\frac{d^2V}{dx^2}(x') = -\varepsilon < 0$. Let

$$x'' = \inf\left\{x > x' : \tfrac{d^2V}{dx^2}(x) \geq 0\right\}.$$

Since $\frac{d^2V}{dx^2}$ is continuous by Theorem 4.5.8, it must hold that $x'' > x'$. Suppose that $x'' < \infty$. Since $\frac{d^2V}{dx^2} < 0$ on $[x', x'')$ and both $\left[\gamma_j \frac{dV}{dx}(x) - \lambda\right]^+$ and $\varrho - c(x)$ are non-increasing, (4.5.23) implies that $\frac{d^2V}{dx^2}(x'') \leq \frac{d^2V}{dx^2}(x') < 0$. Thus x'' cannot be finite, from which it follows that $\frac{d^2V}{dx^2}(x) \leq -\varepsilon < 0$ for all $x \in [x', \infty)$, implying that V is not bounded below and we are led to a contradiction. This proves part (a). If c is not bounded, then it is clear from (4.5.23) that since $\frac{dV}{dx} \geq 0$, we must have $\frac{dV}{dx}(x) \to \infty$ as $x \to \infty$, and part (b) follows. □

The proof of Lemma 4.5.11 shows that if (V, ϱ) solves (4.5.19), then V is bounded below, if and only if $\frac{d^2V}{dx^2}(x) \geq 0$ for all $x \in \mathbb{R}_+$. Thus $\mathcal{Q}_\lambda$ defined in (4.5.20a) has an alternate characterization given in the following corollary.

Corollary 4.5.12 *Suppose the running cost c is nondecreasing on $[0, \infty)$ and that it satisfies Assumption 4.5.2. Then, for all $\lambda \in \Lambda$,*

$$\mathcal{Q}_\lambda = \left\{(V, \varrho) \text{ solves } (4.5.19) : \tfrac{d^2V}{dx^2} \geq 0 \text{ on } \mathbb{R}_+\right\}.$$

Comparing (4.5.18) and (4.5.19), a classical application of Lagrange duality [88, p. 224, theorem 1] yields the following:

Lemma 4.5.13 *If c satisfies Assumption 4.5.2 and is nondecreasing on $[0, \infty)$, then for any $\bar{p} \in (0, p_{\max}]$ and $\lambda_{\bar{p}} \in \Lambda(\bar{p})$, we have*

$$\varrho_{\lambda_{\bar{p}}} - \lambda_{\bar{p}}\,\bar{p} = \max_{\lambda \geq 0}\,\{\varrho_\lambda - \lambda\bar{p}\} = J^*(\bar{p})\,. \qquad (4.5.24)$$

Moreover, if λ_0 attains the maximum in $\lambda \mapsto \varrho_\lambda - \lambda\bar{p}$, then $\varrho_{\lambda_0} = J^*(\bar{p}) + \lambda_0\bar{p}$, which implies that $\lambda_0 \in \Lambda(\bar{p})$.

Proof The first part follows directly from [88, p. 224, theorem 1]. To prove that $\lambda_0 \in \Lambda(\bar{p})$, note that since the infimum of $\int_{\mathbb{R}_+ \times \tilde{\mathbb{U}}} c\, d\pi$ is achieved at some $\bar{\pi} \in H(\bar{p})$, then

$$\inf_{\pi \in \mathcal{G}} \int_{\mathbb{R}_+ \times \tilde{\mathbb{U}}} L(x, \tilde{u}, \bar{p}, \lambda_0)\, \pi(dx, d\tilde{u})$$

is attained at $\bar{\pi}$ and $\int_{\mathbb{R}_+\times\tilde{\mathbb{U}}} h(\tilde{u})\,\bar{\pi}(\mathrm{d}x, \mathrm{d}\tilde{u}) = \bar{p}$. Let $\bar{p}'$ be arbitrary and $\bar{\pi}'$ attain the minimum of of $\pi \mapsto \int_{\mathbb{R}_+\times\tilde{\mathbb{U}}} c\,\mathrm{d}\pi$ in $H(\bar{p}')$. We have

$$\begin{aligned}
J^*(\bar{p}) &= \inf_{\pi\in\mathscr{G}} \int_{\mathbb{R}_+\times\tilde{\mathbb{U}}} L(x, \tilde{u}, \bar{p}, \lambda_0)\,\pi(\mathrm{d}x, \mathrm{d}\tilde{u}) \\
&\leq \int_{\mathbb{R}_+\times\tilde{\mathbb{U}}} L(x, \tilde{u}, \bar{p}, \lambda_0)\,\bar{\pi}'(\mathrm{d}x, \mathrm{d}\tilde{u}) \\
&= J^*(\bar{p}') + \lambda_0(\bar{p}' - \bar{p})\,.
\end{aligned}$$

Since this holds for all $\bar{p}'$ it follows that $\lambda_0 \in \Lambda(\bar{p})$. □

Remark 4.5.14 Lemma 4.5.13 furnishes a method for solving (4.5.6). This can be done as follows: With λ viewed as a parameter, we first solve for ϱ_λ which is defined in (4.5.20b). Then, given $\bar{p}$, we obtain the corresponding value of the Lagrange multiplier via the maximization in (4.5.24). The optimal control can then be evaluated using (4.5.22), with $\lambda = \lambda_{\bar{p}}$. Section 4.5.6 contains an example demonstrating this method.

4.5.5 Solution of the HJB equation

In this section we present an explicit solution of the HJB equation (4.5.19). We deal only with the case where the cost function c is nondecreasing and inf-compact. However, the only reason for doing so is in the interest of simplicity and clarity. If c is bounded the optimal control may have less than N threshold points, but other than the need to introduce some extra notation, the solution we outline below for unbounded c holds virtually unchanged for the bounded case. Also, without loss of generality, we assume that $\gamma_1 > \cdots > \gamma_N > 0$.

We parameterize the controls in (4.5.22) by a collection of points

$$\{\hat{x}_1, \dots, \hat{x}_N\} \subset \mathbb{R}_+\,.$$

In other words, if V is the solution (4.5.23), then $\hat{x}_i$ is the least positive number such that $\frac{\mathrm{d}V}{\mathrm{d}x}(\hat{x}_i) \geq \gamma_i^{-1}$. Thus, if we define

$$\mathscr{X}^N := \{\hat{x} = (\hat{x}_1, \dots, \hat{x}_N) \in \mathbb{R}_+^N : \hat{x}_1 < \cdots < \hat{x}_N\}\,,$$

then for each $\hat{x} \in \mathscr{X}^N$, there corresponds a multi-threshold control $v_{\hat{x}}$ of the form

$$(v_{\hat{x}})_i(x) = \begin{cases} p_{\max} & \text{if } x \geq \hat{x}_i\,, \\ 0 & \text{otherwise}\,, \end{cases} \qquad 1 \leq i \leq N\,. \tag{4.5.25}$$

To facilitate expressing the solution of (4.5.23), we need to introduce some new notation. For $i = 1, \dots, N$, define

$$\tilde{\pi}_i := \sum_{j=1}^{i} \pi_i\,, \quad \tilde{\gamma}_i := \sum_{j=1}^{i} \pi_i\gamma_i\,, \quad \Gamma_i := \frac{\tilde{\gamma}_i}{\gamma_i} - \tilde{\pi}_i\,.$$

Note that from (4.5.5), we obtain the identity

$$\alpha_i = \frac{2p_{\max}}{\sigma^2}\tilde{\gamma}_i\,,\quad i = 1,\dots,N\,.$$

For $x, z \in \mathbb{R}_+$, with $z \le x$, we define the functions

$$F_0(\varrho, x) := \varrho x - \int_0^x c(y)\,\mathrm{d}y\,,$$

and for $i = 1,\dots,N$,

$$F_i(\varrho, x, z) := [\varrho + \lambda p_{\max}\Gamma_i]\left(1 - e^{\alpha_i(z-x)}\right) - \alpha_i \int_z^x e^{\alpha_i(z-y)}c(y)\,\mathrm{d}y\,,$$

$$G_i(\varrho, x, z) := \varrho + \lambda p_{\max}\Gamma_i - \alpha_i \int_z^x e^{\alpha_i(z-y)}c(y)\,\mathrm{d}y - e^{\alpha_i(z-x)}c(x)\,.$$

Using the convention $\hat{x}_{N+1} \equiv \infty$, we write the solution of (4.5.23) as

$$\frac{\mathrm{d}V}{\mathrm{d}x}(x) = \begin{cases} \frac{2}{\sigma^2}F_0(\varrho, x) & \text{if } 0 \le x < \hat{x}_1\,, \\ \frac{2}{\sigma^2\alpha_i}e^{\alpha_i(x-\hat{x}_i)}F_i(\varrho, x, \hat{x}_i) + \frac{\lambda}{\gamma_i} & \text{if } x \in [\hat{x}_i, \hat{x}_{i+1})\,,\ i \ge N\,. \end{cases} \tag{4.5.26}$$

In addition, the following boundary conditions are satisfied

$$\begin{aligned} F_0(\varrho, \hat{x}_1) &= \frac{\lambda\sigma^2}{2\gamma_1} = 0\,, \\ F_i(\varrho, \hat{x}_{i+1}, \hat{x}_i) &= \lambda p_{\max}\tilde{\gamma}_i e^{\alpha_i(\hat{x}_i - \hat{x}_{i+1})}\left(\frac{1}{\gamma_{i+1}} - \frac{1}{\gamma_i}\right), \quad 1 \le i < N\,. \end{aligned} \tag{4.5.27}$$

Also, for $i = 1,\dots,N$, we have

$$\frac{\mathrm{d}^2V}{\mathrm{d}x^2}(x) = \frac{2}{\sigma^2}e^{\alpha_i(x-\hat{x}_i)}G_i(\varrho, x, \hat{x}_i)\,,\quad x \in (\hat{x}_i, \hat{x}_{i+1})\,.$$

Since c is monotone, the map

$$x \mapsto \alpha_i \int_z^x e^{\alpha_i(z-y)}c(y)\,\mathrm{d}y + e^{\alpha_i(z-x)}c(x) \tag{4.5.28}$$

is nondecreasing. Moreover, using the fact that c is either inf-compact (or strictly monotone increasing, when bounded), an easy calculation yields

$$G_i(\varrho, x, z) > \lim_{x\to\infty} G_i(\varrho, x, z)\,. \tag{4.5.29}$$

Suppose $\hat{x} \in \mathscr{X}^N$ are the threshold points of a solution (V, ϱ) of (4.5.23). It follows from (4.5.29) that $\lim_{x\to\infty} G_N(\varrho, x, \hat{x}_N) \ge 0$ is a necessary and sufficient condition for $\frac{\mathrm{d}^2V}{\mathrm{d}x^2}(x) \ge 0$, $x \in (\hat{x}_N, \infty)$. This condition translates to

$$\varrho + \lambda p_{\max}\Gamma_N - \alpha_N \int_{\hat{x}_N}^{\infty} e^{\alpha_N(\hat{x}_N - y)}c(y)\,\mathrm{d}y \ge 0\,. \tag{4.5.30}$$

The arguments in the proof of Lemma 4.5.11 actually show that (4.5.30) is sufficient for $\frac{\mathrm{d}^2V}{\mathrm{d}x^2}$ to be nonnegative on $\mathbb{R}_+$. We sharpen this result by showing in Lemma 4.5.15 below that (4.5.30) implies that $\frac{\mathrm{d}^2V}{\mathrm{d}x^2}$ is strictly positive on $\mathbb{R}_+$.

Lemma 4.5.15 *Suppose $\hat{x} \in \mathscr{X}^N$ satisfies* (4.5.27). *If* (4.5.30) *holds, then $\varrho > c(\hat{x}_1)$ and $G_i(\varrho, x, \hat{x}_i) > 0$ for all $x \in [\hat{x}_i, \hat{x}_{i+1}]$, $i = 0, \ldots, N-1$.*

Proof We argue by contradiction. If $\varrho \le c(\hat{x}_1)$, then $G_1(\varrho, \hat{x}_i, \hat{x}_i) \le 0$, hence it is enough to assume that $G_i(\varrho, x, \hat{x}_i) \le 0$ for some $x \in [\hat{x}_i, \hat{x}_{i+1}]$ and $i \in \{1, \ldots, N-1\}$. Then, since (4.5.28) is nondecreasing, we have

$$G_i(\varrho, \hat{x}_{i+1}, \hat{x}_i) \le 0 . \tag{4.5.31}$$

Therefore, since

$$F_i(\varrho, x, \hat{x}_i) = G_i(\varrho, x, \hat{x}_i) + e^{\alpha_i(\hat{x}_i - x)} c(x) - [\varrho + \lambda p_{\max} \Gamma_i] e^{\alpha_i(\hat{x}_i - x)} , \tag{4.5.32}$$

combining (4.5.27) and (4.5.31)–(4.5.32), we obtain

$$c(\hat{x}_{i+1}) - \varrho - \lambda p_{\max} \Gamma_i \ge \lambda \tilde{\gamma}_i \left(\frac{1}{\gamma_{i+1}} - \frac{1}{\gamma_i} \right) ,$$

which simplifies to

$$c(\hat{x}_{i+1}) - \varrho + \lambda p_{\max} \tilde{\pi}_i \ge \lambda p_{\max} \frac{\tilde{\gamma}_i}{\gamma_{i+1}} . \tag{4.5.33}$$

Since

$$\frac{\tilde{\gamma}_i}{\gamma_{i+1}} - \tilde{\pi}_i = \frac{\tilde{\gamma}_{i+1}}{\gamma_{i+1}} - \tilde{\pi}_{i+1} = \Gamma_{i+1} ,$$

(4.5.33) yields

$$\varrho + \lambda p_{\max} \Gamma_{i+1} \le c(\hat{x}_{i+1}) . \tag{4.5.34}$$

Using the monotonicity of $x \mapsto G_{i+1}(\varrho, x, \hat{x}_{i+1})$ together with (4.5.34), we obtain $G_{i+1}(\varrho, x, \hat{x}_{i+1}) \le 0$ for all $x \in [\hat{x}_{i+1}, \hat{x}_{i+2}]$, and iterating this argument, we conclude that $G_N(\varrho, x, \hat{x}_N) \le 0$ for all $x \in (\hat{x}_N, \infty)$, thus contradicting (4.5.30). □

Combining Corollary 4.5.12 with Lemma 4.5.15, yields the following.

Corollary 4.5.16 *Suppose (V, ϱ) satisfies* (4.5.26)–(4.5.27) *for some $\hat{x} \in \mathscr{X}^N$ and $\lambda \in \Lambda$. Then $(V, \varrho) \in \mathscr{Q}_\lambda$, if and only if* (4.5.30) *holds.*

For $\lambda \in \Lambda$, define

$$R_\lambda := \{\varrho \in \mathbb{R}_+ : (V, \varrho) \in \mathscr{Q}_\lambda\} .$$

For each $\varrho \in R_\lambda$, equations (4.5.27) define a map $\varrho \mapsto \hat{x}$, which we denote by $\hat{x}(\varrho)$.

Lemma 4.5.17 *Let $\lambda \in \Lambda$ and suppose $\varrho_0 \in R_\lambda$. With ϱ_λ as defined in* (4.5.20b), *and denoting the left-hand side of* (4.5.30) *by $G_N(\varrho, \infty, \hat{x}_N)$, the following hold:*

(a) *if $\varrho' > \varrho_0$, then $\varrho' \in R_\lambda$ and $G_N(\varrho', \infty, \hat{x}(\varrho')) > 0$;*

(b) *if* $G_N(\varrho_0, \infty, \hat{x}(\varrho_0)) > 0$, *then* $\varrho_0 > \varrho_\lambda$*;*

(c) $R_\lambda = [\varrho_\lambda, \infty)$*, and* ϱ_λ *is the only point in* R_λ *which satisfies*

$$G_N(\varrho_\lambda, \infty, \hat{x}(\varrho_\lambda)) = 0\,.$$

Proof Part (a) follows easily from (4.5.23). Denoting by V_0 and V' the solutions of (4.5.23) corresponding to ϱ_0 and ϱ', respectively, a standard argument shows that

$$\frac{\mathrm{d}^2(V' - V_0)}{\mathrm{d}x^2}(x) \geq \varrho' - \varrho_0 > 0 \qquad \forall x \in \mathbb{R}_+\,,$$

implying

$$\frac{\mathrm{d}V'}{\mathrm{d}x}(x) \geq \frac{\mathrm{d}V_0}{\mathrm{d}x}(x) \qquad \forall x \in \mathbb{R}_+\,. \tag{4.5.35}$$

Hence, since by the definition of $\mathcal{Q}_\lambda$, V_0 is bounded below, the same holds for V', in turn, implying that $(V', \varrho') \in \mathcal{Q}_\lambda$. By (4.5.35), we have $\hat{x}(\varrho') \leq \hat{x}(\varrho_0)$, and since $\hat{x}_N \mapsto G_N(\varrho, \infty, \hat{x}_N)$ is non-increasing and $\varrho' > \varrho_0$, we obtain $G_N(\varrho', \infty, \hat{x}(\varrho')) > 0$.

To prove (b), write (4.5.27) in the form $\tilde{F}(\varrho, \hat{x}) = 0$, with $\tilde{F} \colon \mathbb{R}_+^{N+1} \to \mathbb{R}_+^N$. The map $\tilde{F}$ is continuously differentiable and as a result of Lemma 4.5.15 its Jacobian $D_{\hat{x}}F$ with respect to $\hat{x}$ has full rank at $(\varrho_0, \hat{x}(\varrho_0))$. By the implicit function theorem, there exists an open neighborhood $W(\varrho_0)$ and a continuous map $\hat{x} \colon W(\varrho_0) \to \mathbb{R}_+$, such that $\tilde{F}(\varrho, \hat{x}(\varrho)) = 0$ for all $\varrho \in W(\varrho_0)$. Using the continuity of G_N, we may shrink $W(\varrho_0)$ so that $G_N(\varrho, \infty, \hat{x}(\varrho)) > 0$ for all $\varrho \in W(\varrho_0)$. Hence $W(\varrho_0) \subset R_\lambda$, implying that $\varrho_0 > \varrho_\lambda$.

Part (c) follows directly from (a) and (b). □

Combining Corollary 4.5.10 and Lemma 4.5.15, we obtain the following characterization of the solution to the HJB equation (4.5.19).

Theorem 4.5.18 *Let the running cost c be nondecreasing and inf-compact. Then the threshold points* $(\hat{x}_1, \dots, \hat{x}_N) \in \mathcal{X}^N$ *of the stationary Markov optimal control in* (4.5.25) *and the optimal value* $\varrho_\lambda > 0$ *are the (unique) solution of the set of* $N + 1$ *algebraic equations which is comprised of the equations in* (4.5.27) *and* $G_N(\varrho_\lambda, \infty, \hat{x}(\varrho_\lambda)) = 0$.

4.5.6 Minimizing the mean delay

We specialize the optimization problem to the case $c(x) = x$, which corresponds to minimizing the mean delay.

First consider the case $N = 1$, letting $\alpha \equiv \alpha_1$ and $\hat{x} \equiv \hat{x}_1$. Solving (4.5.19) we obtain

$$\frac{\mathrm{d}V}{\mathrm{d}x}(x) = \frac{2\varrho}{\sigma^2}x - \frac{x^2}{\sigma^2}\,, \quad x \leq \hat{x}\,,$$

with

$$\hat{x} = \varrho - \sqrt{\varrho^2 - \tfrac{\lambda\sigma^2}{\gamma}}\,. \tag{4.5.36}$$

Also, for $x \geq \hat{x}$,

$$\frac{\mathrm{d}V}{\mathrm{d}x}(x) = \frac{2e^{\alpha(x-\hat{x})}}{\sigma^2\alpha}\left(\varrho - \hat{x} - \frac{1}{\alpha}\right) + \frac{2}{\sigma^2\alpha}\left(\varrho - \lambda p_{\max} + x + \frac{1}{\alpha}\right).$$

Therefore, for $x > \hat{x}$,

$$\frac{\mathrm{d}^2V}{\mathrm{d}x^2}(x) = \frac{2}{\sigma^2}\left(\varrho - \hat{x} - \frac{1}{\alpha}\right)e^{\alpha(x-\hat{x})} + \frac{2}{\sigma^2\alpha}. \tag{4.5.37}$$

It follows from (4.5.37) that

$$\varrho_\lambda = \hat{x} + \frac{1}{\alpha}. \tag{4.5.38}$$

By (4.5.36) and (4.5.38),

$$\varrho_\lambda = \sqrt{\tfrac{1}{\alpha^2} + \tfrac{\lambda\sigma^2}{\gamma}}. \tag{4.5.39}$$

Let $\bar{p} \in (0, p_{\max}]$ be given. Applying Lemma 4.5.13, we obtain from (4.5.39)

$$\lambda_{\bar{p}} = \frac{p_{\max}}{2\alpha\bar{p}^2} - \frac{1}{2\alpha p_{\max}}$$

and

$$J^*(\bar{p}) = \frac{1}{2\alpha}\left(\frac{p_{\max}}{\bar{p}} + \frac{\bar{p}}{p_{\max}}\right).$$

Moreover, the threshold point of the optimal control is given by

$$\hat{x} = \frac{1}{\alpha}\left(\frac{p_{\max}}{\bar{p}} - 1\right).$$

Now consider the case $N = 2$. We obtain:

$$\frac{\mathrm{d}V}{\mathrm{d}x}(x) = \frac{2\varrho}{\sigma^2}x - \frac{x^2}{\sigma^2}, \quad x \leq \hat{x}_1, \tag{4.5.40a}$$

$$\begin{aligned}\frac{\mathrm{d}V}{\mathrm{d}x}(x) = {}& \frac{2}{\sigma^2\alpha_1}\left(\varrho - \hat{x}_1 - \frac{1}{\alpha_1}\right)[e^{\alpha_1(x-\hat{x}_1)} - 1] \\ & + \frac{2(x-\hat{x}_1)}{\sigma^2\alpha_1} + \frac{\lambda}{\gamma_1}, \quad \hat{x}_1 \leq x < \hat{x}_2,\end{aligned} \tag{4.5.40b}$$

and for $x \geq \hat{x}_2$,

$$\begin{aligned}\frac{\mathrm{d}V}{\mathrm{d}x}(x) = {}& \frac{2}{\sigma^2\alpha_2}\left(\varrho - \hat{x}_2 - \frac{1}{\alpha_2} + \lambda p_{\max}\pi_1\frac{\gamma_1 - \gamma_2}{\gamma_2}\right) \\ & \times [e^{\alpha_2(x-\hat{x}_2)} - 1] + \frac{2(x-\hat{x}_2)}{\sigma^2\alpha_2} + \frac{\lambda}{\gamma_2}.\end{aligned} \tag{4.5.40c}$$

Since $\frac{\mathrm{d}V}{\mathrm{d}x}(\hat{x}_1) = \frac{\lambda}{\gamma_1}$, we obtain by (4.5.40a),

$$\hat{x}_1 = \varrho - \sqrt{\varrho^2 - \tfrac{\lambda\sigma^2}{\gamma_1}}. \tag{4.5.41}$$

By (4.5.40c), $\frac{d^2V}{dx^2}(x) \geq 0$ for all $x > \hat{x}_2$, if and only if

$$\varrho - \hat{x}_2 - \frac{1}{\alpha_2} + \lambda p_{\max}\pi_1 \frac{\gamma_1 - \gamma_2}{\gamma_2} \geq 0\,.$$

Also, since $\frac{dV}{dx}(\hat{x}_2) = \frac{\lambda}{\gamma_2}$, we obtain from (4.5.40b),

$$\left(\varrho - \hat{x}_1 - \frac{1}{\alpha_1}\right)[e^{\alpha_1(\hat{x}_2 - \hat{x}_1)} - 1] + \hat{x}_2 - \hat{x}_1 = \frac{\sigma^2 \lambda \alpha_1}{2}\left(\frac{1}{\gamma_2} - \frac{1}{\gamma_1}\right).$$

We apply Theorem 4.5.18 to compute the optimal control. Let $\hat{x}_1(\varrho)$ be as in (4.5.41) and

$$\hat{x}_2(\varrho) := \hat{x}_1(\varrho) + \sqrt{\varrho^2 - \tfrac{\lambda\sigma^2}{\gamma_1}} - \frac{1}{\alpha_2} + \lambda p_{\max}\pi_1 \frac{\gamma_1 - \gamma_2}{\gamma_2}\,.$$

Then ϱ_λ is the solution of

$$\left(\sqrt{\varrho^2 - \tfrac{\lambda\sigma^2}{\gamma_1}} - \frac{1}{\alpha_1}\right) e^{\alpha_2(\hat{x}_2(\varrho) - \hat{x}_1(\varrho))} + \left(\frac{1}{\alpha_1} - \frac{1}{\alpha_2}\right) = 0\,.$$

This completely specifies an optimal stationary control.

4.6 Bibliographical note

Section 4.2. This is based on [31, 38].

Section 4.3. This is based on [36].

Section 4.5. This is from [123].

5

Controlled Switching Diffusions

In this chapter we extend the results of Chapter 3 to switching diffusions, also referred to as diffusions with a discrete parameter [108]. A controlled switching diffusion is a typical example of a hybrid system which arises in numerous applications of systems with multiple modes or failure modes, such as fault tolerant control systems, multiple target tracking, flexible manufacturing systems etc. The state of the system at time t is given by a pair $(X_t, Y_t) \in \mathbb{R}^d \times \mathbb{S}$, $\mathbb{S} = \{1, 2, \dots, N\}$. The continuous component X_t is governed by a controlled diffusion process with a drift vector and diffusion matrix which depend on the discrete component Y_t. Thus X_t switches from one diffusion path to another as the discrete component Y_t jumps from one state to another. On the other hand, the discrete component Y_t is a controlled Markov chain with a transition matrix depending on the continuous component. The evolution of the process (X_t, Y_t) is governed by the following equations:

$$\mathrm{d}X_t = b(X_t, Y_t, U_t)\,\mathrm{d}t + \sigma(X_t, Y_t)\,\mathrm{d}W_t\,,$$

$$\mathbb{P}\Bigl(Y_{t+\delta t} = j \Bigm| Y_t = i, X_s, Y_s, s \le t\Bigr) = \lambda_i^j(X_t, U_t)\delta t + o(\delta t), \quad i \ne j\,,$$

with $\lambda_i^j \ge 0$ for $i \ne j$, and $\sum_{j=1}^N \lambda_i^j = 0$.

For the most part we follow the structure of Chapter 3. Virtually all of the results for a controlled nondegenerate diffusion can be extended to controlled switching diffusions. From an analytical viewpoint, the study of ergodic control for switching diffusions relies on the extension of the theory of scalar elliptic equations in non-divergence form to coupled cooperative elliptic systems of second order. The key theorems for elliptic systems used in this chapter are summarized in Section 5.2.

5.1 The mathematical model

5.1.1 Construction of a switching diffusion as a Markov process

We construct the controlled switching diffusion $Z = (X, Y)$ as a strong Markov process, continuous from the right, in the phase space $\mathbb{R}^d \times \mathbb{S}$, and action space $\mathbb{U}$, a compact metric space, on a probability space $(\Omega, \mathfrak{F}, \mathbb{P})$. For the time being, let $u \in \mathbb{U}$ be fixed. Let τ be the first exit time of the discrete component Y from the initial state $Y_0 = y$. On the set $\{\tau > t\}$ the exit time τ satisfies $\tau \circ \theta_t = \tau - t$. Using this property, we can verify that for each $y \in \mathbb{S}$, the function $Q_y(t, x, A) = \mathbb{P}_{x,y}(\tau > t,\ X_t \in A)$, $A \in \mathscr{B}(\mathbb{R}^d)$, is a transition function, i.e., a sub-probability kernel that satisfies the Chapman-Kolmogorov equations. Indeed

$$\begin{aligned} Q_y(t+s, x, A) &= \mathbb{P}_{x,y}(\tau > t+s,\ X_{t+s} \in A) \\ &= \mathbb{E}_{x,y}[(\mathbb{I}_A(X_s)\,\mathbb{I}\{\tau > s\}) \circ \theta_t] \\ &= \int_{\mathbb{R}^d} Q_y(t, x, \mathrm{d}x')\,\mathbb{E}_{x',y}\left[\mathbb{I}_A(X_s)\,\mathbb{I}\{\tau > s\}\right] \\ &= \int_{\mathbb{R}^d} Q_y(t, x, \mathrm{d}x') Q_y(s, x', A)\,. \end{aligned}$$

The transition function $Q_y(t, x, A)$, indexed by the parameter y, is associated with a process governed on $[0, \tau)$ by the Itô equation:

$$\mathrm{d}X_t = b(X_t, y, u)\,\mathrm{d}t + \sigma(X_t, y)\,\mathrm{d}W_t\,, \tag{5.1.1}$$

with

$$\begin{aligned} b &= \left[b^1, \ldots, b^d\right]^{\mathsf{T}} : \mathbb{R}^d \times \mathbb{S} \times \mathbb{U} \to \mathbb{R}^d, \\ \sigma &= \left[\sigma^{ij}\right] : \mathbb{R}^d \times \mathbb{S} \to \mathbb{R}^{d \times d}\,. \end{aligned}$$

At the same time, the distribution of τ is specified as follows.

$$\mathbb{P}_{x,y}\left(\tau > t \,\middle|\, \mathfrak{F}_t^X\right) = \exp\left(-\int_0^t \lambda(X_s, y, u)\,\mathrm{d}s\right). \tag{5.1.2}$$

At the jump time τ the transitions of the discrete component Y are governed by a transition function $\hat{Q}(i, j; x, u)$, having a parametric dependence on $(x, u) \in \mathbb{R}^d \times \mathbb{U}$, and satisfying $\hat{Q}(i, i; x, u) = 0$ for all $i \in \mathbb{S}$. In other words, we require that the process satisfies

$$\mathbb{P}\left(Y_\tau = j \,\middle|\, \mathfrak{F}_{\tau-}^Z\right) = \hat{Q}(Y_{\tau-}, j; X_\tau, u)\,, \tag{5.1.3}$$

where $\mathfrak{F}_t^Z$ is the right-continuous completion of $\sigma\{(X_s, Y_s) : s \le t\}$. Let $\mathfrak{p}$ be a Poisson random measure on $\mathbb{R} \times \mathbb{R}_+$, which is independent of W, with intensity $\mathbb{E}\left[\mathfrak{p}(\mathrm{d}\xi \times \mathrm{d}t)\right] = \mathfrak{m}(\mathrm{d}\xi)\,\mathrm{d}t$, where $\mathfrak{m}$ denotes the Lebesgue measure on $\mathbb{R}$. Let

$$\{\Delta_{ij}(x, u) : i \neq j\,,\ i, j \in \mathbb{S}\}$$

be consecutive, with respect to lexicographic ordering, left closed, right open intervals of the real line, with $\Delta_{ij}(x, u)$ having length $\lambda(x, i, u)\hat{Q}(i, j; x, u)$. Also define the function $g : \mathbb{R}^d \times \mathbb{S} \times \mathbb{U} \times \mathbb{R} \to \mathbb{R}$ by

$$g(x, i, u, \xi) = \begin{cases} y - i & \text{if } \xi \in \Delta_{iy}(x, u)\,, \\ 0 & \text{otherwise.} \end{cases} \tag{5.1.4}$$

The jump times and transitions in (5.1.2) – (5.1.3) may be realized by the stochastic integral

$$\mathrm{d}Y_t = \int_{\mathbb{R}} g(X_t, Y_{t-}, u, \xi)\, \mathfrak{p}(\mathrm{d}\xi \times \mathrm{d}t)\,. \tag{5.1.5}$$

Equations (5.1.1) and (5.1.5) describe the process Z.

Let

$$\begin{aligned} \lambda_y^j(x,u) &:= \lambda(x,y,u)\hat{Q}(y,j;x,u)\,, \quad j \neq y\,, \\ \lambda_y^y(x,u) &:= -\lambda(x,y,u)\,. \end{aligned} \tag{5.1.6}$$

By (5.1.6), $\sum_j \lambda_y^j = 0$.

5.1.2 The controlled switching diffusion model

We make the following assumptions on the parameters.

Assumption 5.1.1 Let $\gamma\colon (0,\infty) \to (0,\infty)$ be a positive function that plays the role of a parameter.

(i) On every ball B_R the functions $b(x,y,u)$, $\sigma^{ij}(x,y)$ and $\lambda_y^j(x,u)$ are continuous and Lipschitz in x, with a Lipschitz constant $\gamma(R)$ uniformly with respect to u.

(ii) Also $b(x,y,u)$ and $\sigma(x,y)$ have linear growth in x, that is

$$|b(x,y,u)| + \|\sigma(x,y)\| \le \gamma(1)(1+|x|) \qquad \forall (x,y,u) \in \mathbb{R}^d \times \mathbb{S} \times \mathbb{U}\,.$$

(iii) σ_{ij} is uniformly elliptic, on every ball B_R, i.e., with $a := \frac{1}{2}\sigma\sigma^{\mathsf{T}}$, we have

$$\sum_{i,j=1}^{d} a^{ij}(x,y)\xi_i\xi_j \ge \gamma^{-1}(R)|\xi|^2 \qquad \forall (x,y) \in B_R \times \mathbb{S}\,,$$

for all $\xi = (\xi_1,\ldots,\xi_d) \in \mathbb{R}^d$.

Remark 5.1.2 By Assumption 5.1.1 (i), λ_y^j are locally bounded. This implies that there is no accumulation of an infinite time of jumps at the discrete component in any finite time interval. Indeed, in view of (5.1.2), the number of jumps in any interval $[0,t]$ is a.s. finite.

The controlled, switching diffusion process constructed in Section 5.1.1 can be modeled via the following Itô stochastic differential equations.

$$\begin{aligned} \mathrm{d}X_t &= b(X_t, Y_t, U_t)\,\mathrm{d}t + \sigma(X_t, Y_t)\,\mathrm{d}W_t\,, \\ \mathrm{d}Y_t &= \int_{\mathbb{R}} g(X_t, Y_{t-}, U_t, \xi)\, \mathfrak{p}(\mathrm{d}\xi \times \mathrm{d}t)\,, \end{aligned} \tag{5.1.7}$$

for $t \ge 0$, with g defined in (5.1.4), where

(a) X_0 and Y_0 are prescribed $\mathbb{R}^d$-valued and $\mathbb{S}$-valued random variables, respectively.

(b) W is a d-dimensional standard Wiener process, and $\mathfrak{p}$ is a Poisson random measure on $\mathbb{R} \times \mathbb{R}_+$, with intensity $\mathfrak{m}(\mathrm{d}z) \times \mathrm{d}t$.

(c) X_0, Y_0, W, and $\mathfrak{p}$ are independent.

(d) U is a $\mathbb{U}$-valued process with measurable sample paths satisfying the non-anticipativity property that the σ-fields

$$\sigma\{X_0, Y_0, U_s, W_s, \mathfrak{p}(A, B) : A \in \mathscr{B}([0, s]),\ B \in \mathscr{B}(\mathbb{R}),\ s \le t\},$$

and

$$\sigma\{W_s - W_t, \mathfrak{p}(A, B) : A \in \mathscr{B}([s, \infty)),\ B \in \mathscr{B}(\mathbb{R}),\ s \ge t\}.$$

are independent for each $t \in \mathbb{R}$.

If $(W, \mathfrak{p}, X_0, Y_0, U)$ satisfying (a)–(d) above are given on a prescribed probability space $(\Omega, \mathfrak{F}, \mathbb{P})$, then, under Assumption 5.1.1, equation (5.1.7) admits an a.s. unique strong solution with $X \in C([0, \infty); \mathbb{R}^d)$, $Y \in \mathcal{D}([0, \infty); \mathbb{S})$, where $\mathcal{D}([0, \infty); \mathbb{S})$ is the space of right-continuous functions on $[0, \infty)$ with left limits taking values in $\mathbb{S}$. This follows readily using the Picard iterations. As also defined in Chapter 2, a control process U satisfying (d) is called admissible, and the set of all admissible controls is denoted by $\mathfrak{U}$.

Definition 5.1.3 If $\mathcal{X}(\mathbb{R}^d)$ is a vector space of real functions over $\mathbb{R}^d$, we adopt the notation $\mathcal{X}(\mathbb{R}^d \times \mathbb{S})$ to indicate the space $(\mathcal{X}(\mathbb{R}^d))^N$, endowed with the product topology. For example,

$$L^p(\mathbb{R}^d \times \mathbb{S}) := \left\{f : \mathbb{R}^d \times \mathbb{S} \to \mathbb{R} : f(\cdot, i) \in L^p(\mathbb{R}^d),\ i \in \mathbb{S}\right\}$$

and similarly we define $C^k(\mathbb{R}^d \times \mathbb{S})$, $\mathscr{W}^{k,p}_{\mathrm{loc}}(\mathbb{R}^d \times \mathbb{S})$, etc. We denote the components of such a function f, either as $f_y(x)$ or $f(x, y)$. If $\|\cdot\|_{\mathcal{X}(\mathbb{R}^d)}$ is a norm on $\mathcal{X}(\mathbb{R}^d)$, then the corresponding norm on $\mathcal{X}(\mathbb{R}^d \times \mathbb{S})$ is defined by

$$\|f\|_{\mathcal{X}(\mathbb{R}^d \times \mathbb{S})} = \sum_{y \in \mathbb{S}} \|f_y\|_{\mathcal{X}(\mathbb{R}^d)}\,.$$

5.1.3 The extended controlled generator of the Markov semigroup

For $\varphi \in C^2(\mathbb{R}^d \times \mathbb{S})$, define $\mathcal{L}^u$ and Π^u by

$$(\mathcal{L}^u \varphi)(x, y) := \sum_{i,j=1}^{d} a^{ij}(x, y) \frac{\partial^2 \varphi}{\partial x_i \partial x_j}(x, y) + \sum_{j=1}^{d} b^j(x, y, u) \frac{\partial \varphi}{\partial x_j}(x, y)\,, \tag{5.1.8a}$$

$$\begin{aligned}(\Pi^u \varphi)(x, y) &:= -\lambda(x, y, u)\varphi(x, y) + \lambda(x, y, u) \sum_{j \ne y} \varphi(x, j) \hat{Q}(y, j; x, u) \\ &= -\lambda^y_y(x, u)\varphi(x, y) + \sum_{j \ne y} \lambda^j_y(x, u)\varphi(x, j)\,. \end{aligned} \tag{5.1.8b}$$

The last equality in (5.1.8b) follows from (5.1.6).

Lemma 5.1.4 *Let $\varphi \in C_c^2(\mathbb{R}^d \times \mathbb{S})$. Then for $t > 0$,*

$$\begin{aligned}\mathbb{E}_z^{U_s}[\varphi(Z_t)] - \varphi(z) &= \mathbb{E}_z \int_0^t \left[\mathcal{L}^{U_s}\varphi(Z_s) + \Pi^{U_s}\varphi(Z_s)\right] \mathrm{d}s \\ &= \mathbb{E}_z \int_0^t \left[\mathcal{L}^{U_s}\varphi(Z_s) + \sum_{j\in\mathbb{S}} \lambda_{Y_s}^j(X_s, U_s)\varphi(Z_s)\right] \mathrm{d}s\,. \end{aligned} \tag{5.1.9}$$

Proof Let $\{\tau_1, \dots, \tau_\ell\}$ denote the times of jumps of Y in the interval $[0, t]$, and set $\tau_0 = 0$, and $\tau_{\ell+1} = t$. Applying the Itô formula, we obtain, for $k = 0, \dots, \ell$,

$$\begin{aligned}\varphi(X_{\tau_{k+1}}, Y_{\tau_k}) - \varphi(X_{\tau_k}, Y_{\tau_k}) &= \int_{\tau_k}^{\tau_{k+1}} \mathcal{L}^{U_s}\varphi(X_s, Y_{\tau_k})\,\mathrm{d}s \\ &\quad + \int_{\tau_k}^{\tau_{k+1}} \langle \nabla\varphi(X_s, Y_{\tau_k}), \sigma(X_s, Y_{\tau_k})\,\mathrm{d}W_s\rangle\,. \end{aligned} \tag{5.1.10}$$

Adding (5.1.10) over k from 0 to ℓ, we obtain

$$\begin{aligned}\varphi(Z_t) &= \varphi(z) + \int_0^t \mathcal{L}^{U_s}\varphi(Z_s)\,\mathrm{d}s + \int_0^t \langle \nabla\varphi(Z_s), \sigma(Z_s)\,\mathrm{d}W_s\rangle \\ &\qquad + \sum_{k=1}^{\ell} [\varphi(X_{\tau_k}, Y_{\tau_k}) - \varphi(X_{\tau_k}, Y_{\tau_k -})] \\ &= \varphi(z) + \int_0^t \mathcal{L}^{U_s}\varphi(Z_s)\,\mathrm{d}s + \int_0^t \langle \nabla\varphi(Z_s), \sigma(Z_s)\,\mathrm{d}W_s\rangle \\ &\quad + \int_0^t \int_{\mathbb{R}} [\varphi(X_s, Y_s + g(Z_s, U_s, \xi)) - \varphi(Z_s)]\mathfrak{p}(\mathrm{d}\xi \times \mathrm{d}t)\,. \end{aligned} \tag{5.1.11}$$

Since

$$\Pi^u\varphi(x, y) = \int_{\mathbb{R}} [\varphi(x, y + g(x, y, u, \xi)) - \varphi(x, y)]\,\mathfrak{m}(\mathrm{d}\xi)\,,$$

taking expectation in (5.1.11), and using (5.1.6) we obtain (5.1.9). □

Definition 5.1.5 Let $a_y^{ij}(x) := a^{ij}(x, y)$ and $b_y^i(x, u) := b^i(x, y, u)$. Also for $u \in \mathbb{U}$ and $\varphi \in C^2(\mathbb{R}^d)$, we define

$$\begin{aligned}\mathcal{L}_y^u\varphi(x) &:= a_y^{ij}(x)\partial_{ij}\varphi(x) + b_y^i(x, u)\partial_i\varphi(x)\,, \\ \tilde{\mathcal{L}}_y^u\varphi(x) &:= a_y^{ij}(x)\partial_{ij}\varphi(x) + b_y^i(x, u)\partial_i\varphi(x) + \lambda_y^y(x, u)\varphi(x)\,. \end{aligned}$$

The operator $\tilde{\Pi}^u$ is defined by

$$\tilde{\Pi}^u f(x, y) := \sum_{j\neq y} \lambda_y^j(x, u) f(x, j)\,, \quad u \in \mathbb{U}\,.$$

Combining the above definitions, for $f \in C^2(\mathbb{R}^d \times \mathbb{S})$, we obtain

$$\begin{aligned}\mathcal{A}^u f(x, y) &:= \mathcal{L}_y^u f(x, y) + \sum_{j\in\mathbb{S}} \lambda_y^j(x, u) f(x, j) \\ &= \tilde{\mathcal{L}}_y^u f(x, y) + \tilde{\Pi}^u f(x, y)\,, \quad u \in \mathbb{U}\,. \end{aligned}$$

Equivalently, we view the operator $\mathcal{A}^u$ as acting on functions in $C^2(\mathbb{R}^d)^N$, and with Λ denoting the matrix $[\lambda_i^j]$, we represent $\mathcal{A}$ as

$$\mathcal{A}^u f(x) = \mathcal{L}^u f(x) + \Lambda^u f(x)\,, \tag{5.1.12}$$

where we adopt the natural vector notation for $f : \mathbb{R}^d \times \mathbb{S} \to \mathbb{R}^N$, i.e., represent it as $f = (f_1, \dots, f_N)$, with $f_i(x) = f(x, i)$. In expanded form (5.1.12) is written as

$$\begin{aligned}(\mathcal{A}^u f)_k(x) &= (\mathcal{L}^u f)_k(x) + (\Lambda^u f)_k(x)\\ &= \mathcal{L}^u_k f_k(x) + \sum_{j\in\mathbb{S}} \lambda_k^j(x,u) f_j(x)\\ &= \tilde{\mathcal{L}}^u_k f_k(x) + (\tilde{\Pi}^u f)_k(x)\,, \quad k \in \mathbb{S}\,, \quad u \in \mathbb{U}\,.\end{aligned}$$

We adopt the relaxed control framework, and for $v \in \mathscr{P}(\mathbb{U})$, we define

$$b_v(x,y) := \int_{\mathbb{U}} b(x,y,u) v(\mathrm{d}u \mid x,y)\,, \qquad \lambda^j_{y,v}(x) := \int_{\mathbb{U}} \lambda^j_y(x,u) v(\mathrm{d}u \mid x,y)\,,$$

and

$$\mathcal{A}^v f(x,y) = \int_{\mathbb{U}} \mathcal{A}^u f(x,y)\, v(\mathrm{d}u \mid x,y)\,. \tag{5.1.13}$$

We also use the vector notation $b_v = (b_{1,v}, \dots, b_{N,v})$, with $b_{y,v}(x) := b_v(x,y)$.

All these definitions also hold for $f \in \mathscr{W}^{2,p}_{\mathrm{loc}}(\mathbb{R}^d \times \mathbb{S})$.

Remark 5.1.6 If we vectorize $v \in \mathfrak{U}_{\mathrm{sm}}$ by defining $v_y(\mathrm{d}u \mid x) := v(\mathrm{d}u \mid x,y)$, then $v = (v_1, \dots v_N) \in (\mathscr{P}(\mathbb{U}))^N$. It is important to note that $b_{y,v} = b_{y,v_y}$, $\lambda^j_{y,v} = \lambda^j_{y,v_y}$ for all $j \in \mathbb{S}$, and therefore,

$$(\mathcal{A}^v f)_k = \mathcal{L}^{v_k}_k f_k(x) + \sum_{j\in\mathbb{S}} \lambda^j_{k,v_k}(x) f_j(x) \qquad \forall k \in \mathbb{S}\,.$$

Under a stationary Markov control $v \in \mathfrak{U}_{\mathrm{sm}}$, a direct extension of the results in [117] shows that (5.1.7) admits a pathwise unique strong solution $Z = (X, Y)$ which is a Feller process with extended controlled generator $\mathcal{A}^v$ (compare with Theorem 2.2.12). As usual we let $P^v : \mathbb{R}_+ \times \mathbb{R}^d \times \mathbb{S} \to \mathscr{P}(\mathbb{R}^d \times \mathbb{S})$ denote its transition function. Also $\mathbb{P}^v_{x,i}$ and $\mathbb{E}^v_{x,i}$ denote the probability measure and the expectation operator, respectively, on the canonical space, of the process (X, Y) starting at $(x, i) \in \mathbb{R}^d \times \mathbb{S}$.

5.2 Cooperative elliptic systems

The model in (5.1.7) gives rise to the class of elliptic operators in (5.1.13), with $v \in \mathfrak{U}_{\mathrm{sm}}$ appearing as a parameter. These take the form of a coupled (coupling in the zeroth order term) elliptic system of operators, with coupling coefficients which are nonnegative (cooperative). In the interest of economy of notation we have introduced a convenient parameterization in Assumption 5.1.1, via the function γ which suppresses the need for distinct Lipschitz constants and growth bounds. We use the same parameterization to define a class of elliptic operators as follows.

Definition 5.2.1 Let $\gamma\colon (0,\infty) \to (0,\infty)$ be a positive function that plays the role of a parameter. We let $\mathfrak{A}(\gamma)$ denote the class of operators $\mathcal{A}$ which take the following form: with $\varphi = (\varphi_1, \dots, \varphi_N) \in \mathscr{W}^{2,p}_{\mathrm{loc}}(\mathbb{R}^d \times \mathbb{S})$

$$(\mathcal{A}\varphi)_k(x) = \tilde{\mathcal{L}}_k \varphi_k(x) + \sum_{j \neq k} \lambda^j_k(x)\varphi_j(x)\,, \tag{5.2.1}$$

and satisfying

(i) $\tilde{\mathcal{L}}_k \in \mathfrak{L}(\gamma)$ (see Definition A.1.1).
(ii) $\|\lambda^j_k\|_{L^\infty(B_R)} \le \gamma(R)$, $\lambda^j_k \ge 0$ for $k \neq j$, and $\sum_j \lambda^j_k \le 0$.[1]

Comparing Assumption 5.1.1 to Definition 5.2.1, it is clear that for the model in (5.1.7), under Assumption 5.1.1, there corresponds a function $\gamma\colon (0,\infty) \to (0,\infty)$ such that $\mathcal{A}^v \in \mathfrak{A}(\gamma)$ for all $v \in \mathfrak{U}_{\mathrm{sm}}$. Thus any property which holds uniformly over solutions involving the operators in $\mathfrak{A}(\gamma)$, also holds uniformly over all $\mathcal{A}^v$, $v \in \mathfrak{U}_{\mathrm{sm}}$.

The ergodic behavior of $Z = (X, Y)$, depends heavily on the coupling coefficients $\{\lambda^j_i\}$. Recall that a matrix $[M_{ij}] \in \mathbb{R}^{N\times N}$ is called irreducible provided that for any pair of non-empty sets I, J which form a partition of $\mathbb{S}$, there exist $i_0 \in I$ and $j_0 \in J$, such that $M_{i_0 j_0} \neq 0$.

Definition 5.2.2 If $D \subset \mathbb{R}^d$ is a bounded domain and $\mathcal{A} \in \mathfrak{A}(\gamma)$, then with $\{\lambda^j_{i,\mathcal{A}}\}$ denoting the coefficients that correspond to $\mathcal{A}$, define

$$\hat{\lambda}^j_{i,\mathcal{A}}(D) := \big\|\lambda^j_{i,\mathcal{A}}\big\|_{L^1(D)}$$

and let $\hat{\Lambda}_{\mathcal{A}}(D)$ denote the $N \times N$ matrix $[\hat{\lambda}^j_{i,\mathcal{A}}(D) : i \neq j]$.

We say that $\mathcal{A} \in \mathfrak{A}(\gamma)$ is *fully coupled in a bounded domain* $D \subset \mathbb{R}^d$ if $\hat{\Lambda}_{\mathcal{A}}(D)$ is irreducible, and that it is *fully coupled* if it is fully coupled in some bounded domain $D \subset \mathbb{R}^d$. We also say that a class of elliptic operators $\mathfrak{A}' \subset \mathfrak{A}(\gamma)$ is θ-uniformly fully coupled in D, with $\theta > 0$ a constant, if for each $\mathcal{A} \in \mathfrak{A}'$ and any non-empty set $I \subset \mathbb{S}$ there exist $i_0 \in I$ and $j_0 \in I^c$ such that

$$\hat{\lambda}^{j_0}_{i_0,\mathcal{A}}(D) \ge \theta\,.$$

Definition 5.2.3 Analogously to Definition 5.2.2, with $v = (v_1, \dots v_N) \in \mathfrak{U}_{\mathrm{sm}}$ as a parameter we let

$$\tilde{\lambda}^j_{i,v}(B_R) := \int_{B_R} \big|\lambda^j_i(x, v_i(x))\big|\,\mathrm{d}x\,, \quad i, j \in \mathbb{S}\,,$$

and define the matrix

$$\tilde{\Lambda}^v(B_R) := [\tilde{\lambda}^j_{i,v}(B_R) : i \neq j]\,, \quad v = (v_1, \dots, v_N) \in \mathfrak{U}_{\mathrm{sm}}\,.$$

We say that the controlled switching diffusion in (5.1.7) is *fully coupled in* B_R if $\tilde{\Lambda}^v(B_R)$ is irreducible for all $v \in \mathfrak{U}_{\mathrm{sm}}$. We also say that the controlled switching diffusion in (5.1.7) is *fully coupled* if it is fully coupled in some ball B_R.

[1] Note that this amounts to a more general framework than we need.

Comparing Definition 5.2.3 to Definition 5.2.2, we note that if (5.1.7) is fully coupled in B_R, then then the collection $\{\mathcal{A}^v : v \in \mathfrak{U}_{\mathrm{sm}}\}$ is θ-uniformly coupled in B_R for some constant $\theta > 0$. Thus, if (5.1.7) is fully coupled, then there exist $R_0 > 0$, $\theta > 0$ such that

$$\|\lambda^j_{i,v}\|_{L^1(B_{R_0})} \geq \theta M_{ij}(v), \quad i \neq j, \quad \forall v \in \mathfrak{U}_{\mathrm{sm}},$$

where $M(v) \in \{0,1\}^{N\times N}$ is an irreducible matrix. Note also that the switching diffusion is fully coupled unless there exists a non-empty $I \subset \mathbb{S}$ such that for each $i \in I$,

$$\min_{u\in\mathbb{U}} \lambda^j_i(x,u) = 0 \text{ a.e.} \qquad \forall j \in I^c.$$

It is rather straightforward to show that since $\lambda^j_i \geq 0$ for $i \neq j$, the weak and strong maximum principles in Theorems A.2.2–A.2.3 hold for the elliptic system in (5.2.1).

An extension of the Alexandroff–Bakelman–Pucci maximum principle for elliptic systems in the class $\mathfrak{A}$ is summarized in the following theorem [42].

Theorem 5.2.4 (ABP estimate for elliptic systems) *Let $D \subset \mathbb{R}^d$ be a bounded domain. There exists a constant C_A, depending only on d, N, D, and γ, such that if $\varphi \in \mathscr{W}^{2,d}_{\mathrm{loc}}(D\times\mathbb{S}) \cap C(\bar{D}\times\mathbb{S})$ satisfies $\mathcal{A}\varphi \geq -f$, with $\mathcal{A} \in \mathfrak{A}(\gamma)$, then*

$$\sup_D \Bigl(\max_{i\in\mathbb{S}} \varphi_i\Bigr) \leq \sup_{\partial D} \Bigl(\max_{i\in\mathbb{S}} \varphi_i^+\Bigr) + C_A \Bigl\|\max_{i\in\mathbb{S}} f_i^+\Bigr\|_{L^d(D)}.$$

Moreover, if $\mathcal{A}\varphi \leq f$, then

$$-\inf_D \Bigl(\min_{i\in\mathbb{S}} \varphi_i\Bigr) \leq \sup_{\partial D} \Bigl(\max_{i\in\mathbb{S}} \varphi_i^-\Bigr) + C_A \Bigl\|\max_{i\in\mathbb{S}} f_i^+\Bigr\|_{L^d(D)}.$$

It is evident that Lemma A.2.5 and Theorem A.2.9 hold for $\mathcal{A} \in \mathfrak{A}$, since they hold component-wise.

For a C^2 bounded domain $D \subset \mathbb{R}^d$, $g \in C(\partial D\times\mathbb{S})$, and $f \in L^p(D\times\mathbb{S})$, $p \geq d$, we address the Dirichlet problem,

$$\mathcal{A}\varphi = f \quad \text{in } D\times\mathbb{S}, \qquad \varphi = g \quad \text{on } \partial D\times\mathbb{S}. \tag{5.2.2}$$

Consider the sequence of functions $\{\varphi^{(n)} : n \geq 0\}$, defined by $f^{(0)} \equiv 0$, and inductively for $n > 0$, as the solution of

$$\begin{aligned} \tilde{\mathcal{L}}_k \varphi_k^{(n)} &= f_k - \sum_{j\neq k} \lambda^j_k \varphi_j^{(n-1)} \quad \text{in } D,\\ \varphi_k^{(n)} &= g_k \quad \text{on } \partial D \end{aligned}$$

for $k \in \mathbb{S}$. By the strong maximum principle $\varphi_k^{(n+1)} \geq \varphi_k^{(n)}$ for all $n \geq 0$, and hence, since $\mathcal{A}\varphi^{(n)} \geq f$ for $n \in \mathbb{N}$ by Theorem 5.2.4, $\varphi^{(n)}$ is bounded in $D\times\mathbb{S}$ uniformly over $n \in \mathbb{N}$. By Lemma A.2.5, each component $\varphi_k^{(n)}$ is bounded in $\mathscr{W}^{2,p}(D')$ for any subdomain $D' \subset D$. It is then straightforward to show that $\varphi^{(n)}$ converges to some $\varphi \in \mathscr{W}^{2,p}_{\mathrm{loc}}(D\times\mathbb{S})$ uniformly on compact subsets of $D\times\mathbb{S}$ and that φ solves the Dirichlet problem (5.2.2). By the weak maximum principle this solution is unique. We have proved the following:

Theorem 5.2.5 *Let D be a bounded C^2 domain in $\mathbb{R}^d$, $\mathcal{A} \in \mathfrak{A}(\gamma)$, $g \in C(\partial D\times\mathbb{S})$, and $f \in L^p(D\times\mathbb{S})$, $p \geq d$. Then* (5.2.2) *has a unique solution $\varphi \in \mathscr{W}^{2,p}_{\mathrm{loc}}(D\times\mathbb{S})\cap C(\bar{D}\times\mathbb{S})$. Moreover, if $g \equiv 0$, then $\varphi \in \mathscr{W}^{2,p}(D \times \mathbb{S}) \cap \mathscr{W}^{1,p}_0(D \times \mathbb{S})$ and we have the estimate*

$$\|\varphi\|_{\mathscr{W}^{2,p}(D\times\mathbb{S})} \leq C_0'\|f\|_{L^d(D\times\mathbb{S})}$$

for some constant $C_0' = C_0'(d, p, D, \gamma)$.

In view of Remark 5.1.6 the same approach via approximating solutions can be followed for the quasilinear problem to obtain the analog of Theorem 3.5.3.

We could also consider Dirichlet problems on $D = (D_1, \dots, D_N) \subset \mathbb{R}^d \times \mathbb{S}$, where each D_i is a C^2 domain in $\mathbb{R}^d$. For the problem to be well-posed the boundary condition should be placed on $\cup_k \bar{D}_k\times\mathbb{S}\setminus D$. Consider for example a problem with $\mathbb{S} = \{1, 2\}$ and D a bounded C^2 domain in $\mathbb{R}^d$. The function $\varphi_i(x) = \mathbb{P}_{x,i}(\uptau(D\times\mathbb{S}) > \uptau(\mathbb{R}^d\times\{1\}))$ is a solution of

$$\tilde{\mathcal{L}}_1\varphi_1 = 0 \quad \text{in } D\,, \qquad \varphi_1 = 1 \quad \text{on } \partial D\,, \quad \varphi_2 = 0 \quad \text{on } \mathbb{R}^d\,,$$

while $\psi_i(x) = \mathbb{E}_{x,i}\,[\uptau(D \times \{1\})]$ is a solution of

$$\tilde{\mathcal{L}}_1\psi_1 = -1 \quad \text{in } D\,, \qquad \psi_1 = 0 \quad \text{on } \partial D\,, \quad \psi_2 = 0 \quad \text{on } \bar{D}\,.$$

Definition 5.2.6 For $\delta > 0$ and D a bounded domain, let $\mathbf{\mathfrak{K}}(\delta, D)$ denote the class of functions $f = (f_1, \dots, f_N)\colon D \times \mathbb{S} \to \mathbb{R}$, with $f_i \in \mathfrak{K}(\delta, D)$ for all $i \in \mathbb{S}$ (see Definition A.2.11).

Theorem 5.2.7 *Suppose $\mathfrak{A}' \subset \mathfrak{A}(\gamma)$ is θ-uniformly fully coupled in $B_R \subset \mathbb{R}^d$. There exists a constant $\tilde{C}_A = \tilde{C}_D(d, N, \gamma, R, \theta, \delta)$ such that if $\varphi \in \mathscr{W}^{2,p}(B_R\times\mathbb{S})\cap\mathscr{W}^{1,p}_0(B_R\times\mathbb{S})$ satisfies $\mathcal{A}\varphi = -f$ in $B_R \times \mathbb{S}$, $\varphi = 0$ on $\partial B_R \times \mathbb{S}$, with $f \in \mathbf{\mathfrak{K}}(\delta, B_R)$ and $\mathcal{A} \in \mathfrak{A}'$, then*

$$\inf_{B_{R/2}} \Bigl(\min_{i\in\mathbb{S}} \varphi_i\Bigr) \geq \tilde{C}_A \sum_{i\in\mathbb{S}} \|f_i\|_{L^1(B_R)}\,.$$

Harnack's inequality for cooperative elliptic systems can be stated as follows [7].

Theorem 5.2.8 (Harnack's inequality for elliptic systems) *Suppose $\mathfrak{A}' \subset \mathfrak{A}(\gamma)$ is θ-uniformly fully coupled in B_R. There exists a constant $C_H = C_H(d, N, R, \gamma, \theta, \delta)$, such that if $\varphi \in \mathscr{W}^{2,d}(B_{2R} \times \mathbb{S})$ satisfies $\mathcal{A}\varphi = -f$ and $\varphi \geq 0$ in B_{2R}, with $f \in \mathbf{\mathfrak{K}}(\delta, B_{2R})$ and $\mathcal{A} \in \mathfrak{A}'$, then*

$$\varphi_i(x) \leq C_H\varphi_j(y) \qquad \forall x, y \in B_R\,,\ \forall i, j \in \mathbb{S}\,.$$

Theorem 5.2.9 *Under a stationary Markov control $v \in \mathfrak{U}_{\mathrm{sm}}$, the solution $Z = (X, Y)$ to* (5.1.7) *has a strong Feller resolvent. Moreover, the resolvent kernel is mutually absolutely continuous with respect to the Lebesgue measure in $\mathbb{R}^d$.*

Proof For $R > 0$ let $\uptau_R$ denote the first exit time of Z from $D_R \times \mathbb{S}$. Consider the Dirichlet problem

$$\mathcal{A}^v\varphi = -f \quad \text{in } B_R\,, \qquad \varphi = 0 \quad \text{on } \partial B_R\,. \tag{5.2.3}$$

A straightforward application of Dynkin's formula together with the ABP estimate (Theorem 5.2.4) for the elliptic system, yields a constant C_R satisfying

$$\mathbb{E}^v_{x,i}\left|\int_0^{t\wedge\tau_R} f(Z_s)\,\mathrm{d}s\right| \le C_R\|f\|_{L^d(B_R\times\mathbb{S})} \qquad \forall f\in L^d(B_R\times\mathbb{S})\,.$$

Thus

$$f\mapsto\int_0^t \mathbb{E}^v_{x,i}[f(Z_s)\,\mathbb{I}\{s<\tau_R\}]\,\mathrm{d}s$$

defines a bounded linear functional on $L^d(B_R\times\mathbb{S})$. Invoking the Riesz representation theorem there exists a function $g^v_R(t,x,i,\cdot)\in L^{(d+1)/d}(B_R\times\mathbb{S})$ such that

$$\int_0^t \mathbb{E}^v_{x,i}[f(Z_s)\,\mathbb{I}\{s<\tau_R\}]\,\mathrm{d}s = \int_{\mathbb{R}^d}\sum_{k\in\mathbb{S}} g^v_R(t,x,i,y,k)f_k(y)\,\mathrm{d}y\,. \tag{5.2.4}$$

Fix $f=\mathbb{I}_{\{A,k\}}$ for $A\in\mathscr{B}(B_R)$ and $k\in\mathbb{S}$. Clearly, g^v_R is increasing in R to some limit g^v and letting $R\to\infty$, (5.2.4) yields

$$\int_0^t P^v(s,x,i,A,k)\,\mathrm{d}s = \int_A g^v(t,x,i,y,k)\,\mathrm{d}y\,.$$

It follows that

$$Q_\alpha(x,i,\cdot,k) = \alpha\int_0^\infty \mathrm{e}^{-\alpha t}P^v(t,x,i,\cdot,k)\,\mathrm{d}t$$

is absolutely continuous with respect to the Lebesgue measure.

Next, let A be a bounded Borel set in $B_{R_0}\times\mathbb{S}$ for some $R_0>0$, and f^n a bounded sequence of continuous functions such that

$$\lim_{n\to\infty}\sum_{k\in\mathbb{S}}\int_{\mathbb{R}^d}|\mathbb{I}_A(x,k)-f^n_k(x)|\,\mathrm{d}x = 0\,.$$

In order to simplify the notation we let P^v denote the $N\times N$ matrix with elements $\mathrm{P}^v_{ij}(t,x,A) = P^v(t,x,i,A_j,j)$, where A_j is the jth element of the natural partition of $A\in\mathscr{B}(\mathbb{R}^d\times\mathbb{S})$ as $A=\cup_{j\in\mathbb{S}}(A_j,j)$. Correspondingly, we view $\mathbb{I}_A$ is an N-vector. Let $\mathrm{P}^v_{i,\cdot}$ denote the ith row of P^v. Also, for a function $g\colon\mathbb{R}^d\times\mathbb{S}\to\mathbb{R}$, let $|g| := \sum_{i\in\mathbb{S}}|g_i|$. If M is a bound for $\{f^n\}$, then for any $R>0$, and with $Z_s=(X_s,Y_s)$, we have

$$\begin{aligned}
\left|\int_0^t \mathrm{P}^v_{i,\cdot}(s,x,A)\,\mathrm{d}s - \int_0^t\int_{\mathbb{R}^d}\mathrm{P}^v_{i,\cdot}(s,x,\mathrm{d}y)f^n(y)\,\mathrm{d}s\right|
&\le \mathbb{E}^v_{x,i}\int_0^t |\mathbb{I}_A(Z_s)-f^n(Z_s)|\,\mathrm{d}s\\
&\le \mathbb{E}^v_{x,i}\int_0^{t\wedge\tau_R}|\mathbb{I}_A(Z_s)-f^n(Z_s)|\,\mathrm{d}s + t(1+M)\,\mathbb{P}^v_{x,i}(\tau_R<t)\\
&\le C_R\left\|\mathbb{I}_A-f^n\right\|_{L^d(B_R\times\mathbb{S})} + t(1+M)\,\mathbb{P}^v_{x,i}(\tau_R<t)\,.
\end{aligned} \tag{5.2.5}$$

With $x \in B_{R_0}$ and $R > R_0$, we first take limits in (5.2.5) as $n \to \infty$, and then take limits as $R \to \infty$. Since

$$\mathbb{P}^v_{x,i}(\tau_R < t) = \mathbb{P}^v_{x,i}\Bigl(\sup_{s \le t} |X_s| > R\Bigr) \le R^{-2}\, \mathbb{E}_{x,i}|X_t|^2 ,$$

the right-hand side of (5.2.5) tends to zero uniformly in $(x, i) \in B_{R_0} \times \mathbb{S}$. Therefore

$$\sup_{(x,i)\in B_{R_0}\times\mathbb{S}} \left| \int_0^t \mathrm{P}^v_{i,\cdot}(s, x, A)\,\mathrm{d}s - \int_0^t \int_{\mathbb{R}^d} \mathrm{P}^v_{i,\cdot}(s, x, \mathrm{d}y) f(y)\,\mathrm{d}s \right| \xrightarrow[n\to\infty]{} 0 ,$$

thus establishing that $x \mapsto \int_0^t \mathrm{P}^v_{i,\cdot}(s, x, A)$ is the uniform limit of continuous functions on each ball in $\mathbb{R}^d$, and hence also continuous. Lastly, an application of Dynkin's formula to the Dirichlet problem (5.2.3), in combination with Theorems 5.2.4 and 5.2.7, shows that for any Borel set A with positive Lebesgue measure, and any $x \in \mathbb{R}^d$, $\int_0^t \mathrm{P}^v_{i,\cdot}(s, x, A)\,\mathrm{d}s > 0$ for some $t > 0$, and hence $Q_\alpha(x, i, A) > 0$. □

5.3 Recurrence and ergodicity of fully coupled switching diffusions

In this section we assume that the switching diffusion (5.1.7) is fully coupled in some ball B_{R_0}.

Definition 5.3.1 For $D \times J \subset \mathbb{R}^d \times \mathbb{S}$, we let $\tau(D \times J)$ denote the first exit time of (X_t, Y_t) from the set $D \times J$. Also, we use the abbreviation $\tau(D)$ for $\tau(D \times \mathbb{S})$. As usual $\breve{\tau}$ denotes the first recurrence time to a set, i.e.,

$$\breve{\tau}(D \times J) := \tau((D \times J)^c) ,$$

and we use the analogous abbreviation $\breve{\tau}(D)$ for $\breve{\tau}(D \times \mathbb{S})$.

Lemma 5.3.2 *Let D_1 and D_2 be two open balls in $\mathbb{R}^d$, satisfying $D_1 \Subset D_2$. Then*

$$0 < \inf_{\substack{(x,i)\in \bar{D}_1\times\mathbb{S}\\ v\in\mathfrak{U}_{\mathrm{sm}}}} \mathbb{E}^v_{x,i}[\tau(D_2)] \le \sup_{\substack{(x,i)\in \bar{D}_1\times\mathbb{S}\\ v\in\mathfrak{U}_{\mathrm{sm}}}} \mathbb{E}^v_{x,i}[\tau(D_2)] < \infty , \tag{5.3.1a}$$

$$\sup_{(x,i)\in\partial D_2\times\mathbb{S}} \mathbb{E}^v_{x,i}[\tau(D_1^c)] < \infty \qquad \forall v \in \mathfrak{U}_{\mathrm{ssm}} . \tag{5.3.1b}$$

Proof Let h be the unique solution in $\mathscr{W}^{2,p}(D_2 \times \mathbb{S}) \cap \mathscr{W}^{1,p}_0(D_2 \times \mathbb{S})$, $p \ge 1$, of

$$\mathcal{A}^v h = -1 \quad \text{in } D_2 \times \mathbb{S} , \qquad h = 0 \quad \text{on } \partial D_2 \times \mathbb{S} .$$

By Dynkin's formula,

$$h(x, i) = \mathbb{E}^v_{x,i}[\tau(D_2)] \qquad \forall x \in D_2 .$$

The positive lower bound in (5.3.1a) follows by Theorem A.2.12, noting that h_i satisfies $\tilde{\mathcal{L}}_i h_i \le -1$ on D_2.

Let $\tau_i \equiv \tau(\mathbb{R}^d \times \{i\})$. For the upper bound in (5.3.1a), we use

$$\mathbb{E}^v_{x,i}[\tau(D_2)] = \mathbb{E}^v_{x,i}[\tau(D_2)\,\mathbb{I}\,\{\tau_i > \tau(D_2)\}] + \mathbb{E}^v_{x,i}[\tau(D_2)\,\mathbb{I}\,\{\tau_i \le \tau(D_2)\}]\,. \tag{5.3.2}$$

The first term on the right-hand side of (5.3.2) is dominated by $\mathbb{E}^v_{x,i}[\tau(D_2) \wedge \tau_i]$, and therefore, it is bounded by some constant M_0. Let φ be the unique solution in $\mathscr{W}^{2,p}(D_2 \times \mathbb{S}) \cap \mathscr{W}^{1,p}_0(D_2 \times \mathbb{S})$, $p \ge 1$, of the Dirichlet problem

$$\begin{aligned} \mathcal{A}^v \varphi &= 0 \quad \text{in } D_2 \times \{i\}\,, \\ \varphi &= 1 \quad \text{on } \partial D_2 \times \{i\}\,, \\ \varphi &= 0 \quad \text{on } D_2 \times \{i\}^c\,. \end{aligned}$$

Then φ satisfies $\tilde{\mathcal{L}}^v_i \varphi_i(x) = 0$ for $x \in D_2$ and $i \in \mathbb{S}$, and since $\varphi_i(x) = \mathbb{P}^v_{x,i}(\tau_i > \tau(D_2))$, it follows by the strong maximum principle (and Harnack's inequality) that for some $\rho \in (0, 1)$,

$$\sup_{\substack{(x,i)\in \bar{D}_2\times\mathbb{S}\\ v\in\mathfrak{U}_{\mathrm{sm}}}} \mathbb{P}^v_{x,i}(\tau_i > \tau(D_2)) \ge \rho\,.$$

Hence we write the second term on the right-hand side of (5.3.2) as

$$\begin{aligned} \mathbb{E}^v_{x,i}\left[\tau(D_2)\,\mathbb{I}\,\{\tau_i \le \tau(D_2)\}\right] &= \mathbb{E}^v_{x,i}\Big[\mathbb{E}^v_{x,i}\left[\tau(D_2)\,\mathbb{I}\,\{\tau_i \le \tau(D_2)\} \mid \mathfrak{F}_{\tau_i}\right]\Big] \\ &= \mathbb{E}^v_{x,i}\left[\mathbb{I}\,\{\tau_i \le \tau(D_2)\}\, \mathbb{E}^v_{X_{\tau_i},Y_{\tau_i}}[\tau(D_2)]\right] \\ &= \mathbb{P}^v_{x,i}(\tau_i \le \tau(D_2)) \sup_{\substack{(x,i)\in \bar{D}_2\times\mathbb{S}\\ v\in\mathfrak{U}_{\mathrm{sm}}}} \mathbb{E}^v_{x,i}[\tau(D_2)] \\ &\le (1-\rho) \sup_{\substack{(x,i)\in \bar{D}_2\times\mathbb{S}\\ v\in\mathfrak{U}_{\mathrm{sm}}}} \mathbb{E}^v_{x,i}[\tau(D_2)]\,. \end{aligned} \tag{5.3.3}$$

By (5.3.2) – (5.3.3)

$$\sup_{\substack{(x,i)\in \bar{D}_2\times\mathbb{S}\\ v\in\mathfrak{U}_{\mathrm{sm}}}} \mathbb{E}^v_{x,i}[\tau(D_2)] \le M_0 + (1-\rho) \sup_{\substack{(x,i)\in \bar{D}_2\times\mathbb{S}\\ v\in\mathfrak{U}_{\mathrm{sm}}}} \mathbb{E}^v_{x,i}[\tau(D_2)]\,,$$

and the result follows since $\rho > 0$.

Let $\{R_n : n \in \mathbb{N}\} \subset \mathbb{R}_+$ be an increasing, divergent sequence, and let $g^{(n)}$ be the unique solution in $\mathscr{W}^{2,p}((B_{R_n} \setminus \bar{D}_1) \times \mathbb{S}) \cap \mathscr{W}^{1,p}_0((B_{R_n} \setminus \bar{D}_1) \times \mathbb{S})$, $p \ge 1$, of the Dirichlet problem

$$\mathcal{A}^v g^{(n)} = -1 \quad \text{in } (B_{R_n} \setminus \bar{D}_1) \times \mathbb{S}\,, \qquad g^{(n)} = 0 \quad \text{on } (\partial B_{R_n} \cup \partial D_1) \times \mathbb{S}\,.$$

If $x_0 \in \partial D_2$ and $v \in \mathfrak{U}_{\mathrm{ssm}}$, then $\mathbb{E}^v_{x_0,i}[\tau(D^c_1)] < \infty$. Since

$$g^{(n)}_i(x_0) = \mathbb{E}^v_{x_0,i}\left[\tau(D^c_1) \wedge \tau(D_{R_n})\right] \le \mathbb{E}^v_{x_0,i}\left[\tau(D^c_1)\right],$$

by Harnack's inequality the increasing sequence $f^{(n)} = g^{(n)} - g^{(1)}$ of $\mathcal{A}^v$-harmonic functions is bounded locally in $D^c_1 \times \mathbb{S}$, and hence approaches a limit as $n \to \infty$, which is an $\mathcal{A}^v$-harmonic function on $D_1^c \times \mathbb{S}$. Therefore $g = \lim_{n\to\infty} g^{(n)}$ is a bounded function on $\partial D_2 \times \mathbb{S}$ and clearly $g_i(x) = \mathbb{E}^v_{x,i}[\tau(D_1^c)]$. Property (5.3.1b) follows. □

Let D_1 and D_2 be two open balls in $\mathbb{R}^d$, satisfying $D_1 \Subset D_2$. Let $\hat{\tau}_0 = 0$ and for $k = 0, 1, \dots$, define inductively an increasing sequence of stopping times by

$$\hat{\tau}_{2k+1} = \min\{t \geq \hat{\tau}_{2k} : X_t \in D_2^c\},$$

$$\hat{\tau}_{2k+2} = \min\{t \geq \hat{\tau}_{2k+1} : X_t \in D_1\}.$$

Invariant measures for switching diffusions are characterized in the same manner as in Theorems 2.6.7 and 2.6.9. Namely if $\tilde{\mu} \in \mathcal{P}(\partial D_1 \times \mathbb{S})$ denotes the unique stationary probability distribution of $(\tilde{X}_n, \tilde{Y}_n) := (\tilde{X}_{\hat{\tau}_{2n}}, \tilde{Y}_{\hat{\tau}_{2n}})$ under $v \in \mathfrak{U}_{\mathrm{ssm}}$, then $\eta \in \mathcal{P}(\mathbb{R}^d \times \mathbb{S})$ defined by

$$\int_{\mathbb{R}^d\times\mathbb{S}} f(x,y)\,\eta(\mathrm{d}x,\mathrm{d}y) = \frac{\int_{\partial D_1\times\mathbb{S}} \mathbb{E}^v_{x,y}\left[\int_0^{\hat{\tau}_2} f(X_t,Y_t)\,\mathrm{d}t\right]\tilde{\mu}(\mathrm{d}x,\mathrm{d}y)}{\int_{\partial D_1\times\mathbb{S}} \mathbb{E}^v_{x,y}[\hat{\tau}_2]\,\tilde{\mu}(\mathrm{d}x,\mathrm{d}y)}\,.$$

for $f \in C_b(\mathbb{R}^d \times \mathbb{S})$ is the unique invariant probability measure of (X, Y), under $v \in \mathfrak{U}_{\mathrm{ssm}}$.

It is convenient to use vector notation for measures in $\mathcal{P}(\mathbb{R}^d \times \mathbb{S})$. Specifically, for $\eta \in \mathcal{P}(\mathbb{R}^d \times \mathbb{S})$, we define the vector-valued measure

$$\vec{\eta} = (\eta_1, \dots, \eta_N) \in \mathcal{P}(\mathbb{R}^d \times \mathbb{S}),$$

where $\eta_y(\cdot) := \eta(\cdot \times \{y\})$ is a sub-probability measure on $\mathbb{R}^d$. Thus, provided the map $f = (f_1, \dots, f_N)^{\mathsf{T}} : \mathbb{R}^d \to \mathbb{R}^N$ is integrable under $\vec{\eta}$, we define the pairing $\langle f, \vec{\eta}\rangle$ by

$$\begin{aligned}\int_{\mathbb{R}^d} \langle f(x), \vec{\eta}(\mathrm{d}x)\rangle &:= \sum_{y\in\mathbb{S}} \int_{\mathbb{R}^d} f_y(x)\,\eta_y(\mathrm{d}x)\\ &= \int_{\mathbb{R}^d\times\mathbb{S}} f(x,y)\,\eta(\mathrm{d}x,\mathrm{d}y)\,.\end{aligned}$$

Invariant probability measures are also characterized as in (2.6.40), namely that $\eta \in \mathcal{P}(\mathbb{R}^d \times \mathbb{S})$ is an invariant probability measure for the process associated with $\mathcal{A}^v$, if and only if

$$\int_{\mathbb{R}^d} \langle \mathcal{A}^v f(x), \vec{\eta}(\mathrm{d}x)\rangle = 0 \qquad \forall f \in \mathcal{C}\,,$$

where $\mathcal{C}$ is any dense set in $(C_c^2(\mathbb{R}^d))^N$.

As usual, we use $\mathcal{H}$ and $\mathcal{G}$ to denote the set of invariant probability measures and ergodic occupation measures respectively. When convenient, we use the vector notation $\vec{\pi}$ for $\pi \in \mathcal{G}$. Also, for $f : \mathbb{R}^d \times \mathbb{U} \to \mathbb{R}^N$, the pairing $\langle f, \vec{\pi}\rangle$ is analogously defined by

$$\int_{\mathbb{R}^d\times\mathbb{U}} \langle f(x,u), \vec{\pi}(\mathrm{d}x,\mathrm{d}u)\rangle := \sum_{y\in\mathbb{S}} \int_{\mathbb{R}^d\times\mathbb{U}} f_y(x,u)\,\pi_y(\mathrm{d}x,\mathrm{d}u)\,.$$

The same characterization as in Lemma 3.2.2 applies, and hence the argument in Lemma 3.2.3 shows that $\mathscr{G}$ is a closed and convex subset of $\mathscr{P}(\mathbb{R}^d \times \mathbb{S} \times \mathbb{U})$.

Regularity of invariant probability measures is the topic of the next section.

5.3.1 Regularity of invariant probability measures

Lemma 5.3.3 *Let B be an open ball in $\mathbb{R}^d$ and μ a finite Borel measure on B, with $\mu(B) > 0$, and suppose that μ is singular with respect to the Lebesgue measure. Let ρ be a nonnegative mollifier supported on the unit ball centered at the origin, and for $\varepsilon > 0$, define*

$$\rho_\varepsilon(x) := \varepsilon^{-d} \rho\left(\tfrac{x}{\varepsilon}\right), \quad \textit{and} \quad \varphi_\varepsilon(x) := \int_B \rho_\varepsilon(x-y)\mu(\mathrm{d}y)\,.$$

Then, for any $p > 1$,

$$\int_B |\varphi_\varepsilon(x)|^p \,\mathrm{d}x \xrightarrow[\varepsilon\to 0]{} \infty\,.$$

Proof Since μ is a Borel measure and $\mu(B) > 0$, there exists a compact set $K \subset B$, with $\mu(K) > 0$, and Lebesgue measure $|K| = 0$. Let K^ε denote the ε-neighborhood of K, i.e., $K^\varepsilon = \{x \in B : \mathfrak{d}(x, K) < \varepsilon\}$, where $\mathfrak{d}$ denotes Euclidean distance. Define

$$\tilde\varphi_\varepsilon(x) := \int_K \rho_\varepsilon(x-y)\mu(\mathrm{d}y)\,,$$

and note that $\tilde\varphi_\varepsilon$ is supported on K^ε, and $\tilde\varphi_\varepsilon \le \varphi_\varepsilon$ on B. Let $\mu_{\tilde\varphi_\varepsilon}$ denote the distribution of $\tilde\varphi_\varepsilon$, i.e.,

$$\mu_{\tilde\varphi_\varepsilon}(t) = \Bigl|\{x \in B : \tilde\varphi_\varepsilon(x) > t\}\Bigr|\,, \qquad t > 0\,.$$

We argue by contradiction. Suppose that for some $p > 1$,

$$\varliminf_{\varepsilon\to 0} \int_B |\tilde\varphi_\varepsilon(x)|^p \,\mathrm{d}x \le M < \infty\,. \tag{5.3.4}$$

On the one hand,

$$\mu_{\tilde\varphi_\varepsilon}(t) \le t^{-p} \int_B |\tilde\varphi_\varepsilon(x)|^p \,\mathrm{d}x\,, \tag{5.3.5}$$

while on the other $\mu_{\tilde\varphi_\varepsilon} \le |K^\varepsilon|$. Hence, by (5.3.4) – (5.3.5), since $|K^\varepsilon| \to 0$ as $\varepsilon \to 0$, using dominated convergence

$$\begin{aligned}
\varliminf_{\varepsilon\to 0} \int_B \tilde\varphi_\varepsilon(x)\,\mathrm{d}x &= \varliminf_{\varepsilon\to 0} \int_0^\infty \mu_{\tilde\varphi_\varepsilon}(t)\,\mathrm{d}t \\
&\le M \lim_{\varepsilon\to 0} \int_0^\infty (|K^\varepsilon| \wedge t^{-p})\,\mathrm{d}t \\
&= 0\,.
\end{aligned} \tag{5.3.6}$$

However, using Fubini's theorem, we obtain $\int_B \tilde\varphi_\varepsilon(x)\,\mathrm{d}x = \mu(K) > 0$ for all $\varepsilon > 0$, which contradicts (5.3.6). □

The operator $\mathcal{A}$ takes the form

$$(\mathcal{A}\varphi)_k = \tilde{\mathcal{L}}_k\varphi_k + \sum_{\ell\neq k}\lambda_k^\ell\varphi_\ell$$
$$\tilde{\mathcal{L}}_k = a_k^{ij}\partial_{ij} + b_k^i\partial_i + \lambda_k^k\,,$$

with a_k^{ij} locally Lipschitz and the rest of the coefficients are in $L^\infty_{\mathrm{loc}}(\mathbb{R}^d)$. Moreover, $\lambda_k^\ell \geq 0$ for $k \neq \ell$, and $\sum_\ell \lambda_k^\ell = 0$.

If $\vec{\mu} = (\mu_1, \dots, \mu_N) \in \mathscr{P}(\mathbb{R}^d \times \mathbb{S})$, we adopt the notation

$$\langle\varphi(x), \vec{\mu}(\mathrm{d}x)\rangle = \sum_{k=1}^N \varphi_k(x)\mu_k(\mathrm{d}x)\,.$$

Theorem 5.3.4 *Suppose μ is a Borel probability measure on $\mathbb{R}^d \times \mathbb{S}$ satisfying*

$$\int_{\mathbb{R}^d} \langle\mathcal{A}f(x), \vec{\mu}(\mathrm{d}x)\rangle = 0 \qquad \forall f \in C_c^2(\mathbb{R}^d \times \mathbb{S})\,.$$

Then the measure μ is absolutely continuous with respect to the Lebesgue measure. Let $\hat{\Lambda} \in \{0,1\}^{N\times N}$ be defined by

$$\hat{\Lambda}_{ij} := \begin{cases} 0 & \text{if } i = j \text{ or } \lambda_i^j = 0 \text{ a.e.,} \\ 1 & \text{otherwise.} \end{cases}$$

Then, provided $\hat{\Lambda}$ is an irreducible matrix, ψ is strictly positive on $\mathbb{R}^d \times \mathbb{S}$.

Proof Let $\mathbb{S}_\mu = \{k \in \mathbb{S} : \mu_k(\mathbb{R}^d) > 0\}$. Decompose $\mu_k = \beta_k\mu_k' + (1-\beta_k)\mu_k''$, with $\mu_k' \perp \mathfrak{m}$ and $\mu_k'' \prec \mathfrak{m}$, and suppose $\beta_{\hat{k}} > 0$ for some $\hat{k} \in \mathbb{S}_\mu$. Select $R > 0$ large enough such that $\mu_{\hat{k}}'(B_R) > 0$, and for $\varepsilon \in (0, R)$ define

$$\bar{\rho}_\varepsilon(x) = \frac{(\mathrm{e}^{-1} - \mathrm{e}^{-\frac{2R}{\varepsilon}})^{-1}}{\varepsilon}\int_\varepsilon^{2R} \mathrm{e}^{-\frac{r}{\varepsilon}}\rho_r(x)\,\mathrm{d}r\,.$$

With "$*$" denoting convolution, since $\partial_i(h * \bar{\rho}_\varepsilon) = (\partial_i h) * \bar{\rho}_\varepsilon$ and $\bar{\rho}_\varepsilon$ is symmetric with respect to the origin, using Fubini's theorem, we obtain

$$\begin{aligned}\int b_k^i(y)\partial_i(f_k * \bar{\rho}_\varepsilon)(y)\,\mu_k(\mathrm{d}y) &= \int b_k^i(y)\left(\int \partial_i f_k(x)\bar{\rho}_\varepsilon(y-x)\,\mathrm{d}x\right)\mu_k(\mathrm{d}y) \\ &= \int\left(\int b_k^i(y)\bar{\rho}_\varepsilon(x-y)\,\mu_k(\mathrm{d}y)\right)\partial_i f_k(x)\,\mathrm{d}x\,.\end{aligned}$$

Thus we can move the mollifier $\bar{\rho}_\varepsilon$ outside the operators ∂_{ij} and ∂_i to obtain

$$\int_{\mathbb{R}^d} \langle\mathcal{A}(f * \bar{\rho}_\varepsilon)(y), \vec{\mu}(\mathrm{d}y)\rangle = \int_{\mathbb{R}^d} \langle\mathcal{A}^\varepsilon f(x), \vec{1}_{\mathbb{S}_\mu}\rangle\,\mathrm{d}x = 0 \tag{5.3.7}$$

for all $f \in C_c^2(\mathbb{R}^d \times \mathbb{S})$, where $\vec{1}_{\mathbb{S}_\mu}$ denotes the function in $\mathbb{R}^d \times \mathbb{S}$ which is equal to 1 on $\mathbb{R}^d \times \mathbb{S}_\mu$ and 0 on its complement. The operator $\mathcal{A}^\varepsilon$ in (5.3.7) is given by

$$(\mathcal{A}^\varepsilon f)_k = \mathcal{L}_k^\varepsilon f_k + \sum_{\ell \neq k} \lambda_{k,\varepsilon}^\ell f_\ell \,,$$
$$\mathcal{L}^\varepsilon = a_{k,\varepsilon}^{ij}\,\partial_{ij} + b_{k,\varepsilon}^i\,\partial_i + \lambda_{k,\varepsilon}^k \,,$$

where the coefficients $a_{k,\varepsilon}^{ij}$, $b_{k,\varepsilon}^i$ and $\lambda_{k,\varepsilon}^\ell$ of $\mathcal{L}_k^\varepsilon$ are given by

$$a_{k,\varepsilon}^{ij}(x) := \int_{\mathbb{R}^d} a^{ij}(y)\bar{\rho}_\varepsilon(x-y)\mu_k(\mathrm{d}y)\,,$$
$$b_{k,\varepsilon}^{i}(x) := \int_{\mathbb{R}^d} b^{i}(y)\bar{\rho}_\varepsilon(x-y)\mu_k(\mathrm{d}y)\,,$$
$$\lambda_{k,\varepsilon}^{\ell}(x) := \int_{\mathbb{R}^d} \lambda_k^{\ell}(y)\bar{\rho}_\varepsilon(x-y)\mu_k(\mathrm{d}y)\,, \quad \ell \in \mathbb{S}\,.$$

Define

$$\psi_k^\varepsilon(x) := \int_{\mathbb{R}^d} \bar{\rho}_\varepsilon(x-y)\mu_k(\mathrm{d}y)\,, \quad k \in \mathbb{S}_\mu\,,$$

and note that $\psi_k^\varepsilon > 0$ for all $k \in \mathbb{S}_\mu$, on B_{2R}. We let

$$\tilde{a}_{k,\varepsilon}^{ij} := \frac{a_{k,\varepsilon}^{ij}}{\psi_k^\varepsilon}\,, \qquad \tilde{b}_\varepsilon^i := \frac{b_{k,\varepsilon}^i}{\psi_k^\varepsilon}\,, \qquad \tilde{\lambda}_{k,\varepsilon}^\ell := \frac{\lambda_{k,\varepsilon}^\ell}{\psi_k^\varepsilon}\,, \qquad k \in \mathbb{S}_\mu\,,$$

and for $k \in \mathbb{S}_\mu$, we define

$$(\tilde{\mathcal{A}}^\varepsilon f)_k := \tilde{\mathcal{L}}_k^\varepsilon f_k + \sum_{\ell \neq k} \tilde{\lambda}_{k,\varepsilon}^\ell f_\ell \,,$$
$$\tilde{\mathcal{L}}_k^\varepsilon := \tilde{a}_{k,\varepsilon}^{ij}\,\partial_{ij} + \tilde{b}_{k,\varepsilon}^i\,\partial_i + \tilde{\lambda}_{k,\varepsilon}^k \,.$$

Then we write (5.3.7) as

$$\int_{\mathbb{R}^d} \langle \psi^\varepsilon(x), \tilde{\mathcal{A}}^\varepsilon f(x)\rangle\,\mathrm{d}x = \int_{\mathbb{R}^d} \langle (\tilde{\mathcal{A}}^\varepsilon)^*\,\psi^\varepsilon(x), f(x)\rangle\,\mathrm{d}x = 0 \qquad \forall f \in C_c^2(B_{2R} \times \mathbb{S})\,,$$

which yields $((\tilde{\mathcal{A}}^\varepsilon)^*\,\psi^\varepsilon)_k = 0$ in B_{2R} for all $k \in \mathbb{S}$, where

$$((\tilde{\mathcal{A}}^\varepsilon)^*\,\psi^\varepsilon)_k = \begin{cases} \partial_i(\tilde{a}_{k,\varepsilon}^{ij}\partial_j\psi_k^\varepsilon + (\partial_j\tilde{a}_{k,\varepsilon}^{ij} - \tilde{b}_{k,\varepsilon}^i)\psi_k^\varepsilon) + \sum_{\ell\in\mathbb{S}_\mu} \tilde{\lambda}_{\ell,\varepsilon}^k\psi_\ell^\varepsilon & \text{for } k \in \mathbb{S}_\mu\,, \\ \sum_{\ell\in\mathbb{S}_\mu} \tilde{\lambda}_{\ell,\varepsilon}^k\psi_\ell^\varepsilon & \text{for } k \notin \mathbb{S}_\mu\,. \end{cases} \tag{5.3.8}$$

On B_{2R}, the family $\{(\tilde{\mathcal{A}}^\varepsilon)^* : \varepsilon \in (0,1)\}$ has bounded coefficients (uniformly in ε), and shares the same ellipticity constant with $\mathcal{A}^*$. Also $\tilde{a}_{k,\varepsilon}^{ij}$ inherits the Lipschitz constant of a_k^{ij}. Next, write (5.3.8) for $k \in \mathbb{S}_\mu$ as

$$\hat{\mathcal{L}}_k^\varepsilon\psi_k^\varepsilon = -\sum_{\substack{\ell\in\mathbb{S}_\mu \\ \ell\neq k}} \tilde{\lambda}_{\ell,\varepsilon}^k\psi_k^\varepsilon\,, \tag{5.3.9}$$

where $\hat{\mathcal{L}}_k^\varepsilon$ is the divergence form operator

$$\hat{\mathcal{L}}_k^\varepsilon f := \partial_i(\tilde{a}_{k,\varepsilon}^{ij}\partial_j f + (\partial_j\tilde{a}_{k,\varepsilon}^{ij} - \tilde{b}_{k,\varepsilon}^i)f) + \tilde{\lambda}_{k,\varepsilon}^k f\,.$$

Since μ is a probability measure, using Fubini's theorem, we obtain

$$|B_R| \inf_{B_R} \psi_k^\varepsilon \le \sum_{k\in\mathbb{S}_\mu} \int_{\mathbb{R}^d} \psi_k^\varepsilon(x)\,\mathrm{d}x = 1\,. \tag{5.3.10}$$

Since ψ_k^ε is a supersolution for $\hat{\mathcal{L}}_k^\varepsilon$, i.e., satisfies $\hat{\mathcal{L}}_k^\varepsilon \psi_k^\varepsilon \le 0$, a standard estimate [67, theorem 8.18, p. 194] asserts that for all p such that $1 \le p < \frac{n}{n-2}$, there exists a constant C_p, depending on p (and R), such that

$$\|\psi_k^\varepsilon\|_{L^p(2B_R)} \le C_p \inf_{B_R} \psi_k^\varepsilon\,. \tag{5.3.11}$$

By (5.3.10) – (5.3.11), $\|\psi_k^\varepsilon\|_{L^p(2B_R)} \le C_p |B_R|^{-1}$ for all $\varepsilon > 0$, which, in turn, implies by Lemma 5.3.3 that $\mu \prec \mathfrak{m}$ for all $k \in \mathbb{S}_\mu$.

By the strong maximum principle $\psi_k > 0$, unless it is identically zero on $\mathbb{R}^d$. Note that as $\varepsilon \to 0$, then $\psi_\ell^\varepsilon \to \psi_\ell$ and $\tilde{\lambda}_{\ell,\varepsilon}^k \to \lambda_\ell^k$ for $\ell \in \mathbb{S}_\mu$. Hence, if $\mathbb{S}_\mu \ne \mathbb{S}$, (5.3.8) yields

$$\sum_{\ell\in\mathbb{S}_\mu} \lambda_\ell^k \psi_\ell = 0\,, \quad \forall k \notin \mathbb{S}_\mu\,.$$

However, this implies that $\lambda_\ell^k = 0$ for all $(\ell, k) \in \mathbb{S}_\mu \times \mathbb{S}_\mu^c$ and hence $\hat{\Lambda}$ must be reducible. □

5.4 Existence of an optimal control

Let $c\colon \mathbb{R}^d \times \mathbb{S} \times \mathbb{U} \mapsto \mathbb{R}$ be a continuous function bounded from below (without loss of generality we assume it is nonnegative). As usual, the ergodic control problem seeks to minimize a.s. over all admissible controls the functional

$$\limsup_{t\to\infty} \frac{1}{t} \int_0^t \bar{c}(X_s, Y_s, U_s)\,\mathrm{d}s\,, \tag{5.4.1}$$

or, in a more restricted sense, the functional

$$\limsup_{t\to\infty} \frac{1}{t} \int_0^t \mathbb{E}^U \left[\bar{c}(X_s, Y_s, U_s)\right]\,\mathrm{d}s\,. \tag{5.4.2}$$

We let ϱ^* denote the infimum of (5.4.2) over $\mathfrak{U}_{\mathrm{sm}}$.

The results of Section 3.2 carry over to switching diffusions in a straightforward manner. Existence of optimal controls can be demonstrated under the assumption of near-monotonicity, which takes the form

$$\liminf_{|x|\to\infty} \min_{u\in\mathbb{U}} c(x, y, u) > \varrho^* \qquad \forall y \in \mathbb{S}\,, \tag{5.4.3}$$

or the assumption that $\mathscr{G}$ is compact. Under either of these two assumptions, the infimum of (5.4.1) or (5.4.2) over $\mathfrak{U}_{\mathrm{sm}}$ is attained at some $v^* \in \mathfrak{U}_{\mathrm{sd}}$. Moreover, ϱ^* is the infimum over $\mathfrak{U}$ of (5.4.2), and under (5.4.3) also the infimum over $\mathfrak{U}$ of (5.4.1). In the stable case, i.e., $\mathfrak{U}_{\mathrm{sm}} = \mathfrak{U}_{\mathrm{ssm}}$, an additional requirement, namely the tightness of the empirical measures defined in (3.4.5), is needed to assert that ϱ^* is the infimum over $\mathfrak{U}$ of (5.4.1).

5.5 HJB equations

We assume that $c\colon \mathbb{R}^d \times \mathbb{S} \times \mathbb{U}$ is locally Lipschitz continuous in its first argument uniformly in $u \in \mathbb{U}$.

Define

$$J_\alpha^U(x,y) := \mathbb{E}_{x,y}^U \left[\int_0^\infty \mathrm{e}^{-\alpha t} \bar{c}(X_t, Y_t, U_t)\, \mathrm{d}t \right], \quad U \in \mathfrak{U}$$

and $V_\alpha(x,y) := \inf_{U\in\mathfrak{U}} J_\alpha^U(x,y)$. Then by analogy to Theorem 3.5.6, V_α is the minimal nonnegative solution in $C^2(\mathbb{R}^d \times \mathbb{S})$ of

$$\min_{u\in\mathbb{U}} \left[\mathcal{A}^u V_\alpha(x,y) + c(x,y,u)\right] = \alpha V_\alpha(x,y)\,, \tag{5.5.1}$$

and $v \in \mathfrak{U}_{\mathrm{sm}}$ is α-discounted optimal if and only if v is a measurable selector from the minimizer in (5.5.1).

We proceed to derive the HJB equations for the ergodic control problem, via the vanishing discount method. First, under (5.4.3), V_α attains its infimum in some compact subset of $\mathbb{R}^d \times \mathbb{S}$ for all $\alpha > 0$ and, moreover, Lemma 3.6.1 holds. Next, the Harnack inequality in Theorem 5.2.8 permits us to extend Lemma 3.6.3 to switching diffusions, while the ABP estimate in Theorem 5.2.4 allows us to extend the convergence results in Lemma 3.6.4 to cooperative elliptic systems. In this manner Theorem 3.6.6, Lemmas 3.6.8 and 3.6.9, and Theorem 3.6.10 can be extended to switching diffusions. We summarize these as follows:

Theorem 5.5.1 *Suppose* (5.4.3) *holds. Then*

(a) *there exists a unique, up to a constant, function $V \in C^2(\mathbb{R}^d \times \mathbb{S})$, which is bounded below in $\mathbb{R}^d \times \mathbb{S}$, such that*

$$\min_{u\in\mathbb{U}} \left[\mathcal{A}^u V(x,y) + c(x,y,u)\right] = \varrho \tag{5.5.2}$$

holds for some constant $\varrho \le \varrho^*$;

(b) *a stationary Markov control is optimal if and only if it is measurable a selector from the minimizer of* (5.5.2).

Turning to the stable case, the results of Theorem 3.7.4 and Theorem 3.7.6 do hold. We summarize these in the following form.

Theorem 5.5.2 *There exist a constant C_0 depending only on the radius $R > 0$ such that, for all $v \in \mathfrak{U}_{\mathrm{ssm}}$ and $\alpha \in (0,1)$,*

$$\begin{aligned} \left\| J_\alpha^v - J_\alpha^v(0,1) \right\|_{\mathscr{W}^{2,p}(B_R\times\mathbb{S})} &\le \frac{C_0}{\eta_v(B_{2R}\times\mathbb{S})} F(R,c,v)\,, \\ \sup_{B_R\times\mathbb{S}} \alpha J_\alpha^v &\le C_0 F(R,c,v)\,, \end{aligned} \tag{5.5.3}$$

where

$$F(R,c,v) := \left(\|c_v\|_{L^\infty(B_{4R}\times\mathbb{S})} + \frac{\varrho_v}{\eta_v(B_{2R}\times\mathbb{S})} \right).$$

Moreover, if $\varrho_{\hat{v}} < \infty$ for some $\hat{v} \in \mathfrak{U}_{\mathrm{sm}}$, then (5.5.3) *still holds if J_α^v is replaced by V_α.*

Theorem 5.5.2 permits us to use the vanishing discount method to obtain the counterparts of Theorems 3.7.11–3.7.12, Corollary 3.7.13 and Theorems 3.7.14–3.7.15.

5.6 The general case

In this section we relax the assumption that the switching diffusion is fully coupled. As a result, under a stationary Markov control the process is not necessarily irreducible. We say that j is accessible from i under $v \in \mathfrak{U}_{\mathrm{sm}}$ and denote this by $i \overset{v}{\rightsquigarrow} j$, if there there exist $\{i_1, \dots, i_k\} \subset \mathbb{S}$ such that for some $R > 0$,

$$\tilde{\lambda}^{i_1}_{i,v}(B_R)\tilde{\lambda}^{i_2}_{i_1,v}(B_R)\cdots\tilde{\lambda}^{j}_{i_k,v}(B_R) > 0\,,$$

or equivalently if

$$\mathbb{P}^v_{x,i}\Big(\breve{\tau}(\mathbb{R}^d \times \{j\}) < \infty\Big) > 0\,.$$

If $i \overset{v}{\rightsquigarrow} j$ and $j \overset{v}{\rightsquigarrow} i$, then we say that i and j communicate and denote it by $i \overset{v}{\leftrightsquigarrow} j$. If i communicates with all states that are accessible from it, then we say that i is $\mathbb{S}$-recurrent, otherwise we call it $\mathbb{S}$-transient. The equivalence classes under $\overset{v}{\leftrightsquigarrow}$ are referred to as $\mathbb{S}$-communicative classes. An $\mathbb{S}$-recurrent ($\mathbb{S}$-transient) class of states refers to an $\mathbb{S}$-communicative class of $\mathbb{S}$-recurrent ($\mathbb{S}$-transient) states. We let $\mathfrak{R}^v$ and $\mathfrak{T}^v$ denote the sets of $\mathbb{S}$-recurrent and $\mathbb{S}$-transient states under $v \in \mathfrak{U}_{\mathrm{sm}}$, respectively.

We modify the definition of stability. We say that a control $v \in \mathfrak{U}_{\mathrm{sm}}$ is stable if for any bounded domain $D \in \mathbb{R}^d$, $\mathbb{E}^v_{x,i}[\tau_{D^c}] < \infty$ for all $(x, i) \in \mathbb{R}^d \times \mathbb{S}$. The set of all stable controls is denoted by $\mathfrak{U}_{\mathrm{ssm}}$. Unless $\mathcal{A}$ is fully coupled, it is not generally the case that the switching diffusion is positive recurrent under a stable control. However, it is straightforward to show that if $v \in \mathfrak{U}_{\mathrm{ssm}}$, then $\mathbb{E}^v_{x,i}[\breve{\tau}(D \times \mathfrak{R}^v)] < \infty$ for any bounded domain D and all $i \in \mathbb{S}$. Indeed, select bounded $D_2, D_1 \Supset D$, and note that since $v \in \mathfrak{U}_{\mathrm{ssm}}$, we have $\mathbb{E}^v_{x,i}\,[\tau(D_1^c)] < \infty$. By Harnack's inequality

$$\inf_{(x,j)\in\partial D_1 \times \mathbb{S}} \mathbb{P}_{x,j}(\breve{\tau}(D \times \mathfrak{R}^v) < \tau(D_2)) > 0\,.$$

The claim then follows as in the proof of Theorem 2.6.10 (b) ⇒ (a).

Let $\mathfrak{n}_v$ denote the number of $\mathbb{S}$-recurrent classes under $v \in \mathfrak{U}_{\mathrm{sm}}$ and

$$\mathfrak{R}^v = \mathfrak{R}^v_1 \cup \mathfrak{R}^v_2 \cup \cdots \cup \mathfrak{R}^v_{\mathfrak{n}_v}$$

be the partition of $\mathfrak{R}^v$ into its $\mathbb{S}$-recurrent classes. It is evident that the restriction of $\mathcal{A}^v$ to any $\mathbb{S}$-recurrent class $\mathfrak{R}^v_k$ defines a fully coupled, cooperative elliptic system. Therefore if $v \in \mathfrak{U}_{\mathrm{ssm}}$, there is a unique probability measure η^k_v in $\mathscr{P}(\mathbb{R}^d \times \mathfrak{R}^v_k)$ which is invariant for the diffusion restricted to $\mathbb{R}^d \times \mathfrak{R}^v_k$. It is also evident that the invariant measures for the switching diffusion under v are supported on $\mathbb{R}^d \times \mathfrak{R}^v$ and take the form

$$\eta_v = \left(\lambda_1\eta^1_v, \lambda_2\eta^2_v, \dots, \lambda_{\mathfrak{n}_v}\eta^{\mathfrak{n}_v}_v\right),$$

where $\lambda_i \in \mathbb{R}_+$ and $\lambda_1 + \cdots + \lambda_{\mathfrak{n}_v} = 1$. The analogous property holds for the ergodic occupation measures π_v. Recall the definition of $\mathscr{H}_{\mathcal{U}}$ and $\mathscr{G}_{\mathcal{U}}$ on p. 89, for $\mathcal{U} \subset \mathfrak{U}_{\mathrm{ssm}}$.

Suppose $\mathscr{H}_{\mathcal{U}}$ is tight. Then one can show that the set-valued map $v \mapsto \eta_v$ from $\bar{\mathcal{U}}$ to $\mathscr{H}_{\bar{\mathcal{U}}}$ is upper semicontinuous under the total variation norm topology, while the set-valued map $v \mapsto \pi_v$ from $\bar{\mathcal{U}}$ to $\mathscr{G}_{\bar{\mathcal{U}}}$ is upper semicontinuous in $\mathscr{P}(\mathbb{R}^d \times \mathbb{S} \times \mathbb{U})$. This extends Lemma 3.2.6.

Let $v \in \mathfrak{U}_{\mathrm{ssm}}$. Choose an arbitrary $\mathfrak{j}_k \in \mathfrak{R}^v_k$, $k = 1, \dots, \mathfrak{n}_v$, and let $\mathfrak{J}$ denote the collection of these points, i.e., $\mathfrak{J} = \{\mathfrak{j}_k : 1 \le k \le \mathfrak{n}_v\}$. With $\breve{\tau}_{\mathfrak{J}}$ denoting as usual the first recurrence time of the set $\mathbb{R}^d \times \mathfrak{J}$, let

$$\hat{J}^v_\alpha(x, y) := \mathbb{E}^v_{x,y}[J^v_\alpha(0, Y_{\breve{\tau}_{\mathfrak{J}}})] .$$

It is evident that $\hat{J}^v_\alpha(x, y) = J^v_\alpha(0, \mathfrak{j}_k)$ for all $y \in \mathfrak{R}^v_k$, and that $\hat{J}^v_\alpha$ solves $\mathcal{A}^v \hat{J}^v_\alpha = 0$ on $\mathbb{R}^d \times \mathbb{S}$. Let

$$\bar{V}_\alpha(x, y) := J^v_\alpha(x, y) - \hat{J}^v_\alpha(x, y) .$$

Using vector notation, it follows that $\bar{V}_\alpha$ satisfies

$$\mathcal{A}^v \bar{V}_\alpha + c_v = \alpha \bar{V}_\alpha + \alpha \hat{J}^v_\alpha .$$

Moreover, one can show that for any $R > 0$, the functions $\bar{V}_\alpha$ and $\alpha \hat{J}^v_\alpha$ are bounded in $\mathscr{W}^{2,p}(B_R \times \mathbb{S})$, uniformly in $\alpha \in (0, 1)$. Taking limits as $\alpha \to 0$ we obtain a solution pair $(\hat{V}_v, \varrho_v) \in (\mathscr{W}^{2,p}_{\mathrm{loc}}(\mathbb{R}^d \times \mathbb{S}))^2$ of

$$\begin{aligned} \mathcal{A}^v \hat{V}_v + c_v &= \varrho_v , \\ \mathcal{A}^v \varrho_v &= 0 . \end{aligned}$$

Moreover, the function ϱ_v satisfies

$$\varrho_v(x, y) = \lim_{T \to \infty} \frac{1}{T} \mathbb{E}^v_{x,y} \left[\int_0^T c(X_t, Y_t, U_t)\, \mathrm{d}t \right] .$$

The ergodic cost ϱ_v is constant on $\mathbb{R}^d \times \mathfrak{R}^v_k$ for each recurrent class $\mathfrak{R}^v_k$.

Using the same methodology as in Chapter 3, but with some important differences in technical details, one can show that the HJB equation takes the form

$$\begin{aligned} \min_{u \in \mathbb{U}} \, [\mathcal{A}^u V(x, y) + c(x, y, u)] &= \varrho(x, y) , \\ \min_{u \in \mathbb{U}} \, [\mathcal{A}^u \varrho(x, y)] &= 0 . \end{aligned} \tag{5.6.1}$$

The pair of equations in (5.6.1) is analogous to *Howard's equations* for finite state controlled Markov chains [6, 53].

5.7 An example

Suppose there is one machine producing a single commodity. We assume that the demand rate is a constant $d > 0$. Let the machine state S_t take values in $\{0, 1\}$, $S_t = 0$

or 1, according as the machine is down or functional. We model S_t as a continuous time Markov chain with generator

$$\begin{bmatrix} -\lambda_0 & \lambda_0 \\ \lambda_1 & -\lambda_1 \end{bmatrix},$$

where λ_0 and λ_1 are positive constants corresponding to the infinitesimal rates of repair and failure respectively. The inventory X_t is governed by the Itô equation

$$\mathrm{d}X_t = (U_t - d)\,\mathrm{d}t + \sigma\,\mathrm{d}W_t\,, \tag{5.7.1}$$

where $\sigma > 0$, U_t is the production rate and W_t is a one-dimensional Wiener process independent of S_t. The last term in (5.7.1) can be interpreted as *sales return*, *inventory spoilage*, *sudden demand fluctuations*, etc. A negative value of X_t represents backlogged demand. The production rate is constrained by

$$U_t \in \begin{cases} \{0\} & \text{if } S_t = 0\,, \\ [0, r] & \text{if } S_t = 1. \end{cases}$$

The cost function is given by

$$c(x) = \frac{c^+ + c^-}{2}|x| + \frac{c^+ - c^-}{2}x\,,$$

with c^+ and c^- positive constants, Thus c is near-monotone. We show that a certain hedging-point Markov control is stable.

The HJB equations in this case are

$$\begin{aligned} &\frac{\sigma^2}{2}V_0''(x) - dV_0'(x) - \lambda_0 V_0(x) + \lambda_0 V_1(x) + c(x) = \varrho \\ &\frac{\sigma^2}{2}V_1''(x) + \min_{u\in[0,r]}\,[(u-d)V_1'(x)] + \lambda_1 V_0(x) + \lambda_1 V_1(x) + c(x) = \varrho\,. \end{aligned} \tag{5.7.2}$$

The results in this chapter ensure the existence of a C^2 solution (V, ϱ^*) of (5.7.2), where ϱ^* is the optimal cost. Using the convexity of c, it can be shown that V_i is convex for each i. Hence, there exists an x^* such that

$$\begin{aligned} V_1'(x) &\le 0 \qquad \text{for } x \le x^*\,, \\ V_1'(x) &\ge 0 \qquad \text{for } x \ge x^*\,. \end{aligned} \tag{5.7.3}$$

It follows, from (5.7.3), that the value of u which minimizes $(u - d)V_1'$ is

$$u - \begin{cases} r & \text{if } x < x^*\,, \\ 0 & \text{if } x > x^*. \end{cases}$$

Since $V_1'(x^*) = 0$, any $u \in [0, r]$ minimizes $(u - d)V_1'(x^*)$. Therefore the action $u \in [0, r]$ can be chosen arbitrarily at $x = x^*$. To be specific, we let $u(x^*) = d$, i.e., we produce at the level that meets the demand exactly. Thus the following precise stable Markov control is optimal

$$v^*(x,0) = 0, \qquad v^*(x,1) = \begin{cases} r & \text{if } x < x^*\,, \\ d & \text{if } x = x^*\,, \\ 0 & \text{if } x > x^*\,. \end{cases} \tag{5.7.4}$$

Provided that the set of stable precise Markov controls is non-empty, the stability of the optimal control (5.7.4) follows. We show next that the zero-inventory control v given by

$$v(x,0) = 0, \qquad v(x,1) = \begin{cases} r & \text{if } x \le 0\,, \\ 0 & \text{if } x > 0\,, \end{cases}$$

is stable if and only if

$$\frac{(r-d)}{\lambda_1} > \frac{d}{\lambda_0}\,. \tag{5.7.5}$$

The condition in (5.7.5) is in accord with intuition. Note that λ_0^{-1} and λ_1^{-1} are the mean sojourn times of the chain in states 0 and 1 respectively. In state 0 the mean inventory depletes at a rate d while in state 1 it builds up at a rate $(r-d)$. Thus, if (5.7.5) is satisfied, one would expect the zero-inventory control to stabilize the system.

The density φ of the invariant probability measure η_v can be obtained by solving the adjoint system

$$(\mathcal{L}^v)^*\varphi = 0\,, \tag{5.7.6}$$

subject to

$$\varphi_i(x) > 0, \qquad \sum_{i\in\{0,1\}} \int_{\mathbb{R}} \varphi_i(x)\,\mathrm{d}x = 1\,. \tag{5.7.7}$$

Define

$$\tilde{\lambda}_0 := \frac{2\lambda_0}{\sigma^2}, \quad \tilde{\lambda}_1 := \frac{2\lambda_1}{\sigma^2}, \quad \tilde{d} := \frac{2d}{\sigma^2} \quad \text{and} \quad \tilde{r} := \frac{2r}{\sigma^2}\,.$$

Then (5.7.6) is equivalent to

$$\begin{aligned} &\left.\begin{aligned} \varphi_0''(x) + \tilde{d}\varphi_0'(x) - \tilde{\lambda}_0\varphi_0(x) + \tilde{\lambda}_1\varphi_1(x) &= 0 \\ \varphi_1''(x) + \tilde{d}\varphi_1'(x) - \tilde{\lambda}_1\varphi_1(x) + \tilde{\lambda}_0\varphi_0(x) &= 0 \end{aligned}\right\} \quad \text{for } x > 0\,, \\ &\left.\begin{aligned} \varphi_0''(x) + \tilde{d}\varphi_0'(x) - \tilde{\lambda}_0\varphi_0(x) + \tilde{\lambda}_1\varphi_1(x) &= 0 \\ \varphi_1''(x) - (\tilde{r} - \tilde{d})\varphi_1'(x) - \tilde{\lambda}_1\varphi_1(x) + \tilde{\lambda}_0\varphi_0(x) &= 0 \end{aligned}\right\} \quad \text{for } x < 0\,. \end{aligned} \tag{5.7.8}$$

A solution of (5.7.8), subject to the constraint (5.7.7), exists if and only if (5.7.5) holds and takes the form:

$$\varphi(x) = \begin{pmatrix} \varphi_0(x) \\ \varphi_1(x) \end{pmatrix} = \begin{cases} a_1 \begin{pmatrix} \tilde{\lambda}_1 \\ \tilde{\lambda}_0 \end{pmatrix} e^{-s_1 x} + a_2 \begin{pmatrix} -\tilde{\lambda}_1 \\ \tilde{\lambda}_1 \end{pmatrix} e^{-s_2 x} & \text{for } x \geq 0, \\ a_3 \begin{pmatrix} \tilde{\lambda}_1 \\ -\psi(s_3) \end{pmatrix} e^{s_3 x} + a_4 \begin{pmatrix} -\tilde{\lambda}_1 \\ \psi(s_4) \end{pmatrix} e^{s_4 x} & \text{for } x < 0, \end{cases}$$

where

$$\psi(s) = s^2 + \tilde{d}s - \tilde{\lambda}_0 ,$$
$$s_1 = \tilde{d} ,$$
$$s_2 = \frac{\tilde{d}}{2} + \frac{1}{2}[\tilde{d}^2 + 4(\tilde{\lambda}_0 + \tilde{\lambda}_1)]^{1/2} ,$$

and s_3, s_4 are the positive roots of the polynomial

$$s^3 - (\tilde{r} - 2\tilde{d})s^2 - [(\tilde{r} - \tilde{d})\tilde{d} + \tilde{\lambda}_0 + \tilde{\lambda}_1]s + [(\tilde{r} - \tilde{d})\tilde{\lambda}_0 - \tilde{d}\tilde{\lambda}_1)] ,$$

ordered by $0 < s_3 < s_4$. Also, the coefficients $\{a_1, a_2, a_3, a_4\}$ are given by:

$$a_1 = \frac{1}{\Delta}\Bigl\{\frac{(s_4 - s_3)s_2}{\tilde{\lambda}_0 + \tilde{\lambda}_1} + \frac{s_4 + s_2}{s_3 + \tilde{d}} - \frac{s_3 + s_2}{s_4 + \tilde{d}}\Bigr\},$$
$$a_2 = \frac{1}{\Delta}\frac{(s_4 - s_3)s_2}{\tilde{\lambda}_0 + \tilde{\lambda}_1},$$
$$a_3 = \frac{1}{\Delta}\frac{s_4 + s_2}{s_3 + \tilde{d}},$$
$$a_4 = \frac{1}{\Delta}\frac{s_3 + s_2}{s_4 + \tilde{d}},$$
$$\Delta = \frac{(s_4 - s_3)(s_2 - \tilde{d})}{\tilde{d}} + \frac{\tilde{\lambda}_0 + \tilde{\lambda}_1}{\tilde{d}}\Bigl\{\frac{s_4 + s_2}{s_3} - \frac{s_3 + s_2}{s_4}\Bigr\}.$$

Note that if φ^* denotes the density of the invariant probability measure corresponding to a hedging-point control as in (5.7.4), then

$$\varphi^*(x) = \varphi(x - x^*) .$$

Given a convex cost function, the ergodic cost $\varrho(x^*)$ corresponding to such a control can be readily computed and is a convex function of the threshold value x^*.

5.8 Bibliographical note

This chapter is based on [64, 65] and contains several enhancements

6

Controlled Martingale Problems

6.1 Introduction

In this chapter, we present an abstract treatment of the ergodic control problem in a very broad framework. Later on the results are specialized to situations like degenerate diffusions and partially observed diffusions. The framework is, in fact, applicable to several other infinite dimensional problems as well, see Section 6.2.1.

In Section 6.2, we formally define a controlled martingale problem. As one might expect, this extends the martingale characterization of solutions of stochastic differential equations for diffusions, viz., that for $f \in C_b^2(\mathbb{R}^d)$,

$$f(X_t) - \int_0^t \mathcal{L}^{U_s} f(X_s)\,\mathrm{d}s\,, \quad t \geq 0\,,$$

is a martingale. We specify the conditions to be satisfied by the candidate controlled extended generator in this more abstract framework and provide several examples. In this chapter, we denote the controlled extended generator by $\mathcal{A}$.

In Chapter 3, we repeatedly used the fact that if $\int \mathcal{A}f\,\mathrm{d}\pi = 0$ for a sufficiently rich class of functions, then π must be an ergodic occupation measure, i.e., it disintegrates as $\pi(\mathrm{d}x, \mathrm{d}u) = \eta_v(\mathrm{d}x)v(\mathrm{d}u \mid x)$, where η_v is the unique invariant probability measure under the relaxed Markov control $v(\mathrm{d}u \mid x)$. This fact was proved using the theory of elliptic PDEs at our disposal. Unfortunately this characterization is no longer straightforward in the more general case we analyze in Section 6.3, forcing us to take a more convoluted route. As alluded to in Section 1.3, we begin with bounded operators $\tilde{\mathcal{A}}_n$ constructed from the resolvent of $\mathcal{A}$, which approximate $\mathcal{A}$. This allows us to construct a stationary Markov chain X^n with controlled extended generator $\tilde{\mathcal{A}}_n$ and marginal μ. The process X^n changes states at the click of an exponential clock with rate n. We establish the tightness of X^n and extract a limiting process X as $n \to \infty$, that is stationary, with marginal μ, and with controlled extended generator $\mathcal{A}$.

Another key fact used in Chapter 3 was that the extreme points of ergodic occupation measures correspond to stable stationary Markov controls. Once again the proof of this property was facilitated by the theory of elliptic PDEs which ensures the existence of a density for the invariant probability measure with respect to the Lebesgue measure, under any stable stationary Markov control. In the more abstract setup of this chapter, we see in Section 6.4 that not only is the characterization of extremal solutions much more difficult, but also the results are weaker. One considers, for a fixed initial law, equivalence classes of joint state-control processes whose marginals at time t agree for almost all t. Passing to the compact and convex set of these equivalence classes, one argues that the extremal solutions are Markov. What is still missing though is a characterization of when the extremal solutions are time-homogeneous Markov.

Not surprisingly, the existence results for optimal controls in suitable classes thereof are correspondingly weaker than those in Chapter 3. In Section 6.5 we introduce counterparts of the near-monotonicity and stability conditions of Chapter 3, and under these we assert the existence of an optimal state-control process that is either ergodic (but not necessarily Markov), or Markov (but not necessarily stationary). The optimal control, however, may be taken to be a stationary Markov control regardless.

In Section 6.6, we recall the formulation of the ergodic control problem as that of minimizing a linear functional over the closed convex set of ergodic occupation measures, and observe that it is an infinite dimensional linear program. An important spin-off is the dual linear program in function spaces, which is reminiscent of the maximal subsolution formulation of the HJB equations. Moreover, there is no duality gap.

In Section 6.7 we present the Krylov selection procedure, originally a scheme for extracting a Markov family out of non-unique solutions to the martingale problem for different initial conditions. We recall here this procedure for the infinite horizon discounted cost. It consists of successive minimization of a countable family of secondary discounted costs for every initial condition, over the non-empty compact and convex set of optimal laws. In other words, we successively minimize each of the sequences of costs over the set of minimizers of the one that precedes it. The family of costs is chosen so that the intersection of the nested decreasing set of minimizing laws is a singleton for each initial condition, and the collection of the above laws constitutes a Markov family.

The level of generality as well as the flavor of our arguments in the early sections is essentially that of Ethier and Kurtz [55] for the (uncontrolled) martingale problem, and a good familiarity with Chapters 1–4 of [55] will ease the passage through what follows.

6.2 The controlled martingale problem

We begin our development of the abstract controlled martingale problem by introducing some notation. We consider processes taking values in a Polish space E (state space), which together with its Borel σ-field $\mathscr{B}(E)$ forms a measurable space. As usual, $\mathcal{B}(E)$ denotes the space of bounded measurable maps $E \to \mathbb{R}$.

We say that a sequence $\{f_k\}_{k\in\mathbb{N}} \subset \mathcal{B}(E)$ converges to $f \in \mathcal{B}(E)$ in a bounded and pointwise sense, and we denote this convergence by $f_k \xrightarrow{\text{bp}} f$, if

$$\sup_{k\in\mathbb{N}} \sup_{x\in E} |f_k(x)| < \infty \quad \text{and} \quad f_k(x) \xrightarrow[k\to\infty]{} f(x) \qquad \forall x \in E\,.$$

A set $B \subset \mathcal{B}(E)$ is *bp-closed* if $f_k \in B$ for all $k \in \mathbb{N}$, and $f_k \xrightarrow{\text{bp}} f$ together imply that $f \in B$. Define the bp-closure (C) for $C \in \mathcal{B}(E)$ to be the smallest bp-closed set containing C.

Let $\mathbb{U}$ and $\mathscr{U}$ be as in Section 2.3, and $\mathcal{D}([0,\infty); E)$ as usual be the space of r.c.l.l. paths $[0,\infty) \to E$, with the Skorohod topology. Recall that $\mathscr{U}$ is the space of measurable maps $[0,\infty) \to \mathscr{P}(\mathbb{U})$ with the coarsest compact metrizable topology that renders continuous each of the maps

$$U \in \mathscr{U} \mapsto \int_0^T g(t) \int_{\mathbb{U}} h(u) U_t(\mathrm{d}u)\,\mathrm{d}t$$

for all $T > 0$, $g \in L^2[0,T]$, and $h \in C_b(\mathbb{U})$. A sequence f_n in $\mathcal{D}([0,\infty); E)$ converges to f if for each $T > 0$, there exist monotone continuous maps $\lambda_n^T : [0,T] \to [0,T]$ with $\lambda_n^T(0) = 0$ and $\lambda_n^T(T) = T$ such that

$$\sup_{t\in[0,T]} |\lambda_n^T(t) - t| \xrightarrow[n\to\infty]{} 0\,,$$

$$\sup_{t\in[0,T]} d(f_n(\lambda_n^T(t)), f(t)) \xrightarrow[n\to\infty]{} 0\,.$$

This topology is separable and metrizable by a complete metric (the Skorohod metric) which renders $\mathcal{D}([0,\infty); E)$ Polish (see p. 3).

Definition 6.2.1 Let $\mathcal{A}$ be a linear operator with domain $\mathscr{D}(\mathcal{A}) \subset C_b(E)$ and range $\mathscr{R}(\mathcal{A}) \subset C_b(E \times \mathbb{U})$. Let $\nu \in \mathscr{P}(E)$.

An $E \times \mathbb{U}$-valued process $\{(X_t, U_t) : 0 \le t < \infty\}$ defined on a probability space $(\Omega, \mathfrak{F}, \mathbb{P})$ is said to be a solution to the *controlled martingale problem* for $(\mathcal{A}, \nu)$ with respect to a filtration $\{\mathfrak{F}_t : t \ge 0\}$ if

(i) (X, U) is progressively measurable with respect to $\{\mathfrak{F}_t\}$;
(ii) $\mathscr{L}(X_0) = \nu$;
(iii) for all $f \in \mathscr{D}(\mathcal{A})$,

$$f(X_t) - \int_0^t \mathcal{A} f(X_s, U_s)\,\mathrm{d}s$$

is an $(\mathfrak{F}_t)$-martingale.

Correspondingly, an $E \times \mathscr{P}(\mathbb{U})$-valued process (X, U) defined on a probability space $(\Omega, \mathfrak{F}, \mathbb{P})$ is said to be a solution to the *relaxed controlled martingale problem* for $(\mathcal{A}, \nu)$ with respect to a filtration $\{\mathfrak{F}_t : t \ge 0\}$ if

(i) (X, U) is progressively measurable with respect to $\{\mathfrak{F}_t\}$;
(ii) $\mathscr{L}(X_0) = \nu$;

(iii) for all $f \in \mathscr{D}(\mathcal{A})$,

$$f(X_t) - \int_0^t \int_{\mathbb{U}} \mathcal{A}f(X_s, u)U_s(\mathrm{d}u)\,\mathrm{d}s \tag{6.2.1}$$

is an $(\mathfrak{F}_t)$-martingale.

In most of what follows, $\{\mathfrak{F}_t\}$ (assumed to satisfy the usual conditions of completeness and right-continuity) and ν are understood from the context and not mentioned explicitly. We define an operator $\bar{\mathcal{A}}\colon \mathscr{D}(\mathcal{A}) \to C_b(E \times \mathscr{P}(\mathbb{U}))$ by

$$\bar{\mathcal{A}}f(x, v) = \int_{\mathbb{U}} \mathcal{A}f(x, u)\, v(\mathrm{d}u)\,, \quad f \in \mathscr{D}(\mathcal{A})\,,\ x \in E\,,\ v \in \mathscr{P}(\mathbb{U})\,.$$

Hence the expression in (6.2.1) can be written as

$$f(X_t) - \int_0^t \bar{\mathcal{A}}f(X_s, U_s)\,\mathrm{d}s\,.$$

We often use the notation $\bar{\mathcal{A}}^v f(x) = \bar{\mathcal{A}}f(x, v)$. Then with $v \in \mathscr{P}(\mathbb{U})$ treated as a parameter, $\bar{\mathcal{A}}^v\colon \mathscr{D}(\mathcal{A}) \to C_b(E)$. Analogous notation is used for $\mathcal{A}$.

We impose the following conditions on $\mathcal{A}$:

(A6.1) There exists a countable set $\{g_k\} \subset \mathscr{D}(\mathcal{A})$ such that

$$\{(g, \mathcal{A}g) : g \in \mathscr{D}(\mathcal{A})\} \subset \text{bp-closure}\ \{(g_k, \mathcal{A}g_k) : k \ge 1\}\,.$$

(A6.2) $\mathscr{D}(\mathcal{A})$ is an algebra that separates points in E and contains constant functions. Furthermore, $\mathcal{A}\mathbf{1} = 0$ where $\mathbf{1}$ is the constant function identically equal to one.

(A6.3) For each $u \in \mathbb{U}$ and $x \in E$, there exists an r.c.l.l. solution to the martingale problem [55, chapter 4] for $(\mathcal{A}^u, \delta_x)$, where δ_x is the Dirac measure at x.

Condition (A6.3) implies that the operator $\mathcal{A}^u$ is dissipative for all $u \in \mathbb{U}$ [55, Proposition 4.3.5, p. 178], i.e.,

$$\|(\lambda I - \mathcal{A}^u)f\| \ge \lambda\|f\| \qquad \forall f \in \mathscr{D}(\mathcal{A}^u)\,,\ \forall \lambda > 0\,.$$

6.2.1 Examples

Example 6.2.2 The controlled diffusion governed by (2.2.1). Here $\mathcal{A} = \mathcal{L}$, and $\mathscr{D}(\mathcal{A}) = C_0^2(\mathbb{R}^d)$.

Example 6.2.3 Let H be a real, separable Hilbert space and X an H-valued controlled diffusion described by

$$X_t = X_0 + \int_0^t b(X_s, U_s)\,\mathrm{d}s + \int_0^t \sigma(X_s)\,\mathrm{d}W_s\,, \quad t \ge 0\,,$$

where $b\colon H \times \mathbb{U} \to H$ is Lipschitz in its first argument uniformly with respect to the second, $\sigma\colon H \to L^2(H, H)$ is also Lipschitz, and W is an H-valued cylindrical Wiener process independent of X_0. Here, $L^2(H, H)$ denotes the space of Hilbert–Schmidt operators on H with the Hilbert–Schmidt norm $\|\cdot\|_{\mathrm{HS}}$. Fix a complete orthonormal system $\{e_i : i \geq 1\}$ in $L^2(H, H)$ and define $Q_n \colon H \to \mathbb{R}^n$ by

$$Q_n(x) := (\langle x, e_1 \rangle, \ldots, \langle x, e_n \rangle) .$$

Let

$$\mathscr{D}(\mathcal{A}) = \left\{ f \circ Q_n : f \in C_0^2(\mathbb{R}^n) \,,\; n \geq 1 \right\} \subset C_b(H) \tag{6.2.2}$$

and define $\mathcal{A} \colon \mathscr{D}(\mathcal{A}) \to C_b(H \times \mathbb{U})$ by

$$\mathcal{A}(f \circ Q_n)(x, u) := \sum_{i=1}^{n} \langle b(x, u), e_i \rangle \frac{\partial f}{\partial y_i} \circ Q_n(x) + \frac{1}{2} \sum_{i,j=1}^{n} \langle \sigma^*(x) e_i, \sigma^*(x) e_j \rangle \frac{\partial^2 f}{\partial y_i \partial y_j} \circ Q_n(x) .$$

Example 6.2.4 Let H and H' be separable Hilbert spaces and let $\mathbb{U}$ be the unit closed ball of H' with the weak topology. Consider the H-valued controlled stochastic evolution equation (interpreted in the mild sense)

$$\mathrm{d}X_t = -\mathcal{L} X_t \,\mathrm{d}t + (F(X_t) + BU_t) \,\mathrm{d}t + \mathrm{d}W_t \,,$$

where $-\mathcal{L}$ is the infinitesimal generator of a differentiable semigroup of contractions on H such that $\mathcal{L}^{-1}$ is a bounded self-adjoint operator with discrete spectrum, $F \colon H \to H$ is bounded Lipschitz, $B \colon H' \to H$ is bounded and linear, W is an H-valued Wiener process independent of X_0 with incremental covariance given by a trace class operator R, and U is as in Example 6.2.3. Let $\{e_i\}$ denote the unit norm eigenfunctions of $\mathcal{L}^{-1}$, with $\{\lambda_i^{-1}\}$ the corresponding eigenvalues. This is a complete orthonormal system of H. Let $\mathscr{D}(\mathcal{A})$ be as in (6.2.2), and define the operator $\mathcal{A} \colon \mathscr{D}(\mathcal{A}) \to C_b(H \times \mathbb{U})$ by

$$\mathcal{A}(f \circ Q_n)(x, u) := \sum_{i=1}^{n} \langle e_i, (F(x) + Bu - \lambda_i x) \rangle \frac{\partial f}{\partial y_i} \circ Q_n(x) + \frac{1}{2} \sum_{i,j=1}^{n} \langle e_i, Re_j \rangle \frac{\partial^2 f}{\partial y_i \partial y_j} \circ Q_n(x) .$$

Example 6.2.5 In case of partially observed diffusions, the measure-valued process of the conditional law of the state given the observations and its *unnormalized* counterpart are also examples of controlled martingale problems. These are discussed in detail in Chapter 8.

6.3 Ergodic occupation measures

Let (X, U) be a stationary solution to the relaxed controlled martingale problem for $\mathcal{A}$, i.e., a solution to the relaxed controlled martingale problem with the additional proviso that the pair (X, U) defines a stationary process. Define the *ergodic occupation measure* $\pi \in \mathscr{P}(E \times \mathbb{U})$ by

$$\int_{E\times\mathbb{U}} f\,\mathrm{d}\pi = \mathbb{E}\left[\int_{\mathbb{U}} f(X_t, u) U_t(\mathrm{d}u)\right], \quad f \in C_b(E \times \mathbb{U}). \tag{6.3.1}$$

Then for $f \in \mathscr{D}(\mathcal{A})$ and $t > 0$,

$$\begin{aligned} 0 &= \mathbb{E}\left[f(X_t)\right] - \mathbb{E}\left[f(X_0)\right] \\ &= \int_0^t \mathbb{E}\left[\bar{\mathcal{A}} f(X_s, U_s)\right] \mathrm{d}s \\ &= t \int_{E\times\mathbb{U}} \mathcal{A} f\,\mathrm{d}\pi, \end{aligned}$$

implying that

$$\int_{E\times\mathbb{U}} \mathcal{A} f\,\mathrm{d}\pi = 0 \qquad \forall f \in \mathscr{D}(\mathcal{A}). \tag{6.3.2}$$

The main result in this section is given in Theorem 6.3.6 below, which extends Lemma 3.2.2 for nondegenerate diffusions to show that (6.3.2) completely characterizes all ergodic occupation measures for the controlled martingale problem. The proof consists of approximating the operator $\mathcal{A}$ by a sequence of operators $\tilde{\mathcal{A}}_n$, and using condition (A6.3) to establish that there exists a stationary solution to the controlled martingale problem for $(\tilde{\mathcal{A}}_n, \pi)$, i.e., a stationary pair (X, U) that solves the controlled martingale problem. Note that $\pi = \mathscr{L}((X_t, U_t))$ for all $t \geq 0$. Then we proceed to establish convergence of this solution in probability and use a monotone class argument to prove that the limit is a solution to the relaxed controlled martingale problem for $\mathcal{A}$.

Definition 6.3.1 We define the following objects:

(i) Define a sequence of operators

$$\tilde{\mathcal{A}}_n \colon \mathscr{D}(\tilde{\mathcal{A}}_n) := \mathscr{R}(I - n^{-1}\mathcal{A}) \to C_b(E \times \mathbb{U}), \quad n \in \mathbb{N},$$

by

$$\tilde{\mathcal{A}}_n g = n[(I - n^{-1}\mathcal{A})^{-1} - I]g \qquad \forall g \in \mathscr{R}(I - n^{-1}\mathcal{A}).$$

(ii) Let $M \subset C_b(E \times E \times \mathbb{U})$ denote the linear space of functions of the form

$$F(x, y, u) = \sum_{i=1}^{m} f_i(x) g_i(y, u) + f(y, u),$$

with $f_i \in C_b(E)$, $g_i \in \mathscr{D}(\tilde{\mathcal{A}}_n)$ for $i = 1, \dots, m$, and $f \in C_b(E \times \mathbb{U})$.

(iii) Fix $n \in \mathbb{N}$. Define a linear functional $\Lambda \colon M \to \mathbb{R}$ by

$$\Lambda F = \int_{E\times\mathbb{U}} \left[\sum_{i=1}^{m} f_i(x)[(I - n^{-1}\mathcal{A})^{-1} g_i](x) + f(x, u)\right] \pi(\mathrm{d}x, \mathrm{d}u). \tag{6.3.3}$$

Since by (A6.3) $\mathcal{A}$ is dissipative, $\tilde{\mathcal{A}}_n$ is a well-defined bounded operator for each $n \in \mathbb{N}$, satisfying, for $f \in \mathscr{D}(\mathcal{A})$ and $f_n := (I - n^{-1}\mathcal{A})f$,

$$\|f_n - f\| \to 0 \quad \text{and} \quad \|\tilde{\mathcal{A}}_n f_n - \mathcal{A}f\| \to 0 \quad \text{as } n \to \infty\,.$$

In fact, $\tilde{\mathcal{A}}_n f_n = \mathcal{A}f$. Also, if $g \in \mathscr{D}(\tilde{\mathcal{A}}_n)$, then $g = (I - n^{-1}\mathcal{A})f$ for some $f \in \mathscr{D}(\mathcal{A})$, and thus (6.3.2) implies

$$\int_{E\times\mathbb{U}} \tilde{\mathcal{A}}_n g\,\mathrm{d}\pi = \int_{E\times\mathbb{U}} \mathcal{A}f\,\mathrm{d}\pi = 0 \qquad \forall n \in \mathbb{N}\,.$$

Lemma 6.3.2 *The functional* Λ *in* (6.3.3) *is well defined.*

Proof Suppose that F admits two different representations,

$$F = \sum_{i=1}^{m} f_i(x)g_i(y,u) + f(y,u) = \sum_{i=1}^{m} f_i'(x)g_i'(y,u) + f'(y,u)\,.$$

Then

$$f'(y,u) - f(y,u) = \sum_{i=1}^{m}\big(f_i(x)g_i(y,u) - f_i'(x)g_i'(y,u)\big)\,, \tag{6.3.4}$$

in particular $f' - f \in \mathscr{D}(\tilde{\mathcal{A}}_n)$. Thus with Λ' defined as in (6.3.3) relative to functions $f' \in C_b(E \times \mathbb{U})$, $\{f_i'\} \subset C_b(E)$ and $\{g_i'\} \subset \mathscr{D}(\tilde{\mathcal{A}}_n)$,

$$\begin{aligned}\Lambda F - \Lambda' F = \int_{E\times\mathbb{U}}\Bigg[&\sum_{i=1}^{m}\big(f_i(x)[(I - n^{-1}\mathcal{A})^{-1}g_i](x) + f(x,u)\big)\\ &- \big(f_i'(x)[(I - n^{-1}\mathcal{A})^{-1}g_i'](x) + f'(x,u)\big)\Bigg]\pi(\mathrm{d}x,\mathrm{d}u)\,.\end{aligned} \tag{6.3.5}$$

From (6.3.4),

$$\begin{aligned}&[(I - n^{-1}\mathcal{A})^{-1}f'](y) - [(I - n^{-1}\mathcal{A})^{-1}f](y)\\ &\qquad = \sum_{i=1}^{m}\Big(f_i(x)[(I - n^{-1}\mathcal{A})^{-1}g_i](y) - f_i'(x)[(I - n^{-1}\mathcal{A})^{-1}g_i'](y)\Big)\,.\end{aligned}$$

In particular,

$$\begin{aligned}&[(I - n^{-1}\mathcal{A})^{-1}f'](x) - [(I - n^{-1}\mathcal{A})^{-1}f](x)\\ &\qquad = \sum_{i=1}^{m}\Big(f_i(x)[(I - n^{-1}\mathcal{A})^{-1}g_i](x) - f_i'(x)[(I - n^{-1}\mathcal{A})^{-1}g_i'](x)\Big)\,.\end{aligned} \tag{6.3.6}$$

Thus by (6.3.4) – (6.3.6), with π_1 denoting the marginal of π on E,

$$\Lambda F - \Lambda' F = \int_E (I - n^{-1}\mathcal{A})^{-1}(f' - f)\,\mathrm{d}\pi_1 + \int_{E\times\mathbb{U}} (f - f')\,\mathrm{d}\pi\,. \tag{6.3.7}$$

For $h \in \mathscr{D}(\tilde{\mathcal{A}}_n)$, $h = (I - n^{-1}\mathcal{A})q$ for some $q \in \mathscr{D}(\mathcal{A})$. Now,

$$\int_{E\times\mathbb{U}} [(I - n^{-1}\mathcal{A})q]\,\mathrm{d}\pi = \int_E q\,\mathrm{d}\pi_1 \qquad \forall q \in \mathscr{D}(\mathcal{A})\,.$$

That is,

$$\int_E [(I - n^{-1}\mathcal{A})^{-1}h]\,\mathrm{d}\pi_1 = \int_{E\times\mathbb{U}} h\,\mathrm{d}\pi \qquad \forall h \in \mathcal{D}(\tilde{\mathcal{A}}_n)\,. \tag{6.3.8}$$

Thus the right-hand side of (6.3.7) is zero. □

Next we show that Λ is a positive functional of norm 1. We first need a technical lemma. For economy of notation we extend the definition of operators to vector-valued functions in the standard manner. For example, for $\boldsymbol{f} = (f_1, \dots, f_m)$ with $f_i \in \mathcal{D}(\mathcal{A})$, we adopt the streamlined notation $\mathcal{A}\boldsymbol{f} = (\mathcal{A}f_1, \dots, \mathcal{A}f_m)$.

Lemma 6.3.3 *Let $\varphi\colon \mathbb{R}^m \to \mathbb{R}$, $m \ge 1$, be convex and continuously differentiable and $f_i \in \mathcal{D}(\mathcal{A})$, $i = 1, \dots, m$, satisfy $\varphi(\boldsymbol{f}) \in \mathcal{D}(\mathcal{A})$, where $\boldsymbol{f} = (f_1, \dots, f_m)$. Then*

$$\mathcal{A}\varphi(\boldsymbol{f}) \ge \nabla\varphi(\boldsymbol{f}) \cdot \mathcal{A}\boldsymbol{f}\,.$$

Proof Let X be an r.c.l.l. solution to the martingale problem for $(\mathcal{A}^u, \delta_x)$, $u \in \mathbb{U}$, $x \in E$. Then by the convexity of φ,

$$\begin{aligned}\mathbb{E}\left[\int_0^t \mathcal{A}^u\varphi(\boldsymbol{f}(X_s))\,\mathrm{d}s\right] &= \mathbb{E}\left[\varphi(\boldsymbol{f}(X_t))\right] - \varphi(\boldsymbol{f}(x)) \\ &\ge \nabla\varphi(\boldsymbol{f}(x)) \cdot \Big(\mathbb{E}\left[\boldsymbol{f}(X_t)\right] - \boldsymbol{f}(x)\Big) \\ &= \nabla\varphi(\boldsymbol{f}(x)) \cdot \mathbb{E}\left[\int_0^t \mathcal{A}^u \boldsymbol{f}(X_s)\,\mathrm{d}s\right]\end{aligned}$$

for $t > 0$. Divide by t and let $t \downarrow 0$ to conclude the proof. □

Lemma 6.3.4 *The functional Λ defined in* (6.3.3) *satisfies*

(a) $|\Lambda F| \le \|F\|$*;*
(b) $\Lambda 1 = 1$*;*
(c) $\Lambda F \ge 0$*, whenever* $F \ge 0$*.*

Proof Let $\alpha_k = \big\|(I - n^{-1}\mathcal{A})h_k\big\|$, with $h_k \in \mathcal{D}(\mathcal{A})$ for $1 \le k \le m$, and let φ be a polynomial on $\mathbb{R}^m$ that is convex on $\prod_{i=1}^m [-\alpha_i, \alpha_i]$. Since $\mathcal{D}(\mathcal{A})$ is an algebra, we have $\varphi(\boldsymbol{h}) \in \mathcal{D}(\mathcal{A})$. By Lemma 6.3.3,

$$\mathcal{A}\varphi(\boldsymbol{h}) \ge \nabla\varphi(\boldsymbol{h}) \cdot \mathcal{A}\boldsymbol{h}\,.$$

Therefore

$$\begin{aligned}\varphi((I - n^{-1}\mathcal{A})\boldsymbol{h}) &\ge \varphi(\boldsymbol{h}) - n^{-1}\nabla\varphi(\boldsymbol{h}) \cdot \mathcal{A}\boldsymbol{h} \\ &\ge \varphi(\boldsymbol{h}) - n^{-1}\mathcal{A}\varphi(\boldsymbol{h})\,.\end{aligned}$$

Thus for $g_i = (I - n^{-1}\mathcal{A})h_i \in \mathcal{D}(\tilde{\mathcal{A}}_n)$, $1 \le i \le m$,

$$\int_{E\times\mathbb{U}} \varphi(\boldsymbol{g})\,\mathrm{d}\pi \ge \int_{E\times\mathbb{U}} \varphi\Big((I - n^{-1}\mathcal{A})^{-1}\boldsymbol{g}\Big)\,\mathrm{d}\pi\,. \tag{6.3.9}$$

Since a convex $\varphi\colon \mathbb{R}^m \to \mathbb{R}$ can be approximated uniformly on compacts by convex polynomials, (6.3.9) holds for all convex φ. Now let

$$\varphi(r_1, \dots, r_m) := \sup_{x \in E} \left(\sum_{i=1}^{m} f_i(x) r_i \right). \tag{6.3.10}$$

Then φ is convex and using (6.3.9) – (6.3.10), we obtain

$$\begin{aligned}
\Lambda F &= \int_{E\times\mathbb{U}} \left[\sum_{i=1}^{m} f_i(x)[(I - n^{-1}\mathcal{A})^{-1} g_i](x) + f(x,u) \right] \pi(\mathrm{d}x, \mathrm{d}u) \\
&\le \int_{E\times\mathbb{U}} \left[\varphi((I - n^{-1}\mathcal{A})^{-1} g)(x) + f(x,u) \right] \pi(\mathrm{d}x, \mathrm{d}u) \\
&\le \int_{E\times\mathbb{U}} [\varphi(g)(x,u) + f(x,u)]\, \pi(\mathrm{d}x, \mathrm{d}u) \\
&= \int_{E\times\mathbb{U}} \left[\sup_{y\in E} \left(\sum_{i=1}^{m} f_i(y) g_i(x,u) \right) + f(x,u) \right] \pi(\mathrm{d}x, \mathrm{d}u) \le \|F\|.
\end{aligned}$$

Also, $-\Lambda F = \Lambda(-F) \le \|-F\| = \|F\|$; so $|\Lambda F| \le \|F\|$, proving part (a). Observe that $\Lambda 1 = 1$ and for $F \ge 0$,

$$\|F\| - \Lambda F = \Lambda\left(\|F\| - F\right) \le \left\|(\|F\| - F)\right\| \le \|F\|,$$

implying that $\Lambda F \ge 0$. This completes the proof. □

By the Hahn–Banach theorem, Λ extends to a bounded, positive real functional on $C_b(E \times E \times \mathbb{U})$, satisfying $\|\Lambda\| = 1$. Define

$$\begin{aligned}
F_h(x,y,u) &:= h(x), \quad h \in C_b(E), \\
F^f(x,y,u) &:= f(y,u), \quad f \in C_b(E\times\mathbb{U}).
\end{aligned}$$

Note that

$$\Lambda F_h = \int_E h\, \mathrm{d}\pi_1 \quad \text{and} \quad \Lambda F^f = \int_{E\times\mathbb{U}} f\, \mathrm{d}\pi. \tag{6.3.11}$$

Although the Riesz representation theorem cannot be invoked here since the space $E{\times}E{\times}\mathbb{U}$ is not assumed to be compact, we use Daniell's theorem to obtain an integral representation of Λ. This is established as part of the proof of the next lemma.

Lemma 6.3.5 *For each $\pi \in \mathscr{P}(E\times\mathbb{U})$ satisfying (3.6.3), and each $n \in \mathbb{N}$, there exists a stationary solution (X^n, U^n) to the martingale problem for $(\tilde{\mathcal{A}}_n, \pi)$.*

Proof First we show that there exists a $\nu \in \mathscr{P}(E\times E\times\mathbb{U})$ such that

$$\Lambda F = \int_{E\times E\times\mathbb{U}} F\, \mathrm{d}\nu \qquad \forall F \in C_b(E\times E\times\mathbb{U}). \tag{6.3.12}$$

By [94, Theorem II.5.7], there exists a unique finitely additive measure ν on the Borel field of $E\times E\times\mathbb{U}$ that satisfies the above. Fix $\varepsilon > 0$. Since E is Polish, we can choose a compact set $K \subset E$ such that

$$\pi_1(K) \ge 1 - \varepsilon \quad \text{and} \quad \pi(K\times\mathbb{U}) \ge 1 - \varepsilon.$$

Define $\tilde{K} = K \times K \times \mathbb{U}$. Then

$$\begin{aligned}\nu(\tilde{K}^c) &\leq \nu(K^c \times E \times \mathbb{U}) + \nu(E \times K^c \times \mathbb{U})\\ &\leq \pi_1(K^c) + \pi(K^c \times \mathbb{U})\\ &\leq 2\varepsilon\,.\end{aligned}$$

Let $\{F_n : n \in \mathbb{N}\} \subset C_b(E\times E\times\mathbb{U})$, such that $F_n \downarrow 0$. Then for each $\delta > 0$, $\{F_n \geq \delta\}\cap\tilde{K}$, $n \in \mathbb{N}$, is a decreasing sequence of compact sets with empty intersection. By the finite intersection property, there exists $n_0 \geq 1$ such that $\{F_n \geq \delta\} \cap \tilde{K} = \varnothing$ for all $n \geq n_0$. Thus if $n \geq n_0$, we have

$$\Lambda F_n = \int_{E\times E\times\mathbb{U}} F_n \,\mathrm{d}\nu \leq \int_{\tilde{K}} F_n \,\mathrm{d}\nu + \int_{\tilde{K}^c} F_n \,\mathrm{d}\nu \leq \delta + 2\varepsilon\|F_1\|\,.$$

Since this holds for all $\delta > 0$ and $\varepsilon > 0$, we obtain

$$\Lambda F_n \to 0 \quad \text{as } n \to \infty\,.$$

By Daniell's theorem [93, prop. II.7.1], there exists a unique σ-additive measure, denoted by ν again, on the Borel σ-field $\mathscr{B}(E \times E \times \mathbb{U})$, satisfying (6.3.12). We may disintegrate ν as

$$\nu(\mathrm{d}x, \mathrm{d}y, \mathrm{d}u) = \pi_1(\mathrm{d}x)\,\eta(\mathrm{d}y, \mathrm{d}u \mid x) \tag{6.3.13}$$

for a measurable $\eta\colon x \in E \mapsto \eta(\mathrm{d}y, \mathrm{d}u \mid x) \in \mathscr{P}(E \times \mathbb{U})$ [32, pp. 39–40]. Then for $g \in \mathscr{D}(\tilde{\mathcal{A}}_n)$ and $f \in C_b(E)$, using (6.3.3), (6.3.8) and (6.3.13) we obtain, for $F(x, y, u) := f(x)g(y, u)$,

$$\begin{aligned}\Lambda F &= \int_{E\times E\times\mathbb{U}} f(x)g(y, u)\,\nu(\mathrm{d}x, \mathrm{d}y, \mathrm{d}u)\\ &= \int_E f(x)\left[\int_{E\times\mathbb{U}} g(y, u)\,\eta(\mathrm{d}y, \mathrm{d}u \mid x)\right]\pi_1(\mathrm{d}x)\\ &= \int_{E\times\mathbb{U}} f(x)[(I - n^{-1}\mathcal{A})^{-1}g](x)\,\pi(\mathrm{d}x, \mathrm{d}u)\\ &= \int_E f(x)[(I - n^{-1}\mathcal{A})^{-1}g](x)\,\pi_1(\mathrm{d}x)\,.\end{aligned}$$

Therefore

$$\int_{E\times\mathbb{U}} g(y, u)\,\eta(\mathrm{d}y, \mathrm{d}u \mid x) = [(I - n^{-1}\mathcal{A})^{-1}g](x) \quad \pi_1\text{-a.s.} \tag{6.3.14}$$

Let $\{(Y(k), W(k)) : k \geq 0\}$ be a Markov chain on $E \times \mathbb{U}$ with initial distribution π and transition kernel η. Then using (6.3.11), with $f = \mathbb{I}_B$ for $B \in \mathscr{B}(E \times \mathbb{U})$, we conclude that

$$\int_E \eta(x, B)\,\pi_1(\mathrm{d}x) = \nu(E \times B) = \pi(B) \qquad \forall B \in \mathscr{B}(E \times \mathbb{U})\,.$$

It follows that $(Y(k), W(k))$ is a stationary chain. Let V^n a Poisson process with rate n, and independent of $(Y(k), W(k))$. Define

$$X_t^n = Y(V_t^n), \quad U_t^n = W(V_t^n), \quad t \geq 0.$$

Let $\{\mathfrak{F}_t^n\}$ be the right-continuous completion of the filtration

$$\mathfrak{G}_t^n = \sigma\left(X_s^n, \int_a^b f(U_\tau^n)\,\mathrm{d}\tau : s \leq t,\ 0 \leq a \leq b \leq t,\ f \in C_b(\mathbb{U})\right).$$

Direct verification using (6.3.14) shows that for $g \in \mathcal{D}(\tilde{\mathcal{A}}_n)$,

$$g(Y(k), W(k)) - \sum_{j=0}^{k-1} n^{-1}\tilde{\mathcal{A}}_n g(Y(j), W(j))$$

is a $(\sigma(Y(i), W(i) : i \leq k))$-martingale. It follows as in Ethier and Kurtz [55, pp. 163–164] that (X^n, U^n) is a stationary solution to the martingale problem for $(\tilde{\mathcal{A}}_n, \pi)$ for every $n \geq 1$. □

The proof of the theorem that follows is rather lengthy; therefore we split it into several steps.

Theorem 6.3.6 *For each $\pi \in \mathcal{P}(E \times \mathbb{U})$ satisfying*

$$\int_{E\times\mathbb{U}} \mathcal{A}f\,\mathrm{d}\pi = 0 \qquad \forall f \in \mathcal{D}(\mathcal{A}),$$

there exists a stationary solution (X, U) to the relaxed controlled martingale problem for $\mathcal{A}$ such that π is the associated ergodic occupation measure.

Proof Let (X^n, U^n) be the stationary solution to the martingale problem for $\tilde{\mathcal{A}}_n$ constructed in Lemma 6.3.5.

Step 1. We first show that $(X^n, \delta_{U^n}) \Rightarrow (X, U)$ for some $E \times \mathcal{P}(\mathbb{U})$-valued process (X, U).

Let $\{g_m\}$ be the countable collection in condition (A6.1) of Section 6.2. Set

$$a_m = \|g_m\| \quad \text{and} \quad \hat{E} = \prod_{i=1}^{\infty}[-a_m, a_m].$$

Define $g\colon E \to \hat{E}$ by $g(x) = (g_1(x), g_2(x), \dots)$. Since $\mathcal{D}(\mathcal{A})$ separates points of E and vanishes nowhere, so does $\{g_m\}$. Thus g is one-to-one and continuous. It follows that $g(E)$ is Borel and $g^{-1}\colon g(E) \to E$ measurable [94, corollary I.3.3]. Fix an element $e \in E$ and set $g^{-1}(x) = e$ for $x \notin g(E)$. For $f \in \mathcal{D}(\mathcal{A})$ and $f_n = (I - n^{-1}\mathcal{A})f$, $n \geq 1$, let

$$\xi_n(t) = f_n(X_t^n) \quad \text{and} \quad \varphi_n(t) = \tilde{\mathcal{A}}_n f_n(X_t^n, U_t^n), \quad t \geq 0,\ n \geq 1.$$

Then

$$\xi_n(t) - \int_0^t \varphi_n(s)\,\mathrm{d}s, \quad t \geq 0,$$

is a martingale relative to the natural filtration of (X^n, U^n). Recall that $\|f - f_n\| \to 0$ and $\|\mathcal{A}f - \tilde{\mathcal{A}}_n f_n\| \to 0$. Hence, we can use [55, theorem 3.9.4, p. 145] to claim that the laws of $f(X^n)$ are tight and therefore relatively compact in $\mathscr{P}(\mathcal{D}([0,\infty); R))$. It follows that the laws of $g(X^n)$ are relatively compact in

$$\mathscr{P}(\mathcal{D}([0,\infty); \hat{E})) \approx \mathscr{P}\Big(\textstyle\prod_{i=1}^{\infty} \mathcal{D}([0,\infty); [-a_i, a_i])\Big).$$

Since $\mathscr{U}$ is a compact metric space, so is $\mathscr{P}(\mathscr{U})$, by Prohorov's theorem. As a result, $\{\mathscr{L}(g(X^n), \delta_{U^n})\}$ is relatively compact in $\mathscr{P}(\mathcal{D}([0,\infty); \hat{E}) \times \mathscr{U})$. Suppose that it converges to $\mathscr{L}(Z, U) \in \mathscr{P}(\mathcal{D}([0,\infty); \hat{E}) \times \mathscr{U})$ along some subsequence, which is also denoted by $\{n\}$. By Skorohod's theorem [32, pp. 23–24], we may construct all these processes (to be precise, their replicas in law) on a suitable probability space such that

$$(g(X^n), \delta_{U^n}) \to (Z, U) \quad \text{a.s. in } \mathcal{D}([0,\infty); \hat{E}) \times \mathscr{U}\,.$$

Since $\mathscr{L}(g(X_t^n)) = \pi_1 \circ g^{-1}$ for all t and n, it follows that $\mathscr{L}(Z_t) = \pi_1 \circ g^{-1}$ for all t, implying that $\mathbb{P}(Z_t \in g(E)) = 1$. Let $X_t = g^{-1}(Z_t)$, $t \geq 0$. Then $(X^n, \delta_{U^n}) \Rightarrow (X, U)$.

Step 2. We now prove that $X_t^n \to X_t$ in probability for all $t \geq 0$.

Note that

$$g_j(X_t^n) g_k(X_t) \to g_j(X_t) g_k(X_t) \quad \text{a.s.}, \quad \forall j, k \in \mathbb{N}\,.$$

Let $H \subset C_b(E \times E)$ denote the algebra generated by the family of functions

$$\{h\colon E \times E \to \mathbb{R} : h(x,y) = g_j(x) g_k(y)\,,\ j, k \in \mathbb{N}\}\,.$$

Since $\{g_k\}$ separates points in E, it follows that H separates points in $E \times E$. Also, for $h \in H$,

$$h(X_t^n, X_t) \xrightarrow{\mathbb{P}} h(X_t, X_t)\,.$$

Since $\{\mathscr{L}(X_t^n, X_t) : n \geq 1\}$ is trivially relatively compact, we can choose, for any given $\delta > 0$, a compact set $K \subset E \times E$ such that

$$\mathbb{P}((X_t^n, X_t) \in K) \geq 1 - \delta \qquad \forall n \in \mathbb{N}\,. \tag{6.3.15}$$

Let ρ be a complete metric on E. Restrict ρ and H to K and view H as a subset of $C(K)$ with (K, ρ) a compact metric space. By the Stone–Weierstrass theorem, H is dense in $C(K)$. Thus we can find $\{h_k\} \subset H$ such that

$$\sup_{x,y \in K} \big|\rho(x,y) - h_k(x,y)\big| \xrightarrow[k\to\infty]{} 0\,. \tag{6.3.16}$$

Let $\varepsilon > 0$, and define

$$\rho_n := \rho(X_t^n, X_t), \qquad G_n := \{(X_t^n, X_t) \in K\}\,, \quad n \in \mathbb{N}\,.$$

Using (6.3.15), we obtain

$$\begin{aligned}\mathbb{P}(\rho_n > \varepsilon) &\le \mathbb{P}(\{\rho_n > \varepsilon\} \cap G_n) + \delta \\ &\le \mathbb{P}\Bigl(\bigl\{\bigl|\rho_n - h_k(X_t^n, X_t)\bigr| > \tfrac{\varepsilon}{3}\bigr\} \cap G_n\Bigr) \\ &\qquad + \mathbb{P}\Bigl(\bigl\{\bigl|h_k(X_t^n, X_t) - h_k(X_t, X_t)\bigr| > \tfrac{\varepsilon}{3}\bigr\} \cap G_n\Bigr) \\ &\qquad\qquad + \mathbb{P}\Bigl(\bigl\{|h_k(X_t, X_t)| > \tfrac{\varepsilon}{3}\bigr\} \cap G_n\Bigr) + \delta\,. \end{aligned} \tag{6.3.17}$$

By (6.3.16) we can choose k such that

$$\sup_{x,y\in K} |h_k(x,y) - \rho(x,y)| < \frac{\varepsilon}{3}\,.$$

In particular $\sup_{x\in K} \bigl|h_k(x,x)\bigr| < \frac{\varepsilon}{3}$. Then (6.3.17) yields

$$\mathbb{P}(\rho_n > \varepsilon) \le \mathbb{P}\Bigl(\bigl\{\bigl|h_k(X_t^n, X_t) - h_k(X_t, X_t)\bigr| > \tfrac{\varepsilon}{3}\bigr\} \cap G_n\Bigr) + \delta\,.$$

Thus

$$\limsup_{n\to\infty}\ \mathbb{P}(\rho(X_t^n, X_t) > \varepsilon) \le \delta\,.$$

Since $\delta > 0$ was arbitrary, the claim follows.

Step 3. It follows that, for $h \in C_b(E \times \mathbb{U})$,

$$\bigl|h(X_t^n, U_t^n) - h(X_t, U_t^n)\bigr| \xrightarrow[n\to\infty]{\mathbb{P}} 0 \qquad \forall t \ge 0\,.$$

Also, the topology of $\mathscr{U}$ implies

$$\int_0^t h(X_s, U_s^n)\,\mathrm{d}s \to \int_0^t \int_{\mathbb{U}} h(X_s, u) U_s(\mathrm{d}u)\,\mathrm{d}s \quad \mathbb{P}\text{-a.s.},\ \forall t \ge 0\,.$$

Thus

$$\int_0^t h(X_s^n, U_s^n)\,\mathrm{d}s \xrightarrow{\mathbb{P}} \int_0^t \int_{\mathbb{U}} h(X_s, u) U_s(\mathrm{d}u)\,\mathrm{d}s \qquad \forall t \ge 0\,.$$

Let $f \in \mathscr{D}(\mathcal{A})$, and $f_n = (I - n^{-1}\mathcal{A})f$ for $n \ge 1$. Also, let $h_i \in C_b(E \times \mathbb{U})$ for $i = 1, \dots, m$. With $0 \le t_1 < \cdots < t_{m+1}$ and $0 \le \delta_i \le t_i$ for $1 \le i \le m$, we have

$$\begin{aligned}&\mathbb{E}\biggl[\Bigl(f(X_{t_{m+1}}) - f(X_{t_m}) - \int_{t_m}^{t_{m+1}} \int_{\mathbb{U}} \mathcal{A}f(X_s, u) U_s(\mathrm{d}u)\,\mathrm{d}s\Bigr) \times \\ &\qquad\qquad \prod_{i=1}^m \int_{t_i-\delta_i}^{t_i} \int_{\mathbb{U}} h_i(X_{t_i}, u) U_s(\mathrm{d}u)\,\mathrm{d}s\biggr] \\ &= \lim_{n\to\infty} \mathbb{E}\biggl[\Bigl(f_n(X_{t_{m+1}}^n) - f_n(X_{t_m}^n) - \int_{t_m}^{t_{m+1}} \tilde{\mathcal{A}}_n f_n(X_s^n, U_s^n)\,\mathrm{d}s\Bigr) \times \\ &\qquad\qquad \prod_{i=1}^m \int_{t_i-\delta_i}^{t_i} h_i(X_{t_i}^n, U_s^n)\,\mathrm{d}s\biggr] = 0\,,\end{aligned}$$

because the expectation in the middle expression is always zero by the martingale property of

$$f_n(X_t^n) - \int_0^t \bar{\mathcal{A}}_n f_n(X_s^n, U_s^n)\,\mathrm{d}s\,, \quad t \geq 0\,,\ n \geq 1\,.$$

By a standard monotone class argument, it follows that (X, U) is a solution to the relaxed controlled martingale problem for $\mathcal{A}$. Furthermore, since (X^n, U^n) are stationary, so is (X, U), and for $t \geq 0$, $\delta > 0$ and $h \in C_b(E \times \mathbb{U})$,

$$\begin{aligned}\int_t^{t+\delta} \mathbb{E}\left[\int_{\mathbb{U}} h(X_s, u) U_s(\mathrm{d}u)\right] \mathrm{d}s &= \lim_{n\to\infty} \int_t^{t+\delta} \mathbb{E}\left[h(X_s^n, U_s^n)\right]\,\mathrm{d}s \\ &= \delta \int h\,\mathrm{d}\pi\,,\end{aligned}$$

implying (6.3.1) for a.e. t, where the qualification "a.e." may be dropped by suitably modifying U_t on a Lebesgue-null set of t. This completes the proof of the theorem. □

The relaxed control process U in the stationary solution of the relaxed martingale problem can be taken to be a Markov control, as the following corollary indicates.

Corollary 6.3.7 *The relaxed control process U in Theorem 6.3.6 may be taken to be of the form $U_t = v(X_t)$, $t \in \mathbb{R}$, for a measurable $v\colon E \to \mathscr{P}(\mathbb{U})$.*

Proof Disintegrate π as $\pi(\mathrm{d}x, \mathrm{d}u) = \pi_1(\mathrm{d}x)\hat{v}(\mathrm{d}u \mid x)$, and define $v\colon E \to \mathscr{P}(\mathbb{U})$ by $v(x) = \hat{v}(\mathrm{d}u \mid x)$, which is unique π_1-a.s. We use the symbol u for the generic element of $\mathscr{P}(\mathbb{U})$. Now apply Theorem 6.3.6 to the operator $\bar{\mathcal{A}}$ with $\bar{\pi} \in \mathscr{P}(E \times \mathscr{P}(\mathbb{U}))$ defined as $\bar{\pi}(\mathrm{d}x, \mathrm{du}) = \pi_1(\mathrm{d}x)\delta_{v(x)}(\mathrm{du})$ replacing π. Then by the foregoing there exists an $E \times \mathscr{P}(\mathscr{P}(\mathbb{U}))$-valued stationary solution $(X, \bar{U})$ to the relaxed controlled martingale problem for $\bar{\mathcal{A}}$ with respect to some filtration $\{\mathcal{F}_t\}$, such that

$$\mathbb{E}\left[g(X_t) \int_{\mathscr{P}(\mathbb{U})} h(\mathrm{u})\bar{U}_t(\mathrm{du})\right] = \int_{E\times\mathscr{P}(\mathbb{U})} g(x)\,h(\mathrm{u})\,\bar{\pi}(\mathrm{d}x, \mathrm{du})$$

for $g \in C_b(E)$, $h \in C(\mathscr{P}(\mathbb{U}))$, and $t \in \mathbb{R}$. Define a $\mathscr{P}(\mathscr{P}(\mathbb{U}))$-valued stationary process $\tilde{U}$ by

$$\int_{\mathscr{P}(\mathbb{U})} h(\mathrm{u})\tilde{U}_t(\mathrm{du}) = \mathbb{E}\left[\int_{\mathscr{P}(\mathbb{U})} h(\mathrm{u})\bar{U}_t(\mathrm{du}) \,\Big|\, \mathfrak{F}_t^X\right], \qquad t \in \mathbb{R}$$

for all h in a countable dense subset of $C(\mathscr{P}(\mathbb{U}))$, and $\{\mathfrak{F}_t^X\}$ the natural filtration of X, i.e., $\mathfrak{F}_t^X$ is the right-continuous completion of $\sigma\{X_s : -\infty < s \leq t\}$. Then it is easy to see that $(X, \tilde{U})$ is a stationary solution to the relaxed controlled martingale problem for $\bar{\mathscr{A}}$, with respect to $\{\mathfrak{F}_t^X\}$, satisfying

$$\begin{aligned}\mathbb{E}\left[g(X_t)\int_{\mathscr{P}(\mathbb{U})} h(\mathrm{u})\tilde{U}_t(\mathrm{du})\right] &= \int_{E\times\mathscr{P}(\mathbb{U})} g(x)\,h(\mathrm{u})\,\bar{\pi}(\mathrm{d}x,\mathrm{du})\\ &= \int_{E\times\mathscr{P}(\mathbb{U})} g(x)\,h(\mathrm{u})\,\pi_1(\mathrm{d}x)\,\delta_{v(x)}(\mathrm{du})\\ &= \int_E g(x)\,h(v(x))\,\pi_1(\mathrm{d}x)\\ &= \mathbb{E}\left[g(X_t)\,h(v(X_t))\right]\end{aligned}$$

for $g \in C_b(E)$, $h \in C(\mathscr{P}(\mathbb{U}))$, and $t \in \mathbb{R}$. Then

$$\mathbb{E}\left[\int_{\mathscr{P}(\mathbb{U})} h(\mathrm{u})\tilde{U}_t(\mathrm{du}) \,\Big|\, X_t\right] = h(v(X_t)) \quad \text{a.s.,} \qquad \forall t \ge 0\,,\ \forall h \in C(\mathscr{P}(\mathbb{U}))\,.$$

In particular,

$$\mathbb{E}\left[\int_{\mathscr{P}(\mathbb{U})} h^2(\mathrm{u})\tilde{U}_t(\mathrm{du}) \,\Big|\, X_t\right] = h^2(v(X_t)) \quad \text{a.s.} \quad \forall t \ge 0\,,$$

and hence

$$\begin{aligned}\mathbb{E}\left[\int_{\mathscr{P}(\mathbb{U})} \left[h(\mathrm{u}) - h(v(X_t))\right]^2 \tilde{U}_t(\mathrm{du}) \,\Big|\, X_t\right] &= \mathbb{E}\left[\int_{\mathscr{P}(\mathbb{U})} h^2(\mathrm{u})\tilde{U}_t(\mathrm{du}) \,\Big|\, X_t\right]\\ &\quad - 2\,\mathbb{E}\left[\int_{\mathscr{P}(\mathbb{U})} h(\mathrm{u})\tilde{U}_t(\mathrm{du}) \,\Big|\, X_t\right] h(v(X_t)) + h^2(v(X_t)) = 0\,.\end{aligned}$$

Thus $\tilde{U}_t = \delta_{v(X_t)}$ a.s., proving the claim. □

6.4 Extremal solutions

The main result of this section is the following: If we identify the processes (X, U) whose one-dimensional marginals agree for a.e. t, then the resulting equivalence classes form a closed, convex set whose extreme points correspond to Markov processes X. The idea of the proof is simple. For $T > 0$ we consider the regular conditional law of $X \circ \theta_T$ given $X_{[0,T]}$. This can be thought of as a three-step process. First pick an X_T according to its law. Then pick $X_{[0,T]}$ from the set of continuous functions $[0, T] \to E$ according to its regular conditional law given X_T. Finally, pick $X \circ \theta_T$ according to its regular conditional law given $X_{[0,T]}$ and X_T, i.e., given $X_{[0,T]}$. Suppose that instead of following the second step we pick a single function $x_{[0,T]} : \mathbb{R}^d \to C([0, T]; E)$ with $x_T = X_T$, measurably depending on X_T, and then pick $X \circ \theta_T$ according to its regular conditional law given X_T and $x_{[0,T]}$, with the same functional dependence on the latter as in the above procedure. For each realization of X_T, the first procedure yields a conditional law of $X \circ \theta_T$ given $X_{[0,T]}$, which is a mixture of what one would get in the second procedure. Thus the aforementioned equivalence classes in the first case are mixtures of those in the second. What follows makes this intuition precise.

We consider solutions (X, U) of the relaxed controlled martingale problem for $(\mathcal{A}, \nu)$ with a fixed $\nu \in \mathscr{P}(E)$. Let $\Gamma_\nu \subset \mathscr{P}(\mathcal{D}([0,\infty); E) \times \mathscr{U})$ be the set of all laws $\mathscr{L}(X, U)$ such that (X, U) solves the relaxed controlled martingale problem for $\mathcal{A}$, and satisfies $\mathscr{L}(X_0) = \nu$. Then $\mathscr{L}(X, U) \in \Gamma_\nu$ is completely characterized by

$$\mathbb{E}\left[f(X_0)\right] = \int_E f \, \mathrm{d}\nu , \quad f \in C_b(E), \tag{6.4.1}$$

and for $0 \le t_0 < \cdots < t_{m+1}$, $0 < \delta_i < t_i$, $1 \le i \le m$, $h_1, \dots, h_m \in C_b(E \times \mathbb{U})$ and $f \in \mathscr{D}(\bar{\mathcal{A}})$

$$\mathbb{E}\left[\left(f(X_{t_{m+1}}) - f(X_{t_m}) - \int_{t_m}^{t_{m+1}} \bar{\mathcal{A}} f(X_s, U_s)\,\mathrm{d}s\right) \times \prod_{i=1}^{m} \int_{t_i-\delta_i}^{t_i} \int_{\mathbb{U}} h_i(X_{t_i}, u) U_s(\mathrm{d}u)\,\mathrm{d}s\right] = 0 . \tag{6.4.2}$$

Since (6.4.1) – (6.4.2) are preserved under convex combinations and limits with respect to the topology of $\mathscr{P}(\mathcal{D}([0,\infty); E) \times \mathscr{U})$, we conclude that Γ_ν is closed and convex. We make the following additional assumption.

(A6.4) For every $\eta > 0$ and $T > 0$, there exists a compact set $K_{\eta,T} \subset E$ such that

$$\inf_U \mathbb{P}\left(X(t) \in K_{\eta,T}, \ \forall\, t \in [0, T]\right) \ge 1 - \eta .$$

Lemma 6.4.1 *Under (A6.1) – (A6.4), Γ_ν is compact in $\mathscr{P}(\mathcal{D}([0,\infty); E) \times \mathscr{U})$.*

Proof Since $\mathscr{U}$ and therefore also $\mathscr{P}(\mathscr{U})$ is compact, it suffices to prove that $\{\mathscr{L}(X) : \mathscr{L}(X, U) \in \Gamma_\nu\}$ is compact. Note that for each $t \ge 0$ and $f \in \mathscr{D}(\bar{\mathcal{A}})$,

$$\sup_{\Gamma_\nu} \mathbb{E}\left[\left(\int_0^t \|\bar{\mathcal{A}} f(X_s, U_s)\|^p \,\mathrm{d}s\right)^{1/p}\right] < \infty , \quad 1 \le p < \infty .$$

Hence by [55, theorem 9.4, p. 145], $\{\mathscr{L}(f(X))\}$ is tight for all $f \in \mathscr{D}(\mathcal{A})$. It follows that $\{\mathscr{L}(g(X)) : \mathscr{L}(X, U) \in \Gamma_\nu\}$, with g as in the proof of Theorem 6.3.6, is tight in $\mathscr{P}(\mathcal{D}([0,\infty); \hat{E}))$. The claim now follows from (A6.4) and [55, theorem 9.1, p. 142] as in Theorem 6.3.6. □

Let $\{\mathfrak{F}_t^{X,U} : t \ge 0\}$ denote the natural filtration of (X, U), and for some $t \ge 0$, let $\mathfrak{G}_t$ be a sub-σ-field of $\mathfrak{F}_t^{X,U}$ containing $\sigma(X_t)$. The following lemma is proved along the lines of Lemma 2.3.7.

Lemma 6.4.2 *The regular conditional law of $(X \circ \theta_t, U \circ \theta_t)$ given $\mathfrak{G}_t$ is a.s. equal to the law of some (X', U') which is a solution to the relaxed controlled martingale problem for $(\mathcal{A}, \delta_{X_t})$.*

6.4.1 A technical lemma

This section concerns an important characterization of extreme points of the set of probability measures on a product space with a given marginal.

Let S_1 and S_2 be Polish spaces with Borel σ-fields $\mathfrak{S}_1$ and $\mathfrak{S}_2$, respectively, and $\mu \in \mathscr{P}(S_1 \times S_2)$. We disintegrate μ as

$$\mu(\mathrm{d}s_1, \mathrm{d}s_2) = \mu_1(\mathrm{d}s_1)\,\mu_2^1(\mathrm{d}s_2 \mid s_1),$$

where $\mu_1 \in \mathscr{P}(S_1)$ and $\mu_2^1 \colon S_1 \to \mathscr{P}(S_2)$ are respectively the marginal on S_1 and the regular conditional law on S_2, the latter specified μ_1-a.s. uniquely. For $\xi \in \mathscr{P}(S_1)$, define

$$\mathscr{P}_\xi(S_1 \times S_2) := \{\mu \in \mathscr{P}(S_1 \times S_2) : \mu_1 = \xi\},$$

$$\mathscr{P}_\xi^\delta(S_1 \times S_2) := \{\mu \in \mathscr{P}(S_1 \times S_2) : \mu_1 = \xi \text{ and } \mu_2^1 \text{ is } \mu_1\text{-a.s. Dirac}\}.$$

Lemma 6.4.3 *If*

$$\mu(\mathrm{d}x, \mathrm{d}y) = \xi(\mathrm{d}x)\,\nu(\mathrm{d}y \mid x) \in \mathscr{P}_\xi(S_1 \times S_2) \setminus \mathscr{P}_\xi^\delta(S_1 \times S_2),$$

then there exist Borel $A \subset S_1$ and $f \in C_b(S_2)$ such that $\xi(A) > 0$ and for all $x \in A$, f is not a constant $\nu(x)$-a.s.

Proof Let $\{f_i\} \subset C_b(S_2)$ be a countable set that separates points of S_2. Suppose that for ξ-a.s. x, f_i is $\nu(x)$-a.s. a constant for all i. Then $\nu(x)$ is a Dirac measure for such x, contradicting the hypothesis $\mu \notin \mathscr{P}_\xi^\delta(S_1 \times S_2)$. Thus there exists a Borel $A' \subset S_1$ such that $\xi(A') > 0$ and for $x \in A'$, f_i is not a constant $\nu(x)$-a.s. for some $i \in \mathbb{N}$. Let

$$A_i = \{x \in S_1 : f_i \text{ is not a constant } \nu(x)\text{-a.s.}\}, \qquad i \in \mathbb{N}.$$

Then A_i is the complement of

$$\bigcap_{k\in\mathbb{N}} \Bigl\{x \in S_1 : \int g_k f_i \,\mathrm{d}\nu(x) = \int g_k \,\mathrm{d}\nu(x) \int f_i \,\mathrm{d}\nu(x)\Bigr\}$$

for $\{g_k\} \subset C_b(S_2)$, chosen so that $\int g_k \,\mathrm{d}m_1 = \int g_k \,\mathrm{d}m_2\ \forall k \in \mathbb{N}$ implies $m_1 = m_2$ for any finite signed measures m_1, m_2 on S_2. Therefore A_i is measurable. We may set $A' = \cup_i A_i$. Then $\xi(A') > 0$ implies that $\xi(A_{i_0}) > 0$ for some i_0 and the claim follows with $A = A_{i_0}$, $f = f_{i_0}$. □

We now come to the main result of this section. For $q \in \mathscr{P}(S_2)$ and $f \in C_b(S_2)$ let b_q be the least number such that

$$q(\{x : f(x) > b_q\}) \le \tfrac{1}{2}.$$

Let

$$A_1 = \{x : f(x) < b_q\}, \quad A_2 = \{x : f(x) > b_q\}, \quad A_3 = \{x : f(x) = b_q\},$$

and $\delta \in [0, 1]$ such that

$$q(A_1) + \delta\, q(A_3) = q(A_2) + (1 - \delta)\, q(A_3) = \tfrac{1}{2},$$

setting $\delta = 0$ when $q(A_3) = 0$. Define

$$\alpha_1(q) = 2(I_{A_1} + \delta I_{A_3})\,q\,,$$
$$\alpha_2(q) = 2(I_{A_2} + (1-\delta)I_{A_3})\,q\,.$$

By [49, theorem 2.1], verifying that $\alpha_1, \alpha_2 \colon \mathscr{P}(S_2) \to \mathscr{P}(S_2)$ are measurable is equivalent to verifying that for a Borel $A \subset S_2$, the maps $q \mapsto \alpha_i(q)(A)$, $i = 1, 2$, are measurable. Consider, say, $i = 1$. Then

$$\alpha_1(q)(A) = 2q(A \cap A_1) + 2\delta q(A \cap A_3)\,,$$

where δ, A_1, A_3 depend on q through their dependence on b_q. It is easy to verify that $b \mapsto q(A \cap \{x : f(x) < b\})$ and $b \mapsto q(A \cap \{x : f(x) = b\})$ are measurable. Thus it suffices to prove that $q \mapsto b_q$ is measurable. But for $c \in \mathbb{R}$,

$$\{q : b_q > c\} = \{q : q(\{x : f(x) > c\}) > \tfrac{1}{2}\}\,,$$

which is measurable by [49, theorem 2.1]. This establishes the measurability of α_1, α_2. Note that

$$q = \frac{\alpha_1(q) + \alpha_2(q)}{2}\,.$$

Lemma 6.4.4 *$\mathscr{P}^{\delta}_{\xi}(S_1 \times S_2)$ is the set of extreme points of $\mathscr{P}_{\xi}(S_1 \times S_2)$.*

Proof Define

$$Q := \mathscr{P}_{\xi}(S_1 \times S_2)\,,$$
$$Q_\delta := \mathscr{P}^{\delta}_{\xi}(S_1 \times S_2)\,,$$

and let Q_e stand for the set of extreme points of Q. Let $\mu(\mathrm{d}x, \mathrm{d}y) = \xi(\mathrm{d}x)\nu(\mathrm{d}y \mid x)$, and suppose $\mu \notin Q_\delta$. Pick a Borel $A \subset S_1$ and $f \in C_b(S_2)$ as in Lemma 6.4.3. Then $\alpha_1(\nu(\cdot \mid x)) \neq \alpha_2(\nu(\cdot \mid x))$ for $x \in A$. Define

$$\mu_i(\mathrm{d}x, \mathrm{d}y) = \xi(\mathrm{d}x)\,\alpha_i(\nu(\mathrm{d}y \mid x))\,, \quad i = 1, 2\,.$$

Then $\mu = \frac{\mu_1 + \mu_2}{2}$, $\mu_1 \neq \mu_2$, implying $\mu \notin Q_e$. Thus $Q_e \subset Q_\delta$. Conversely, let $\mu = \frac{\mu_1 + \mu_2}{2}$ for $\mu \in Q_\delta$, $\mu_1 \neq \mu_2$ in Q. Then it follows from the decomposition

$$\mu_i(\mathrm{d}x, \mathrm{d}y) = \xi(\mathrm{d}x)\,\nu_i(\mathrm{d}y \mid x)\,, \quad i = 1, 2\,,$$

that $\nu_1(\cdot \mid x)$ and $\nu_2(\cdot \mid x)$ must differ for x in a set of strictly positive ξ-measure. For such x

$$\nu(\cdot \mid x) = \frac{\nu_1(\cdot \mid x) + \nu_2(\cdot \mid x)}{2}$$

cannot be Dirac, a contradiction. Thus $Q_\delta \subset Q_e$. □

6.4.2 Marginal classes

Definition 6.4.5 Define an equivalence relation "∼" on Γ_V by:

$$\mathscr{L}(X,U) \sim \mathscr{L}(X',U') \iff \mathscr{L}(X_t,U_t) = \mathscr{L}(X'_t,U'_t)$$

for almost all t. The equivalence class under $\sim$ that contains $\mathscr{L}(X,U)$ is denoted by $\langle\!\langle \mathscr{L}(X,U) \rangle\!\rangle$ and is called a *marginal class*. Let Γ'_V denote the set of marginal classes with the quotient topology inherited from Γ_V. We extend the definition of this equivalence relation to the space $\mathfrak{M}_s$ of finite signed measures on $\mathcal{D}([0,\infty);E) \times \mathscr{U}$ with weak*-topology as follows. Two elements of $\mathfrak{M}_s$ are equivalent if their images under the map

$$(x_{[0,\infty)}, v_{[0,\infty)}) \to (x_t, v_t)$$

agree a.e. in $t \in \mathbb{R}_+$.

Remark 6.4.6 Since v is only measurable, the map $v_{[0,\infty)} \to v_t$ needs to be defined with some care. Recall the identification of $v \in \mathscr{U}$ with an $\alpha = (\alpha_1, \alpha_2, \dots) \in B^\infty$ from Section 2.3. If t is a Lebesgue point for all α_i's, let

$$\bar{\alpha}_i(t) = \lim_{\Delta \to 0} \frac{1}{2\Delta} \int_{t-\Delta}^{t+\Delta} \alpha_i(s)\,\mathrm{d}s\,, \quad i \geq 1\,,$$

and set $v_t = \varphi^{-1}(\bar{\alpha}_1(t), \bar{\alpha}_2(t), \dots)$ for φ as in Section 2.3. For any other t, set v_t equal to an arbitrary element of $\mathscr{P}(\mathbb{U})$. This defines $v_{[0,\infty)} \mapsto v_t$ as a measurable map for $t \geq 0$.

The set Γ'_V may now be viewed as a compact, convex subset of $\mathfrak{M}'_s$, the space of equivalence classes in $\mathfrak{M}_s$ under $\sim$, with the quotient topology. We call the equivalence class $\langle\!\langle \mathscr{L}(X,U) \rangle\!\rangle \in \Gamma'_V$ *extremal* if it is an extreme point of Γ'_V. It turns out that every member of an extremal element of Γ'_V is a Markov process. This result is stated as Theorem 6.4.16 at the end of this section. First, we need to develop some necessary intermediate results.

Let $\{h_i : i \in \mathbb{N}\}$ be a countable subset of $C_b(E \times \mathscr{P}(\mathbb{U}))$ that separates points of $\mathscr{P}(E \times \mathscr{P}(\mathbb{U}))$. For $i \in \mathbb{N}$ and $\alpha \in (0,\infty)$ define $h_{i,\alpha} \colon \Gamma_V \to \mathbb{R}$ by

$$\Gamma_V \ni \mathscr{L}(X,U) \mapsto \mathbb{E}\left[\int_0^\infty \mathrm{e}^{-\alpha t} h_i(X_t, U_t)\,\mathrm{d}t\right].$$

This map is constant on marginal classes and therefore may be viewed as a map $\Gamma'_V \to \mathbb{R}$. The following lemma follows easily from the injectivity of the Laplace transform on R_+ and the choice of $\{h_i\}$.

Lemma 6.4.7 *If $\mu_1, \mu_2 \in \Gamma'_V$ satisfy $h_{i,\alpha}(\mu_1) = h_{i,\alpha}(\mu_2)$ for all $i \in \mathbb{N}$ and rational $\alpha > 0$, then $\mu_1 = \mu_2$.*

For $x \in E$, $i \in \mathbb{N}$ and positive constants α and ε, we define the subset $H^\varepsilon_{i,\alpha}(x)$ of $\Gamma'_{\delta_x} \times \Gamma'_{\delta_x}$ by

$$H^\varepsilon_{i,\alpha}(x) := \{(\eta_1, \eta_2) \in \Gamma'_{\delta_x} \times \Gamma'_{\delta_x} : |h_{i,\alpha}(\eta_1) - h_{i,\alpha}(\eta_2)| \geq \varepsilon\}$$

and let

$$H_x := \bigcup_{i\in\mathbb{N},\, \alpha>0,\, \varepsilon>0} H^{\varepsilon}_{i,\alpha}(x)\,.$$

Since H_x remains unchanged if the union in its definition is taken over all positive rationals α and ε, it follows that it is in fact a countable union. By Lemma 6.4.7, the set H_x is the complement of the diagonal in $\Gamma'_{\delta_x} \times \Gamma'_{\delta_x}$. Let $\vartheta\colon \mathfrak{M}'_s \times \mathfrak{M}'_s \to \mathfrak{M}'_s$, denote the map

$$\vartheta(\eta_1, \eta_2) = \frac{\eta_1 + \eta_2}{2}\,.$$

It is clear that $\eta \in \Gamma'_{\delta_x}$ is extremal if and only if $\eta \notin \vartheta(H_x)$.

Fix $\mathscr{L}(X, U) \in \Gamma_\nu$. Let $p(\mathrm{d}y \mid x)$ denote the regular conditional law of (X, U) given $X_0 = x$, defined ν-a.s. uniquely. By Lemma 6.4.2, we may assume that $p(\mathrm{d}y \mid x) \in \Gamma_{\delta_x}$ for each x.

Lemma 6.4.8 *Suppose that $\langle\!\langle p(\mathrm{d}y \mid x)\rangle\!\rangle$ is not an extreme point of Γ'_{δ_x} for all x in a set of positive ν-measure. Then there exist a relatively compact set $C \subset E$ with $\nu(C) > 0$, and positive constants $i \in \mathbb{N}$ and $\alpha, \varepsilon \in \mathbb{Q} \cap (0, \infty)$ such that*

$$\langle\!\langle p(\mathrm{d}y \mid x)\rangle\!\rangle \in \vartheta(H^{\varepsilon}_{i,\alpha}(x)) \qquad \forall x \in C\,.$$

Proof By hypothesis,

$$\nu(\{x : \langle\!\langle p(\mathrm{d}y \mid x)\rangle\!\rangle \in \vartheta(H_x)\}) > 0\,.$$

Therefore, for some i, α, ε as in the statement of the lemma,

$$\nu(\{x : \langle\!\langle p(\mathrm{d}y \mid x)\rangle\!\rangle \in \vartheta(H^{\varepsilon}_{i,\alpha}(x))\}) > 0\,.$$

Since E is Polish, every probability measure on E is tight by the Oxtoby–Ulam theorem [24, theorem 1.4, p. 10] and hence it assigns mass 1 to a countable union of compact sets. Therefore there exists a relatively compact set

$$C \subset \{x : \langle\!\langle p(\mathrm{d}y \mid x)\rangle\!\rangle \in \vartheta(H^{\varepsilon}_{i,\alpha}(x))\}$$

which has positive ν-measure. □

Define $\Delta \subset \mathscr{P}(\mathcal{D}([0,\infty); E) \times \mathscr{U})$ by

$$\Delta := \overline{\bigcup_{x\in C} \Gamma_{\delta_x}}\,.$$

The arguments of Lemma 6.4.1 can be adapted to prove that Δ is compact. We denote by $\Delta' \subset \mathfrak{M}'_s$ the corresponding (compact) set of marginal classes.

Lemma 6.4.9 *If $\langle\!\langle \mathscr{L}(X, U)\rangle\!\rangle$ is an extreme point of Γ'_ν, then for ν-a.s. x, $\langle\!\langle p(\mathrm{d}y \mid x)\rangle\!\rangle$ is an extreme point of Γ'_{δ_x}.*

Proof Suppose not. Let $C \subset E$, and i, α and ε as in Lemma 6.4.8. Define

$$G := \{(\eta_1, \eta_2) \in \mathfrak{M}'_s \times \mathfrak{M}'_s : |h_{i,\alpha}(\eta_1) - h_{i,\alpha}(\eta_2)| \geq \varepsilon\}.$$

For $x \in C$, let

$$K_x := \{(\eta_1, \eta_2) \in H^{\varepsilon}_{i,\alpha}(x) : \langle\!\langle p(\mathrm{d}y \mid x) \rangle\!\rangle = \tfrac{\eta_1+\eta_2}{2}\}.$$

By Lemma 6.4.8, $K_x \neq \varnothing$. It is also compact, since it is clearly closed and it is a subset of $H^{\varepsilon}_{i,\alpha}(x)$ which is compact. Let $G' \subset \Delta' \times \Delta'$ be closed, therefore compact. Note that

$$\{x \in C : K_x \cap G' \neq \varnothing\} = \{x \in C : \langle\!\langle p(\mathrm{d}y \mid x) \rangle\!\rangle \in \vartheta(G \cap G')\}. \tag{6.4.3}$$

Since G is closed, $G \cap G'$ and thus $\vartheta(G \cap G')$ is compact. The map $x \mapsto p(\mathrm{d}y \mid x)$ is measurable, and therefore, $x \mapsto \langle\!\langle p(\mathrm{d}y \mid x) \rangle\!\rangle$ is also measurable. Therefore the set in (6.4.3) is measurable. From this, we conclude that the map $x \mapsto K_x \subset \Delta' \times \Delta'$ is measurable and therefore weakly measurable in the sense of [118, p. 862], in view of the remarks in [118, paragraph 5, p. 863]. By [118, theorem 4.1, p. 867], there exists a measurable map $x \mapsto (\langle\!\langle \eta_{1x} \rangle\!\rangle, \langle\!\langle \eta_{2x} \rangle\!\rangle) \in \Delta' \times \Delta'$ such that

$$\langle\!\langle p(\mathrm{d}y \mid x) \rangle\!\rangle = \frac{\eta_{1x} + \eta_{2x}}{2}$$

and

$$|h_{i,\alpha}(\eta_{1x}) - h_{i,\alpha}(\eta_{2x})| \geq \varepsilon.$$

Define

$$C^+ := \{x \in B : h_{i,\alpha}(\eta_{1x}) - h_{i,\alpha}(\eta_{2x}) \geq \varepsilon\},$$
$$C^- := \{x \in B : h_{i,\alpha}(\eta_{1x}) - h_{i,\alpha}(\eta_{2x}) \leq -\varepsilon\}.$$

Since $\nu(C) > 0$ and $C = C^+ \cup C^-$, it follows that $\max(\nu(C^+), \nu(C^-)) > 0$. Suppose that $\nu(C^+) > 0$ (if not, replace C^+ by C^-). Define, for $i = 1, 2$,

$$\bar{\eta}_{ix} := \begin{cases} \eta_{ix} & \text{if } x \in C^+, \\ p(\mathrm{d}y \mid x) & \text{otherwise,} \end{cases}$$

and

$$\mu_i(\mathrm{d}x, \mathrm{d}y) := \nu(\mathrm{d}x)\, \bar{\eta}_{ix}(\mathrm{d}y).$$

Since $\nu(C^+) > 0$, $\langle\!\langle \mu_1 \rangle\!\rangle \neq \langle\!\langle \mu_2 \rangle\!\rangle$. Clearly, $\langle\!\langle \mathscr{L}(X, U) \rangle\!\rangle = \frac{\langle\!\langle \mu_1 \rangle\!\rangle + \langle\!\langle \mu_2 \rangle\!\rangle}{2}$. Since $\bar{\eta}_{1x}, \bar{\eta}_{2x} \in \Gamma_{\delta_x}$ for each x, we have $\langle\!\langle \mu_1 \rangle\!\rangle, \langle\!\langle \mu_2 \rangle\!\rangle \in \Gamma'_\nu$. Thus $\langle\!\langle \mathscr{L}(X, U) \rangle\!\rangle$ is not an extreme point of Γ'_ν. This contradiction establishes the claim. □

Two solutions of the relaxed martingale problem can be concatenated as shown in the following lemma.

Lemma 6.4.10 *Let (X, U) and (X', U') be two solutions to the relaxed controlled martingale problem for $\mathcal{A}$ such that*

$$\mathscr{L}(X_T, U_T) = \mathscr{L}(X'_0, U'_0),$$

for some $T > 0$. Then there exists a solution $(\tilde{X}, \tilde{U})$ to the relaxed controlled martingale problem for $\mathcal{A}$ such that

$$\mathscr{L}(\tilde{X}_t, \tilde{U}_t) = \begin{cases} \mathscr{L}(X_t, U_t) & \text{if } t \in [0, T), \\ \mathscr{L}(X'_{t-T}, U'_{t-T}) & \text{if } t \geq T. \end{cases}$$

Proof Let $\Omega = \mathcal{D}([0, \infty); E) \times \mathscr{U}$, and $\mathscr{B}(\Omega)$ its Borel σ-field. Define $(\tilde{X}, \tilde{U})$ to be the canonical process on $(\Omega, \mathfrak{B}(\Omega))$, i.e., $(\tilde{X}, \tilde{U})$ evaluated at $\omega = (x, y) \in \Omega$ is given by $\tilde{X} = x$, $\tilde{U} = y$. Then specifying a probability measure $\mathbb{P}$ on $(\Omega, \mathscr{B}(\Omega))$ is equivalent to prescribing the law of $(\tilde{X}, \tilde{U})$. Let Π_1, Π_2 map $\omega = (x, y) \in \Omega$ to its restriction on $[0, T)$, $[T, \infty)$, respectively. Let $\mathscr{L}(\tilde{X}, \tilde{U})$ be such that

$$\mathscr{L}(\Pi_1(\tilde{X}, \tilde{U})) = \mathscr{L}(\Pi_1(X, U)),$$

and the regular conditional law of $\Pi_2(\tilde{X}, \tilde{U})$ given $\Pi_1(\tilde{X}, \tilde{U})$ is the same as the regular conditional law of $\Pi_2(\tilde{X}, \tilde{U})$ given $(\tilde{X}_T, \tilde{U}_T)$, which in turn is the same as the regular conditional law of (X', U') given (X'_0, U'_0), prescribed a.s. with respect to

$$\mathscr{L}(X'_0, U'_0) = \mathscr{L}(X_T, U_T) = \mathscr{L}(\tilde{X}_T, \tilde{U}_T).$$

The law $\mathscr{L}(\tilde{X}, \tilde{U})$ as defined meets our requirements by construction. □

Definition 6.4.11 We call $(\tilde{X}, \tilde{U})$ defined in Lemma 6.4.10 the *T-concatenation* of (X, U) and (X', U').

Lemma 6.4.12 *Let $\langle\!\langle \mathscr{L}(X, U) \rangle\!\rangle$ be an extreme point of Γ'_ν. Then $\langle\!\langle \mathscr{L}(X \circ \theta_T, U \circ \theta_T) \rangle\!\rangle$ is an extreme point of Γ'_{ν_T}, where $\nu_T = \mathscr{L}(X_T)$.*

Proof Suppose not. Then there exist $a \in (0, 1)$, $\langle\!\langle \mathscr{L}(X^i, U^i) \rangle\!\rangle \in \Gamma'_{\nu_T}$, $i = 1, 2$, such that

$$\langle\!\langle \mathscr{L}(X^1, U^1) \rangle\!\rangle \neq \langle\!\langle \mathscr{L}(X^2, U^2) \rangle\!\rangle$$

and

$$\langle\!\langle \mathscr{L}(X \circ \theta_T, U \circ \theta_T) \rangle\!\rangle = a \langle\!\langle \mathscr{L}(X^1, U^1) \rangle\!\rangle + (1 - a) \langle\!\langle \mathscr{L}(X^2, U^2) \rangle\!\rangle.$$

By Lemma 6.4.10 there exists a pair of processes $(\bar{X}^i, \bar{U}^i)$, $i = 1, 2$, which are solutions to the relaxed controlled martingale problem for $\mathcal{A}$, such that for $i = 1, 2$,

$$\mathscr{L}(\bar{X}^i_t, \bar{U}^i_t) = \begin{cases} \mathscr{L}(X_t, U_t) & \text{if } t \in [0, T], \\ \mathscr{L}(X^i_{t-T}, U^i_{t-T}) & \text{if } t > T. \end{cases}$$

Then

$$\langle\!\langle \mathscr{L}(\bar{X}^1, \bar{U}^1) \rangle\!\rangle \neq \langle\!\langle \mathscr{L}(\bar{X}^2, \bar{U}^2) \rangle\!\rangle$$

in Γ'_ν and

$$\langle\!\langle \mathscr{L}(X,U)\rangle\!\rangle = a\langle\!\langle \mathscr{L}(\bar{X}^1,\bar{U}^1)\rangle\!\rangle + (1-a)\langle\!\langle \mathscr{L}(\bar{X}^2,\bar{U}^2)\rangle\!\rangle\,,$$

contradicting the extremality of $\langle\!\langle \mathscr{L}(X,U)\rangle\!\rangle$. The claim follows. □

Let $(S_1,\mathfrak{S}_1)$ and $(S_2,\mathfrak{S}_2)$ be measurable spaces and $f\colon S_1 \to S_2$ a measurable map. We denote the image of $\mu \in \mathscr{P}(S_1)$ under f by $f_*(\mu)$. In other words, $f_*\colon \mathscr{P}(S_1) \to \mathscr{P}(S_2)$ and $f_*(\mu)(A) = \mu(f^{-1}(A))$ for $A \in \mathfrak{S}_2$. Similarly, we denote by $f_*(\mathcal{P})$ the image of $\mathcal{P} \subset \mathscr{P}(S_1)$ under f.

Henceforth, let $\langle\!\langle \mathscr{L}(X,U)\rangle\!\rangle$ be an extreme point of Γ'_ν. Fix $T > 0$, and Let x_T denote the restriction of x to $[0,T]$. Define

$$\beta\colon \mathcal{D}([0,T];E) \to E \times \mathcal{D}([0,T];E)$$

by

$$\beta(x_T) = \bigl(x(T), x_T\bigr)\,.$$

We introduce the notation:

$$\begin{aligned}
\mu_T &= \mathscr{L}(X_{[0,T]})\,,\\
\nu_T &= \mathscr{L}(X_T)\,,\\
\mathcal{Y} &= \{\gamma\colon E \to \mathcal{D}([0,T];E) : \gamma \text{ is measurable and } \gamma(x)(T) = x\,,\ \forall x \in E\}\,,\\
M_T &= \{\gamma_*(\nu_T) \in \mathscr{P}(\mathcal{D}([0,T];E)) : \gamma \in \mathcal{Y}\}\,.
\end{aligned}$$

Lemma 6.4.13 *μ_T (respectively, $\beta_*(\mu_T)$) is the barycenter of some probability measure supported on M_T (respectively, $\beta_*(M_T)$).*

Proof Concerning μ_T, the property follows by Lemma 6.4.4 and Choquet's theorem (Theorem 1.5.7). The claim for $\beta_*(\mu_T)$ follows from the fact that the barycentric representation is preserved under β_*. □

Let q be any version of the regular conditional law of $(X \circ \theta_T, U \circ \theta_T)$ given $(X_T, X_{[0,T]})$. Then q takes the form

$$q(\cdot \mid x,y) \in \mathscr{P}(\mathcal{D}([0,\infty);E) \times \mathscr{U})\,, \qquad \text{for } (x,y) \in E \times \mathcal{D}([0,T];E)\,.$$

Let

$$\Psi = \mathscr{P}(E \times \mathcal{D}([0,T];E) \times (\mathcal{D}([0,\infty);E) \times \mathscr{U}))$$

and set $\psi = \mathscr{L}(X_T, X_{[0,T]}, (X \circ \theta_T, U \circ \theta_T)) \in \Psi$. Thus ψ takes the form

$$\psi(\mathrm{d}x,\mathrm{d}y,\mathrm{d}z) = \beta_*(\mu_T)(\mathrm{d}x,\mathrm{d}y)\,q(\mathrm{d}z \mid x,y)\,.$$

Let $\Psi^\delta \subset \Psi$ be the set of measures ρ of the form

$$\rho(\mathrm{d}x,\mathrm{d}y,\mathrm{d}z) = \nu_T(\mathrm{d}x)\,\delta_{\gamma(x)}(\mathrm{d}y)\,q(\mathrm{d}z \mid x,y)$$

for some $\gamma \in \mathcal{Y}$, where $\delta_{\gamma(x)}$ denotes the Dirac measure at $\gamma(x)$. By Lemma 6.4.13, ψ is the barycenter of a probability measure ξ_1 on Ψ^δ.

Lemma 6.4.14 *With ξ_1-probability 1, the measure $\Phi \in \mathscr{P}(\mathcal{D}([0,\infty); E) \times \mathscr{U})$ defined by*

$$\int f(z)\,\Phi(dz) = \iint \nu_T(dx)\, q(dz \mid x, \gamma(x))\, f(z)$$

for $f \in C_b(\mathcal{D}([0,\infty); E) \times \mathscr{U})$, is in Γ_{ν_T}. Furthermore, $q(dz \mid x, \gamma(x))$ can be chosen to be in Γ_{δ_x} by choosing an appropriate version.

Proof By Lemma 6.4.2, $q(dz \mid x, y) \in \Gamma_{\delta_x}$, $\beta_*(\mu_T)$-a.s. Hence

$$q(dz \mid x, y) \in \Gamma_{\delta_x}, \quad \nu_T(dx)\,\delta_{\gamma(x)}(dy)\text{-a.s.} \quad \text{for } \xi_1\text{-a.s. } \rho\,.$$

The claim follows. □

Denote γ, Φ above as γ_ρ, Φ_ρ to make explicit their dependence on ρ. Recall that ψ is the barycenter of a probability measure ξ_1 on Ψ^δ.

Lemma 6.4.15 *For an extremal $\mathscr{L}(X, U)$, and x outside a set of zero ν_T-measure,*

$$\langle\!\langle q(dz \mid x, \gamma_\rho(x)) \rangle\!\rangle = \langle\!\langle p(dz \mid x) \rangle\!\rangle \quad \xi_1\text{-a.s. } \rho\,.$$

Proof Let $\bar{\Phi} = \mathscr{L}(X \circ \theta_T, U \circ \theta_T)$. Then for $f \in C_b(\mathcal{D}([0,\infty); E) \times \mathscr{U})$,

$$\int f(z)\,\bar{\Phi}(dz) = \int\int \beta_*(\mu_T)(dx, dy)\, q(dz \mid x, y)\, f(z)\,.$$

By Lemma 6.4.12, $\langle\!\langle \bar{\Phi} \rangle\!\rangle$ is an extreme point of Γ'_{ν_T}. Disintegrate $\bar{\Phi}$ as

$$\bar{\Phi}(dx, dz) = \nu_T(dx)\, p(dz \mid x)\,,$$

where $p(dz \mid x)$ is the regular conditional law of $(X \circ \theta_T, U \circ \theta_T)$ given $X_T = x$. By Lemma 6.4.9, $\langle\!\langle p(dz \mid x) \rangle\!\rangle$ is an extreme point of Γ'_{δ_x} for ν_T-a.s. x. Then

$$\begin{aligned}\bar{\Phi}(dx, dz) &= \nu_T(dx)\, p(dz \mid x)\\ &= \int \xi_1(d\rho)\,\nu_T(dx)\, q(dz \mid x, \gamma_\rho(x))\\ &= \nu_T(dx) \int \xi_1(d\rho)\, q(dz \mid x, \gamma_\rho(x))\,,\end{aligned}$$

so that for ν_T-a.s. x, $p(dz \mid x)$ is the barycenter of a probability measure on

$$\{q(dz \mid x, \gamma_\rho(x)) : \rho \in \Psi^\delta\}$$

and, in turn, $\langle\!\langle p(dz \mid x) \rangle\!\rangle$ is the barycenter of a probability measure on

$$\{\langle\!\langle q(dz \mid x, \gamma_\rho(x)) \rangle\!\rangle : \rho \in \Psi^\delta\}\,.$$

For x outside a set of zero ν_T-measure outside which the foregoing properties hold and $\langle\!\langle p(dz \mid x) \rangle\!\rangle$ is extremal in Γ'_{δ_x}, we must have

$$\langle\!\langle p(dz \mid x) \rangle\!\rangle = \langle\!\langle q(dz \mid x, \gamma_\rho(x)) \rangle\!\rangle$$

for ξ_1-a.s. ρ, thus proving the claim. □

Theorem 6.4.16 *Every representative of an extremal element of Γ'_ν is a Markov process.*

Proof Fix $t > 0$ and let $\hat{p}(\mathrm{d}z \mid x)$ and $\hat{q}(\mathrm{d}z \mid x, y)$ denote the images of $p(\mathrm{d}z \mid x)$ and $q(\mathrm{d}z \mid x, y)$, respectively, under the map

$$(x_{[0,\infty)}, y_{[0,\infty)}) \in \mathcal{D}([0,\infty); E) \times \mathscr{U} \mapsto x(t) \in E\,.$$

By Lemma 6.4.15,

$$\hat{p}(\mathrm{d}z \mid x) = \hat{q}(\mathrm{d}z \mid x, \gamma_\rho(x)) \quad \nu_T\text{-a.s. } x\,, \text{ and } \xi_1\text{-a.s. } \rho\,,$$

i.e., the right-hand side is independent of ρ, ξ_1-a.s. Thus we have

$$\begin{aligned}
\mathscr{L}(X_T, X_{[0,T]}, X_{T+t}) &= \beta_*(\mu_T)(\mathrm{d}x, \mathrm{d}y)\,\hat{q}(\mathrm{d}z \mid x, y) \\
&= \int \xi_1(\mathrm{d}\rho)\,\nu_T(\mathrm{d}x)\,\delta_{\gamma_\rho(x)}(\mathrm{d}y)\,\hat{q}(\mathrm{d}z \mid x, \gamma_\rho(x)) \\
&= \nu_T(\mathrm{d}x)\,\eta(\mathrm{d}y \mid x)\,\hat{p}(\mathrm{d}z \mid x)\,,
\end{aligned}$$

where

$$\eta(\mathrm{d}y \mid x) = \int \xi_1(\mathrm{d}\rho)\,\delta_{\gamma_\rho(x)}(\mathrm{d}y)\,.$$

Thus X_{T+t}, $X_{[0,T]}$ are conditionally independent given X_T. Given the arbitrary choice of T, t, the claim follows. □

6.5 Existence results

We now state and prove an assortment of results concerning the existence of optimal controls in various classes of controls. These are direct consequences of the theory so far.

Let $c\colon E \times \mathbb{U} \to \mathbb{R}_+$ be a prescribed running cost function. Transforming into the relaxed control framework, we define $\bar{c}\colon E \times \mathscr{P}(\mathbb{U}) \to \mathbb{R}_+ \cup \{\infty\}$ by

$$\bar{c}(x, v) = \int c(x, u) v(\mathrm{d}u) \qquad \forall (x, v) \in E \times \mathscr{P}(\mathbb{U})\,.$$

Consider the associated ergodic control problem, in the average formulation where one seeks to minimize

$$\limsup_{t\to\infty} \frac{1}{t} \int_0^t \mathbb{E}\,[\bar{c}(X_s, U_s)]\;\mathrm{d}s\,. \tag{6.5.1}$$

Note that for (X, U) a stationary solution to the relaxed controlled martingale problem for $\mathcal{A}$, (6.5.1) takes the form

$$F(\pi) := \int_{E\times\mathbb{U}} c\,\mathrm{d}\pi$$

for some ergodic occupation measure π. This motivates looking at the optimization problem: Minimize F on $\mathcal{G}$, the set of ergodic occupation measures. We assume that $F(\pi) < \infty$ for at least one $\pi \in \mathcal{G}$. Consider the following two alternative conditions:

(A6.5) (near-monotonicity) The map F is inf-compact.

(A6.6) (stability) $\mathcal{G}$ is compact.

Remark 6.5.1 If (A6.5) holds, $\{\pi : F(\pi) < \infty\}$ is σ-compact. But by the Baire category theorem, a Polish space that is not locally compact cannot be σ-compact. Thus F cannot be everywhere finite. Hence the inclusion of "∞" as a possible value for $\bar{c}(x, v)$.

The following is immediate:

Lemma 6.5.2 *Under (A6.5) or (A6.6), F attains its minimum on $\mathcal{G}$.*

As usual, we let ϱ^* denote this minimum.

Corollary 6.5.3 *There exists a stationary solution to the relaxed controlled martingale problem for $\mathcal{A}$ for which the cost is ϱ^*.*

Proof Combine Lemma 6.5.2 and Theorem 6.3.6. □

Recall the mean empirical measures $\bar{\zeta}_t \in \mathcal{P}(E \times \mathbb{U})$, $t \geq 0$, defined by

$$\int f \, \mathrm{d}\bar{\zeta}_t = \frac{1}{t} \int_0^t \mathbb{E}\left[\int_{\mathbb{U}} f(X_s, u) U_s(\mathrm{d}u)\right] \mathrm{d}s \,, \quad f \in C_b(E \times \mathbb{U}) \,,$$

corresponding to a solution (X, U) of the relaxed controlled martingale problem for $\mathcal{A}$. We have the following lemma.

Lemma 6.5.4 *Under (A6.5),*

$$\liminf_{t\to\infty} \frac{1}{t} \int_0^t \mathbb{E}\left[\bar{c}(X_s, U_s)\right] \mathrm{d}s \geq \varrho^* \quad a.s. \tag{6.5.2}$$

Proof The left-hand side of (6.5.2) is

$$\liminf_{t\to\infty} \int c \, \mathrm{d}\bar{\zeta}_t \,.$$

Suppose $\bar{\zeta}_t \to \hat{\zeta} \in \mathcal{P}(E \times \mathbb{U})$ along a subsequence, say, $\{t_n\}$. Let $\{g_k\}$ be the family in assumption (A6.1) of Section 6.2. Then

$$g_k(X_t) - \int_0^t \bar{\mathcal{A}} g_k(X_s, U_s) \, \mathrm{d}s \,, \quad t \geq 0 \,,$$

is a martingale and thus

$$\mathbb{E}\left[g_k(X_t)\right] - \mathbb{E}\left[g_k(X_0)\right] = \int_0^t \mathbb{E}\left[\bar{\mathcal{A}} g_k(X_s, U_s)\right] \mathrm{d}s \,.$$

Divide by t on both sides and let $t \to \infty$ along $\{t_n\}$ to conclude that $\int \mathcal{A} g_k \, \mathrm{d}\hat{\zeta} = 0$, i.e., $\hat{\zeta} \in \mathcal{G}$. Thus

$$\liminf_{n\to\infty} \int c \, \mathrm{d}\bar{\zeta}_{t_n} \geq \inf_{\pi\in\mathcal{G}} \int c \, \mathrm{d}\pi = \varrho^* \,.$$

On the other hand, let $\{t_n\}$, $t_n \uparrow \infty$, be such that $\{\bar{\zeta}_{t_n}\}$ has no limit point in $\mathscr{P}(E \times \mathbb{U})$. Then by (A6.5),

$$\lim_{n\to\infty} \int c \, \mathrm{d}\bar{\zeta}_{t_n} = \infty > \varrho^*,$$

which completes the proof. □

Lemma 6.5.5 *Under (A6.5), there exists a stationary solution which is optimal for the ergodic control problem* (6.5.1). *Under (A6.6), there exists a stationary solution optimal among all stationary solutions. It is also optimal among all solutions under the additional assumption that the mean empirical measures* $\{\bar{\zeta}_t : t \geq 0\}$ *are tight in* $\mathscr{P}(E \times \mathbb{U})$.

This is essentially contained in the foregoing.

Corollary 6.5.6 *In Lemma 6.5.5, "optimal stationary solution" may be replaced by "optimal ergodic solution."*

Proof Consider the ergodic decomposition of an optimal stationary solution. Then the law of the latter is the barycenter of a probability measure on the set of laws of ergodic solutions. The claim then follows by standard arguments. □

Based on the experience with nondegenerate diffusions, we do, however, expect more, viz., an optimal stationary Markov control, an optimal Markov process, etc. As a prelude to such results, consider solutions (X, U), (X^i, U^i), $1 \leq i \leq m$, of the controlled martingale problem such that

$$\langle\!\langle \mathscr{L}(X, U) \rangle\!\rangle = \sum_{i=1}^{m} a_i \langle\!\langle \mathscr{L}(X^i, U^i) \rangle\!\rangle \tag{6.5.3}$$

for some $\{a_i\} \subset (0, 1)$ with $\sum_i a_i = 1$.

Lemma 6.5.7 *If* $U_t = v(X_t)$ *for some measurable* $v \colon E \to \mathscr{P}(\mathbb{U})$ *and* (6.5.3) *holds, then* $U^i_t = v(X^i_t)$ *a.s. for* $1 \leq i \leq m$.

Proof For almost all t (i.e., outside a set of zero Lebesgue measure), the following holds. Let ξ_i, ξ, ψ_i, ψ denote the laws of (X^i_t, U^i_t), (X_t, U_t), X^i_t, X_t respectively for $1 \leq i \leq m$. Disintegrate ξ_i, ξ as

$$\xi_i(\mathrm{d}x, \mathrm{d}u) = \psi_i(\mathrm{d}x)\, q_i(\mathrm{d}u \mid x), \quad 1 \leq i \leq m,$$

$$\xi(\mathrm{d}x, \mathrm{d}u) = \psi(\mathrm{d}x)\, \delta_{v(x)}(\mathrm{d}u).$$

Clearly, $\psi = \sum a_i \psi_i$. Let

$$\Lambda_i = a_i \frac{\mathrm{d}\psi_i}{\mathrm{d}\psi}.$$

Then $\sum_i \Lambda_i = 1$ ψ-a.s., and $\xi = \sum a_i \xi_i$ must disintegrate as

$$\xi(\mathrm{d}x, \mathrm{d}u) = \psi(\mathrm{d}x) \sum_i \Lambda_i(x)\, q_i(\mathrm{d}u \mid x),$$

implying that for ψ-a.s. x,

$$\sum_i \Lambda_i(x)\, q_i(\mathrm{du} \mid x) = \delta_{v(x)}(\mathrm{du})\,.$$

Since a Dirac measure cannot be a convex combination of two or more distinct probability measures, the claim follows. The qualification "almost all t" can be dropped by modifying the U^i's suitably. □

Let (X, U) be a solution to the relaxed controlled martingale problem. By Theorem 6.4.16, $\langle\!\langle \mathscr{L}(X, U)\rangle\!\rangle$ is the barycenter of a probability measure on

$$\{\langle\!\langle \mathscr{L}(X, U)\rangle\!\rangle : X \text{ is a Markov process}\}\,.$$

Thus there exists an $\mathscr{L}(\bar{X}, \bar{U}) \in \langle\!\langle \mathscr{L}(X, U)\rangle\!\rangle$ such that $\mathscr{L}(\bar{X}, \bar{U})$ is the barycenter of a probability measure Ψ on the set

$$H = \{\mathscr{L}(X, U) : X \text{ is a Markov process}\}\,.$$

We may suppose, as in Theorem 2.2.13, that for any $\mathscr{L}(\hat{X}, \hat{U}) \in H$, $\hat{U}_t = q(\hat{X}_t, t)$ a.s. for some $q\colon E \times [0, \infty) \to \mathscr{P}(\mathbb{U})$. Suppose $U_t = v(X_t)$, $t \geq 0$, as before. Then without any loss of generality, we may suppose that $\bar{U}_t = v(\bar{X}_t)$, $t \geq 0$.

Lemma 6.5.8 *For Ψ-a.s. $\hat{\Phi}$ such that $\hat{\Phi} = \mathscr{L}(\hat{X}, \hat{U})$, one has $\hat{U}_t = v(\hat{X}_t)$, $\hat{\Phi}$-a.s. for almost all $t \geq 0$.*

Proof Let $\bar{H}$ denote the closed convex hull of H. Construct the probability measure

$$\psi(\mathrm{d}p, \mathrm{d}x, \mathrm{d}\mathfrak{v}) = \Psi(\mathrm{d}p)\, p(\mathrm{d}x, \mathrm{d}\mathfrak{v}) \in \mathscr{P}(\bar{H} \times \mathcal{D}([0, \infty); E) \times \mathscr{U})\,.$$

Let $(\xi, \tilde{X}, \tilde{U})$ be the canonical random variables on this space. In other words, if

$$\omega = (\omega_1, \omega_2, \omega_3) \in \bar{H} \times \mathcal{D}([0, \infty); E) \times \mathscr{U}\,,$$

then $\xi(\omega) = \omega_1$, $\tilde{X}(\omega) = \omega_2$ and $\tilde{U}(\omega) = \omega_3$. For $t \geq 0$, we let p_t denote the image of $p \in \mathscr{P}(\mathcal{D}([0, \infty); E) \times \mathscr{U})$ under the map $(x_{[0,\infty)}, \mathfrak{v}_{[0,\infty)}) \mapsto (x_{[0,\infty)}, \mathfrak{v}_t)$ (see Remark 6.4.6). Let

$$\psi_t = \mathscr{L}(\xi, \tilde{X}, \tilde{U}_t) = \Psi(\mathrm{d}p)\, p_t(\mathrm{d}x, \mathrm{du})\,.$$

Then for t outside a Lebesgue-null set, the following applies. Since Ψ is supported on H, by the remarks preceding the statement of this lemma, ψ_t must disintegrate as

$$\psi_t(\mathrm{d}p, \mathrm{d}x, \mathrm{du}) = \Psi(\mathrm{d}p)\, \varphi_p(\mathrm{d}x)\, \delta_{f(p, x(t), t)}(\mathrm{du})\,,$$

where φ_p is the regular conditional law of $\tilde{X}$ given ξ, and $f\colon H \times E \times [0, \infty) \to \mathscr{P}(\mathbb{U})$ is a measurable map. Of course, $x(t)$ is the evaluation of $x \in \mathcal{D}([0, \infty); E)$ at t. Thus any $h \in C(\mathscr{P}(\mathbb{U}))$ satisfies

$$\mathbb{E}\left[h(\tilde{U}_t) \mid \xi, \tilde{X}_t\right] = h(f(\xi, \tilde{X}_t, t)) \quad \text{a.s.}$$

for almost all t. Let $\bar{A}_n := [A_{n1}, \ldots, A_{nm_n}]$, $n \geq 1$, be a sequence of finite partitions of H such that

(i) $\bar{A}_{n+1}$ refines $\bar{A}_n$;
(ii) A_{ni} are Borel with $\Psi(A_{ni}) > 0$, $\forall n, i$;
(iii) if $\sigma(\bar{A}_n)$ is the σ-field generated by $\bar{A}_n$ for $n \geq 1$, then $\bigvee_n \sigma(\bar{A}_n)$ is the Borel σ-field of H.

Such a sequence of partitions exists because H is a subset of a Polish space. Let $\mathscr{L}(X^i, U^i)$ be the barycenter of the probability measure

$$\frac{\mathbb{I}_{A_{ni}}(p)\Psi(\mathrm{d}p)}{\Psi(A_{ni})}$$

for $1 \leq i \leq m_n$, which is in $\bar{H}$ by virtue of the convexity of the latter. Then clearly

$$\langle\!\langle \mathscr{L}(\tilde{X}, \tilde{U}) \rangle\!\rangle = \langle\!\langle \mathscr{L}(\bar{X}, \bar{U}) \rangle\!\rangle = \langle\!\langle \mathscr{L}(X, U) \rangle\!\rangle$$

is a convex combination of $\{\langle\!\langle \mathscr{L}(X^i, U^i) \rangle\!\rangle : 1 \leq i \leq m_n\}$. By Lemma 6.5.7, $U^i_t = v(X^i_t)$ a.s. Thus

$$\mathbb{E}\left[h(\tilde{U}_t) \,\middle|\, \tilde{X}, \mathbb{I}_{A_{ni}}(\xi), 1 \leq i \leq m_n\right] = h(v(\tilde{X}_t)) \quad \text{a.s.} \tag{6.5.4}$$

Let $n \to \infty$. By the martingale convergence theorem, the term on the left-hand side of (6.5.4) converges a.s. Moreover, the limit equals $h(f(\xi, \tilde{X}_t, t))$ a.s. for all $t > 0$. Since $h \in C(\mathscr{P}(\mathbb{U}))$ was arbitrary, we have $v(\tilde{X}_t) = f(\xi, \tilde{X}_t, t)$ a.s. Hence

$$\mathbb{E}\left[h(v(\tilde{X}_t)) \mid \xi = p\right] = \mathbb{E}\left[h(f(\xi, \tilde{X}_t, t)) \mid \xi = p\right]$$

for Ψ a.s. p. In other words,

$$v(\hat{X}_t) = f(p, \hat{X}_t, t) \quad \text{a.s.}$$

for almost all t and for Ψ-a.s. $p = \mathscr{L}(\hat{X}, \hat{U})$. The claim follows. □

Corollary 6.5.9 *Suppose that either*

(i) (A6.5) *holds, or*
(ii) (A6.6) *holds together with the condition that* $\{\bar{\zeta}_t, t \geq 0\} \subset \mathscr{P}(E \times \mathbb{U})$ *is tight for all admissible* $\mathscr{L}(X, U)$.

Then there exists an optimal $\mathscr{L}(X, U)$ *such that* X *is a Markov process, with* U *a stationary Markov control.*

Proof Suppose we start in Lemma 6.5.8 with $\mathscr{L}(X, U)$ equal to the optimal ergodic solution guaranteed by Corollary 6.5.6, wherein Corollary 6.3.7 allows us to suppose that $U_t = v(X_t)$ for a measurable $v \colon E \to \mathscr{P}(\mathbb{U})$. Under either condition, we have

$$\liminf_{t \to \infty} \frac{1}{t} \int_0^t \mathbb{E}\left[\bar{c}(X'_s, U'_s)\right] \mathrm{d}s \geq \varrho^* \tag{6.5.5}$$

for any admissible $\mathscr{L}(X', U')$, in particular, those in the support of Ψ. In view of Lemma 6.5.8 and the fact that

$$\lim_{t \to \infty} \frac{1}{t} \int_0^t \mathbb{E}\left[\bar{c}(X_s, U_s)\right] \mathrm{d}s = \varrho^*,$$

by our choice of $\mathscr{L}(X, U)$, it follows that the inequality in (6.5.5) must be an equality for Ψ-a.s. $p = \mathscr{L}(X', U')$ with $U' = v(X')$. The claim follows. □

It should be kept in the mind that the Markov process thus obtained need not be stationary, not even time-homogeneous.

6.6 The linear programming formulation

The foregoing suggests looking at the infinite dimensional linear program:

$$\text{minimize} \quad \int c \, d\mu$$

subject to $\mu \geq 0$, $\int d\mu = 1$ and

$$\int \mathcal{A} f \, d\mu = 0 \qquad \forall f \in \mathscr{D}(\mathcal{A}).$$

To cast this into standard framework, we recall from Anderson and Nash [3] some facts about infinite dimensional linear programs. Two topological vector spaces X, Y are said to form a dual pair it there exists a bilinear form $\langle \cdot, \cdot \rangle \colon X \times Y \to \mathbb{R}$ such that the functions $x \mapsto \langle x, y \rangle$ for $y \in Y$ (respectively $y \mapsto \langle x, y \rangle$ for $x \in X$) separate points of X (respectively Y). Endow X with the coarsest topology, denoted $\sigma(X, Y)$, required to render continuous the maps $x \mapsto \langle x, y \rangle$, $y \in Y$, and endow Y with the dual topology. Let C be the positive cone in X and define the dual cone $C^* \subset Y$ by

$$C^* = \{y \in Y : \langle x, y \rangle \geq 0 \quad \forall x \in C\}.$$

Let Z, W be another dual pair of topological vector spaces. Let $F \colon X \to Z$ be a $\sigma(X, Y) - \sigma(Z, W)$-continuous linear map. Define $F^* \colon W \to X^*$, the algebraic dual of X, by $\langle Fx, w \rangle = \langle x, F^* w \rangle$, $x \in X$, $w \in W$.

The primal linear programming problem then is

$$\begin{aligned} &\text{minimize} \quad \langle x, c \rangle, \\ &\text{subject to} \quad Fx = b, \quad x \in C, \end{aligned}$$

where $b \in Z$, $c \in Y$ are prescribed. Let $\underline{\varrho}$ denote the infimum of $\langle x, c \rangle$ subject to these constraints. The dual problem is

$$\begin{aligned} &\text{maximize} \quad \langle b, w \rangle, \\ &\text{subject to} \quad - F^* w + c \in C^*, \quad w \in W. \end{aligned}$$

Let $\bar{\varrho}$ denote the supremum of $\langle b, w \rangle$ subject to these constraints. From the theory of infinite dimensional linear programming [3], one knows that $\underline{\varrho} \geq \bar{\varrho}$. Let

$$\begin{aligned} K &= \{x \in C : Fx = b\}, \\ D &= \{(Fx, \langle x, c \rangle) : x \in C\}. \end{aligned}$$

We shall use the following result from Anderson and Nash [3, p. 53].

Lemma 6.6.1 *If $K \neq \emptyset$, D is closed and $x \mapsto \langle x, c \rangle$ attains its minimum on K, then $\bar{\varrho} = \underline{\varrho}$.*

That is, under the said conditions, there is no *duality gap*. We assume that there exists a continuous map $h\colon E \to [0, \infty)$ with $\inf_E h > 0$ and $\sup_{(x,u)\in E\times\mathbb{U}} \left|\frac{c(x,u)}{h(x)}\right| < \infty$. For the problem at hand, X is selected as the subspace of $\mathfrak{M}_s(E\times\mathbb{U})$, the finite signed measures on $E\times\mathbb{U}$, satisfying

$$\int h(x)\,|\mu(\mathrm{d}x, \mathrm{d}u)| < \infty\,,$$

and Y as

$$Y := \left\{ f \in C_b(E\times\mathbb{U}) : \sup_{(x,u)\in E\times\mathbb{U}} \left|\tfrac{f(x,u)}{h(x)}\right| < \infty \right\}.$$

These form a dual pair under the bilinear form $\langle \mu, f\rangle_{X,Y} = \int f\,\mathrm{d}\mu$. Also, we let $\bar{W} \equiv \mathcal{D}(\mathcal{A})$, rendered a normed linear space with norm

$$\|f\| = \sup_{x\in E} |f(x)| + \sup_{(x,u)\in E\times\mathbb{U}} |\mathcal{A}f(x,u)|\,,$$

and $\bar{Z} \equiv \bar{W}^*$, the space of bounded linear functionals on $\bar{W}$. Then $Z := \bar{Z}\times\mathbb{R}$ and $W := \bar{W}\times\mathbb{R}$ are a dual pair via the bilinear form

$$\langle z, w\rangle_{Z,W} = \langle \bar{z}, \bar{w}\rangle_{\bar{Z},\bar{W}} + ab\,,$$

where $z = (\bar{z}, a)$, $w = (\bar{w}, b)$. Letting $\mathbf{1}$ denote the constant function identically equal to one, define $F\colon X \to Z$ as

$$F\mu = \left(\lambda_\mu, \int 1\,\mathrm{d}\mu\right),$$

where $\lambda_\mu \in \bar{W}^*$ is defined by

$$\langle \lambda_\mu, f\rangle_{\bar{Z},\bar{W}} = -\int \mathcal{A}f\,\mathrm{d}\mu\,, \qquad f \in \bar{W}\,.$$

The primal problem can now be cast as:

$$\begin{aligned} &\text{minimize} && \langle \mu, c\rangle_{X,Y}\,,\\ &\text{subject to} && F\mu = (\theta, 1), \quad \mu \in X \cap \mathfrak{M}(E\times\mathbb{U})\,, \end{aligned}$$

where θ is the zero element of $\bar{Z}$. The dual problem now takes the form:

$$\begin{aligned} &\text{maximize} && \langle (\theta, 1), (f, a)\rangle_{Z,W} = a\,,\\ &\text{subject to} && \mathcal{A}f - a + c \geq 0\,, \quad f \in \mathcal{D}(\mathcal{A})\,. \end{aligned}$$

Note that $\underline{\varrho}$ is the optimal ergodic cost. Under either (A6.5) or (A6.6), we then have:

Theorem 6.6.2

$$\underline{\varrho} = \sup\left\{ a \in \mathbb{R} : \inf_{u\in\mathbb{U}} [\mathcal{A}f(x,u) + c(x,u)] \geq a\,,\ f \in \mathcal{D}(\mathcal{A}) \right\}.$$

Proof Let $\{\mu_n\} \in B$ be such that

$$F\mu_n = (\lambda_{\mu_n}, 1) \to (\hat{\lambda}, 1)$$

and $\int c \, d\mu_n \to d \in \mathbb{R}$. Under (A6.5) or (A6.6), $\{\mu_n\}$ is relatively compact. Therefore we may let $\mu_n \to \mu$ along a subsequence, which we also label as $\{\mu_n\}$. Then $\int \mathcal{A}f \, d\mu_n \to \int \mathcal{A}f \, d\mu$ for $f \in \mathcal{D}(\mathcal{A})$, implying $F\mu = (\hat{\lambda}, 1)$. Also, $\int c \, d\mu_n \to \int c \, d\mu$. Thus D is closed. The claim now follows from Lemma 6.6.1. □

Theorem 6.6.2 gives a dual characterization of the optimal cost, akin to the maximal subsolution characterization of the value function for finite horizon or discounted infinite horizon control problems [84].

6.7 Krylov's Markov selection

In Section 6.5 we established the existence of an optimal ergodic solution and an optimal Markov, though possibly time-inhomogeneous, solution, both corresponding to a stationary Markov relaxed control. In this section we present Krylov's Markov selection procedure [115, chapter 12]. Originally intended for extracting a Markov family of probability measures satisfying a martingale problem in the presence of non-uniqueness, this procedure was adapted to extract an optimal Markov solution to degenerate controlled diffusions in Haussmann [72] and El Karoui *et al.* [54], following a suggestion of Varadhan. We make use of the Krylov selection later in Section 8.5. We work under assumptions (A6.1) – (A6.4).

Lemma 6.7.1 *Let $\Gamma^0 = \{\Gamma^0(\nu) \subset \Gamma_\nu : \nu \in \mathcal{P}(E)\}$ be such that*

(a) *$\Gamma^0(\nu)$ is compact for all $\nu \in \mathcal{P}(E)$.*
(b) *Γ^0 is closed under conditioning at X_0: if $\Phi = \mathcal{L}(X, U) \in \Gamma^0(\nu)$ then the regular conditional law of Φ given $X_0 = x$, denoted by $\Phi(x)$, is in $\Gamma^0(\delta_x)$ for ν-a.s. x.*
(c) *If $\{\Phi^n : n \in \mathbb{N}\} \subset \Gamma^0(\nu)$ and $\{C_n : n \in \mathbb{N}\} \subset \mathcal{B}(E)$ is a collection of disjoint sets with $\sum_n \nu(C_n) = 1$ then*

$$\bar{\Phi} := \sum_n \int_{C_n} \Phi^n(x) \nu(dx) , \tag{6.7.1}$$

is a member of $\Gamma^0(\nu)$.
(d) *Γ^0 is closed under T-shifts: if $\Phi = \mathcal{L}(X, U) \in \Gamma^0(\nu)$ and $T > 0$, then*

$$\Phi_T := \mathcal{L}(X \circ \theta_T, U \circ \theta_T) \in \Gamma^0(\mathcal{L}(X_T)) .$$

(e) *Γ^0 is closed under T-concatenations: if $\breve{\Phi} \in \Gamma^0(\mathcal{L}(X_T))$ and Φ and T are as in part* (d)*, then the T-concatenation of Φ and $\breve{\Phi}$ is in $\Gamma^0(\nu)$.*

For a constant $\beta > 0$ *and* $f \in C_b(E)$ *define*

$$F_1(\Phi) := \int_0^\infty \mathrm{e}^{-\beta t}\, \mathbb{E}^\Phi\left[f(X_t)\right] \mathrm{d}t\,,$$
$$\Psi_1(\nu) := \inf_{\Phi \in \Gamma^0(\nu)} F_1(\Phi)\,,$$
$$\Gamma^1(\nu) := \left\{\Phi \in \Gamma^0(\nu) : F_1(\Phi) = \Psi_1(\nu)\right\}.$$

Then $\Gamma^1 = \{\Gamma^1(\nu) : \nu \in \mathscr{P}(E)\}$ *satisfies* (a)–(e).

Proof Since $\Phi \mapsto F_1(\Phi)$ is continuous, it follows that $\Gamma^1(\nu)$ is a closed subset of $\Gamma^0(\nu)$, and hence is compact.

We introduce the following notation. We denote $\Psi_1(\delta_x)$ and $\Gamma^i(\delta_x)$, $i = 0, 1$, where δ_x is the Dirac measure at x, by $\Psi_1([x])$ and $\Gamma^i([x])$, respectively. Note then that property (b) holds for Γ^1 if and only if

$$\Psi_1(\nu) = \int_E \Psi_1([x])\, \nu(\mathrm{d}x)\,.$$

It is evident that $\Psi_1(\nu) \geq \int \Psi_1([x])\, \nu(\mathrm{d}x)$. To show the converse inequality, for an arbitrary $\varepsilon > 0$, let $\Phi^n = \mathscr{L}(X^n, V^n) \in \Gamma^0(\nu)$ be such that

$$F_1(\Phi^n) \leq \Psi_1(\nu) + \frac{\varepsilon}{2^n}\,. \tag{6.7.2}$$

By (b) we can find a set $N \subset E$ with $\nu(N) = 0$ such that $\Phi^n(x) \in \Gamma^0([x])$ for all $x \notin N$. Let

$$B_n = \{x \in E : F_1(\Phi^n(x)) < \Psi_1([x]) + \varepsilon\}\,, \quad n \geq 1\,.$$

Then (6.7.2) implies that $\nu(B_n^c) < 2^{-n}$. Define $C_n \subset E$, $n \geq 1$, successively by

$$C_1 = B_1\,, \quad \text{and} \quad C_n = B_n \cap (\cup_{m<n} B_m)^c\,, \quad \text{for } n > 1\,.$$

Then clearly the sets C_n are disjoint and $\nu(\cup_n C_n) = \nu(\cup_n B_n) = 1$. Define $\bar{\Phi}$ by (6.7.1). Then $\bar{\Phi}$ is the law of a process $(\bar{X}, \bar{V})$ such that $\mathscr{L}(\bar{X}_0) = \nu$ and for $x \in C_n$, $n \geq 1$, the regular conditional law of $(\bar{X}, \bar{V})$ given $\bar{X}_0 = x$ is $\Phi^n(x)$. By (c), $\bar{\Phi} \in \Gamma^0(\nu)$. We obtain

$$\begin{aligned} F_1(\bar{\Phi}) &= \sum_n \int_{C_n} F_1(\Phi^n(x))\, \nu(\mathrm{d}x) \\ &\leq \sum_n \int_{C_n} (\Psi_1([x]) + \varepsilon)\, \nu(\mathrm{d}x) \\ &\leq \int_E \Psi_1([x])\, \nu(\mathrm{d}x) + \varepsilon\,. \end{aligned} \tag{6.7.3}$$

Since $\varepsilon > 0$ was arbitrary, (6.7.3) yields

$$\Psi_1(\nu) \leq \int_E \Psi_1([x])\, \nu(\mathrm{d}x)\,,$$

thus completing the proof that (b) holds for Γ^1.

That (c) holds for Γ^1 is evident since, by property (b) which holds for Γ^1, we have

$$\begin{aligned} F_1(\bar{\Phi}) &= \sum_n \int_{C_n} F_1(\Phi^n(x))\,\nu(\mathrm{d}x) \\ &= \sum_n \int_{C_n} \Psi_1([x])\,\nu(\mathrm{d}x) \\ &= \Psi_1(\nu), \end{aligned}$$

and therefore $\bar{\Phi} \in \Gamma^1(\nu)$.

To prove (d) and (e) let $\Phi = \mathscr{L}(X, U) \in \Gamma^1(\nu)$. Decompose $\Psi(\nu)$ as

$$\Psi_1(\nu) = \mathbb{E}^{\Phi}\left[\int_0^T \mathrm{e}^{-\beta s} f(X_s)\,\mathrm{d}s\right] + \int_0^\infty \mathbb{E}^{\Phi}[\mathrm{e}^{-\beta(T+s)} f(X_{T+s})]\,\mathrm{d}s\,. \tag{6.7.4}$$

Since by property (d) for Γ^0,

$$\Phi_T = \mathscr{L}(X \circ \theta_T, U \circ \theta_T) \in \Gamma^0(\mathscr{L}(X_T)),$$

by conditioning at T and using property (b) for Γ^0, we obtain

$$\begin{aligned} \int_0^\infty \mathbb{E}^{\Phi}\left[\mathrm{e}^{-\beta(T+s)} f(X_{T+s})\right] \mathrm{d}s &= \mathbb{E}^{\Phi}\left[\mathrm{e}^{-\beta T}\,\mathbb{E}^{\Phi}\left[\int_0^\infty \mathrm{e}^{-\beta s} f(X_{T+s})\,\mathrm{d}s \,\middle|\, X_T\right]\right] \\ &= \mathbb{E}^{\Phi}\left[\mathrm{e}^{-\beta T} F_1(\Phi_T(X_T))\right]. \end{aligned} \tag{6.7.5}$$

By (6.7.4) – (6.7.5), we have

$$\Psi_1(\nu) = \mathbb{E}^{\Phi}\left[\int_0^T \mathrm{e}^{-\beta s} f(X_s)\,\mathrm{d}s\right] + \mathbb{E}^{\Phi}[\mathrm{e}^{-\beta T} F_1(\Phi_T(X_T))]\,. \tag{6.7.6}$$

Let $\nu' := \mathscr{L}(X_T)$ and select any $\Phi' = \mathscr{L}(X', U') \in \Gamma^1(\nu')$. Hence, by definition,

$$\Psi_1(\nu') = F_1(\Phi') = \int_0^\infty \mathrm{e}^{-\beta s}\,\mathbb{E}^{\Phi'}[f(X'_s)]\,\mathrm{d}s\,. \tag{6.7.7}$$

Let $(\tilde{X}, \tilde{U})$ be the T-concatenation of (X, U) and (X', U'), and set $\tilde{\Phi} := \mathscr{L}(\tilde{X}, \tilde{U})$. By property (e) for Γ^0, $\tilde{\Phi} \in \Gamma^0(\nu)$. This implies that $F_1(\tilde{\Phi}) \geq \Psi_1(\nu)$. Using the definition of $(\tilde{X}, \tilde{U})$, (6.7.7), and property (b) for Γ^1, which was established earlier, we obtain

$$\mathbb{E}^{\tilde{\Phi}}\left[\int_0^\infty \mathrm{e}^{-\beta s} f(\tilde{X}_{T+s})\,\mathrm{d}s \,\middle|\, \tilde{X}_T\right] = \Psi_1([\tilde{X}_T]) \quad \nu'\text{-a.s.} \tag{6.7.8}$$

Therefore, by decomposing $F_1(\tilde{\Phi})$ as in (6.7.4) and conditioning as in (6.7.5), then using (6.7.8), we obtain

$$\begin{aligned} \Psi_1(\nu) \leq F_1(\tilde{\Phi}) &= \mathbb{E}^{\tilde{\Phi}}\left[\int_0^T \mathrm{e}^{-\beta s} f(\tilde{X}_s)\,\mathrm{d}s\right] + \int_0^\infty \mathbb{E}^{\tilde{\Phi}}[\mathrm{e}^{-\beta(T+s)} f(\tilde{X}_{T+s})]\,\mathrm{d}s \\ &= \mathbb{E}^{\Phi}\left[\int_0^T \mathrm{e}^{-\beta s} f(X_s)\,\mathrm{d}s\right] + \mathbb{E}^{\Phi}[\mathrm{e}^{-\beta T}\Psi_1([X_T])]\,. \end{aligned} \tag{6.7.9}$$

Since $\Psi_1([x]) \leq F_1(\Phi_T(x))$, (6.7.6) and (6.7.9) yield

$$\Psi_1([x]) = F_1(\Phi_T(x)) \quad \nu'\text{-a.s.}, \tag{6.7.10}$$

and

$$\Psi_1(\nu) = F_1(\tilde{\Phi}) = \mathbb{E}^{\Phi}\left[\int_0^T e^{-\beta s} f(X_s)\,ds\right] + \mathbb{E}^{\Phi}[e^{-\beta T}\Psi_1([X_T])]. \tag{6.7.11}$$

By (6.7.10)

$$\Psi_1(\nu') = \int_E \Psi_1([x])\,d\nu' = \int_E F_1(\Phi_T(x))\,d\nu' = F_1(\Phi_T),$$

which implies property (d) for Γ^1, while (6.7.11) implies property (e). □

In the next theorem we use the map $\kappa\colon \mathbb{N}\times\mathbb{N} \to \mathbb{N}$ defined by

$$\kappa(i,j) := \tfrac{1}{2}[(i+j-1)^2 + j - i + 1], \quad i,j \geq 1.$$

Note that κ is one-to-one and onto.

Theorem 6.7.2 *Suppose* Γ^0 *satisfies the hypotheses* (a)–(e) *in Lemma 6.7.1. Let* $\{f_i\} \subset C_b(E)$ *be a countable separating class for* $\mathscr{P}(E)$ *and* $\{\beta_j\}$ *be a countable dense set in* $(0,\infty)$. *We define* $F_\ell\colon \Gamma_\nu \mapsto \mathbb{R}$, $\ell \in \mathbb{N}$, *by*

$$F_{\kappa(i,j)}(\Phi) := \int_0^\infty e^{-\beta_i t}\,\mathbb{E}^{\Phi}\left[f_j(X_t)\right] dt.$$

For $i = 1,2,\dots,$ *define inductively*

$$\Psi_i(\nu) := \inf_{\Phi\in\Gamma^{i-1}(\nu)} F_i(\Phi),$$

$$\Gamma^i(\nu) := \left\{\Phi \in \Gamma^{i-1}(\nu) : F_i(\Phi) = \Psi_i(\nu)\right\}.$$

Then for any two elements $\mathscr{L}(X,U)$ *and* $\mathscr{L}(X',U')$ *of* $\Gamma^\infty(\nu) := \cap_i\Gamma^i(\nu)$ *it holds that* $\mathscr{L}(X) = \mathscr{L}(X')$. *Moreover, X is a time-homogeneous Markov process, and U can be taken to be of the form* $U_t = v(X_t)$, $t \geq 0$, *for a measurable* $v\colon E \to \mathscr{P}(\mathbb{U})$.

Proof By Lemma 6.7.1, for fixed $\nu \in \mathscr{P}(E)$, $\{\Gamma^i(\nu) : i \geq 0\}$ is a nested, decreasing family of compact non-empty sets. Therefore $\Gamma^\infty(\nu) \neq \varnothing$. Clearly, Γ^∞ satisfies (a) – (e) of Lemma 6.7.1. Let $\mathscr{L}(X,U)$, $\mathscr{L}(X',U')$ be two elements of $\Gamma^\infty(\nu)$. By Lemma 6.4.7, for bounded measurable $f\colon [0,\infty) \to \mathbb{R}$,

$$\int_0^\infty e^{-\beta_j t} f(t)\,dt = 0 \quad \forall j \in \mathbb{N} \quad \Longrightarrow \quad f(t) = 0 \quad \text{a.e.}$$

Therefore, since $\{f_m\}$ is a separating class for $\mathscr{P}(E)$, X and X' must have the same one-dimensional marginals. The same applies to any two elements of $\Gamma^\infty([x])$, $x \in E$. Define $q(x,t,B) := \mathbb{P}(X_t \in B)$ for $B \in \mathscr{B}(E)$ and any $\mathscr{L}(X,U) \in \Gamma^\infty([x])$. The exact choice of $\mathscr{L}(X,U)$ is immaterial since for any two such elements the marginals of X agree. Since the map $x \mapsto \Gamma^\infty([x])$ is upper semicontinuous and compact-valued, $x \mapsto q(x,t,B)$ may be selected so as to be measurable. It also follows by properties (b), (d) and (e) of Lemma 6.7.1 that $\{q(x,t,\cdot) : x \in E,\ t \geq 0\}$ satisfy the Chapman–Kolmogorov equations. Since properties (b) and (d) hold for $\Gamma^\infty(\nu)$ then for $\mathscr{L}(X,U) \in \Gamma^\infty(\nu)$ the regular conditional law of X_{t+s} for $s > 0$ given $\mathfrak{F}_t^X$ is in $\Gamma^\infty([X_t])$. Moreover, since $\Gamma^\infty([x])$ is a singleton, this conditional law is completely determined by $q(x, t, \cdot)$. Thus for every $\mathscr{L}(X, U) \in \Gamma^\infty(\nu)$, X corresponds to a Markov process with transition kernel $q(x, t, \cdot)$. Since the initial law and the transition kernel completely specify the law of a Markov process, $\mathscr{L}(X)$ is uniquely determined. The last claim follows from Theorem 2.2.13. □

The conclusions of Theorem 6.7.2 can be strengthened. If we modify the hypotheses of Lemma 6.7.1 in analogy to the conditions in Stroock and Varadhan [115, section 12.2] which concerns the uncontrolled case, the Krylov selection extracts a strong Markov process.

6.8 Bibliographical note

Sections 6.2–6.3. These sections largely follow [21], drawing upon [23, 112] for some technical lemmas. See also [110, 81] for analogous results.

Section 6.4. This is based on [30].

Section 6.5. This follows [28, 21, 30].

Section 6.6. This follows [21]. See also [111] for related work.

Section 6.7. For more detailed results on Markov selections we refer the reader to [115, Chapter 12].

7

Degenerate Controlled Diffusions

7.1 Introduction

In this chapter we turn to the study of degenerate controlled diffusions. For the nondegenerate case the theory is more or less complete. This is not the case if the uniform ellipticity hypothesis is dropped. Indeed, the differences between the nondegenerate and the degenerate cases are rather striking. In the nondegenerate case, the state process X is strong Feller under a Markov control. This, in turn, facilitates the study of the ergodic behavior of the process. In contrast, in the degenerate case, under a Markov control, the Itô stochastic differential equation (2.2.1) is not always well posed. From an analytical viewpoint, in the nondegenerate case, the HJB equation is uniformly elliptic and the associated regularity properties benefit its study. The degenerate case, on the other hand, is approached via a particular class of weak solutions known as viscosity solutions. This approach does not yield as satisfactory results as in the case of classical solutions. In fact ergodic control of degenerate diffusions should not be viewed as a single topic, but rather as a class of problems, which are studied under various hypotheses. We first formulate the problem as a special case of a controlled martingale problem and then summarize those results from Chapter 6 that are useful here. Next, in Section 7.3, we study the HJB equations in the context of viscosity solutions for a specific class of problems that bears the name of *asymptotically flat* diffusions. Then in Section 7.4, we turn our attention to a class of diffusions which have a *partial nondegeneracy* property.

7.2 Controlled martingale formulation

The state process X is governed by the Itô stochastic differential equation (2.2.1) under the hypotheses (i)–(iv) on p. 30. Treating $u \in \mathbb{U}$ as a parameter, we let $\mathcal{L}$ $(= \mathcal{L}^u)$ be the operator defined in (2.2.12) with domain $\mathscr{D}(\mathcal{L}) \subset C_b(\mathbb{R}^d)$ and range $\mathscr{R}(\mathcal{L}) \subset C_b(\mathbb{R}^d \times \mathbb{U})$. It is simple to verify that (A6.1)–(A6.4) in Chapter 6, p. 220 and p. 232, hold for $\mathcal{L}$. We adopt the relaxed control framework, using the notation $\bar{b}$, $\bar{c}$ and $\bar{\mathcal{L}}$. By Theorem 6.3.6 the set of ergodic occupation measures $\mathscr{G} \subset \mathscr{P}(\mathbb{R}^d \times \mathbb{U})$

is characterized by

$$\mathscr{G} = \left\{\pi \in \mathscr{P}(\mathbb{R}^d \times \mathbb{U}) : \int_{\mathbb{R}^d\times\mathbb{U}} \mathcal{L}^u f(x)\,\pi(\mathrm{d}x, \mathrm{d}u) = 0 \quad \forall f \in \mathscr{D}(\mathcal{L})\right\}.$$

Clearly, $\mathscr{G}$ is closed. We assume that there exists $\pi \in \mathscr{G}$ such that $\int c\,\mathrm{d}\pi < \infty$. Let $c \in C(\mathbb{R}^d \times \mathbb{U})$ be a prescribed running cost function. In this chapter we seek to minimize over all $U \in \mathfrak{U}$ the average cost

$$\limsup_{t\to\infty} \frac{1}{t} \int_0^t \mathbb{E}^U[\bar{c}(X_s, U_s)]\,\mathrm{d}s. \tag{7.2.1}$$

Let

$$\varrho^* = \inf_{\pi\in\mathscr{G}} \int c\,\mathrm{d}\pi. \tag{7.2.2}$$

The existence of an optimal control is guaranteed under either of the following conditions:

(C7.1) Strong near-monotonicity: the running cost c is inf-compact.

(C7.2) Stability: there exists an inf-compact function $\mathcal{V} \in C^2(\mathbb{R}^d, \mathbb{R}_+)$ such that the map $(x, u) \mapsto -\mathcal{L}^u\mathcal{V}(x)$ is inf-compact.

Note that (C7.1) implies that the sets $\left\{\pi \in \mathscr{G} : \int c\,\mathrm{d}\pi \le k\right\}$, $k \in \mathbb{R}_+$, are compact. Hence hypothesis (A6.5) in Chapter 6 on p. 242 holds. On the other hand, under (C7.2), Lemma 2.5.3 asserts that the mean empirical measures $\{\bar{\zeta}^U_{\nu,t} : U \in \mathfrak{U}\}$ are tight for each $\nu \in \mathscr{P}(\mathbb{R}^d)$, and the argument used in the proof of the same lemma shows that their accumulation points as $t \to \infty$ are in $\mathscr{G}$. Then, in view of Theorem 6.3.6, $\mathscr{G}$ is identical with the the set of accumulation points of the mean empirical measures as $t \to \infty$, and consequently, $\mathscr{G}$ is compact. Thus hypothesis (A6.6) in Chapter 6 on p. 242 holds. Therefore, from the general results in Chapter 6, we have the following theorem.

Theorem 7.2.1 *Under either (C7.1) or (C7.2), there exists an optimal ergodic pair (X, U), where $U \in \mathfrak{U}_{\mathrm{sm}}$. Alternatively, there exists an optimal pair (X, U), such that $U \in \mathfrak{U}_{\mathrm{sm}}$ and X is a Markov process.*

7.3 Asymptotically flat controlled diffusions

In this section we study the ergodic HJB equation in the framework of viscosity solutions for a controlled process governed by (2.2.1). In addition to the standard local Lipschitz and growth assumptions in (2.2.3) and (2.2.4) on p. 31 we assume that the diffusion matrix σ and the running cost c are Lipschitz continuous in x (uniformly in $u \in \mathbb{U}$ for the latter), and we denote their common Lipschitz constant by C_{Lip}. We carry out our program under a stability assumption, which yields *asymptotic flatness*.

We introduce the following notation: for x, z in $\mathbb{R}^d$ define

$$\Delta_z b(x,u) := b(x+z,u) - b(x,u)\,,$$
$$\Delta_z \sigma(x) := \sigma(x+z) - \sigma(x)\,,$$
$$\tilde{a}(x;z) := \Delta_z\sigma(x)\Delta_z\sigma^{\mathsf{T}}(x)\,.$$

Assumption 7.3.1 There exist a symmetric positive definite matrix Q and a constant $r > 0$ such that for $x, z \in \mathbb{R}^d$, with $z \neq 0$, and $u \in \mathbb{U}$,

$$2\Delta_z b^{\mathsf{T}}(x,u)Qz - \frac{|\Delta_z\sigma^{\mathsf{T}}(x)Qz|^2}{z^{\mathsf{T}}Qz} + \operatorname{tr}(\tilde{a}(x;z)Q) \leq -r|z|^2\,. \tag{7.3.1}$$

The following example shows that Assumption 7.3.1 arises naturally.

Example 7.3.2 Let $\mathbb{U} = [0,1]^d$, $b(x,u) = Bx + Du$, $\sigma_1(x) = x$ and $\sigma_i(x) = 0$ for all $i \neq 1$. Here B, D are constant $d \times d$ matrices where all eigenvalues of B have negative real part and σ_j is the jth column of σ. It is well known that there exists a positive definite matrix Q such that $B^{\mathsf{T}}Q + QB = -I$ [106, theorem 7.11, p. 124]. Using this property, we can verify that Assumption 7.3.1 holds. Indeed, for $z \neq 0$,

$$\begin{aligned}
2\Delta_z b^{\mathsf{T}}(x,u)Qz - &\frac{|\Delta_z\sigma^{\mathsf{T}}(x)Qz|^2}{z^{\mathsf{T}}Qz} + \operatorname{tr}(\tilde{a}(x;z)Q) \\
&= z^{\mathsf{T}}(B^{\mathsf{T}}Q + QB)z - \frac{(z^{\mathsf{T}}Qz)^2}{z^{\mathsf{T}}Qz} + z^{\mathsf{T}}Qz \\
&= -|z|^2 - z^{\mathsf{T}}Qz + z^{\mathsf{T}}Qz \\
&= -|z|^2\,.
\end{aligned}$$

Example 7.3.3 Assumption 7.3.1 allows for certain state-dependent diffusion matrices as in Example 7.3.2. If we assume a constant diffusion matrix σ and additive control of the form $b(x,u) = h(x) - x + f(u)$ for some bounded continuous f, asymptotic flatness reduces to

$$\langle h(x) - h(y),\ x - y\rangle \leq \beta|x-y|^2\,, \quad x, y \in \mathbb{R}^d$$

for some $\beta \in (0,1)$. This holds, e.g., if h is a contraction or if $-h$ is monotone, i.e., $\langle h(x) - h(y),\ x - y\rangle \leq 0$.

We now establish asymptotic flatness of the flow under Assumption 7.3.1. In the lemma that follows, the Brownian motion and an admissible control are prescribed on a probability space, and strong solutions starting from distinct initial conditions are compared – see Theorem 2.3.4 for a justification.

Lemma 7.3.4 *Let U be any admissible relaxed control. Let X_t^x be the corresponding solution with initial condition $X_0 = x$. Then under Assumption 7.3.1 there exist constants $C_1 > 0$, $C_2 > 0$, which do not depend on U, such that*

$$\mathbb{E}^U\left|X_t^x - X_t^y\right| \leq \sqrt{\|Q^{-1}\|\,\|Q\|}\exp\left(-\frac{rt}{2\|Q\|}\right)|x-y| \qquad \forall x, y \in \mathbb{R}^d\,. \tag{7.3.2}$$

Proof Consider the Lyapunov function

$$w_\varepsilon(x) = \frac{x^{\mathsf{T}} Q x}{(\varepsilon + x^{\mathsf{T}} Q x)^{1/2}}\,, \qquad \varepsilon > 0\,.$$

For $f \in C^2(\mathbb{R}^d)$ and $u \in \mathbb{U}$ define

$$\tilde{\mathcal{L}}^u f(x; z) = \sum_{i=1}^{d} \Delta_z b^i(x, u) \frac{\partial f}{\partial x_i}(z) + \frac{1}{2} \sum_{i,j=1}^{d} \tilde{a}_{ij}(x; z) \frac{\partial^2 f}{\partial x_i \partial x_j}(z)\,.$$

A simple computation yields,

$$\begin{aligned}
\tilde{\mathcal{L}}^u w_\varepsilon(x; z) &= \frac{\varepsilon + \frac{1}{2} z^{\mathsf{T}} Q z}{(\varepsilon + z^{\mathsf{T}} Q z)^{3/2}} \left[2\Delta_z b^{\mathsf{T}}(x, u) Q z + \operatorname{tr}(\tilde{a}(x; z) Q) \right] \\
&\qquad - \frac{2\varepsilon + \frac{1}{2} z^{\mathsf{T}} Q z}{(\varepsilon + z^{\mathsf{T}} Q z)^{5/2}} |\Delta_z \sigma^{\mathsf{T}}(x) Q z|^2 \\
&= \frac{\varepsilon + \frac{1}{2} z^{\mathsf{T}} Q z}{(\varepsilon + z^{\mathsf{T}} Q z)^{3/2}} \left[2\Delta_z b^{\mathsf{T}}(x, u) Q z - \frac{|\Delta_z \sigma^{\mathsf{T}}(x) Q z|^2}{z^{\mathsf{T}} Q z} + \operatorname{tr}(\tilde{a}(x; z) Q) \right] \\
&\qquad + \varepsilon \frac{\varepsilon - \frac{1}{2} z^{\mathsf{T}} Q z}{(\varepsilon + z^{\mathsf{T}} Q z)^{5/2}} \frac{|\Delta_z \sigma^{\mathsf{T}}(x) Q z|^2}{z^{\mathsf{T}} Q z}\,.
\end{aligned} \tag{7.3.3}$$

Then using Assumption 7.3.1, and the bound

$$\frac{|\Delta_z \sigma^{\mathsf{T}}(x) Q z|^2}{(z^{\mathsf{T}} Q z)^2} \le C_{\mathrm{Lip}}^2\,,$$

where C_{Lip} is the Lipschitz constant of σ, (7.3.3) yields

$$\begin{aligned}
\tilde{\mathcal{L}}^u w_\varepsilon(x; z) &\le -r \frac{\varepsilon + \frac{1}{2} z^{\mathsf{T}} Q z}{\varepsilon + z^{\mathsf{T}} Q z} \frac{|z|^2}{z^{\mathsf{T}} Q z} w_\varepsilon(z) + \varepsilon \frac{\varepsilon - \frac{1}{2} z^{\mathsf{T}} Q z}{(\varepsilon + z^{\mathsf{T}} Q z)^{5/2}} (z^{\mathsf{T}} Q z)\, C_{\mathrm{Lip}}^2 \\
&\le -\frac{r}{2\|Q\|} w_\varepsilon(z) + \sqrt{\varepsilon}\, C_{\mathrm{Lip}}^2\,.
\end{aligned}$$

Let $\tau = \inf\{t \ge 0 : X_t^{x+z} = X_t^x\}$ (possibly $+\infty$). By Dynkin's formula, and letting $C_2 = \frac{r}{2\|Q\|}$, we obtain

$$\begin{aligned}
\mathbb{E}^U \left[w_\varepsilon(X_{t\wedge\tau}^{x+z} - X_{t\wedge\tau}^x) \right] - w_\varepsilon(z) &= \mathbb{E}^U \left[\int_0^{t\wedge\tau} \int_{\mathbb{U}} \tilde{\mathcal{L}}^u w_\varepsilon(X_s^x; X_s^{x+z} - X_s^z) U_s(\mathrm{d}u)\, \mathrm{d}s \right] \\
&\le -C_2\, \mathbb{E}^U \left[\int_0^{t\wedge\tau} w_\varepsilon(X_s^{x+z} - X_s^x)\, \mathrm{d}s \right] + \sqrt{\varepsilon}\, C_{\mathrm{Lip}}^2 t \\
&\le -C_2 \int_0^t \mathbb{E}^U \left[w_\varepsilon(X_{s\wedge\tau}^{x+z} - X_{s\wedge\tau}^x) \right] \mathrm{d}s + \sqrt{\varepsilon}\, C_{\mathrm{Lip}}^2 t\,,
\end{aligned}$$

since, for $t \ge \tau$, $X_t^{x+z} = X_t^x$ a.s. by the pathwise uniqueness of the solution of (2.2.1). Thus by Gronwall's inequality it follows that for any $t \ge 0$,

$$\mathbb{E}^U \left[w_\varepsilon(X_t^{x+z} - X_t^x) \right] \le w_\varepsilon(z) \mathrm{e}^{-C_2 t} + \frac{\sqrt{\varepsilon}\, C_{\mathrm{Lip}}^2}{C_2} (1 - \mathrm{e}^{-C_2 t})\,. \tag{7.3.4}$$

Taking limits as $\varepsilon \to 0$ in (7.3.4), and using monotone convergence and the bound $|z|^2 \le \|Q^{-1}\|(z^{\mathsf{T}}Qz)$, we obtain, with $w_0(z) := (z^{\mathsf{T}}Qz)^{1/2}$,

$$\begin{aligned}\mathbb{E}^U\left|X_t^{x+z} - X_t^x\right| &\le \sqrt{\|Q^{-1}\|}\,\mathbb{E}^U\left[w_0(X_t^{x+z} - X_t^x)\right]\\ &\le \sqrt{\|Q^{-1}\|}\,(z^{\mathsf{T}}Qz)^{1/2}\,\mathrm{e}^{-C_2t}\\ &\le \sqrt{\|Q^{-1}\|\,\|Q\|}\,\mathrm{e}^{-C_2t}|z|\,,\end{aligned}$$

thus establishing (7.3.2). □

Remark 7.3.5 Property (7.3.2) is known as asymptotic flatness.

A variation of the method in the proof of Lemma 7.3.4 yields the following.

Lemma 7.3.6 *Under Assumption 7.3.1 there exist a constant $\delta > 0$ such that, for any compact set $K \subset \mathbb{R}^d$,*

$$\sup_{t\ge0}\,\sup_{x\in K}\,\sup_{U\in\mathfrak{U}}\,\mathbb{E}^U\,|X_t|^{1+\delta} < \infty\,.$$

Proof Using (7.3.1) with $x = 0$ we conclude that there exist positive constants r_1 and r_2 such that

$$2b^{\mathsf{T}}(z,u)Qz - \frac{|\sigma^{\mathsf{T}}(z)Qz|^2}{z^{\mathsf{T}}Qz} + \operatorname{tr}\bigl(a(z)Q\bigr) \le -r_1|z|^2 + r_2 \qquad \forall z\in\mathbb{R}^d \tag{7.3.5}$$

for all $u \in \mathbb{U}$. For positive constants ε and δ, we define

$$w_{\varepsilon,\delta}(x) := \frac{(x^{\mathsf{T}}Qx)^{1+\delta/2}}{(\varepsilon + x^{\mathsf{T}}Qx)^{1/2}}\,.$$

Following the method in the proof of Lemma 7.3.4, using (7.3.5), we deduce that there exist positive constants ε, δ, k_0 and k_1 such that

$$\mathcal{L}^u w_{\varepsilon,\delta}(x) \le -k_1 w_{\varepsilon,\delta}(x) + k_0 \qquad \forall u\in\mathbb{U}\,. \tag{7.3.6}$$

By Lemma 2.5.5 on p. 63 and (7.3.6),

$$\mathbb{E}_x^U\left[w_{\varepsilon,\delta}(X_t)\right] \le \frac{k_0}{k_1} + w_{\varepsilon,\delta}(x)\mathrm{e}^{-k_1t} \qquad \forall x\in\mathbb{R}^d\,,\ \forall U\in\mathfrak{U}\,.$$

The claim follows since $|x|^{1+\delta} \in \mathscr{O}(w_{\varepsilon,\delta})$. □

Under Assumption 7.3.1, the existence results in Theorem 7.2.1 can be improved, as the following theorem shows.

Theorem 7.3.7 *Suppose Assumption 7.3.1 holds. Then there exists $\pi^* \in \mathscr{G}$ satisfying $\varrho^* = \int c\,\mathrm{d}\pi^*$. Let (X^*, U^*) be a stationary relaxed solution to the controlled martingale problem corresponding to π^*. Then for any initial condition X_0, we have*

$$\lim_{T\to\infty}\frac{1}{T}\int_0^T \mathbb{E}^{U^*}\left[\bar{c}(X_s, U_s^*)\right]\mathrm{d}s = \varrho^*\,. \tag{7.3.7}$$

Moreover, for any admissible control $U \in \mathfrak{U}$ and any initial law, the corresponding solution X of (2.2.1) *satisfies*

$$\liminf_{T\to\infty} \frac{1}{T}\int_0^T \mathbb{E}^U[\bar{c}(X_s, U_s)]\, ds \geq \varrho^* .$$

Thus U^* is optimal for any initial law.

Proof First note that by (7.3.6), Assumption 7.3.1 implies (C7.2). Thus there exists $\pi^* \in \mathscr{G}$ satisfying $\varrho^* = \int c\, d\pi^*$. Let (X^*, U^*) be a stationary relaxed solution of the controlled martingale problem corresponding to π^*. Then if the law of (X_0^*, U_0^*) is such that

$$\mathbb{E}\left[\int_{\mathbb{U}} f(X_0^*, u) U_0^*(du)\right] = \int_{\mathbb{R}^d\times\mathbb{U}} f\, d\pi^* \qquad \forall f \in C_b(\mathbb{R}^d \times \mathbb{U}),$$

we obtain

$$\lim_{T\to\infty} \frac{1}{T}\int_0^T \mathbb{E}^{U^*}[\bar{c}(X_s^*, U_s^*)]\, ds = \int c\, d\pi^* = \varrho^* . \tag{7.3.8}$$

Now fix the probability space on which the Wiener process W and U^* are defined, as in Theorem 2.3.4. By (7.3.2) and (7.3.8), for any solution X under U^*, (7.3.7) holds. Since (7.3.6) implies that the mean empirical measures are tight, any limit point of the mean empirical measures is in $\mathscr{G}$ by the proof of Lemma 6.5.4. Thus the second assertion follows. □

7.3.1 The HJB equation

We now study the HJB equation for the ergodic control problem given by

$$\inf_{u\in\mathbb{U}} [\mathcal{L}^u V(x) + c(x, u) - \varrho] = 0 , \tag{7.3.9}$$

where $V\colon \mathbb{R}^d \to \mathbb{R}$ and ϱ is a scalar. Note that (7.3.9) corresponds to precise controls while the functional in (7.2.1) is defined in the space of relaxed controls. Even though the results of Section 3.4 which assert optimality in the class of precise stationary Markov controls are not applicable here, Theorem 2.3.1 on p. 47 does hold and shows that this relaxation is valid.

Solving the HJB equation means finding a suitable pair (V, ϱ) satisfying (7.3.9) in an appropriate sense. In Chapter 3 we studied the nondegenerate case. To obtain analogous results for the degenerate case, we strengthen the regularity assumptions on the diffusion matrix from Lipschitz continuity to $\sigma^{ij} \in C^2(\mathbb{R}^d) \cap C^{0,1}(\mathbb{R}^d)$, $1 \leq i, j \leq d$. We introduce the notion of a viscosity solution of (7.3.9).

Definition 7.3.8 The pair $(V, \varrho) \in C(\mathbb{R}^d) \times \mathbb{R}$ is said to be a viscosity solution of (7.3.9) if for any $\psi \in C^2(\mathbb{R}^d)$

$$\inf_{u\in\mathbb{U}} [\mathcal{L}^u \psi(x) + c(x, u) - \varrho] \geq 0$$

at each local maximum x of $(V - \psi)$, and

$$\inf_{u \in \mathbb{U}} \left[\mathcal{L}^u \psi(x) + c(x,u) - \varrho\right] \le 0$$

at each local minimum x of $(V - \psi)$.

We show that (7.3.9) has a unique viscosity solution in $(V, \varrho) \in C^{0,1}(\mathbb{R}^d) \times \mathbb{R}$, satisfying $V(0) = 0$. To this end we follow the traditional vanishing discount method. Let $\alpha > 0$. For an admissible control $U \in \mathfrak{U}$, as usual, let

$$J_\alpha^U(x) = \mathbb{E}^U \left[\int_0^\infty \mathrm{e}^{-\alpha t} \bar{c}(X_t, U_t)\, \mathrm{d}t \;\middle|\; X_0 = x \right],$$

and V_α denote the α-discounted value function, i.e.,

$$V_\alpha(x) = \inf_{U \in \mathfrak{U}} J_\alpha^U(x)\,.$$

Let

$$C_{\mathrm{pol}}(\mathbb{R}^d) := \{f \in C(\mathbb{R}^d) : f \in \mathscr{O}(1 + |x|^m) \quad \text{for some } m \in \mathbb{N}\}\,.$$

By Lions [85, theorem II.2, pp. 1258–1259], V_α is the unique viscosity solution in $C_{\mathrm{pol}}(\mathbb{R}^d)$ of the HJB equation for the discounted control problem, which takes the form

$$\inf_{u \in \mathbb{U}} \left[\mathcal{L}^u V_\alpha(x) + c(x,u) - \alpha V_\alpha(x)\right] = 0\,.$$

Theorem 7.3.9 *Under Assumption 7.3.1, there exists a viscosity solution* (V, ϱ) *in* $C^{0,1}(\mathbb{R}^d) \times \mathbb{R}$ *to* (7.3.9).

Proof For $x, y \in \mathbb{R}^d$, by Lemma 7.3.4 and the Lipschitz continuity of c, we obtain

$$\begin{aligned}
|V_\alpha(x) - V_\alpha(y)| &\le \sup_{U \in \mathfrak{U}} \mathbb{E}^U \left[\int_0^\infty \mathrm{e}^{-\alpha t} \left| \bar{c}(X_t^x, U_t) - \bar{c}(X_t^y, U_t) \right| \mathrm{d}t \right] \\
&\le C_{\mathrm{Lip}} \sup_{U \in \mathfrak{U}} \int_0^\infty \mathrm{e}^{-\alpha t}\, \mathbb{E}^U \left| X_t^x - X_t^y \right| \mathrm{d}t \\
&\le C_{\mathrm{Lip}} \sqrt{\|Q^{-1}\|\,\|Q\|} \int_0^\infty \mathrm{e}^{-\alpha t} \exp\left(-\frac{rt}{2\|Q\|} \right) |x - y|\, \mathrm{d}t \\
&\le \frac{2}{r} \sqrt{\|Q^{-1}\|}\, \|Q\|^{3/2} C_{\mathrm{Lip}} |x - y|\,.
\end{aligned} \tag{7.3.10}$$

Let $\bar{V}_\alpha(x) = V_\alpha(x) - V_\alpha(0)$. Then $\bar{V}_\alpha$ is the unique viscosity solution in $C_{\mathrm{pol}}(\mathbb{R}^d)$ to

$$\inf_{u \in \mathbb{U}} \left[\mathcal{L}^u \bar{V}_\alpha(x) + c(x,u) - \alpha \bar{V}_\alpha(x) + \alpha V_\alpha(0)\right] = 0\,.$$

Let $\{\alpha_n\}$ be a sequence such that $\alpha_n \to 0$ as $n \to \infty$. From (7.3.10) it follows, using the Ascoli–Arzelà theorem, that $\bar{V}_{\alpha_n}$ converges to a function $V \in C(\mathbb{R}^d)$ uniformly on compact subsets of $\mathbb{R}^d$ along some subsequence of $\{\alpha_n\}$. By Lemma 7.3.6, and the linear growth of $\sup_u c(x,u)$ implied by its Lipschitz continuity, $\alpha V_\alpha(0)$ is bounded in α. Hence along a suitable subsequence of $\{\alpha_n\}$ (still denoted by $\{\alpha_n\}$ by an abuse of notation) $\alpha_n V_{\alpha_n}(0)$ converges to a scalar ϱ as $n \to \infty$. Thus by the stability property of the viscosity solution [85, proposition I.3, p. 1241] it follows that the pair (V,ϱ) is a viscosity solution of (7.3.9). Clearly $V(0) = 0$ and from (7.3.10) it follows that V is Lipschitz continuous. □

Theorem 7.3.10 *Under Assumption 7.3.1, if* $(V,\varrho) \in C^{0,1}(\mathbb{R}^d) \times \mathbb{R}$ *is a viscosity solution of* (7.3.9), *then* $\varrho = \varrho^*$.

Proof Let $\alpha > 0$. Write (7.3.9), as

$$\inf_{u\in\mathbb{U}} \left[(\mathcal{L}^u - \alpha)V(x) + c(x,u) - \varrho + \alpha V(x)\right] = 0\,. \tag{7.3.11}$$

Therefore, by the uniqueness of the viscosity solution of the HJB for the discounted cost problem [85, theorem II.2, pp. 1258–1259] it follows that

$$V(x) = \inf_{U\in\mathfrak{U}} \mathbb{E}^U_x \left[\int_0^\infty \mathrm{e}^{-\alpha t}[\bar{c}(X_t, U_t) - \varrho + \alpha V(X_t)]\,\mathrm{d}t\right]. \tag{7.3.12}$$

Let $\pi \in \mathcal{G}$. Let (X, U) be the stationary solution of the martingale problem corresponding to π. Fix the probability space on which the Wiener process W and U are defined and consider the process X with arbitrary initial condition. By (7.3.12), for any $x \in \mathbb{R}^d$,

$$\alpha V(x) + \varrho \le \alpha \int_0^\infty \mathrm{e}^{-\alpha t}\, \mathbb{E}^U_x [\bar{c}(X_t, U_t)]\,\mathrm{d}t + \alpha^2 \int_0^\infty \mathrm{e}^{-\alpha t}\, \mathbb{E}^U_x [V(X_t)]\,\mathrm{d}t\,.$$

Letting $\alpha \to 0$, and using a Tauberian theorem, which asserts that

$$\limsup_{\alpha\to 0}\, \alpha \int_0^\infty \mathrm{e}^{-\alpha t}\, \mathbb{E}^U_x [\bar{c}(X_t, U_t)]\,\mathrm{d}t \le \limsup_{T\to\infty} \frac{1}{T} \int_0^T \mathbb{E}^U_x [\bar{c}(X_t, U_t)]\,\mathrm{d}t\,,$$

we obtain

$$\varrho \le \int c\,\mathrm{d}\pi\,.$$

Hence $\varrho \le \varrho^*$. To obtain the reverse inequality, let U^α be an optimal relaxed feedback control for the cost criterion in (7.3.12). Let $\mathfrak{F}^{X,U^\alpha}_t$ be the natural filtration of (X, U^α), i.e., the right-continuous completion of $\sigma(X_s, U^\alpha_s : s \le t)$. Then, since by hypothesis V has linear growth, it follows from Lemma 7.3.6 that $V(X_t)$ is uniformly integrable under any $U \in \mathfrak{U}$, and hence

$$\mathrm{e}^{-\alpha t} V(X_t) + \int_0^t \mathrm{e}^{-\alpha s}\bigl(\bar{c}(X_s, U^\alpha_s) - \varrho + \alpha V(X_s)\bigr)\,\mathrm{d}s$$

is an $(\mathfrak{F}^{X,U^\alpha}_t)$-martingale by Corollary 2.7.2. By Lemma 7.3.6 and the Lipschitz continuity, and therefore the linear growth of $\bar{c}$ and V, it is uniformly integrable over $\alpha \in (0, 1)$ and t in any compact subset of $[0, \infty)$. Since the martingale property is preserved under convergence in law and uniform integrability, letting $\alpha \to 0$, we can argue as in the proof of Corollary 2.3.9 to establish that for some relaxed feedback control $\bar{U}$

$$V(X_t) + \int_0^t (\bar{c}(X_s, \bar{U}_s) - \varrho)\,\mathrm{d}s$$

is an $(\mathfrak{F}_t^{X,\bar{U}})$-martingale. Therefore

$$\mathbb{E}^{\bar{U}}\left[V(X_t) + \int_0^t (\bar{c}(X_s, \bar{U}_s) - \varrho)\,\mathrm{d}t\right] = \mathbb{E}^{\bar{U}}\left[V(X_0)\right],$$

and dividing by t we obtain

$$\frac{1}{t}\,\mathbb{E}_x^{\bar{U}}[V(X_t)] + \frac{1}{t}\int_0^t \mathbb{E}_x^{\bar{U}}[\bar{c}(X_s, \bar{U}_s)]\,\mathrm{d}s = \varrho + \frac{1}{t}\,\mathbb{E}^{\bar{U}}\left[V(X_0)\right]. \tag{7.3.13}$$

Lemma 7.3.6 and the linear growth of V implies $t^{-1}\,\mathbb{E}_x^{\bar{U}}[V(X_t)] \to 0$ as $t \to \infty$. Thus letting $t \to \infty$ in (7.3.13) yields

$$\varrho = \lim_{t\to\infty}\frac{1}{t}\int_0^t \mathbb{E}^{\bar{U}}[\bar{c}(X_s, \bar{U}_s)]\,\mathrm{d}s \geq \varrho^*,$$

and the proof is complete. □

Theorem 7.3.11 *Let Assumption 7.3.1 hold, and suppose $(V, \varrho^*) \in C^{0,1}(\mathbb{R}^d) \times \mathbb{R}$ is a viscosity solution of* (7.3.9)*. Define*

$$M_t := V(X_t) + \int_0^t (\bar{c}(X_s, U_s) - \varrho^*)\,\mathrm{d}s, \quad t \geq 0.$$

The following hold:

(i) *if $U \in \mathfrak{U}$, then M is an $(\mathfrak{F}_t^{X,U})$-submartingale;*
(ii) *if for some $U \in \mathfrak{U}$ the process M is an $(\mathfrak{F}_t^{X,U})$-martingale, then U is optimal;*
(iii) *if $U \in \mathfrak{U}$ is optimal and the process (X, U) is stationary with*

$$\mathbb{E}^U\left[\int_{\mathbb{U}} f(X_t, u)U_t(\mathrm{d}u)\right] = \int_{\mathbb{R}^d\times\mathbb{U}} f\,\mathrm{d}\nu, \quad \nu \in \mathscr{P}(\mathbb{R}^d \times \mathbb{U}), \quad t \geq 0,$$

then M is an $(\mathfrak{F}_t^{X,U})$-martingale;
(iv) *if U is optimal and the process (X, U) is ergodic with $\mathscr{L}(X_0) = \mu$, then for any viscosity solution $(\psi, \varrho) \in C^{0,1}(\mathbb{R}^d) \times \mathbb{R}$ of* (7.3.9)*, we have $\varrho = \varrho^*$ and $\psi - V$ is constant on* $\operatorname{supp}(\mu)$*, the support of μ.*

Proof (i) Let $U \in \mathfrak{U}$. By (7.3.11) and Corollary 2.7.2,

$$\mathrm{e}^{-\alpha t}V(X_t) + \int_0^t \mathrm{e}^{-\alpha s}[\bar{c}(X_s, U_s) - \varrho^* + \alpha V(X_s)]\,\mathrm{d}s$$

is an $(\mathfrak{F}_t^{X,U})$-submartingale. Letting $\alpha \to 0$, (i) follows, since the submartingale property with respect to the natural filtration is preserved under weak convergence and uniform integrability.

(ii) Since M is an $(\mathfrak{F}_t^{X,U})$-martingale under $\mathbb{P}^U$, we have

$$\mathbb{E}_x^U[V(X_t)] + \int_0^t \mathbb{E}_x^U[\bar{c}(X_s, U_s) - \varrho^*]\,\mathrm{d}s = V(x).$$

Dividing by t and letting $t \to \infty$, arguing as in the proof of Theorem 7.3.10, we obtain

$$\lim_{t\to\infty} \frac{1}{t} \int_0^t \mathbb{E}_x^U[\bar{c}(X_s, U_s)]\, \mathrm{d}s = \varrho^* .$$

Thus U is optimal.

(iii) By stationarity, $\mathbb{E}^U[M_t] = \int V\,\mathrm{d}\nu$. By (i), M is a submartingale. Suppose M is not a martingale. Then there exists $t > s > 0$ and $A \in \mathfrak{F}_s^{X,U}$ with $\mathbb{P}^U(A) > 0$ such that

$$\mathbb{E}^U[M_t \mid \mathfrak{F}_s^{X,U}] > M_s \quad \text{on } A .$$

Now,

$$\begin{aligned}\int V\,\mathrm{d}\nu &= \mathbb{E}^U[M_t] = \mathbb{E}^U[\mathbb{E}^U[M_t \mid \mathfrak{F}_s^{X,U}]] \\ &= \mathbb{E}^U[\mathbb{E}^U[M_t \mathbb{I}_A \mid \mathfrak{F}_s^{X,U}]] + \mathbb{E}^U[\mathbb{E}^U[M_t \mathbb{I}_{A^c} \mid \mathfrak{F}_s^{X,U}]] \\ &= \mathbb{E}^U[\mathbb{I}_A\, \mathbb{E}^U[M_t \mid \mathfrak{F}_s^{X,U}]] + \mathbb{E}^U[\mathbb{I}_{A^c}\, \mathbb{E}^U[M_t \mid \mathfrak{F}_s^{X,U}]] \\ &> \mathbb{E}^U[M_s] = \int V\,\mathrm{d}\nu ,\end{aligned}$$

which is a contradiction.

(iv) That $\varrho = \varrho^*$ is proved in Theorem 7.3.10. Therefore, analogously to (iii),

$$\psi(X_t) + \int_0^t (\bar{c}(X_s, U_s) - \varrho^*)\, \mathrm{d}s$$

is an $(\mathfrak{F}_t^{X,U})$-martingale. Hence so is $V(X_t) - \psi(X_t)$. Since it is uniformly integrable, it converges μ-a.s., and since $V-\psi$ is continuous, it must equal a constant on $\mathrm{supp}(\mu)$.

□

The following example shows that the viscosity solution of (7.3.9) is not necessarily unique.

Example 7.3.12 With $d = 1$, let

$$c(x) = \frac{(x^2 - 1)^2}{(x^2 + 1)^{3/2}} , \qquad b(x, u) = -(x + u) ,$$

$\sigma = 0$, and $\mathbb{U} = [-1, 1]$. Then the function

$$V(x) := \begin{cases} \displaystyle\int_1^x \frac{(1 - z)^2 (1 + z)}{(z^2 + 1)^{3/2}}\, \mathrm{d}z & \text{if } x \geq 1 , \\ \displaystyle\int_x^1 \frac{(1 + z)^2 (1 - z)}{(z^2 + 1)^{3/2}}\, \mathrm{d}z & \text{if } x < 1 \end{cases}$$

is a viscosity solution to

$$\min_{u\in[-1,1]} \left[b(x, u) \frac{\mathrm{d}V}{\mathrm{d}x}(x) + c(x) \right] = 0 , \quad x \in \mathbb{R} , \tag{7.3.14}$$

and so is $V(-x)$. A third viscosity solution to (7.3.14) is given by

$$\frac{1+\operatorname{sign}(x)}{2}V(x)+\frac{1-\operatorname{sign}(x)}{2}V(-x)\,,\qquad x\in\mathbb{R}\,.$$

All three of these functions are in $C^1(\mathbb{R})$, and are therefore also classical solutions. None of these three pairs differ by a constant, hence they are distinct.

The reason behind the lack of uniqueness in Example 7.3.12 is that the optimal ergodic occupation measure in not unique. The theorem that follows clarifies this issue.

Theorem 7.3.13 *If the optimal ergodic occupation measure π^* is unique, then there is a unique viscosity solution in $(V,\varrho^*)\in C^{0,1}(\mathbb{R}^d)\times\mathbb{R}$ of (7.3.9) satisfying $V(0)=0$.*

Proof Let (V,ϱ) and (ψ,ϱ') be two viscosity solutions in $C^{0,1}(\mathbb{R}^d)\times\mathbb{R}$ of (7.3.9), with $V(0)=\psi(0)=0$. By Theorem 7.3.10, $\varrho=\varrho'=\varrho^*$. Fix $X_0=0$ and let $\bar{U}$ be as in the proof of Theorem 7.3.10. Then

$$V(X_t)+\int_0^t\bigl(\bar{c}(X_s,\bar{U}_s)-\varrho^*\bigr)\,\mathrm{d}s\,,\qquad t\geq 0\,,$$

is a martingale under $\mathbb{P}^{\bar{U}}$. It is also clear from the proof of Theorem 7.3.10 that $\bar{U}$ is optimal. Thus the associated mean empirical measures (see Lemma 2.5.3, p. 62) converge to $\pi^*\in\mathscr{G}$. An argument analogous to the one leading to the martingale property also shows that

$$\psi(X_t)+\int_0^t\bigl(\bar{c}(X_s,\bar{U}_s)-\varrho^*\bigr)\,\mathrm{d}s\,,\qquad t\geq 0\,,$$

is a submartingale. Thus $\psi(X_t)-V(X_t)$ is a submartingale. Since $\psi-V$ has linear growth, it is uniformly integrable. Hence, it converges a.s. By Theorem 7.3.11 (iv), it must converge to the constant value C that $\psi-V$ takes on $\operatorname{supp}(\mu^*)$, where μ^* is the marginal of π^* on $\mathbb{R}^d$. Since $\psi(0)-V(0)=0$, the submartingale property leads to $C\geq 0$. Interchanging the roles of V and ψ, and repeating the argument we deduce $C\leq 0$, hence $C=0$. Using the submartingale property from an arbitrary initial condition $x\in\mathbb{R}^d$, we deduce that $\psi(x)-V(x)\leq 0$, and $V(x)-\psi(x)\leq 0$, respectively. Hence $\psi=V$. □

7.4 Partially nondegenerate controlled diffusions

7.4.1 Preliminaries

In this section we turn to the problem of ergodic control of a reflecting diffusion in a compact domain under the condition of partial nondegeneracy, i.e., when its transition kernel after some time is absolutely continuous with respect to the Lebesgue measure on a part of the state space. The existence of a value function and a *martingale dynamic programming principle* is established by mapping the problem to a discrete time control problem. This approach also extends to partially nondegenerate diffusions on compact manifolds without boundary. The underlying idea may be applicable to other problems as well, in so far as it offers a recipe for a passage from continuous time to discrete time problems and back.

The remainder of this section describes the problem and the key hypothesis of partial nondegeneracy. Section 7.4.2 reviews the associated discrete time control problem. Section 7.4.3 uses these results to develop the martingale approach to the dynamic programming principle for the original problem. Section 7.4.4 considers problems on the entire domain, while Section 7.4.5 summarizes some further extensions. It also establishes some implications to the problem of existence of optimal controls.

We state the model in the relaxed control framework. Let $D \subset \mathbb{R}^d$ for $d \geq 1$ be a bounded domain with a smooth boundary ∂D. Let X be a $\bar{D}$-valued controlled reflecting diffusion governed by

$$\mathrm{d}X_t = \bar{b}(X_t, U_t)\,\mathrm{d}t + \sigma(X_t)\,\mathrm{d}W_t - \gamma(X_t)\,\mathrm{d}\xi_t\,, \quad t \geq 0\,. \tag{7.4.1}$$

In this model

(i) $\bar{b}\colon \bar{D}\times\mathscr{P}(\mathbb{U}) \to \mathbb{R}^d$ is continuous and Lipschitz in its first argument uniformly w.r.t. the second.

(ii) $\sigma = [\sigma^{ij}]_{1\leq i,j\leq d} : \bar{D} \to \mathbb{R}^{d\times d}$ is Lipschitz continuous.

(iii) X_0 is prescribed in law.

(iv) $W = [W_1, \dots, W_d]^\mathsf{T}$ is a d-dimensional standard Brownian motion independent of X_0.

(v) U is a $\mathscr{P}(\mathbb{U})$-valued control process with measurable sample paths satisfying the non-anticipativity condition that for $t \geq s$, $W_t - W_s$ is independent of $\{X_0, W_{s'}, U_{s'} : s' \leq s\}$.

(vi) ξ is an $\mathbb{R}$-valued continuous nondecreasing process ("local time on the boundary") satisfying

$$\xi_t = \int_0^t \mathbb{I}_{\partial D}(X_s)\,\mathrm{d}\xi_s\,. \tag{7.4.2}$$

(vii) There exists $\eta \in C^2(\mathbb{R}^d)$ satisfying $|\nabla\eta| \geq 1$ on ∂D, such that

$$D = \{x \in \mathbb{R}^d : \eta(x) < 0\}\,, \qquad \partial D = \{x \in \mathbb{R}^d : \eta(x) = 0\}\,,$$

and

$$\sup_{u\in\mathbb{U}} \langle \nabla\eta(x), b(x,u)\rangle < 0 \quad \text{on } \partial D\,, \quad \text{whenever } \sigma^\mathsf{T}(x)\nabla\eta(x) = 0\,.$$

(viii) $\gamma\colon \bar{D} \to \mathbb{R}^d$ is a smooth vector field such that, for some $\delta > 0$,

$$\langle\gamma(x), \nabla\eta(x)\rangle \geq \delta > 0 \qquad \forall x \in \partial D\,.$$

Under these assumptions the existence of a unique weak solution of (7.4.1)–(7.4.2) given the joint law of (X_0, W, U) can be established as in [52, 86, 114], which consider the uncontrolled case. In the controlled case, a solution to (7.4.1)–(7.4.2) is a quadruplet (X, U, W, ξ) defined on some probability space such that (X_0, W, U) have the prescribed joint law and (7.4.1)–(7.4.2) hold. Uniqueness of the weak solutions means that if $\mathscr{L}(X_0, W, U)$ is specified the law of the solution gets uniquely fixed. This formulation is equivalent to the usual one in terms of feedback controls, to the extent that it does not alter the set of attainable laws of X modulo a possible enlargement of the underlying probability space. This can be argued as in Theorem 2.3.4. In particular, it should be emphasized that we are *not* fixing (X_0, W, U) as processes on a fixed probability space, but only in law.

Let $\mathfrak{m}$ denote the Lebesgue measure on $\mathbb{R}^d$. We work under the following hypothesis:

Assumption 7.4.1 There exist positive constants T and C_0 and a non-empty open set $G_0 \subset D$ such that, for all $x \in \bar{D}$,

$$\mathbb{P}^U(X_T \in A \cap G_0 \mid X_0 = x) \geq C_0\, \mathfrak{m}(A \cap G_0)$$

for all Borel $A \subset D$ and any choice of $U \in \mathfrak{U}$.

We refer to Assumption 7.4.1 as the *partial nondegeneracy* hypothesis. It should be noted that for unbounded D, this is not a reasonable hypothesis, since the left-hand side would typically converge to zero as $|x| \to \infty$ for any bounded G_0. Therefore the assumption of boundedness of D is essential here. We do, however, consider an appropriate extension to the whole space later.

As an example, consider the scalar case with $D = (-4R, 4R)$ for some $R > 0$. Let $\sigma(x) = 0$ for $|x| = 4R$ and ≥ 1 for $|x| \leq R$. Also, let $b(x, u) = -2x + u$ with $u \in \mathbb{U} = [-R, R]$. It is straightforward to verify Assumption 7.4.1 with $G_0 = (-R, R)$. While this example is rather contrived, it does bring out the essential intuition behind the assumption of partial nondegeneracy: that there be a part of the state space, accessible from anywhere in finite time with positive probability, such that the diffusion is nondegenerate there. This is analogous to the "unichain" condition in Markov decision processes.

We conclude this section by noting that an immediate consequence of Assumption 7.4.1 is that for any $t > 0$,

$$\begin{aligned}\mathbb{P}^U_x(X_{T+t} \in A \cap G_0) &= \mathbb{E}^U_x\big[\mathbb{P}^U_{X_t}(X_T \in A \cap G_0)\big] \\ &\geq C_0\, \mathfrak{m}(A \cap G_0)\,.\end{aligned}$$

7.4.2 The associated discrete time control problem

This section describes an equivalent discrete time control problem. Let $T > 0$ be as in Assumption 7.4.1 and set $\tilde{X}_n = X_{nT}$, $n \geq 0$.

For $x \in \bar{D}$, let $A_x \subset \mathscr{P}(C([0, \infty); \bar{D}) \times \mathscr{U}_T)$ be defined by

$$A_x := \{\mathscr{L}(X_{[0,T]}, U_{[0,T]}) : X_0 = x\,,\ (X, U) \text{ satisfies (7.4.1)}\}\,. \tag{7.4.3}$$

Theorem 7.4.2 *The set A_x is compact.*

Proof Let $(X^n, W^n, U^n, \xi^n, X_0^n)$, $n \geq 1$, satisfy (7.4.1)–(7.4.2) with $X_0^n = x$, on probability spaces $(\Omega^n, \mathfrak{F}^n, \mathbb{P}^n)$. Let $\{f_i\}$ be a countable, dense subset of the unit ball of $C(\mathbb{U})$ and define

$$\alpha_i^n(t) = \int_{\mathbb{U}} f_i(u)\, U_t^n(\mathrm{d}u)\,, \quad t \in [0, T]\,.$$

Let B be the unit ball of $L^\infty([0, T])$ with the topology given by the weak topology of $L^2([0, T])$ relativized to B. Let E be the countable product of replicas of B with the product topology. Let

$$\alpha^n = (\alpha_1^n, \alpha_2^n, \dots)\,.$$

Obviously the sequence (X^n, α^n) is tight in $\mathscr{P}(C([0, T]; \bar{D}) \times E)$. Therefore it contains some subsequence, also denoted by (X^n, α^n), which converges in law to a limit (X, α). By Skorohod's theorem we may assume that $(X^n, \alpha^n : n \geq 1)$ and (X, α) are defined on a common probability space $(\Omega, \mathfrak{F}, \mathbb{P})$ and that this convergence is a.s. on $(\Omega, \mathfrak{F}, \mathbb{P})$. Argue as in Theorem 2.3.2 to show that

$$\alpha_i(t) = \int_{\mathbb{U}} f_i(u)\, U_t(\mathrm{d}u)$$

for a $\mathscr{P}(\mathbb{U})$-valued process U. It remains to show that X satisfies (7.4.1)–(7.4.2) for some $\widetilde{W}$, $\tilde{U}$, $\tilde{\xi}$ on $(\Omega, \mathfrak{F}, \mathbb{P})$. Pick $g \in C^2(D) \cap C^1(\bar{D})$ satisfying

$$\langle \gamma(x), \nabla g(x) \rangle \leq -\delta_1\,, \quad x \in \partial D\,,$$

for some $\delta_1 > 0$. For $t \geq t_0 \geq 0$, and $h \in C_b(C([0, t_0]; \bar{D}); \mathbb{R}_+)$, we have

$$\mathbb{E}\left[\left(g(X_t^n) - g(X_{t_0}^n) - \int_{t_0}^t \mathcal{L}^{U_s^n} g(X_s^n)\, \mathrm{d}s\right) h(X_{[0,t_0]}^n)\right] \geq 0\,. \tag{7.4.4}$$

Since (7.4.4) is preserved under weak convergence, letting $n \to \infty$, it also holds for (X, U). As in Chapter 2 we can replace U by a feedback $\tilde{U}$ (i.e., $\tilde{U}$ is progressively measurable w.r.t. the natural filtration of X). Then

$$g(X_t) - \int_0^t \mathcal{L}^{\tilde{U}_s} g(X_s)\, \mathrm{d}s$$

is a submartingale w.r.t. the natural filtration of X. The rest follows by using the submartingale representation result in Stroock and Varadhan [114, theorem 2.5, pp. 165–166]. □

Lemma 7.4.3 *For each $x \in \bar{D}$, A_x is convex and compact, and the set-valued map $x \to A_x$ is upper semicontinuous (i.e., it has a closed graph).*

Proof Note that (7.4.1)–(7.4.2) is completely characterized in law as follows: if $f \in C^2(D) \cap C^1(\bar{D})$ and $h_t \in C_b(C([0, t]; \bar{D}) \times \mathscr{U}_t)$, $t \geq 0$, then for all $0 \leq s \leq t$,

$$\mathbb{E}\left[\left(f(X_t) - f(X_s) - \int_s^t \bar{\mathcal{L}}^{U_r} f(X_r)\, \mathrm{d}r \right.\right.$$
$$\left.\left. - \int_s^t \frac{\partial f}{\partial \gamma}(X_r)\, \mathrm{d}\xi_r\right) h_s(X_{[0,s]}, U_{[0,s]})\right] = 0 \tag{7.4.5}$$

and

$$\mathbb{E}\left[\left(\xi(t) - \xi(s) - \int_s^t \mathbb{I}_{\partial D}(X_r)\, d\xi_r\right) h_t(X_{[0,t]}, U_{[0,t]})\right] = 0\,. \tag{7.4.6}$$

A standard monotone class argument shows that (7.4.5) is in fact the martingale formulation of (7.4.1), and (7.4.6) coincides with (7.4.2). Now if (7.4.5)–(7.4.6) hold under two probability measures, they do so under any convex combination thereof. It follows that the set of attainable laws of $\mathscr{L}(X, U, \xi)$ for $X_0 = x$, and therefore also the set A_x are convex. By Theorem 7.4.2, A_x is also compact. Consider sequences $x_n \to x_\infty \in \bar{D}$ and $\mathscr{L}(X^n, U^n) \in A_{x_n}$, $n \geq 1$, and suppose without loss of generality that $\mathscr{L}(X^n, U^n) \to \mathscr{L}(X^\infty, U^\infty)$. Using the arguments in the proof of Theorem 7.4.2 we can show that $\mathscr{L}(X^\infty, U^\infty) \in A_{x_\infty}$. Thus the set-valued map $x \to A_x$ is upper semicontinuous. □

Definition 7.4.4 For a pair (X, U) satisfying (7.4.1), with $\nu = \mathscr{L}(X_{[0,t]}, U_{[0,t]})$, and $X_0 = x \in \bar{D}$, we denote by $\tilde{P}^\nu_t(x, \mathrm{d}y) \in \mathscr{P}(\bar{D})$ the law of X_t for $t \geq 0$ parameterized by the control ν. In particular, if we define

$$\mathscr{K} := \{(x, \nu) : x \in \bar{D}\,,\ \nu \in A_x\} \subset \bar{D} \times \mathscr{P}(C([0,T]; \bar{D}) \times \mathscr{U}_T)\,,$$

then $(x, \nu) \mapsto \tilde{P}^\nu_T(x, \mathrm{d}y)$ maps $(x, \nu) \in \mathscr{K}$ to $\mathscr{L}(X_T)$. We define $\tilde{X}_n = X_{nT}$, $n \geq 0$. Then $\{\tilde{X}_n\}$ is a $\bar{D}$-valued controlled Markov chain with transition kernel $\tilde{P}^\nu_T(x, \mathrm{d}y)$. Note that by Lemma 7.4.3, $\mathscr{K}$ is compact.

For the controlled process $\tilde{X}$, we consider the ergodic control problem of minimizing

$$\limsup_{n\to\infty} \frac{1}{n} \sum_{m=0}^{n-1} \mathbb{E}\left[\tilde{c}(\tilde{X}_m, \tilde{Z}_m)\right], \tag{7.4.7}$$

where $\tilde{Z}_n$ is an $A_{\tilde{X}_n}$-valued control for $n \geq 0$ and the running cost function is $\tilde{c}$ defined by

$$\tilde{c}(x, \nu) := \frac{1}{T}\, \mathbb{E}\left[\int_0^T \bar{c}(X_t, U_t)\, \mathrm{d}t \,\middle|\, X_0 = x\right], \quad (x, \nu) \in \mathscr{K}\,,$$

where the expectation is with respect to $\nu = \mathscr{L}(X_{[0,T]}, U_{[0,T]}) \in A_x$.

Theorem 7.4.5 *There exists a unique pair $(\tilde{V}, \tilde{\varrho})$, $\tilde{V} \in C(\bar{D})$ and $\tilde{\varrho} \in \mathbb{R}$, such that $\tilde{V}(x_0) = 0$ for a prescribed $x_0 \in D$ and*

$$\tilde{V}(x) + \tilde{\varrho} = \min_{\nu\in A_x} \left[\tilde{c}(x, \nu) + \int_{\bar{D}} \tilde{P}^\nu_T(x, \mathrm{d}y) \tilde{V}(y)\right] \qquad \forall x \in \bar{D}\,. \tag{7.4.8}$$

Moreover,

(i) *$\tilde{\varrho}$ is the optimal cost for the ergodic control problem corresponding to (7.4.7), and,*

(ii) *for any measurable selection* $\bar{D} \ni x \mapsto v(x) \in A_x$, *such that* $v(x)$ *attains the minimum on the right-hand side of* (7.4.8), *the control given by* $\tilde{Z}_n = v(\tilde{X}_n)$, $n \geq 0$ *is optimal for the discrete time ergodic control problem in* (7.4.7).

Proof By Assumption 7.4.1, we have

$$\left\|\tilde{P}_T^{\nu}(x, \cdot) - \tilde{P}_T^{\nu'}(x', \cdot)\right\|_{\mathrm{TV}} \leq 2(1 - C_0 \mathfrak{m}(G_0)) \tag{7.4.9}$$

for all pairs (x, ν), $(x', \nu') \in \mathcal{K}$. It is straightforward to verify using Lemma 7.4.3 that the map $\mathcal{T}$ defined by

$$\mathcal{T} f(x) := \min_{\nu \in A_x} \left[\tilde{c}(x, \nu) + \int_{\bar{D}} \tilde{P}_T^{\nu}(x, \mathrm{d}y) f(y) \right],$$

maps $C(\bar{D})$ to itself, and that by (7.4.9) it is a span contraction. Hence if $x_0 \in \bar{D}$ is some fixed point and $\mathcal{G}$ is the subspace of $C(\bar{D})$ consisting of those functions that vanish at x_0, the map $\mathcal{T}_0 f := \mathcal{T} f - \mathcal{T} f(x_0)$ is also a span contraction on $\mathcal{G}$. Noting that the span semi-norm is equivalent to the supremum norm on $\mathcal{G}$, since $\|f\|_\infty \leq \mathrm{span}(f) \leq 2\|f\|_\infty$ for all $f \in \mathcal{G}$, it follows that $\mathcal{T}_0 \tilde{V} = \tilde{V}$ for some $\tilde{V} \in \mathcal{G}$. Then $\tilde{V}$ and $\tilde{\varrho} := \mathcal{T}\tilde{V}(x_0)$ solve (7.4.8). The rest follows from standard results in Markov decision processes, which can be found in Hernández-Lerma [73, sections 5.1–5.2]. □

Remark 7.4.6 That at least one such measurable selection exists follows from a standard measurable selection theorem [13].

Lemma 7.4.7 *If* ϱ^* *denotes the optimal cost for the continuous time ergodic control problem defined in* (7.2.2), *then* $\varrho^* = \tilde{\varrho}$.

Proof Fix $\mathcal{L}(X_0)$. We establish a correspondence between the laws of $(\tilde{X}, \tilde{Z})$ and (X, U). Given a specific $\mathcal{L}(X, U)$, let

$$C([0, nT]; \bar{D}) \times \mathcal{U}_{nT} \ni (X_{[0,nT]}, U_{[0,nT]}) \mapsto v_n(X_{[0,nT]}, U_{[0,nT]}) \in A_{X_{nT}}$$

for $n \geq 0$, denote the regular conditional law of

$$\{(X_t, U_t) : t \in [nT, (n+1)T]\},$$

given $(X_{[0,nT]}, U_{[0,nT]})$, and let $\tilde{Z}_n = v_n(X_{[0,nT]}, U_{[0,nT]})$ for $n \geq 0$. Then $(\tilde{X}, \tilde{Z})$ is a pair of processes conforming to the above description. (Note that $\tilde{Z}_n$ is not measurable with respect to $\sigma\{\tilde{X}_m, \tilde{Z}_{m-1} : m \leq n\}$, but this is permitted as long as it does not anticipate the future.) It is clear that the ergodic cost (7.4.7) for $(\tilde{X}, \tilde{Z})$ equals the ergodic cost (7.2.1) for (X, U). Thus $\tilde{\varrho} \leq \varrho^*$. Conversely, let $x \mapsto v(x)$ be a measurable selector from the minimizer on the right-hand side of (7.4.8) and $(\tilde{X}, \tilde{Z})$ with $\mathcal{L}(\tilde{X}_0) = \mathcal{L}(X_0)$ and $\tilde{Z}_n = v(\tilde{X}_n)$, $n \geq 0$, be an optimal pair. Construct on the canonical path space $C([0, \infty); \bar{D}) \times \mathcal{U}$ (with its Borel σ-field) the processes (X, U) by the following inductive procedure: Construct (X_t, U_t), $t \in [0, T]$, such that the regular conditional law of $(X_{[0,T]}, U_{[0,T]})$ given X_0 is $v(X_0)$. At step n, $n \geq 1$, construct (X_t, U_t), $t \in [nT, (n+1)T]$ such that the regular conditional law of $(X_{[nT,(n+1)T]}, U_{[nT,(n+1)T]})$ given $(X_{[0,nT]}, U_{[0,nT]})$ is $v(X_{nT})$. This defines (X, U) such that the ergodic cost (7.2.1) for (X, U) is the same as the ergodic cost (7.4.7) for $(\tilde{X}, \tilde{Z})$, which is $\tilde{\varrho}$. Thus $\varrho^* \le \tilde{\varrho}$. The claim follows. □

Lemma 7.4.8 *For $n \ge 0$ define*

$$M_n := \tilde{V}(\tilde{X}_n) + \sum_{m=0}^{n-1}(\tilde{c}(\tilde{X}_m, \tilde{Z}_m) - \tilde{\varrho}),$$

and $\mathcal{F}_n := \sigma(\tilde{X}_m, \tilde{Z}_m : m \le n)$. Then $(M_n, \mathcal{F}_n)$, $n \ge 0$, is a submartingale. If it is a martingale, $\{\tilde{Z}_n\}$ is optimal. Conversely if $(\tilde{X}, \tilde{Z})$ is an optimal stationary pair, $(M_n, \mathcal{F}_n)$, $n \ge 0$, is a martingale.

Proof That $(M_n, \mathcal{F}_n)$, $n \ge 0$, is a submartingale follows from (7.4.8). If $(M_n, \mathcal{F}_n)$ is a martingale, we have

$$\tilde{V}(\tilde{X}_n) = \left(\tilde{c}(\tilde{X}_n, \tilde{Z}_n) - \tilde{\varrho}\right) + \mathbb{E}\left[\tilde{V}(X_{n+1}) \mid \mathcal{F}_n\right].$$

Sum over $n = 0, 1, \dots, N$, take expectations, divide by N and let $N \to \infty$ to verify that (7.4.7) equals $\tilde{\varrho}$. If the last claim were false, we would have

$$\tilde{V}(\tilde{X}_n) \le \left(\tilde{c}(\tilde{X}_n, \tilde{Z}_n) - \tilde{\varrho}\right) + \mathbb{E}\left[\tilde{V}(\tilde{X}_{n+1}) \mid \mathcal{F}_n\right], \quad n \ge 0,$$

with the inequality being strict with positive probability. Taking expectations and using stationarity, we observe that (7.4.7) strictly exceeds $\tilde{\varrho}$, a contradiction. □

7.4.3 The dynamic programming principle

In this section, we show that the value function $\tilde{V}$ of the dynamic programming equation (7.4.8) for the discrete time control problem also qualifies as the value function for the continuous time problem. First we show that it suffices to consider stationary solutions (X, U) of (7.4.2).

Lemma 7.4.9 *There exists a stationary pair (X, U) which satisfies* (7.4.1) – (7.4.2). *Moreover, any attainable value of* (7.4.7) *is also attainable by a stationary (X, U).*

Proof Let (X, U) be any solution of (7.4.1) – (7.4.2) and define $\mathscr{L}(X^n, U^n)$ as follows: For $f \in C_b(C([0, \infty); \bar{D}) \times \mathscr{U})$,

$$\mathbb{E}\left[f(X^n, U^n)\right] = \frac{1}{n}\int_0^n \mathbb{E}\left[f(X \circ \theta_t, U \circ \theta_t)\right] \mathrm{d}t, \quad n \ge 1.$$

Using the fact that the solution measures $\mathscr{L}(X, U)$ of (7.4.1) – (7.4.2) are characterized by (7.4.5) – (7.4.6), we argue using the convexity assertion of Lemma 7.4.3 that $\mathscr{L}(X^n, U^n)$ is also a solution measure for (7.4.1) – (7.4.2) for $n \ge 1$. Obviously $\{\mathscr{L}(X^n, U^n) : n \ge 1\}$ is relatively compact. Using arguments as in the proof of Theorem 7.4.2, we can show that any limit point thereof is also a solution measure for (7.4.1) – (7.4.2). Also, for any $T > 0$ and f as above,

$$\mathbb{E}\left[f(X^n, U^n)\right] - \mathbb{E}\left[f(X^n \circ \theta_T, U^n \circ \theta_T)\right] \xrightarrow[n \to \infty]{} 0,$$

and thus any limit point of $\mathscr{L}(X^n, U^n)$ as $n \to \infty$ is stationary.

The second assertion follows along the lines of the proof of Lemma 6.5.4 on p. 242. □

Lemma 7.4.10 *For $t > 0$, $x \in \bar{D}$, define*

$$V_t(x) = \inf_{U \in \mathfrak{U}} \mathbb{E}_x^U \left[\int_0^t (\bar{c}(X_s, U_s) - \tilde{\varrho}) \, \mathrm{d}s + \tilde{V}(X_t) \right]. \tag{7.4.10}$$

Then $V_t = \tilde{V}$ for all $t \geq 0$.

Proof Clearly $V_0 = \tilde{V}$. By (7.4.8), $V_T = \tilde{V}$. If $t \in [0, T)$, standard dynamic programming (DP) arguments show that $V_t = V_{nT+t}$ for all $n \geq 0$. Thus, without any loss of generality, it suffices to consider $t \in [0, T]$. By standard DP arguments,

$$\tilde{V}(x) = \inf_{U \in \mathfrak{U}} \mathbb{E}_x^U \left[\int_0^{T-t} (\bar{c}(X_s, U_s) - \tilde{\varrho}) \, \mathrm{d}s + V_t(X_{T-t}) \right]. \tag{7.4.11}$$

Combining (7.4.10) – (7.4.11), DP arguments again lead to

$$V_t(x) = \inf_{U \in \mathfrak{U}} \mathbb{E}_x^U \left[\int_0^{T} (\bar{c}(X_s, U_s) - \tilde{\varrho}) \, \mathrm{d}s + V_t(X_T) \right].$$

Note that the solution $\tilde{V}$ of (7.4.8) is unique up to an additive constant if we drop the requirement $\tilde{V}(x_0) = 0$. Thus $V_t(\cdot) = \tilde{V}(\cdot) + a(t)$ for some $a(t) \in \mathbb{R}$. From (7.4.10) – (7.4.11) and standard DP arguments, we then obtain

$$\tilde{V}(x) = \inf_{U \in \mathfrak{U}} \mathbb{E}_x^U \left[\int_0^{T} (\bar{c}(X_s, U_s) - \tilde{\varrho}) \, \mathrm{d}s + \tilde{V}(X_T) \right] + a(t) + a(T-t).$$

By (7.4.8), we have

$$a(t) + a(T-t) = 0, \quad t \in [0, T). \tag{7.4.12}$$

For a stationary solution (X, U),

$$V_t(X_0) \leq \mathbb{E}_{X_0}^U \left[\int_0^t (\bar{c}(X_s, U_s) - \tilde{\varrho}) \, \mathrm{d}s + \tilde{V}(X_t) \right] \qquad \forall t \geq 0.$$

Taking expectations and using stationarity and the fact that $V_t = \tilde{V} + a(t)$, yields

$$a(t) \leq t \left(\mathbb{E}\left[\bar{c}(X_t, U_t)\right] - \tilde{\varrho} \right).$$

Taking the infimum of the right-hand side over all stationary $\mathscr{L}(X, U)$, we obtain by Lemma 7.4.9, $a(t) \leq 0$. In view of (7.4.12), we conclude that $a(t) = 0$ for all t, which completes the proof. □

We have shown that

$$\tilde{V}(x) = \inf_{U \in \mathfrak{U}} \mathbb{E}_x^U \left[\int_0^t (\bar{c}(X_s, U_s) - \varrho^*) \, \mathrm{d}s + \tilde{V}(X_t) \right], \quad t \geq 0. \tag{7.4.13}$$

The proof of the theorem which follows is exactly analogous to Lemma 7.4.8, using (7.4.13).

Theorem 7.4.11 *The family given by*

$$\left(\tilde{V}(X_t) + \int_0^t (\bar{c}(X_s, U_s) - \varrho^*)\,\mathrm{d}s\,,\ \mathfrak{F}_t^{X,U}\right), \quad t \ge 0\,, \tag{7.4.14}$$

is a submartingale, and if it is a martingale, then (X, U) is an optimal pair. Conversely, if (X, U) is a stationary optimal pair, (7.4.14) is a martingale.

Theorem 7.4.11 constitutes a *martingale DP principle* in the spirit of Rishel [99] and Striebel [113]. The next result provides a converse statement.

Theorem 7.4.12 *Suppose $V' \in C(\bar{D})$ and $\varrho' \in \mathbb{R}$ are such that*

$$\left(V'(X_t) + \int_0^t (\bar{c}(X_s, U_s) - \varrho')\,\mathrm{d}s\,,\ \mathfrak{F}_t^{X,U}\right), \quad t \ge 0\,,$$

is a submartingale under any $U \in \mathfrak{U}$. Then $\varrho' \le \varrho^*$. If in addition it is a martingale for some (X, U), then the latter is optimal and $\varrho' = \varrho^*$. Moreover, in this case, $V' - \tilde{V}$ is constant on the support of $\mathscr{L}(X_0) \in \mathscr{P}(\bar{D})$ for any stationary optimal $\mathscr{L}(X, U)$.

Proof Let (X, U) be a stationary solution. By the submartingale property,

$$\mathbb{E}^U[V'(X_T)] + \int_0^T \left(\mathbb{E}^U[\bar{c}(X_s, U_s)] - \varrho'\right)\mathrm{d}s \ge \mathbb{E}^U[V'(X_0)]\,.$$

By stationarity

$$\varrho' \le \mathbb{E}^U[\bar{c}(X_t, U_t)]\,.$$

Taking the infimum over all stationary $\mathscr{L}(X, U)$ and using Lemma 7.4.9, we obtain $\varrho' \le \varrho^*$. If the submartingale is in fact a martingale under some (X, U), we have

$$\frac{\mathbb{E}^U[V'(X_t)] - \mathbb{E}^U[V'(X_0)]}{t} = \varrho' - \frac{1}{t}\int_0^t \mathbb{E}^U[\bar{c}(X_s, U_s)]\,\mathrm{d}s\,.$$

Letting $t \to \infty$, the ergodic cost (7.4.7) for (X, U) is ϱ'. Then we must have $\varrho' = \varrho^*$ and (X, U) must be optimal. Finally, if $\varrho' = \varrho^*$, it follows that $(V'(X_t) - \tilde{V}(X_t)\,,\ \mathfrak{F}_t^{X,U})$ is a bounded submartingale for any stationary optimal $\mathscr{L}(X, U)$, and therefore converges a.s., which is only possible if it is constant a.s. with respect to the stationary distribution. The last claim now follows in view of the continuity of $\tilde{V}$ and V'. □

7.4.4 Extension to the whole space

In this section we investigate (2.2.1), assuming that the partial nondegeneracy condition, Assumption 7.4.1, now holds on a bounded domain G_0, which without loss of generality is assumed C^2, and that σ is nonsingular in some bounded domain $G \Supset G_0$. We also assume that the running cost is bounded and Lipschitz continuous in x uniformly in $u \in \mathbb{U}$. In order to extend the results of Section 7.3 to the present setup, we need the following stronger stochastic Lyapunov condition:

Assumption 7.4.13 There exists a pair of nonnegative, inf-compact functions

$$\mathcal{V}_1 \in C^2(\mathbb{R}^d) \cap C_{\mathrm{pol}}(\mathbb{R}^d) \quad \text{and} \quad \mathcal{V}_2 \in C^2(\mathbb{R}^d)$$

such that $\mathcal{V}_1 \in \mathfrak{o}(\mathcal{V}_2)$, and for some $\beta > 0$,

$$\mathcal{L}^u \mathcal{V}_1 \le -1 \quad \text{and} \quad \mathcal{L}^u \mathcal{V}_2 \le -\beta \mathcal{V}_2 \quad \text{on } G_0^c, \quad \forall u \in \mathbb{U}.$$

Lemma 7.4.14 *Under Assumption 7.4.13 and provided $\mathbb{E}[\mathcal{V}_2(X_0)] < \infty$, the collection $\{\mathcal{V}_1(X_t) : t \ge 0\}$, with U ranging over all admissible controls is uniformly integrable.*

Proof By (2.5.11),

$$\mathbb{E}^U[\mathcal{V}_2(X_t)] \le \frac{k_0}{\beta} + \mathbb{E}^U[\mathcal{V}_2(X_0)] \qquad \forall t \ge 0,$$

for some $k_0 > 0$. The result follows since $\mathcal{V}_1 \in \mathfrak{o}(\mathcal{V}_2)$. □

We need the following technical lemma.

Lemma 7.4.15 *Under the assumption that σ is nonsingular in $G \Supset G_0$, we have $\tau(G_0^c) = \tau(\bar{G}_0^c)$ a.s.*

Proof By Lemma 2.3.7, we may consider $X_{\tau(\bar{G}_0^c)+t}$ in place of X_t and condition on $X_{\tau(\bar{G}_0^c)}$, so as to assume without any loss of generality that $X_0 = x \in \partial G_0$ (i.e., $\tau(\bar{G}_0^c) = 0$). Thus we need to show that $\tau(G_0^c) = 0$ a.s. Since ∂G_0 is C^2, a C^2 local change of coordinates in a neighborhood of x renders x the origin in $\mathbb{R}^d$ and an open neighborhood B of x intersects ∂G_0 in the unit open ball in the hyperplane $\{(x_1, \dots, x_d) : x_d = 0\}$, centered at x (i.e., the origin). X may no longer satisfy (2.2.1), but locally it is an Itô process with a uniformly nondegenerate diffusion matrix in a small open neighborhood $B \subset G$. It suffices to show that a.s., the dth component of this process takes both positive and negative values in a time interval $[0, \tau(B) \wedge \varepsilon)$ for every $\varepsilon > 0$. To see this, first set the drift term to zero for $t \in (0, \varepsilon)$ by an absolutely continuous change of measure using the Girsanov formula. The dth component is now purely a time-changed Brownian motion for $t \in [0, \tau(B) \wedge \varepsilon)$, for which this fact is well known. □

Select an open G_1 such that $G_0 \Subset G_1 \Subset G$. Define $\{\hat{\tau}_n\}$ as in Lemma 2.6.6 on p. 70, with G_0 and G_1 replacing D_0 and D_1, respectively. Then by Lemma 2.5.5, $\{\hat{\tau}_n\}$ have all their moments finite and bounded uniformly over $\mathfrak{U}$ and over initial values lying in compact subsets of $\mathbb{R}^d$. Let $\tilde{X}_n = X_{\hat{\tau}_{2n}}$. As in Section 7.4.2, $\tilde{X}$ is a controlled Markov chain on ∂G_0 with state dependent control space

$$\tilde{A}_x := \{\mathscr{L}(X_{t\wedge\hat{\tau}_2}, U_{t\wedge\hat{\tau}_2}, t \ge 0) : X_0 = x, \ U \in \mathfrak{U}\}, \quad x \in \partial G_0.$$

Lemma 7.4.16 *The map $\partial G_0 \ni x \mapsto \tilde{A}_x$ is compact, convex-valued and upper semicontinuous.*

Proof Let $(\hat{X}_t, \hat{U}_t) := (X_{t\wedge\hat{\tau}_2}, U_{t\wedge\hat{\tau}_2})$. An element $\mathscr{L}(\hat{X}, \hat{U})$ is specified by its martingale characterization: for all $f \in C_b^2(\mathbb{R}^d)$, $g \in C_b(C([0,s];\mathbb{R}^d) \times \mathscr{U}_{[0,s]})$, and $0 \le s < t$, it holds that

$$\mathbb{E}\left[\left(f(\hat{X}_t) - f(\hat{X}_s) - \int_s^t \mathcal{L}^{U_r} f(\hat{X}_r)\,\mathrm{d}r\right) g(\hat{X}_{[0,s]}, \hat{U}_{[0,s]})\right] = 0\,. \tag{7.4.15}$$

If two probability measures on $\mathscr{P}(C([0,\infty);\mathcal{R}^d) \times \mathscr{U})$ are concentrated on paths stopped at the corresponding $\hat{\tau}_2$, so is their convex combination. Also, (7.4.15) holds under convex combinations. Hence $\tilde{A}_x$ is convex. The set of laws $\mathscr{L}(X, U)$ with $X_0 = x$ is compact by Corollary 2.3.9. Thus $\tilde{A}_x$ is tight. Consider a sequence of solutions (X^n, U^n) with $X_0^n = x$, $n \in \mathbb{N}$, and suppose $\mathscr{L}(X^n, U^n) \to \mathscr{L}(X^\infty, U^\infty)$. By Lemma 7.4.15, $\hat{\tau}_2$ is an a.s. continuous function of the trajectories, while by Lemmas 2.5.1 and 7.4.14 it is uniformly integrable over $U \in \mathfrak{U}$ and initial conditions in a compact set. Thus applying (7.4.15) with $X = X^n$, $U = U^n$, $n \ge 1$, we can pass to the limit as $n \to \infty$ to conclude that (7.4.15) holds for $\mathscr{L}(X^\infty, U^\infty)$. Hence $\tilde{A}_x$ is compact. Upper semicontinuity follows by a similar argument by considering a sequence $\mathscr{L}(X^n, U^n)$ with $X_0^n = x_n \in \partial G_0$, and $x_n \to x_\infty$. □

Let $\tilde{P}^\nu(x, \mathrm{d}y)$ denote the controlled transition kernel for $\tilde{X}$, with $\nu \in \tilde{A}_x$.

Lemma 7.4.17 *The map $(x, \nu) \mapsto \tilde{P}^\nu(x, \mathrm{d}y)$ is continuous.*

Proof Suppose (x_n, ν_n) is a sequence converging to $(x_\infty, \nu_\infty) \in \cup_{x\in\partial G_0}\{x\} \times \tilde{A}_x$ as $n \to \infty$, with $(\tilde{X}^{(n)}, \tilde{Z}^{(n)})$ denoting the corresponding controlled processes. Thus $\tilde{X}_0^{(n)} = x_n$ and $\tilde{Z}_0^{(n)} = \nu_n$ for all n. Arguing as in Lemma 7.4.16 we conclude that $\tilde{X}^{(n)} \to \tilde{X}^{(\infty)}$ in law. □

Arguments analogous to the proofs of Theorem 7.4.5 and Lemma 7.4.7 show that there exists $\tilde{V} \in C(\partial G_0)$, and $\tilde{\varrho} \in \mathbb{R}$ such that

(i) $\tilde{\varrho}$ is the optimal cost for the control problem that seeks to minimize

$$\limsup_{T\to\infty} \frac{1}{T} \int_0^T \mathbb{E}_x^U[\bar{c}(X_t, U_t)]\,\mathrm{d}t\,,$$

or equivalently [104, theorem 1],

$$\limsup_{N\to\infty} \frac{\sum_{n=0}^{N-1} \mathbb{E}\left[\tilde{c}(\tilde{X}_n, \tilde{Z}_n)\right]}{\sum_{n=0}^{N-1} \mathbb{E}\left[\hat{\tau}_{2n+2} - \hat{\tau}_{2n}\right]}$$

over all admissible $\tilde{Z}$, with

$$\tilde{c}(x, \nu) := \mathbb{E}_x^\nu \left[\int_0^{\hat{\tau}_2} \bar{c}(X_t, U_t)\,\mathrm{d}t\right],$$

where the expectation $\mathbb{E}_x^\nu$ is with respect to $\nu = \mathscr{L}(X_{t\wedge\hat{\tau}_2}, U_{t\wedge\hat{\tau}_2},\ t \ge 0) \in \tilde{A}_x$.

(ii) $\tilde{V}$ satisfies [104, theorem 2]

$$\tilde{V}(x) = \min_{\nu\in\tilde{A}_x} \left[\tilde{c}(x, \nu) - \tilde{\varrho}\beta(x, \nu) + \int_{\partial G_0} \tilde{P}^\nu(x, \mathrm{d}y)\tilde{V}(y)\right], \tag{7.4.16}$$

where $\beta(x, \nu) = \mathbb{E}_x^\nu[\hat{\tau}_2]$.

(iii) $\tilde{\varrho} = \varrho^*$ [104, theorem 2].

We extend $\tilde{V}$ to $\mathbb{R}^d$ by defining, for $x \in G_0 \cup \bar{G}_0^c$,

$$\begin{aligned} V(x) &:= \inf_{U \in \mathfrak{U}} \tilde{J}(x, U)\,, \\ \tilde{J}(x, U) &:= \mathbb{E}_x^U \left[\int_0^{\tilde{\tau}_0} (\bar{c}(X_s, U_s) - \varrho^*)\,\mathrm{d}s + \tilde{V}(X_{\tilde{\tau}_0}) \right], \end{aligned} \tag{7.4.17}$$

where $\tilde{\tau}_0 := \tau(G_0 \cup \bar{G}_0^c)$. We then have the following lemma.

Lemma 7.4.18 (i) *V is continuous and the infimum in* (7.4.17) *is attained.*
(ii) $\{V(X_t) : t \geq 0\}$ *is uniformly integrable.*

Proof Note that V takes the form

$$V(x) = \inf_{\breve{A}_x} \mathbb{E}_x^U \left[\int_0^{\tilde{\tau}_0} (\bar{c}(X_s, U_s) - \varrho^*)\,\mathrm{d}s + \tilde{V}(X_{\tilde{\tau}_0}) \right],$$

where

$$\breve{A}_x := \{\mathscr{L}(X_{t \wedge \tilde{\tau}_0}\,,\ U_{t \wedge \tilde{\tau}_0}\,,\ t \geq 0) : X_0 = x\,,\ U \in \mathfrak{U}\}\,, \quad x \in \mathbb{R}^d\,.$$

Arguments directly analogous to the proof of Lemma 7.4.16 show that $x \mapsto \breve{A}_x$ is compact, convex-valued and upper semicontinuous. Therefore in order to complete the proof of the first part of the lemma it remains to show that $(x, U) \mapsto \tilde{J}(x, U)$ is continuous. Indeed, let (X^n, U^n) be a sequence, with $X_0^n = x_n$, which converges to (X^∞, U^∞) in law, while $x_n \to x_\infty \in \mathbb{R}^d$ as $n \to \infty$. By Lemma 7.4.15, $\tilde{\tau}_0$ is an a.s. continuous function of the trajectories. By Assumption 7.4.13,

$$\mathcal{L}^u(\mathcal{V}_1 + \mathcal{V}_2) \leq -1 - \beta \mathcal{V}_2 \quad \text{on } G_0^c\,, \quad \forall u \in \mathbb{U}\,.$$

Moreover, since $\mathcal{V}_1 \in \mathfrak{o}(\mathcal{V}_2)$, there exists $\beta_0 > 0$ such that $1 + \beta \mathcal{V}_2 \geq \beta_0 \mathcal{V}_1$ on G_0^c. Therefore, by Lemma 2.5.1,

$$\mathbb{E}_x^U\,[\tilde{\tau}_0^2] \leq \frac{2}{\beta_0}(\mathcal{V}_1(x) + \mathcal{V}_2(x)) \qquad \forall x \in \bar{G}_0^c\,, \quad \forall U \in \mathfrak{U}\,,$$

and uniform integrability of $\tilde{\tau}_0$ over $U \in \mathfrak{U}$ and initial conditions in a compact subset of $\bar{G}_0^c$ follows. Also, since σ has full rank on $G \Supset G_0$, $\mathbb{E}_x^U\,[\tilde{\tau}_0^2]$ is bounded uniformly in $U \in \mathfrak{U}$ and $x \in G_0$ by Theorem 2.6.1 (b). It follows that $\tilde{\tau}_0$ is uniformly integrable over $U \in \mathfrak{U}$ and initial conditions in a compact subset of $\mathbb{R}^d$. Since c is continuous and bounded and $\tilde{V} \in C(\partial G_0)$, it easily follows that $\tilde{J}(x_n, U^n) \to \tilde{J}(x, U^\infty)$, establishing the continuity of $\tilde{J}$.

For part (ii), note that Lemma 2.5.1, Assumption 7.4.13 and (7.4.17) imply that $V \in \mathscr{O}(\mathcal{V}_1)$. The result then follows by Lemma 7.4.14. □

Fix $X_0 = x$ and consider the admissible control U^* obtained by patching up the minimizer in (7.4.17) defined on $[0, \tilde{\tau}_0]$ with the optimal processes on $[\hat{\tau}_{2n}, \hat{\tau}_{2n+2}]$, $n \geq 0$, obtained from (7.4.16).

Lemma 7.4.19 *Let $0 \le \mathfrak{s} \le \mathfrak{t}$ be two bounded stopping times. Then, with $\mathfrak{s}_n = \mathfrak{s} \wedge \hat{\tau}_{2n}$ and $\mathfrak{t}_n = \mathfrak{t} \wedge \hat{\tau}_{2n+2}$, we have*

$$\mathbb{E}^U \left[\int_{\mathfrak{s}_n}^{\mathfrak{t}_n} (\bar{c}(X_s, U_s) - \varrho^*)\, ds + V(X_{\mathfrak{t}_n}) \,\middle|\, \mathfrak{F}^{X,U}_{\mathfrak{s}_n} \right] \ge V(X_{\mathfrak{s}_n}), \quad \mathbb{P}^U\text{-a.s.}, \tag{7.4.18}$$

for all $n \ge 0$ and $U \in \mathfrak{U}$, with equality when $U = U^*$.

Proof In view of Lemma 7.4.18, this follows by a standard dynamic programming argument. □

Taking expectations in (7.4.18) and letting $n \uparrow \infty$, we obtain

$$\mathbb{E}^U \left[\int_{\mathfrak{s}}^{\mathfrak{t}} (\bar{c}(X_s, U_s) - \varrho^*)\, ds + V(X_{\mathfrak{t}}) \right] \ge \mathbb{E}^U \left[V(X_{\mathfrak{s}}) \right],$$

with equality for $U = U^*$. It then follows that

$$M_t := V(X_t) + \int_0^t (\bar{c}(X_s, U_s) - \varrho^*)\, ds, \quad t \ge 0, \tag{7.4.19}$$

is an $(\mathfrak{F}^{X,U}_t)$-submartingale under $\mathbb{P}^U$, and it is a martingale for $U = U^*$. Following the proof of Theorem 7.3.11 yields the following characterization:

Theorem 7.4.20 *Let M be the process in* (7.4.19). *Then*

(i) *M is an $(\mathfrak{F}^{X,U}_t)$-submartingale under $\mathbb{P}^U$ for all $U \in \mathfrak{U}$;*
(ii) *if M is an $(\mathfrak{F}^{X,U}_t)$-martingale, under some $U \in \mathfrak{U}$, then U is optimal;*
(iii) *if (X, U) is stationary optimal, then M is an $(\mathfrak{F}^{X,U}_t)$-martingale.*

Theorem 7.4.21 *Suppose $\sigma^{ij} \in C^2(\mathbb{R}^d) \cap C^{0,1}(\mathbb{R}^d)$. Then (V, ϱ^*) in* (7.4.17) *is the unique (up to a constant) viscosity solution to* (7.3.9) *in $C_{\mathrm{pol}}(\mathbb{R}^d) \times \mathbb{R}$.*

Proof For $x \in \mathbb{R}^d$ let $B(x)$ be an open ball centered at x, such that the family $\tau(B(x))$ is uniformly integrable over $\{\mathbb{P}^U_y : y \in B(x), U \in \mathfrak{U}\}$. It is always possible to select such a ball. Indeed if $x \in G_1$, and $B(x) \Subset G$, then uniform integrability follows from the nondegeneracy hypothesis in G, while if $x \in G_1^c$ and $B(x) \Subset \bar{G}_0^c$, then uniform integrability follows from the Lyapunov condition in Assumption 7.4.13. By Theorem 7.4.20 and a dynamic programming argument

$$V(y) = \min_{U \in \mathfrak{U}} \mathbb{E}^U_y \left[\int_0^{\tau(B(x))} (\bar{c}(X_s, U_s) - \varrho^*)\, ds + V(X_{\tau(B(x))}) \right] \qquad \forall y \in B(x).$$

Treating the expectation on the right-hand side as a cost functional for a control problem on a bounded domain, it follows that V is the value function and hence is a viscosity solution of (7.3.9) in $B(x)$ for any $x \in \mathbb{R}^d$. Thus (V, ϱ^*) is a viscosity solution of (7.3.9) on $\mathbb{R}^d$. In view of Lemma 7.4.18 and the polynomial growth of $\mathcal{V}_1$ in Assumption 7.4.13, uniqueness follows as in Theorems 7.3.10 and 7.3.13 upon noting that the support of the marginal on $\mathbb{R}^d$ of any optimal ergodic occupation measure corresponding to an ergodic (X, U) must contain G_0. □

7.4.5 Extensions and implications

We can extend our approach to the ergodic control problem to various other systems. For example consider the model in Section 4.4.1 for the controlled diffusion in $\mathbb{R}^d$ with periodic coefficients in the degenerate case

$$dX_t = \bar{b}(X_t, U_t)\,dt + \sigma(X_t)\,dW_t\,. \tag{7.4.20}$$

For each $i = 1, 2, \dots, d$, let $b(x, \cdot)$, $\sigma(x)$ and $c(x, \cdot)$ be periodic in x_i, with period T_i. In such a situation the state space $\mathbb{R}^d$ is viewed as a d-dimensional torus given by

$$\mathbb{T} := \mathbb{R}^d / \Pi_{i=1}^d (T_i \mathbb{Z})\,.$$

Since the torus is a compact manifold without boundary, our method can be applied to treat this problem. In fact our approach is applicable to a wide variety of problems as long as the passage from continuous time to discrete time problem and back is feasible and a dynamic programming principle is available for the discrete time problem.

In the rest of this section, we sketch some implications to the problem of existence of optimal Markov controls. The experience with discrete state or time ergodic control would lead us to expect the existence of an optimal (X, U) where X is a time-homogeneous Markov process and U is a stationary Markov control. This is indeed possible here, as argued below.

Without loss of generality assume that $T_i = 1$, $i = 1, \dots, d$. Let $\tau_0 = 0$ and $\{\tau_n\}$ be successive jump times of a Poisson process with rate 1, independent of all other processes under consideration. Let $\hat{X}_n = X_{\tau_n}$, $n \ge 0$, and for $x \in \mathbb{T}$,

$$\begin{aligned}\hat{A}_x = \{\mathscr{L}(X, U) : (X, U) &\text{ satisfies (7.4.1)–(7.4.2) up to an independent} \\ &\text{killing time } \tau \text{ which is exponential with parameter } 1\}\,.\end{aligned}$$

For $\xi = \mathscr{L}(X, U) \in \hat{A}_x$, let $\hat{p}(x, \xi; dy)$ denote the law of X_τ. Then $\{\hat{X}_n\}$ is a controlled Markov process on $\mathbb{T}$ with control space $\hat{A}_x$ at state x and controlled transition kernel $\hat{p}(\,\cdot\,, \cdot\,; dy)$. Assume that that σ is nonsingular in an open subset $G_0 \subset \mathbb{T}$, and for some $\tilde{C}_0 > 0$, the inequality

$$\mathbb{P}\bigl(\hat{X}_1 \in G_0 \cap A \bigm| \hat{X}_0 = x\bigr) \ge \tilde{C}_0\, \mathfrak{m}(G_0 \cap A)$$

holds for any Borel $A \subset \mathbb{T}$. Now we may repeat the arguments of Section 7.4.2 with minor modifications to obtain the following counterpart of (7.4.8): There exists $\hat{V} \in C(\mathbb{T})$ such that

$$\hat{V}(x) = \min_{\xi \in \hat{A}_x} \left[\hat{c}(x, \xi) - \varrho + \int \hat{p}(x, \xi; dy) \hat{V}(y) \right],$$

where

$$\hat{c}(x, \xi) = \mathbb{E}\left[\int_0^\tau \bar{c}(X_s, U_s)\,ds \Bigm| X_0 = x \right],$$

the expectation being w.r.t. ξ. Now consider the problem of minimizing

$$\mathbb{E}\left[\int_0^\tau (\bar{c}(X_s, U_s) - \varrho)\,\mathrm{d}s + \hat{V}(X_\tau)\right], \tag{7.4.21}$$

where τ is exponential with parameter 1, independent of (X, U). For this problem, a straightforward adaptation of Krylov's Markov selection procedure (see Theorem 6.7.2) allows us to obtain a family $\{\mathbb{P}_x^v : x \in \mathbb{T}\} \subset \mathscr{P}(C([0,\infty); \mathbb{T}))$ corresponding to a control $v\colon \mathbb{T} \to \mathscr{P}(\mathbb{U})$ such that:

(i) for each x, $P_x^v = \mathscr{L}(X)$ for X satisfying (7.4.1)–(7.4.2) with $U_t = v(X_t)$;
(ii) for each x, the law $\mathscr{L}(X, v)$ is optimal for the problem of minimizing (7.4.21);
(iii) the set $\{\mathbb{P}_x^v : x \in \mathbb{T}\}$ is a strong Markov family. In particular, it satisfies the Chapman–Kolmogorov equations.

It is straightforward to show that for an optimal solution $(X_t, v(X_t),\ t \geq 0)$ as above

$$\left(\hat{V}(X_{t\wedge\tau}) + \int_0^{t\wedge\tau} [\bar{c}(X_s, v(X_s)) - \varrho]\,\mathrm{d}s\,,\ \mathfrak{F}_{t\wedge\tau}^X\right)$$

is a martingale. By concatenating such optimal solutions on $[\tau_n, \tau_{n+1}]$, $n \geq 0$, we construct a process (X, U) such that

$$\left(\hat{V}(X_t) + \int_0^t [\bar{c}(X_s, v(X_s)) - \varrho]\,\mathrm{d}s,\ \mathfrak{F}_t^X\right)$$

is a martingale and (7.4.1)–(7.4.2) holds with $U_t = v(X_t)$ for all $t \geq 0$. Arguing as in Theorem 7.4.12, we conclude that this is an optimal solution.

Theorem 7.4.22 *For the controlled diffusion with periodic coefficients in* (7.4.20), *under the assumption that σ is not everywhere singular, there exists an optimal time-homogeneous Markov solution corresponding to a stationary Markov control. This solution inherits the periodic structure of the problem data.*

Theorem 7.4.22 uses only the martingale dynamic programming principle, so it also applies to the asymptotically flat and partially degenerate (on a bounded set as well as the whole space) cases studied earlier. Contrast this with the existence results proved in Sections 6.5 and 7.2 for the ergodic control problem for degenerate diffusions in $\mathbb{R}^d$ or more general state spaces. Two kinds of results are available: one can show the existence of an optimal stationary solution X corresponding to a stationary Markov control, but X is not guaranteed to be a Markov process. Alternatively, one has the existence of an optimal X that is a Markov process with U a stationary Markov control, but X is not guaranteed to be either a time-homogeneous Markov process or a stationary process. Thus both results fall short of the ideal. Here, one is able to get some improvement on the latter insofar as one has an optimal time-homogeneous Markov process X corresponding to a stationary control, but this X has not been proven to be stationary.

7.5 Bibliographical note

Section 7.2. The existence results are based on the corresponding results for the controlled martingale problem in [21].

Section 7.3. This is based on [11], with some corrections.

Section 7.4. This is based on [40], with several modifications and extensions. For related work see [8].

8

Controlled Diffusions with Partial Observations

8.1 Introduction

In this chapter we study the ergodic control problem for diffusions with partial observations. The system state is governed by the Itô stochastic differential equation

$$dX_t = b(X_t, U_t)\,dt + \sigma(X_t)\,dW_t\,, \quad t \geq 0\,, \tag{8.1.1}$$

where b, σ and W are as in Section 2.2. The process X in this context is often referred to as the signal process and is unavailable for control purposes. Instead one has to base the control decisions upon a related *observations process* Y described by

$$dY_t = h(X_t)\,dt + dW'_t\,, \quad Y_0 = 0\,,$$

where $h\colon \mathbb{R}^d \to \mathbb{R}^m$, and W' is a standard m-dimensional Wiener process independent of (W, X_0). In the simplest formulation, the controller is required to select a control process U_t, $t \geq 0$, adapted to the natural filtration (i.e., the right-continuous completion of the σ-fields) generated by Y. These are the so-called *strict-sense admissible controls*. We shall expand the class of allowable controls later on. The objective is of course to minimize the ergodic cost

$$\limsup_{T\to\infty} \frac{1}{T}\, \mathbb{E}^U \left[\int_0^T \bar{c}(X_t, U_t)\,dt\right],$$

over the strict-sense admissible controls $U \in \mathfrak{U}$.

It so happens that even for relatively tame cost functionals such as the finite horizon cost, the existence of strict-sense admissible optimal controls remains an open issue. This has motivated a relaxation of this class to the so-called *wide-sense admissible* controls within which an existence result can be established [60]. We defer the definition of this class till after the appropriate machinery has been introduced.

Standard stochastic control methodology suggests that the correct state for this control problem should be the *sufficient statistics* given the observed quantities. The simplest choice at time t in this case (though not unique, as we shall soon see) is the regular conditional law π_t of X_t given the observed quantities $\{Y_s, U_s : s \leq t\}$. (Although for strict-sense admissible controls $\{U_s : s \leq t\}$ is completely specified when $\{Y_s : s \leq t\}$ is, we mention them separately in view of future relaxation of this requirement.) Note that π is a $\mathscr{P}(\mathbb{R}^d)$-valued process. Its evolution is given by the equations of nonlinear filtering described in the next section. These express π as a $\mathscr{P}(\mathbb{R}^d)$-valued controlled Markov process controlled by U. While the state space has now become more complex than the original, the big advantage of this formulation is that it is now a *completely observed* control problem, equivalent to the original partially observed control problem. This separates the two issues of estimation and control: the equations of nonlinear filtering represent a Bayesian state estimation scheme, which is followed by the pure control problem of controlling a completely observed Markov process, viz., the conditional law that is the output of the estimation scheme. For this reason the equivalent problem of controlling the conditional laws is called the *separated* control problem.

In rare cases (this includes the all-important Linear–Quadratic–Gaussian or LQG problem), one is able to find finite dimensional sufficient statistics. In other words, π_t is characterized completely by finitely many scalar processes. These then serve as an equivalent state description, reducing the problem to a finite dimensional controlled diffusion. For the LQG problem, the conditional law in question is Gaussian and hence completely characterized by the conditional mean and conditional covariance matrix. The former is described by the finite dimensional linear diffusion described by the Kalman–Bucy filter, and the latter by the associated (deterministic) Riccati equation. This should not come as a surprise – recall that when estimating one of two jointly Gaussian random variables in terms of the other, the best least squares estimate is the conditional mean, given as an affine function of the observed value of the latter random variable, and the corresponding conditional variance is deterministic.

We make the following assumptions throughout:

(i) b and σ satisfy (2.2.3) with a uniform Lipschitz constant over $\mathbb{R}^d$.
(ii) b and σ are bounded and $\sigma\sigma^{\mathsf{T}}$ is uniformly elliptic.
(iii) $h \in C^2(\mathbb{R}^d; \mathbb{R}^m)$ with linear growth for itself and its first and second order partial derivatives.
(iv) The running cost function $c \colon \mathbb{R}^d \times U \to \mathbb{R}$ is continuous and bounded from below.

8.2 Controlled nonlinear filters

Let $\{\mathcal{F}_t : t \geq 0\}$ and $\{\bar{\mathcal{F}}_t : t \geq 0\}$ denote the natural filtrations of (X, Y, U) and (Y, U), respectively. We may assume that $\mathfrak{F} = \bigvee_{t \geq 0} \mathcal{F}_t$, where $(\Omega, \mathfrak{F}, \mathbb{P})$ is the underlying probability space. We next derive the evolution equation for the probability measure-valued process π. For this purpose, define

$$\bar{\Lambda}_t := \exp\left(-\int_0^t \langle h(X_s), \mathrm{d}W'_s\rangle - \tfrac{1}{2}\int_0^t |h(X_s)|^2 \mathrm{d}s\right).$$

From our assumptions on b, σ, and h above, it follows by Theorem 2.2.2 that, for any $T > 0$,

$$\mathbb{E}_x^U\left[\int_0^T |h(X_s)|^2 \mathrm{d}s\right] < \infty,$$

uniformly over $\mathfrak{U}$ and x in compact subsets of $\mathbb{R}^d$. Then by Portenko's criterion (2.2.37) it follows that $\mathbb{E}[\bar{\Lambda}_t] = 1$ for all $t \geq 0$, and $(\bar{\Lambda}_t, \mathcal{F}_t), t \geq 0$, is an exponential martingale. Thus one can consistently define a new probability measure $\mathbb{P}_0$ on $(\Omega, \mathcal{F})$ as follows: if $\mathbb{P}_t, \mathbb{P}_{0t}$ denote the restrictions of $\mathbb{P}, \mathbb{P}_0$ to $(\Omega, \mathcal{F}_t)$, then

$$\frac{\mathrm{d}\mathbb{P}_{0t}}{\mathrm{d}\mathbb{P}_t} = \bar{\Lambda}_t .$$

From the martingale property, it follows that $\mathbb{P}_{0t}|_{\mathcal{F}_s} = \mathbb{P}_{0s}$ for $t > s$, whence the definition is consistent. Let $\mathbb{E}_0$ denote the expectation operator under $\mathbb{P}_0$. By the Cameron–Martin–Girsanov theorem, it follows that under $\mathbb{P}_0$, Y is a Wiener process independent of (W, X_0). One can also verify that

$$\Lambda_t := \frac{\mathrm{d}\mathbb{P}_t}{\mathrm{d}\mathbb{P}_{0t}} = \exp\left(\int_0^t \langle h(X_s), \mathrm{d}Y_s\rangle - \tfrac{1}{2}\int_0^t |h(X_s)|^2 \mathrm{d}s\right).$$

Furthermore, $\mathbb{E}_0[\Lambda_t] = 1$ for all $t \geq 0$, and $(\Lambda_t, \mathcal{F}_t), t \geq 0$, is an exponential martingale under $\mathbb{P}_0$. This change of measure facilitates our definition of wide-sense admissible controls.

Definition 8.2.1 We say that U is a *wide-sense admissible control* if under $\mathbb{P}_0$, the following holds:

(C8.1) $Y \circ \theta_t - Y_t$ is independent of $(\{U_s, Y_s : s \leq t\}, W, X_0)$.

We let $\mathfrak{U}_{\mathrm{ws}}$ denote the class of wide-sense admissible controls.

That is, the control does not anticipate the future increments of the Wiener process Y. Note that this includes strict-sense admissible controls. Since the specification of wide-sense admissible controls is in law, a wide-sense admissible control as above may be identified with the joint law of (Y, U). That is, the set $\mathfrak{U}_{\mathrm{ws}}$ can be identified with the set of probability measures on $C([0,\infty);\mathbb{R}^m) \times \mathscr{U}$ such that the marginal on $C([0,\infty);\mathbb{R}^m)$ is the Wiener measure and (C8.1) holds. Since independence is preserved under convergence in $\mathscr{P}(C([0,\infty);\mathbb{R}^m) \times \mathscr{U})$, it follows that this set is closed. Since $\mathscr{U}$ is compact and the marginal on $C([0,\infty);\mathbb{R}^m)$ is fixed as the Wiener measure, it is also tight and therefore compact. Strict-sense controls correspond to the situation where the regular conditional law on $\mathscr{U}$ given the Wiener trajectory in $C([0,\infty);\mathbb{R}^m)$ is Dirac, say δ_U, with the process U adapted to the natural filtration of the Wiener process.

From now on we use the notation $\mu(f)$ to denote $\int f\,\mathrm{d}\mu$ for a function f and a nonnegative measure μ. Also, let $\mathfrak{M}(\mathbb{R}^d)$ denote the space of finite nonnegative measures on $\mathbb{R}^d$ equipped with the coarsest topology that renders continuous the maps $\mu \in \mathfrak{M}(\mathbb{R}^d) \to \mu(f)$ for bounded continuous f. This is the same as the weak* topology on the space of finite signed measures relativized to $\mathfrak{M}(\mathbb{R}^d)$. Define the $\mathfrak{M}(\mathbb{R}^d)$-valued process p of the so-called *unnormalized conditional laws* of X_t given $\{\bar{\mathcal{F}}_t : t \geq 0\}$ by

$$p_t(f) := \mathbb{E}_0\left[f(X_t)\Lambda_t \mid \bar{\mathcal{F}}_t\right], \quad f \in C_b(\mathbb{R}^d).$$

Thus $p_0 = \pi_0$. By applying Itô's formula to $f(X_t)\Lambda_t$, taking conditional expectations, and using the fact that Y is adapted to $\{\bar{\mathcal{F}}_t\}$, one obtains the evolution equation for the process p:

$$p_t(f) = \pi_0(f) + \int_0^t p_s(\mathcal{L}^{U_s} f)\,\mathrm{d}s + \int_0^t \langle p_s(fh), \mathrm{d}Y_s\rangle \tag{8.2.1}$$

for $f \in C_b^2(\mathbb{R}^d)$.[1] The evolution equation in (8.2.1) is called the (controlled) *Duncan–Mortensen–Zakai equation* and displays p as an $\mathfrak{M}(\mathbb{R}^d)$-valued Markov process controlled by U. It is a linear stochastic PDE driven by the Wiener process Y when viewed under $\mathbb{P}_0$ (which is usually the case).

Observe that for $f \in C_b(\mathbb{R}^d)$,

$$\begin{aligned}
\pi_t(f) &= \mathbb{E}\left[f(X_t) \mid \bar{\mathcal{F}}_t\right] \\
&= \frac{\mathbb{E}_0\left[f(X_t)\Lambda_t \mid \bar{\mathcal{F}}_t\right]}{\mathbb{E}_0\left[\Lambda_t \mid \bar{\mathcal{F}}_t\right]} \qquad (8.2.2)\\
&= \frac{p_t(f)}{p_t(1)}, \qquad (8.2.3)
\end{aligned}$$

where "1" denotes the constant function identically equal to 1. Equation (8.2.2) is called the Kallianpur–Striebel formula and is an abstract Bayes formula. Equation (8.2.3) establishes a relationship between p and π and justifies the terminology *unnormalized conditional law* for p. Applying the Itô formula to (8.2.3) and organizing the resulting terms, one obtains the evolution equation for π:

$$\pi_t(f) = \pi_0(f) + \int_0^t \pi_s(\mathcal{L}^{U_s}(f))\,\mathrm{d}s + \int_0^t \langle \pi_s(fh) - \pi_s(f)\pi_s(h), \mathrm{d}I_s\rangle, \tag{8.2.4}$$

for $f \in C_b^2(\mathbb{R}^d)$. Here the products of a scalar and a vector are interpreted as componentwise products and

$$\begin{aligned}
I_t &:= Y_t - \int_0^t \pi_s(h)\,\mathrm{d}s \\
&= W'_t + \int_0^t (h(X_s) - \pi_s(h))\,\mathrm{d}s, \quad t \geq 0, \qquad (8.2.5)
\end{aligned}$$

is the so-called *innovations process*. Using the second expression in (8.2.5), it is easily verified that I_t is an $\{\mathcal{F}_t\}$-adapted process. Under $\mathbb{P}$, it is a continuous path zero mean martingale with quadratic variation $\langle I\rangle_t = t$, $t \geq 0$. (The latter is so because it differs from W' by a bounded variation term.) By Levy's characterization

[1] Conditioning inside the stochastic integral is justified by a standard approximation argument using piecewise constant approximations of the integrand.

of Wiener processes [19], it follows that it is a Wiener process under $\mathbb{P}$. Thus under $\mathbb{P}$, (8.2.4) is a stochastic PDE driven by the Wiener process I and displays π as a $\mathscr{P}(\mathbb{R}^d)$-valued controlled Markov process, with control U. This is the (controlled) Fujisaki–Kallianpur–Kunita equation of nonlinear filtering. It should be mentioned that, in the uncontrolled case, the innovations process I generates the same filtration as Y [2]. This is expected from its interpretation as the process that captures the *innovations*, or incremental new information from the observations. In fact, for the discrete time model, the innovations process can be viewed as the output of a Gram–Schmidt orthonormalization procedure in the space of square-integrable random variables applied to the observations process.

The equation

$$\begin{aligned} p_t(1) &= \pi_0(1) + \int_0^t \langle p_s(h), \mathrm{d}Y_s\rangle \\ &= 1 + \int_0^t p_s(1)\langle \pi_s(h), \mathrm{d}Y_s\rangle \end{aligned}$$

has the unique solution

$$p_t(1) = \exp\left(\int_0^t \langle \pi_s(h), \mathrm{d}Y_s\rangle - \frac{1}{2}\int_0^t |\pi_s(h)|^2 \mathrm{d}s\right),$$

whence

$$p_t(f) = \pi_t(f) \exp\left(\int_0^t \langle \pi_s(h), \mathrm{d}Y_s\rangle - \frac{1}{2}\int_0^t |\pi_s(h)|^2 \mathrm{d}s\right), \quad f \in C_b(\mathbb{R}^d). \tag{8.2.6}$$

This specifies the process p in terms of the process π. Thus the two are interconvertible, qualifying p as an alternative state variable for posing the control problem.

We introduce yet another equivalent process, the Clark–Davis pathwise filter. This is the $\mathfrak{M}(\mathbb{R}^d)$-valued process ν defined by

$$\nu_t(f) = \mathbb{E}_0\left[e^{-\langle Y_t, h(X_t)\rangle} \Lambda_t f(X_t) \,\middle|\, \bar{\mathcal{F}}_t\right], \quad f \in C_b(\mathbb{R}^d).$$

Then

$$\begin{aligned} \nu_t(f) &= p_t\left(e^{-\langle Y_t, h(\cdot)\rangle} f\right), \\ p_t(f) &= \nu_t\left(e^{\langle Y_t, h(\cdot)\rangle} f\right). \end{aligned} \tag{8.2.7}$$

Thus this is yet another equivalent state variable. In order to derive its evolution equation, define

$$\tilde{L}_s^U f := \mathcal{L}^{U_s} f - \langle \nabla f, \sigma(x)\sigma^{\mathsf{T}}(x)J^{\mathsf{T}}(x)Y_s\rangle,$$

$$q(x, y, u) := -\langle y, J(x)b(x, u) + \ell(x)\rangle - \frac{1}{2}|h(x)|^2 + \frac{1}{2}\langle y, J(x)\sigma(x)\sigma^{\mathsf{T}}(x)J^{\mathsf{T}}(x)y\rangle,$$

where

$$J := \text{the Jacobian matrix of } h\,,$$

$$H_i := \text{the Hessian matrix of } h_i\,, \quad 1 \le i \le m\,,$$

$$\ell_i := \frac{1}{2} tr\left(\sigma^{\mathsf{T}} H_i \sigma\right), \quad 1 \le i \le m\,,$$

$$\ell := [\ell_1, \dots, \ell_m]^{\mathsf{T}}\,.$$

Then using the definition of ν, (8.2.1), and the Kunita–Itô formula for composition of semimartingales [80, equation (1.2)], it follows that

$$\nu_t(f) = \pi_0(f) + \int_0^t \nu_s(\tilde{L}_s^U f)\,\mathrm{d}s + \int_0^t \nu_s(q(\,\cdot\,, Y_s, U_s)f)\,\mathrm{d}s \tag{8.2.8}$$

for $f \in C_b^2(\mathbb{R}^d)$. This is the evolution equation for ν, known as the Clark–Davis pathwise filter. It stands apart from (8.2.1) and (8.2.4) in that it is not a stochastic PDE, but a deterministic PDE (i.e., a PDE not involving a stochastic integral term) with trajectories of Y, U featuring as random parameters. Note that

$$\nu_t(f) = \tilde{E}_0\left[f(X_t)\exp\left(\int_0^t q(X_s, Y_s, U_s)\,\mathrm{d}s\right)\right], \quad f \in C_b(\mathbb{R}^d), \tag{8.2.9}$$

where X is the time-inhomogeneous diffusion with extended generator $\tilde{\mathcal{L}}_s^U$. The latter is parameterized by the random parameters Y and U. Also, $\tilde{\mathbb{E}}_0$ denotes the expectation taken over the law of X when Y and U are treated as fixed parameters (we use $\tilde{\mathbb{P}}_0$ to denote the corresponding probabilities). For b and σ bounded and $h \in C_b^2(\mathbb{R}^d)$, ν_t, $t > 0$, is mutually absolutely continuous w.r.t. the Lebesgue measure on $\mathbb{R}^d$ [62]. By (8.2.3) and (8.2.6), (8.2.7) it follows that π_t, p_t, ν_t for $t > 0$ are mutually absolutely continuous w.r.t. each other and therefore w.r.t. the Lebesgue measure.

An alternative derivation of (8.2.8) goes as follows:

$$\begin{aligned}\nu_t(f) &= p_t\left(e^{-\langle Y_t, h(\cdot)\rangle} f\right)\\ &= \mathbb{E}_0\left[\exp\left(-\langle Y_t, h(X_t)\rangle + \int_0^t \langle h(X_s), \mathrm{d}Y_s\rangle - \frac{1}{2}\int_0^t |h(X_s)|^2 \mathrm{d}s\right) f(X_t) \,\middle|\, \bar{\mathcal{F}}_t\right],\end{aligned}$$

for $f \in C_b(\mathbb{R}^d)$. Apply the integration by parts formula to $\int_0^t \langle h(X_s), \mathrm{d}Y_s\rangle$ and then change the measure $\mathbb{E}_0\left[\,\cdot \mid \bar{\mathcal{F}}_t\right]$ to $\tilde{E}_0$ (treating Y and U as fixed parameters) to obtain (8.2.9). Then (8.2.8) follows from (8.2.9) by Itô's formula. In fact, (8.2.9) is the *Feynman–Kac formula* associated with the controlled extended generator $\tilde{L}_s^U$ and the potential q.

Concerning the uniqueness of solutions to the evolution equations for π, p and ν, with prescribed (Y, U), it follows from (8.2.3), (8.2.6) and (8.2.7) that it suffices to prove uniqueness for any one of them. For the controlled case considered here, ν appears to be the most convenient. See Haussmann [71] for one such result. For bounded b and σ, and h in $C_b^2(\mathbb{R}^d)$, uniqueness follows from standard results for the Cauchy problem for parabolic equations [83]. We do not address the uniqueness issue for unbounded data. See Xiong [124] and references therein for an extensive account of well-posedness results for *uncontrolled* nonlinear filters.

The final important fact about these processes that we need is the following conversion formulas.

Theorem 8.2.2 *For $f \in C_b(\mathbb{R}^d \times U)$ and (X, U) as above,*

$$\begin{aligned}\mathbb{E}\left[f(X_t, U_t)\right] &= \mathbb{E}\left[\pi_t(f(\cdot\,, U_t))\right]\\ &= \mathbb{E}_0\left[p_t(f(\cdot\,, U_t))\right]\\ &= \mathbb{E}_0\left[\nu_t\left(f(\cdot\,, U_t)e^{\langle Y_t, h(\cdot)\rangle}\right)\right].\end{aligned}$$

Proof The first and the last equalities are clear. For the second, we have

$$\begin{aligned}\mathbb{E}\left[\pi_t(f(\cdot\,, U_t))\right] &= \mathbb{E}_0\left[f(\cdot\,, U_t)\Lambda_t)\right]\\ &= \mathbb{E}_0\left[p_t(f(\cdot\,, U_t))\right]. \qquad \square\end{aligned}$$

Theorem 8.2.2 allows us to express the ergodic cost in various equivalent ways, viz.,

$$\begin{aligned}\limsup_{T\to\infty} \frac{1}{T}\int_0^T \mathbb{E}\left[c(X_t, U_t)\right]\mathrm{d}t &= \limsup_{T\to\infty} \frac{1}{T}\int_0^T \mathbb{E}\left[\pi_t(c(\cdot\,, U_t))\right]\,\mathrm{d}t\\ &= \limsup_{T\to\infty} \frac{1}{T}\int_0^T \mathbb{E}_0\left[p_t(c(\cdot\,, U_t))\right]\,\mathrm{d}t\\ &= \limsup_{T\to\infty} \frac{1}{T}\int_0^T \mathbb{E}_0\left[\nu_t\left(c(\cdot\,, U_t)e^{\langle Y_t, h(\cdot)\rangle}\right)\right]\mathrm{d}t.\end{aligned}$$

The three expressions on the right-hand side are convenient for use with π, p, and ν as the state process with $\mathbb{P}$, $\mathbb{P}_0$, and $\mathbb{P}_0$ the operative probability measure, respectively. We shall stick to the first of these possibilities in what follows.

8.3 The separated control problem

Let $\tilde{c}\colon \mathscr{P}(\mathbb{R}^d) \times \mathbb{U} \to \mathbb{R}$ be defined by

$$\tilde{c}(\pi, u) := \int_{\mathbb{R}^d} c(x, u)\,\pi(\mathrm{d}x)\,.$$

Consider the separated control problem of controlling the process π governed by (8.2.4) so as to minimize the ergodic cost

$$\limsup_{T\to\infty} \frac{1}{T}\int_0^T \mathbb{E}\left[\tilde{c}(\pi_t, U_t)\right]\,\mathrm{d}t \tag{8.3.1}$$

over all $U \in \mathfrak{U}_{\mathrm{ws}}$.

We first verify that this is a special instance of the controlled martingale problem studied in Chapter 6. To this end, let

$$\mathscr{D}(\mathcal{A}) = \Bigl\{ f \in C_b(\mathscr{P}(\mathbb{R}^d)) : f(\mu) = g\bigl(\textstyle\int f_1\,\mathrm{d}\mu, \dots, \int f_n\,\mathrm{d}\mu\bigr) \quad \forall \mu\,, \text{ for some } n \ge 1,$$
$$g \in C_c^\infty(\mathbb{R}^n), \text{ and } f_1, \dots, f_n \in \mathscr{D}(\mathcal{L}) \Bigr\},$$

where $\mathcal{L} = \mathcal{L}^u$ is as usual, with $u \in \mathbb{U}$ treated as a parameter. Define the operator $\mathcal{A}\colon \mathscr{D}(\mathcal{A}) \to C_b(\mathscr{P}(\mathbb{R}^d) \times \mathbb{U})$ by

$$\mathcal{A}f(\mu, u) = \sum_{i=1}^{n} \frac{\partial g}{\partial x_i} \bigl(\textstyle\int f_1\,\mathrm{d}\mu, \dots, \int f_n\,\mathrm{d}\mu\bigr) \mu(\mathcal{L}f_i(\,\cdot\,, u))$$
$$+ \frac{1}{2} \sum_{i,j=1}^{n} \frac{\partial^2 g}{\partial x_i \partial x_j} \bigl(\textstyle\int f_1\,\mathrm{d}\mu, \dots, \int f_n\,\mathrm{d}\mu\bigr)$$
$$\langle \mu(f_i h) - \mu(f_i)\mu(h),\, \mu(f_j h) - \mu(f_j)\mu(h) \rangle\,.$$

Under the hypotheses on the drift and diffusion coefficients of X, the assumptions (A6.1)–(A6.3) of Chapter 6 on p. 220 do hold for $\mathcal{A}$. We claim that assumption (A6.4) on p. 232 also holds. To prove this, let $f \in C^2(\mathbb{R}^d)$ be an inf-compact function with bounded first and second partial derivatives, and let $K > 0$ be a bound of $|\mathcal{L}^u f|$. For any $T > 0$, (8.2.4) yields

$$\sup_{t \in [0,T]} |\pi_t(f)| \le K(1 + T) + \sup_{t \in [0,T]} \left| \int_0^t \langle \varphi_s, \mathrm{d}I_s \rangle \right| ,$$

where $\varphi_s = \pi_s(fh) - \pi_s(f)\pi_s(h)$. Thus for any $M > 0$,

$$\inf_{U \in \mathfrak{U}} \mathbb{P}^U \biggl(\sup_{t \in [0,T]} |\pi_t(f)| > M \biggr) \le \inf_{U \in \mathfrak{U}} \mathbb{P}^U \biggl(\sup_{t \in [0,T]} \left| \int_0^t \langle \varphi_s, \mathrm{d}I_s \rangle \right| > M + K(1 + T) \biggr)$$
$$\le \frac{C(T)}{(M + K(1 + T))^2}\,,$$

for some constant $C(T)$. Here, the last inequality follows from (2.2.11) and (2.2.16a) after a routine calculation. Since $\bigl\{\mu \in \mathscr{P}(\mathbb{R}^d) : \int f\,\mathrm{d}\mu < M\bigr\}$ is compact in $\mathscr{P}(\mathbb{R}^d)$ for any inf-compact function f and any $M > 0$, this verifies (A6.4).

Additionally we consider one of the conditions (C7.1) or (C7.2) on p. 254. We need the following lemma:

Lemma 8.3.1 *Let E be a Polish space, and for $\xi \in \mathscr{P}(\mathscr{P}(E))$, define $\bar{\xi} \in \mathscr{P}(E)$ by:*

$$\bar{\xi}(A) = \int_{\mathscr{P}(E)} \nu(A)\, \xi(\mathrm{d}\nu)\,, \quad A \in \mathscr{B}(E)\,.$$

For $B \subset \mathscr{P}(\mathscr{P}(E))$, define $\bar{B} := \{\bar{\xi} : \xi \in B\} \subset \mathscr{P}(E)$. Then $B \subset \mathscr{P}(\mathscr{P}(E))$ is tight if and only if $\bar{B} \subset \mathscr{P}(E)$ is tight.

Proof Suppose $\bar{B}$ is tight but B is not. By Urysohn's theorem [20, proposition 7.2, p. 106] E can be embedded densely and homeomorphically into a compact subset $\bar{E}$ of the Hilbert cube $[0,1]^\infty$. Let Ψ denote a map that identifies a $\mu \in \mathscr{P}(E)$ with the $\bar{\mu} \in \mathscr{P}(\bar{E})$ that assigns zero mass to $\check{E} := \bar{E} - E$. Then Ψ is a continuous injection of $\mathscr{P}(E)$ into $\mathscr{P}(\bar{E})$. Since B is not tight, we can find a sequence $\{\xi_n\} \subset B$ that has no limit point in $\mathscr{P}(\mathscr{P}(E))$. But by the compactness of $\bar{E}$ and therefore of $\mathscr{P}(\bar{E})$, and in turn of $\mathscr{P}(\mathscr{P}(\bar{E}))$, $\xi_n \to \xi^*$ along a subsequence, for some $\xi^* \in \mathscr{P}(\mathscr{P}(\bar{E}))$. Then it must be the case that

$$\xi^*(\{\nu \in \mathscr{P}(\bar{E}) : \nu(\check{E}) > 0\}) > 0 .$$

Therefore $\bar{\xi}^*(\check{E}) > 0$, contradicting the tightness of $\bar{B}$, since $\bar{\xi}_n \to \bar{\xi}^*$. Thus B must be tight. Conversely, if B is tight, for any $\varepsilon > 0$, there exists a tight $K_\varepsilon \subset \mathscr{P}(E)$ such that $\xi(K_\varepsilon) > 1 - \frac{\varepsilon}{2}$ for all $\xi \in B$. Let G_ε be a compact set in E such that $\nu(G_\varepsilon) > 1 - \frac{\varepsilon}{2}$ for all $\nu \in K_\varepsilon$, which is possible by the tightness of K_ε. Then

$$\begin{aligned}
\bar{\xi}(G_\varepsilon) &= \int_{\mathscr{P}(E)} \nu(G_\varepsilon)\,\xi(\mathrm{d}\nu) \\
&\geq \int_{\mathscr{P}(E)} \mathbb{I}_{K_\varepsilon}(\nu)\nu(G_\varepsilon)\,\xi(\mathrm{d}\nu) \\
&\geq \left(1 - \tfrac{\varepsilon}{2}\right)\xi(K_\varepsilon) \\
&\geq \left(1 - \tfrac{\varepsilon}{2}\right)^2 \\
&\geq 1 - \varepsilon .
\end{aligned}$$

Tightness of $\bar{B}$ follows. □

We use the relaxed control framework as before, thus replacing $\mathbb{U}$ by $\mathscr{P}(\mathbb{U})$ and replacing b, c, $\mathcal{A}$ and $\mathcal{L}$ by $\bar{b}$, $\bar{c}$, $\bar{\mathcal{A}}$ and $\bar{\mathcal{L}}$, respectively. Let

$$\mathscr{G} := \left\{\mu \in \mathscr{P}(\mathscr{P}(\mathbb{R}^d) \times \mathbb{U}) : \int \mathcal{A} f \,\mathrm{d}\mu = 0 ,\ \forall f \in \mathscr{D}(\mathcal{A})\right\}$$

denote the set of all ergodic occupation measures. Note that under a stationary solution (π, U) corresponding to $\mu \in \mathscr{G}$, (8.3.1) equals $\int \tilde{c}\,\mathrm{d}\mu$. We assume that this is finite for at least one such μ. Let $\varrho^* = \inf_{\mu \in \mathscr{G}} \int \tilde{c}\,\mathrm{d}\mu$. The set $\mathscr{G}$ is clearly closed and convex. Under (C7.1), $\left\{\mu \in \mathscr{G} : \int \tilde{c}\,\mathrm{d}\mu \leq k\right\}$ for $k \in (0, \infty)$ are compact. Thus condition (A6.5) of Chapter 6 on p. 242 holds. Let (π, U) denote a stationary solution to the controlled martingale problem for $\bar{\mathcal{A}}$ and X the state process in the background.

Define the mean empirical measures $\{\bar{\zeta}_t : t > 0\}$ by

$$\begin{aligned}
\int_{\mathbb{R}^d \times \mathbb{U}} f \,\mathrm{d}\bar{\zeta}_t &:= \frac{1}{t}\int_0^t \mathbb{E}\left[\pi_s(f(\cdot\,, U_s))\right]\mathrm{d}s \\
&= \frac{1}{t}\int_0^t \mathbb{E}\left[f(X_s, U_s)\right]\mathrm{d}s , \quad f \in C_b(\mathbb{R}^d \times \mathbb{U}) .
\end{aligned}$$

Lemma 8.3.2 *Under (C7.2), the family $\{\bar{\zeta}_t : t \geq t_0\}$ is tight for any $t_0 > 0$.*

Proof This follows from Lemmas 2.5.3 and 8.3.1. □

Our main result of this section now follows from the results of Section 6.5.

Theorem 8.3.3 *Under either (C7.1) or (C7.2), the partially observed ergodic control problem has an optimal Markov solution and an optimal ergodic solution.*

8.4 Split chain and the pseudo-atom

We next consider the problem of dynamic programming. With this in mind, in this section we describe the Athreya–Ney–Nummelin split chain and pseudo-atom construction along the lines of [91, chapter 5], adapted to the controlled Markov process framework. (The original construction is for the uncontrolled case.) This is used in Section 8.5 to derive the martingale dynamic programming principle for the separated control problem. Let E be a Polish space endowed with its Borel σ-field $\mathfrak{E}$, $x \mapsto A_x \subset \mathbb{U}$ an upper semicontinuous set-valued map. Suppose $\{X_n\}$ is a controlled Markov process on E with an associated control process $\{Z_n\}$, where Z_n takes values in the compact metric space A_{X_n}. Thus

$$\mathbb{P}(X_{n+1} \in A \mid X_m, Z_m,\ m \le n) = p(X_n, Z_n; A)\,, \quad A \in \mathfrak{E}\,, \quad \forall n \in \mathbb{N}\,.$$

The (controlled) transition kernel p is said to satisfy a *minorization condition* if there exist $B \in \mathfrak{E}$, $\delta > 0$, and $\nu \in \mathscr{P}(E)$ with $\nu(B) = 1$, such that

$$p(x, u; A) \ge \delta \nu(A)\, \mathbb{I}_B(x) \qquad \forall A \in \mathfrak{E}\,. \tag{8.4.1}$$

Introduce the notation: for $A \in \mathfrak{E}$, let $A_0 = A\times\{0\}$, $A_1 = A\times\{1\}$. Define $E^* = E\times\{0, 1\}$ and for $\mu \in \mathscr{P}(E)$, define $\mu^* \in \mathscr{P}(E^*)$ by:

$$\mu^*(A_0) = (1 - \delta)\mu(A \cap B) + \mu(A \cap B^c)\,,$$
$$\mu^*(A_1) = \delta\mu(A \cap B)$$

for $A \in \mathfrak{E}$. For a measurable $f\colon E \times U \to \mathbb{R}$, define $f^*\colon E^* \times U \to \mathbb{R}$ by

$$f^*((x, i), u) := f(x, u) \qquad \forall x, u\,.$$

The *split chain* is a controlled Markov process $\{X_n^* = (\hat{X}_n, i_n)\}$ on E^* with the associated control process $\{Z_n^*\}$ and the transition kernel $\hat{p}((x, i), u; \mathrm{d}y)$ defined as follows: for $(x, i) \in E^*$,

$$\hat{p}((x,i), u; \mathrm{d}y) = \begin{cases} p^*(x, u; \mathrm{d}y) & \text{if } (x, i) \in E_0 \setminus B_0\,, \\ \frac{1}{1-\delta}(p^*(x, u; \mathrm{d}y) - \delta\nu^*(\mathrm{d}y)) & \text{if } (x, i) \in B_0\,, \\ \nu^*(\mathrm{d}y) & \text{if } (x, i) \in E_1\,. \end{cases}$$

Likewise, for $A \in \mathfrak{E}$, the initial law satisfies

$$\mathbb{P}((\hat{X}_0, i_0) \in A_0) = (1 - \delta)\,\mathbb{P}(X_0 \in A \cap B) + \mathbb{P}(X_0 \in A \cap B^c)\,,$$
$$\mathbb{P}((\hat{X}_0, i_0) \in A_1) = \delta\,\mathbb{P}(X_0 \in A \cap B)\,.$$

Finally, the control process is prescribed in law by

$$\mathbb{P}(Z_n^* \in D \mid X_m^*, Z_k^*, m \le n, k < n) = \mathbb{P}(Z_n \in D \mid X_m, Z_k, m \le n, k < n).$$

Intuitively, the dynamics may be described as follows:

(a) If $\hat{X}_n = x \in B$, $Z_n = u$, and $i_n = 0$, then $\hat{X}_{n+1} = y$ according to

$$\frac{1}{1-\delta}(p(x, u; \mathrm{d}y) - \delta\nu(\mathrm{d}y)).$$

Moreover, if $y \in B$, then $i_{n+1} = 1$ with probability δ. Otherwise, $i_{n+1} = 0$.

(b) If $\hat{X}_n = x \in B$ and $i_n = 1$, then $\hat{X}_{n+1} = y \in B$ according to $\nu(\mathrm{d}y)$, and $i_{n+1} = 0$ or 1 with probability $1 - \delta$ or δ, respectively.

(c) If $\hat{X}_n = x \notin B$, $Z_n = u$, and $i_n = 0$, then $\hat{X}_{n+1} = y$ according to $p(x, u; \mathrm{d}y)$, and i_{n+1} evolves as in (a).

(d) The set $B^c \times \{1\}$ is never visited.

The key point to note is that a transition out of $B \times \{1\}$ occurs with a probability distribution independent of x and u. This makes it an *atom* in the sense of Meyn and Tweedie [91], albeit a controlled version thereof. Furthermore, if $\{X_n\}$ is φ-irreducible [91, chapter 4] and $\varphi(B) > 0$, then $\varphi^*(B \times \{1\}) > 0$, as can be easily verified. That is, the atom is *accessible* in the terminology of *ibid.* As this is an atom for the split chain and not for the original chain, it is called a *pseudo-atom*. The important fact about the split chain which is valuable to us is the following:

Theorem 8.4.1 *The processes $\{X_n\}$ and $\{\hat{X}_n\}$ agree in law.*

Proof For $A \in \mathfrak{E}$,

$$\begin{aligned}
&\mathbb{P}(\hat{X}_{n+1} \in A \mid \hat{X}_m, Z_m^*, m \le n) \\
&\quad = \mathbb{P}(\hat{X}_{n+1} \in A \mid \hat{X}_m, Z_m^*, m \le n, i_n = 0)\,\mathbb{P}(i_n = 0 \mid \hat{X}_m, Z_m^*, m \le n) \\
&\qquad + \mathbb{P}(\hat{X}_{n+1} \in A \mid \hat{X}_m, Z_m^*, m \le n, i_n = 1)\,\mathbb{P}(i_n = 1 \mid \hat{X}_m, Z_m^*, m \le n).
\end{aligned}$$

If $\hat{X}_n \in B$, the right-hand side equals

$$\frac{1}{1-\delta}(p(\hat{X}_n, Z_n^*; A) - \delta\nu(A)) \times (1 - \delta) + \delta\nu(A) = p(\hat{X}_n, Z_n^*; A).$$

If $\hat{X}_n \notin B$, it equals $p(\hat{X}_n, Z_n^*, A)$ anyway. The claim now follows by a simple induction using our specification of the laws of X_0^* and $\{Z_n^*\}$. □

We are interested in the special case when $E = \mathbb{R}^d$ and B is open with compact closure. Let

$$\tau_0 := \min\{n \ge 0 : X_n \in B\}.$$

Suppose there exist nonnegative inf-compact functions $\mathcal{V}$ and g on $\mathbb{R}^d$ such that $\mathcal{V}$ is locally bounded, $g \geq 1$ on B^c and

$$\int_{\mathbb{R}^d} p(x,u;\mathrm{d}y)\mathcal{V}(y) - \mathcal{V}(x) \leq -g(x) \qquad \forall x \in B^c\,. \tag{8.4.2}$$

The implication of (8.4.2) that interests us is:

Lemma 8.4.2 *The stopping time τ_0 is uniformly integrable over the controls Z and over all initial conditions $X_0 = x$ belonging to any compact subset of $\mathbb{R}^d$.*

Proof Let $K \subset \mathbb{R}^d$ be a compact set. For $R > 0$, define

$$T'_R := \sum_{n=1}^{\tau_0} \mathbb{I}_{B_R}(X_n) \quad \text{and} \quad T''_R := \sum_{n=1}^{\tau_0} \mathbb{I}_{B_R^c}(X_n)\,.$$

By the optional sampling theorem,

$$\mathbb{E}_x^Z\left[\sum_{n=0}^{\tau_0\wedge N}\left(\mathcal{V}(X_{n+1}) - \int_{\mathbb{R}^d} p(X_n,Z_n;\mathrm{d}y)\mathcal{V}(y)\right)\right] = 0\,, \quad N \geq 1\,.$$

Hence, the Lyapunov condition (8.4.2) implies that

$$\begin{aligned}\mathcal{V}(x) - \mathbb{E}_x^Z\left[\mathcal{V}(X_{\tau_0\wedge N+1})\right] &= \mathbb{E}_x^Z\left[\sum_{n=1}^{\tau_0\wedge N}\left(\mathcal{V}(X_n) - \int_{\mathbb{R}^d} p(X_n,Z_n;\mathrm{d}y)\mathcal{V}(y)\right)\right]\\ &\geq \mathbb{E}_x^Z\left[\sum_{n=1}^{\tau_0\wedge N} g(X_n)\right].\end{aligned}$$

Letting $N \uparrow \infty$ and using Fatou's lemma, we obtain

$$\mathbb{E}_x^Z\left[\tau_0\right] \leq \mathbb{E}_x^Z\left[\sum_{n=1}^{\tau_0} g(X_n)\right] \leq \mathcal{V}(x) \qquad \forall x \in B^c\,,$$

and all admissible Z. Thus

$$\mathbb{E}_x^Z[T''_R] \leq \frac{\mathcal{V}(x)}{\inf_{B_R^c} g}\,,$$

which implies that for each $\varepsilon > 0$ there exists $R_\varepsilon > 0$ such that if $R > R_\varepsilon$, then $\mathbb{E}_x^Z\left[T''_R\right] < \varepsilon$, for all admissible Z and $x \in K$. Also, using the strong Markov property with $\check{\tau}_R$ denoting the first exit time from B_R^c, we obtain

$$\mathbb{E}_x^Z[T'_R] = \mathbb{E}_x^Z\left[\mathbb{E}_{X_{\check{\tau}_R}}^Z[T'_R]\right] \leq \sup_{x'\in B_R} \mathbb{E}_{x'}^Z[\tau_0] \leq \sup_{x'\in B_R} \mathcal{V}(x')\,,$$

which implies that for each $R > 0$, there exists a constant $M_1(R) > 0$ such that

$\mathbb{E}_x^Z[T'_R] \le M_1(R)$, for all admissible Z and $x \in \mathbb{R}^d$. On the other hand, we have that

$$\begin{aligned}
\frac{1}{2}\mathbb{E}_x^Z[T'_R(T'_R-1)] &= \mathbb{E}\left[\sum_{n=1}^{T'_R}\left(T'_R - \sum_{k=1}^{n}\mathbb{I}_{B_R}(X_k)\right)\right] \\
&= \mathbb{E}_x^Z\left[\sum_{n=1}^{\infty}\left(T'_R - \sum_{k=1}^{n}\mathbb{I}_{B_R}(X_k)\right)\mathbb{I}\{T'_R \ge n\}\right] \\
&= \mathbb{E}_x^Z\left[\sum_{n=1}^{\infty}\mathbb{I}\{T'_R \ge n\}\,\mathbb{E}_x^Z\left[\sum_{k=n+1}^{\tau_0}\mathbb{I}_{B_R}(X_k)\,\middle|\,\mathfrak{F}_n\right]\right] \\
&= \mathbb{E}_x^Z\left[\sum_{n=1}^{\infty}\mathbb{I}\{T'_R \ge n\}\,\mathbb{E}_{X_n}^Z[T'_R]\right] \\
&\le M_1(R)\,\mathbb{E}_x^Z[T'_R] \le M_1^2(R)\,,
\end{aligned}$$

which in turn implies that, if $M_2(R) := M_1(R)(1+2M_1(R))$, then $\mathbb{E}_x^Z[(T'_R)^2] \le M_2(R)$, for all admissible Z and $x \in K$. Thus, using the Cauchy–Schwartz, Markov and Chebyshev inequalities, for any $R > R_\varepsilon$, we obtain

$$\begin{aligned}
\mathbb{E}_x^Z[\tau_0\,\mathbb{I}\{\tau_0 > 2t\}] &\le \mathbb{E}_x^Z[T'_R\,\mathbb{I}\{T'_R > t\}] + \mathbb{E}_x^Z[T'_R\,\mathbb{I}\{T''_R > t\}] + \mathbb{E}_x^Z[T''_R] \\
&\le \sqrt{\mathbb{E}_x^Z[(T'_R)^2]\,\mathbb{P}_x^Z(T'_R > t)} + \sqrt{\mathbb{E}_x^Z[(T'_R)^2]\,\mathbb{P}_x^Z(T''_R > t)} + \mathbb{E}_x^Z[T''_R] \\
&\le \frac{\mathbb{E}_x^Z[(T'_R)^2]}{t} + \frac{\sqrt{\mathbb{E}_x^Z[(T'_R)^2]\,\mathbb{E}_x^Z[T''_R]}}{\sqrt{t}} + \mathbb{E}_x^Z[T''_R] \\
&< \frac{M_2(R)}{t} + \frac{\sqrt{\varepsilon M_2(R)}}{\sqrt{t}} + \varepsilon\,, \qquad \forall x \in K\,,
\end{aligned}$$

for all admissible Z, and the result follows. □

Lemma 8.4.3 *Let*

$$\tau^* = \min\{n \ge 0 : X_n^* \in B \times \{1\}\}\,.$$

Then τ^* is uniformly integrable over the controls Z^* and over x belonging to any compact subset A of $\mathbb{R}^d$.

Proof Let

$$\begin{aligned}
\tilde{\tau}_0 &= \min\{n \ge 0 : X_n^* \in B \times \{0,1\}\}\,, \\
\tilde{\tau}_m &= \min\{n > \tilde{\tau}_{m-1} : X_n^* \in B \times \{0,1\}\}\,, \qquad m \ge 1\,.
\end{aligned}$$

Define

$$K := \sup_{x\in B}\,\sup_{Z^*}\,\mathbb{E}[\tilde{\tau}_0 \mid X_0 = x]\,,$$

and let $\{\mathfrak{F}_k\}$ denote the natural filtration of (X^*, Z^*). Also, let $\mathbb{E}_{x,i}$ denote the conditional expectation given $X_0^* = (x, i)$. In view of Lemma 8.4.2, without loss of generality, we suppose that $x \in B$, i.e., $\mathbb{E}_{x,i}[\hat{\tau}_0] = 0$. Then, for $m_0 \in \mathbb{N}$, we have

$$\begin{aligned}
\mathbb{E}_{x,i}\left[\tau^* \mathbb{I}\{\tau^* > T\}\right] &\le \sum_{m=0}^{\infty} \mathbb{E}_{x,i}\left[\tilde{\tau}_m \mathbb{I}\{\tau^* = \tilde{\tau}_m\} \mathbb{I}\{\tilde{\tau}_m > T\}\right] \\
&\le \sum_{m=0}^{m_0} \mathbb{E}_{x,i}\left[\tilde{\tau}_m \mathbb{I}\{\tilde{\tau}_m > T\}\right] + \sum_{m=m_0+1}^{\infty} \mathbb{E}_{x,i}\left[\tilde{\tau}_m \mathbb{I}\{\tau^* = \tilde{\tau}_m\}\right] \\
&= \sum_{m=0}^{m_0} \mathbb{E}_{x,i}\left[\tilde{\tau}_m \mathbb{I}\{\tilde{\tau}_m > T\}\right] + \mathbb{E}_{x,i}\left[\tilde{\tau}_{m_0} \mathbb{I}\{\tau^* > \tilde{\tau}_{m_0}\}\right] \\
&\quad + \mathbb{E}_{x,i}\left[\sum_{m=m_0+1}^{\infty} \sum_{k=m_0}^{m-1} (\tilde{\tau}_{k+1} - \tilde{\tau}_k) \mathbb{I}\{\tau^* = \tilde{\tau}_m\}\right]. \qquad (8.4.3)
\end{aligned}$$

Using the filtering property of conditional expectation by conditioning first at $\mathfrak{F}_{\tilde{\tau}_{k+1}}$ and then at $\mathfrak{F}_{\tilde{\tau}_k}$ we obtain

$$\begin{aligned}
\mathbb{E}_{x,i}\left[\tilde{\tau}_{m_0} \mathbb{I}\{\tau^* > \tilde{\tau}_{m_0}\}\right] &= \mathbb{E}_{x,i}\left[\sum_{k=0}^{m_0-1} (\tilde{\tau}_{k+1} - \tilde{\tau}_k) \mathbb{I}\{\tau^* > \tilde{\tau}_k\} \mathbb{I}\{\tau^* > \tilde{\tau}_{m_0}\}\right] \\
&= \mathbb{E}_{x,i}\left[\sum_{k=0}^{m_0-1} (\tilde{\tau}_{k+1} - \tilde{\tau}_k) \mathbb{I}\{\tau^* > \tilde{\tau}_k\} \mathbb{P}(\tau^* > \tilde{\tau}_{m_0} \mid \mathfrak{F}_{\tilde{\tau}_{k+1}})\right] \\
&\le \sum_{k=0}^{m_0-1} (1-\delta)^{m_0-k-1} \mathbb{E}_{x,i}\left[\mathbb{I}\{\tau^* > \tilde{\tau}_k\} \mathbb{E}\left[\tilde{\tau}_{k+1} - \tilde{\tau}_k \mid \mathfrak{F}_{\tilde{\tau}_k}\right]\right] \\
&\le \sum_{k=0}^{m_0-1} K(1-\delta)^{m_0-k-1} \mathbb{P}_{x,i}(\tau^* > \tilde{\tau}_k) \\
&\le K m_0 (1-\delta)^{m_0-1}. \qquad (8.4.4)
\end{aligned}$$

Also,

$$\begin{aligned}
\mathbb{E}_{x,i}\left[\sum_{m=m_0+1}^{\infty} \sum_{k=m_0}^{m-1} (\tilde{\tau}_{k+1} - \tilde{\tau}_k) \mathbb{I}\{\tau^* = \tilde{\tau}_m\}\right] &= \mathbb{E}_{x,i}\left[\sum_{k=m_0}^{\infty} \sum_{m=k+1}^{\infty} (\tilde{\tau}_{k+1} - \tilde{\tau}_k) \mathbb{I}\{\tau^* = \tilde{\tau}_m\}\right] \\
&= \sum_{k=m_0}^{\infty} \mathbb{E}_{x,i}\left[\mathbb{E}\left[(\tilde{\tau}_{k+1} - \tilde{\tau}_k) \mathbb{I}\{\tau^* > \tilde{\tau}_k\} \mid \mathfrak{F}_{\tilde{\tau}_k}\right]\right] \\
&= \sum_{k=m_0}^{\infty} \mathbb{E}_{x,i}\left[\mathbb{I}\{\tau^* > \tilde{\tau}_k\} \mathbb{E}\left[\tilde{\tau}_{k+1} - \tilde{\tau}_k \mid \mathfrak{F}_{\tilde{\tau}_k}\right]\right] \\
&\le K \sum_{k=m_0}^{\infty} (1-\delta)^k \\
&= \frac{K}{\delta} (1-\delta)^{m_0}. \qquad (8.4.5)
\end{aligned}$$

Using (8.4.4) and (8.4.5) in (8.4.3) we obtain

$$\mathbb{E}_{x,i}\left[\tau^* \mathbb{I}\{\tau^* > T\}\right] \le \sum_{m=1}^{m_0} \mathbb{E}_{x,i}\left[\tilde{\tau}_m \mathbb{I}\{\tilde{\tau}_m > T\}\right] + K\left(m_0 + \frac{1-\delta}{\delta}\right)(1-\delta)^{m_0-1}. \quad (8.4.6)$$

Given $\varepsilon > 0$ we first select $m_0 = m_0(\varepsilon)$ so that the second term on the right-hand side of (8.4.6) does not exceed $\varepsilon/2$, and then select T large enough so that the first term on the right-hand side of (8.4.6) has the same bound. Then $\mathbb{E}_{x,i}\left[\tau^* \mathbb{I}\{\tau^* > T\}\right] \le \varepsilon$, and the proof is complete. □

Theorem 8.4.4 *Under* (8.4.2), *the process* $\{X_n^*\}$ *has a unique invariant probability measure given by*

$$\eta(A) = \frac{\mathbb{E}\left[\sum_{m=0}^{\tau^*-1} \mathbb{I}\{X_m^* \in A^*\} \mid X_0^* \in B \times \{1\}\right]}{\mathbb{E}\left[\tau^* \mid X_0^* \in B \times \{1\}\right]}, \qquad A \in \mathfrak{E}.$$

Proof From the strong law of large numbers, for $f \in C_b(\mathbb{R}^d)$,

$$\lim_{N\to\infty} \frac{1}{N}\sum_{m=0}^{N} f^*(X_m^*) = \lim_{N\to\infty} \frac{\frac{1}{N}\sum_{m=0}^{N}\sum_{n=\tilde{\tau}_m}^{\tilde{\tau}_{m+1}-1} f^*(X_n^*)}{\frac{1}{N}\sum_{m=0}^{N}(\tilde{\tau}_{m+1} - \tilde{\tau}_m)}$$

$$= \frac{\mathbb{E}\left[\sum_{m=0}^{\tau^*-1} f^*(X_m^*) \mid X_0^* \in B \times \{1\}\right]}{\mathbb{E}\left[\tau^* \mid X_0^* \in B \times \{1\}\right]}.$$

The claim follows from Theorem 8.4.1. □

8.5 Dynamic programming

In this section we derive a martingale dynamic programming principle for the separated ergodic control problem. We make the following additional assumption:

(A8.1) There exist nonnegative inf-compact functions $\mathcal{V} \in C^2(\mathbb{R}^d) \cap C_{\mathrm{pol}}(\mathbb{R}^d)$ and $g \in C(\mathbb{R}^d)$ such that for some positive constants C and R, we have

$$\mathcal{L}^u\mathcal{V}(x) \le -g(x) + C\mathbb{I}_{B_R}(x) \qquad \forall u \in \mathbb{U}.$$

We claim that without loss of generality $\mathcal{V}$ can be assumed to satisfy

$$\lim_{t\to\infty} \frac{\mathbb{E}^U\left[\mathcal{V}(X_t)\right]}{t} = 0 \qquad \forall U \in \mathfrak{U}. \quad (8.5.1)$$

Indeed, let g' be a nonnegative inf-compact function satisfying $g' \in \mathfrak{o}(g)$ and $\mathcal{V}'$ be the minimal nonnegative solution in $C^2(B_R^c)$ of $\max_{u\in\mathbb{U}} \mathcal{L}^u\mathcal{V}'(x) = -g'(x)$ on B_R^c and $\mathcal{V}' = 0$ on ∂B_R. Extending the definition of $\mathcal{V}'$ in B_R and adding a constant to make it nonnegative, the resulting function, which is also denoted by $\mathcal{V}'$, is in $C^2(\mathbb{R}^d)$ and satisfies, for some constant C', $\mathcal{L}^u\mathcal{V}'(x) \le -g'(x) + C'\mathbb{I}_{B_R}(x)$ for all $u \in \mathbb{U}$. By Lemma 3.7.2 (i), $\mathcal{V}' \in \mathfrak{o}(\mathcal{V})$. Therefore $\mathcal{V}' \in C_{\mathrm{pol}}(\mathbb{R}^d)$. Since (3.7.13) – (3.7.14) used in the proof of Lemma 3.7.2 (ii) also hold over any admissible $U \in \mathfrak{U}$, it follows that $\mathcal{V}'$ satisfies $\lim_{t\to\infty} \frac{\mathbb{E}^U[\mathcal{V}'(X_t)]}{t} = 0$ under any $U \in \mathfrak{U}$. This proves the claim.

Consider the discrete time controlled Markov chain $\{X_n : n \ge 0\}$ with associated state-dependent action spaces A_x defined as in (7.4.3) and the transition kernel $\tilde{P}^\nu(x, \mathrm{d}y)$, defined as in Definition 7.4.4 with $t = 1$ and $\bar{D}$ replaced by $\mathbb{R}^d$. Define

$$\hat{g}(x) := \inf_{U\in\mathfrak{U}} \mathbb{E}_x^U \left[\int_0^1 g(X_t)\,\mathrm{d}t \right], \quad x \in \mathbb{R}^d .$$

By (2.6.4), it follows that $\hat{g}$ is inf-compact. Then

$$\int_{\mathbb{R}^d} \tilde{P}^\nu(x, \mathrm{d}y)\mathcal{V}(y) - \mathcal{V}(x) \le -\hat{g}(x) \qquad \forall x \in \mathbb{R}^d , \quad \forall \nu \in A_x .$$

Thus (8.4.2) and therefore also Lemmas 8.4.2 and 8.4.3 hold for $\{X_n : n \ge 0\}$.

Define

$$\mathscr{P}_0(\mathbb{R}^d) = \left\{ \mu \in \mathscr{P}(\mathbb{R}^d) : \int |x|^m \mu(\mathrm{d}x) < \infty , \ m \in \mathbb{N} \right\} .$$

This is a closed subset of $\mathscr{P}(\mathbb{R}^d)$. By Theorem 2.2.2,

$$\mathbb{E}\left[|X_0|^{2m}\right] < \infty \Longrightarrow \mathbb{E}\left[|X_t|^{2m}\right] < \infty \qquad \forall t \ge 0 .$$

Therefore $\int |x|^{2m}\, \pi_t(\mathrm{d}x) < \infty$ a.s. We assume that $\mathbb{E}\left[|X_0|^{2m}\right] < \infty$ for all $m \in \mathbb{N}$, thereby viewing π as a $\mathscr{P}_0(\mathbb{R}^d)$-valued process.

We use the vanishing discount approach. For simplicity of exposition, we assume that the running cost function c is bounded. We also assume that it is Lipschitz continuous. The discounted cost under a $U \in \mathfrak{U}_{\mathrm{ws}}$ and initial law π_0 takes the form

$$J_\alpha^U(\pi_0) = \mathbb{E}_{\pi_0}^U \left[\int_0^\infty \mathrm{e}^{-\alpha t} \bar{c}(X_t, U_t)\,\mathrm{d}t \right] ,$$

where $\alpha > 0$ is the discount factor. Here the explicit dependence on U and π_0 is indicated on the expectation operator. Define the associated discounted value function

$$V_\alpha(\pi_0) = \inf_{U\in\mathfrak{U}_{\mathrm{ws}}} J_\alpha^U(\pi_0) .$$

The dynamic programming principle for this problem, proved by standard arguments [28, pp. 121–122], is as follows.

Theorem 8.5.1 (i) *For any $t > 0$,*

$$V_\alpha(\pi) = \inf_{U\in\mathfrak{U}_{\mathrm{ws}}} \mathbb{E}^U \left[\int_0^t e^{-\alpha s} \bar{\bar{c}}(\pi_s, U_s)\,\mathrm{d}s + e^{-\alpha t} V_\alpha(\pi_t) \,\middle|\, \pi_0 = \pi \right] ,$$

where

$$\bar{\bar{c}}(\pi, U) := \int_{\mathbb{U}} \left[\int_{\mathbb{R}^d} c(x, u)\, \pi(\mathrm{d}x) \right] U(\mathrm{d}u) .$$

(ii) *The process $\int_0^t e^{-\alpha s} \bar{\bar{c}}(\pi_s, U_s)\,\mathrm{d}s + e^{-\alpha t} V_\alpha(\pi_t)$, $t \ge 0$, is a submartingale w.r.t. the natural filtration of π, U, and is a martingale if and only if U is optimal for the α-discounted control problem.*

Next we derive a uniform bound for $|V_\alpha(\hat{\pi}) - V_\alpha(\tilde{\pi})|$ over $\alpha > 0$ for any pair $\hat{\pi}, \tilde{\pi} \in \mathscr{P}_0(\mathbb{R}^d)$. This requires that we be able to construct two processes with different initial conditions on a common probability space, with the same control *process* that is wide-sense admissible for both initial conditions, and optimal for one of them. Let $\xi \in \mathfrak{U}_{\mathrm{ws}}$. Define

$$\Omega' = C([0,\infty);\mathbb{R}^d) \times C([0,\infty);\mathbb{R}^d) \times C([0,\infty);\mathbb{R}^m) \times \mathscr{U} \times C([0,\infty);\mathbb{R}^m) \times \mathbb{R}^d \times \mathbb{R}^d$$

and let $\mathfrak{G}$ denote its product Borel σ-field. For $n \geq 1$, let κ^n denote the Wiener measure on $C([0,\infty);\mathbb{R}^n)$. Define a probability measure $\mathbb{P}'_0$ on $(\Omega', \mathfrak{G})$ by

$$\mathbb{P}'_0(\mathrm{d}b, \mathrm{d}b', \mathrm{d}y, \mathrm{d}u, \mathrm{d}y', \mathrm{d}x, \mathrm{d}x') = \kappa^d(\mathrm{d}b)\, \kappa^d(\mathrm{d}b')\, \xi(\mathrm{d}y, \mathrm{d}u)\, \kappa^m(\mathrm{d}y')\, \Pi(\mathrm{d}x, \mathrm{d}x'),$$

where $\Pi(\mathrm{d}x, \mathrm{d}x') \in \mathscr{P}(\mathbb{R}^d \times \mathbb{R}^d)$ satisfies $\Pi(\mathrm{d}x, \mathbb{R}^d) = \hat{\pi}(\mathrm{d}x)$ and $\Pi(\mathbb{R}^d, \mathrm{d}x') = \tilde{\pi}(\mathrm{d}x')$. Let $\omega' = (b, b', y, u, y', x_0, x'_0)$ denote a generic element of Ω'. Define on $(\Omega', \mathfrak{G}, \mathbb{P}'_0)$ the canonical random variables $U(\omega') = u \in \mathscr{U}$ and

$$\begin{aligned}
\hat{B}(\omega') &= b \in C([0,\infty);\mathbb{R}^d), & \tilde{B}(\omega') &= b' \in C([0,\infty);\mathbb{R}^d), \\
\hat{Y}(\omega') &= y \in C([0,\infty);\mathbb{R}^m), & \tilde{Y}(\omega') &= y' \in C([0,\infty);\mathbb{R}^m), \\
\hat{X}_0(\omega') &= x_0 \in \mathbb{R}^d, & \tilde{X}_0(\omega') &= x'_0 \in \mathbb{R}^d.
\end{aligned}$$

Henceforth we do not explicitly denote the ω'-dependence, as is customary. Let $\hat{X}$, $\tilde{X}$ denote the solutions of (8.1.1) when W is replaced by $\hat{B}$, $\tilde{B}$ with initial conditions $\hat{X}_0$, $\tilde{X}_0$, respectively. Note that $\hat{X}_0$, $\tilde{X}_0$ are *not* independent. Let $\mathfrak{G}_t$ denote the right-continuous completion of

$$\sigma\Big(\{\hat{B}_s, \tilde{B}_s, \hat{Y}_s, \tilde{Y}_s, U_s : s \leq t\}, \hat{X}_0, \tilde{X}_0\Big)$$

for $t \geq 0$. Then $\mathfrak{G} = \bigvee_{t\geq 0} \mathfrak{G}_t$. Define a new probability measure $\mathbb{P}'$ on $(\Omega', \mathfrak{G})$ as follows: If $\mathbb{P}'_t$, $\mathbb{P}'_{0t}$ denote the restrictions of $\mathbb{P}$, $\mathbb{P}_0$, respectively, to $\mathfrak{G}_t$, then

$$\frac{\mathrm{d}\mathbb{P}'_t}{\mathrm{d}\mathbb{P}'_{0t}} = \exp\left(\int_0^t \langle h(\hat{X}_s), \mathrm{d}\hat{Y}_s\rangle + \int_0^t \langle h(\tilde{X}_s), \mathrm{d}\tilde{Y}_s\rangle - \frac{1}{2}\int_0^t \left(\big|h(\hat{X}_s)\big|^2 + \big|h(\tilde{X}_s)\big|^2\right)\mathrm{d}s\right).$$

Then on $(\Omega', \mathfrak{G}, \mathbb{P}')$, X and $\tilde{X}$ are processes governed by a common control U.

Lemma 8.5.2 *The control U is wide-sense admissible for both the processes $\hat{X}$ and $\tilde{X}$, with initial laws $\hat{\pi}$ and $\tilde{\pi}$, respectively.*

Proof Define probability measures $\hat{\mathbb{P}}, \tilde{\mathbb{P}}$ on $(\Omega', \mathfrak{G})$ by: if $\hat{\mathbb{P}}_t, \tilde{\mathbb{P}}_t$ denote their respective restrictions to $\mathfrak{G}_t$ for $t \geq 0$, then

$$\frac{\mathrm{d}\hat{\mathbb{P}}_t}{\mathrm{d}\mathbb{P}'_{0t}} = \exp\left(\int_0^t \langle h(\tilde{X}_s), d\tilde{Y}_s\rangle\rangle - \frac{1}{2}\int_0^t |h(\tilde{X}_s)|^2\,\mathrm{d}s\right),$$

$$\frac{\mathrm{d}\tilde{\mathbb{P}}_t}{\mathrm{d}\mathbb{P}'_{0t}} = \exp\left(\int_0^t \langle h(\hat{X}_s), d\hat{Y}_s\rangle\rangle - \frac{1}{2}\int_0^t |h(\hat{X}_s)|^2\mathrm{d}s\right).$$

Then

$$\frac{d\mathbb{P}'_t}{d\hat{\mathbb{P}}_t} = \exp\left(\int_0^t \langle h(\hat{X}_s), d\hat{Y}_s\rangle\rangle - \frac{1}{2}\int_0^t |h(\hat{X}_s)|^2\, ds\right),$$

$$\frac{d\mathbb{P}'_t}{d\tilde{\mathbb{P}}_t} = \exp\left(\int_0^t \langle h(\tilde{X}_s), d\tilde{Y}_s\rangle\rangle - \frac{1}{2}\int_0^t |h(\tilde{X}_s)|^2 ds\right).$$

Under $\hat{\mathbb{P}}$, $\hat{Y}$ is a Brownian motion independent of $\hat{X}_0, \hat{B}$ and for $t \geq 0$, $\hat{Y}_{t+\cdot} - \hat{Y}_t$ is independent of $\{\hat{X}_0, \hat{B}, (U_s, \hat{Y}_s,\ s \leq t)\}$. Similarly under $\tilde{\mathbb{P}}$, $\tilde{Y}$ is a Brownian motion independent of $\tilde{X}_0, \tilde{B}$ and for $t \geq 0$, $\tilde{Y}_{t+\cdot} - \tilde{Y}_t$ is independent of $\{\tilde{X}_0, \tilde{B}, (U_s, \tilde{Y}_s,\ s \leq t)\}$. The claim follows. □

Let $\hat{\pi}$ and $\tilde{\pi}$ denote the $\mathscr{P}_0(\mathbb{R}^d)$-valued processes of conditional laws of $\hat{X}$ and $\tilde{X}$, respectively, given observations $\hat{Y}$, respectively $\tilde{Y}$, and the control U up to the present time.

In the foregoing, we choose ξ to be optimal for the discounted cost problem with initial law $\hat{\pi}$. Consider the discrete time process $\bar{X}_n = (\tilde{X}_n, \hat{X}_n)$, $n \geq 0$. In view of the nondegeneracy assumption on σ, the minorization condition (8.4.1) holds for $\bar{X}$ relative to any bounded domain $B \in \mathbb{R}^d \times \mathbb{R}^d$, ν the normalized Lebesgue measure on B, and a suitable $\delta > 0$.

Let X^* denote the corresponding split chain and C^* the pseudo-atom. Define

$$\tau := \min\,\{n \geq 0 : \bar{X}^*_n \in C^*\}\,,$$

the coupling time at the pseudo-atom. By (A8.1), the function $\bar{\mathcal{V}}\colon \mathbb{R}^{2d} \to \mathbb{R}$, defined by $\bar{\mathcal{V}}(x,y) := \mathcal{V}(x) + \mathcal{V}(y)$ also satisfies the same Lyapunov condition relative to some ball B_R, with C, R and g appropriately defined. Therefore, by Lemma 2.5.1, $\mathbb{E}_{x,y}\left[\tau(B^c_R)\right] \in \mathscr{O}(\bar{\mathcal{V}}(x,y))$. Hence it is integrable with respect to $\Pi \in \mathscr{P}_0(\mathbb{R}^{2d})$. Then for any $T > 0$ and $\varepsilon > 0$,

$$\begin{aligned}
V_\alpha(\tilde{\pi}) - V_\alpha(\hat{\pi}) &\leq \mathbb{E}\left[\int_0^\infty e^{-\alpha t}\left(\bar{\bar{c}}(\tilde{\pi}_t, U_t) - \bar{\bar{c}}(\hat{\pi}_t, U_t)\right) dt\right] \\
&= \mathbb{E}\left[\int_0^\infty e^{-\alpha t}\left(\bar{c}(\tilde{X}_t, U_t) - \bar{c}(\hat{X}_t, U_t)\right) dt\right] \\
&= \mathbb{E}\left[\int_0^\tau e^{-\alpha t}\left(\bar{c}(\tilde{X}_t, U_t) - \bar{c}(\hat{X}_t, U_t)\right) dt\right] \\
&\leq K_1 \sqrt{T} \int_0^T \left(\mathbb{E}\,|\hat{X}_t - \tilde{X}_t|^2\right)^{1/2} dt + K_1\, \mathbb{E}\left[(\tau - T)^+\right] \\
&\leq K_2(T)\left(\mathbb{E}\,|\hat{X}_0 - \tilde{X}_0|^2\right)^{1/2} + K_1\, \mathbb{E}\left[(\tau - T)^+\right] \\
&\leq K_2(T)\,(\rho_w(\hat{\pi}, \tilde{\pi}) + \varepsilon) + K_1\, \mathbb{E}\left[(\tau - T)^+\right]\,,
\end{aligned} \tag{8.5.2}$$

where ρ_w is the Wasserstein metric (see p. 72), K_1 is a bound for c and also a common Lipschitz constant for c, $\mathrm{tr}(\sigma)$ and b in its first argument, and $K_2(T)$ is a continuous function of T. In (8.5.2), the first inequality follows from our choice of ξ, the first equality follows by deconditioning, the second equality follows from the fact that the conditional laws of $(\hat{X} \circ \theta_\tau, U \circ \theta_\tau)$ and $(\tilde{X} \circ \theta_\tau, U \circ \theta_\tau)$ given $\mathcal{F}_\tau$ coincide,[2] the second inequality from the boundedness and Lipschitz continuity of c, the third inequality follows from Lemma 2.2.5 and Remark 2.2.6, and the last (fourth) inequality follows from the definition of ρ_w by an appropriate choice of the joint law Π for the pair $(\hat{X}_0, \tilde{X}_0)$. By interchanging the roles of $\hat{\pi}$ and $\tilde{\pi}$, we get a symmetric inequality. Combined with the above, it yields

$$|V_\alpha(\tilde{\pi}) - V_\alpha(\hat{\pi})| \le K_2(T)\,(\rho_w(\hat{\pi}, \tilde{\pi}) + \varepsilon) + K_1\,\mathbb{E}\left[(\tau - T)^+\right].$$

Recall that $x \mapsto \sup_{U \in \mathfrak{U}_{\mathrm{ws}}} \mathbb{E}^U_x[\tau] \in \mathcal{O}(\bar{\mathcal{V}})$. Hence there exists a constant $c > 0$ such that, for any compact set $K \subset \mathbb{R}^{2d}$,

$$\sup_{U \in \mathfrak{U}_{\mathrm{ws}}} \mathbb{E}^U_\Pi\left[(\tau - T)^+\right] \le c \int_{K^c} \bar{\mathcal{V}}(x)\,\Pi(\mathrm{d}x) + \sup_{x \in K} \sup_{U \in \mathfrak{U}_{\mathrm{ws}}} \mathbb{E}^U_x\left[(\tau - T)^+\right]. \tag{8.5.3}$$

For $\varepsilon > 0$, choose a suitable K to make the first term on the right-hand side of (8.5.3) less than $\varepsilon/2$. By Lemma 8.4.2, τ is uniformly integrable over all $U \in \mathfrak{U}_{\mathrm{ws}}$ and all initial conditions $x = (\tilde{x}, \hat{x})$ in a compact set. Thus $\mathbb{E}^U_x\left[(\tau - T)^+\right]$ can be made less than $\varepsilon/2$ uniformly over x in compacta and $U \in \mathfrak{U}_{\mathrm{ws}}$, by selecting a sufficiently large T. It follows that the second term on the right-hand side of (8.5.2) can be made arbitrarily small by choosing T large enough, uniformly over Π in a compact subset of $\mathscr{P}(\mathbb{R}^{2d})$. Fix $\pi^* \in \mathscr{P}_0(\mathbb{R}^d)$. It follows by (8.5.2), that $V_\alpha(\tilde{\pi}) - V_\alpha(\pi^*)$ is bounded and equicontinuous with respect to the metric ρ_w on compact subsets of $\mathscr{P}(\mathbb{R}^d)$. Since $\mathscr{P}_0(\mathbb{R}^d)$ is a σ-compact subset of $\mathscr{P}(\mathbb{R}^d)$, it follows that the family $\{V_\alpha(\tilde{\pi}) - V_\alpha(\pi^*) : \alpha \in (0,1)\}$ is locally bounded and locally equicontinuous in $\mathscr{P}_0(\mathbb{R}^d)$.

It is also clear that the quantity $\alpha V_\alpha(\pi^*)$ is bounded uniformly in $\alpha \in (0,1)$ by the bound on c. Invoking the Ascoli–Arzelà and the Bolzano–Weierstrass theorems, we may now drop to a subsequence as $\alpha \downarrow 0$, along which $\bar{V}_\alpha$ converges to some V in $C(\mathscr{P}_0(\mathbb{R}^d))$ and $\alpha \bar{V}_\alpha(\pi^*)$ converges to a constant $\varrho \in \mathbb{R}$. Letting $T = 0$ in (8.4.6), by Lemmas 8.4.2 and 8.4.3 we have that $\mathbb{E}_{x,i}[\tau^*] \in \mathcal{O}(\mathbb{E}_x[\tau_0])$. With $\mathcal{V}$ as in (A8.1), let $\tilde{\mathcal{V}}(\pi) := \int \mathcal{V}(x)\,\pi(\mathrm{d}x)$ for $\pi \in \mathscr{P}_0(\mathbb{R}^d)$. In view of the foregoing, we then have $V_\alpha - V_\alpha(\pi^*) \in \mathcal{O}(\tilde{\mathcal{V}})$, and hence $V \in \mathcal{O}(\tilde{\mathcal{V}})$. In view of (8.5.1), we then have

$$\lim_{t \to \infty} \frac{\mathbb{E}^U\left[V(\pi_t)\right]}{t} = 0 \qquad \forall U \in \mathfrak{U}. \tag{8.5.4}$$

The following now follows by familiar arguments from Theorem 7.3.11 and Sections 7.4.3 and 7.4.4.

Theorem 8.5.3 *Under (A8.1), the constant ϱ above is uniquely characterized as the optimal ergodic cost for the ergodic control problem under partial observations and the pair $(V, \varrho) \in (C(\mathscr{P}_0(\mathbb{R}^d)) \cap \mathcal{O}(\tilde{\mathcal{V}})) \times \mathbb{R}$ satisfies the dynamic programming*

[2] Here we have to explicitly use the correspondence between the continuous time controlled diffusion and its discrete time skeleton as spelt out in the proof of Lemma 7.4.7.

equation

$$V(\tilde{\pi}) = \inf_{U \in \mathfrak{U}_{\mathrm{ws}}} \mathbb{E}^U \left[\int_0^t (\bar{\tilde{c}}(\pi_s, U_s) - \varrho)\, \mathrm{d}s + V(\pi_t) \,\middle|\, \pi_0 = \tilde{\pi} \right], \quad t \geq 0. \tag{8.5.5}$$

In particular,

$$\left(V(\pi_t) - \int_0^t (\bar{\tilde{c}}(\pi_s, U_s) - \varrho)\, \mathrm{d}s,\ \bar{\mathcal{F}}_t \right), \quad t \geq 0, \tag{8.5.6}$$

is a submartingale and if it is a martingale, then the pair (π, U) *is optimal. In the converse direction, if* (π, U) *is a stationary optimal pair, then* (8.5.6) *is a martingale. Moreover, if* $(V', \varrho') \in (C(\mathscr{P}_0(\mathbb{R}^d)) \cap \mathcal{O}(\tilde{\mathcal{V}})) \times \mathbb{R}$ *is another pair satisfying* (8.5.5)*, then* $\varrho' = \varrho$ *and* $V' = V$ *on the support of* $\mathscr{L}(\pi_0)$ *for any stationary optimal* (π, U)*.*

In view of Theorem 8.5.3, Theorem 8.3.3 can be improved as follows:

Theorem 8.5.4 *Under either (C7.1) or (C7.2), the partially observed ergodic control problem admits an optimal time-homogeneous Markov, or optimal ergodic solution.*

The strengthening from "Markov" to "time-homogeneous Markov" goes exactly as in Section 7.4.5.

Remark 8.5.5 Note that the foregoing was carried out under the relaxation of the original control problem over strict-sense admissible controls to the larger class of (relaxed) wide-sense admissible controls. To say that this is a valid relaxation, we need to argue that the infimum of the cost attained over either strict-sense admissible controls or wide-sense admissible controls is the same. While this is quite standard for the simpler costs such as finite horizon or infinite horizon discounted costs, it is far from obvious for the ergodic cost. We first sketch the scenario for the former. Let (X, U) be an optimal pair where U is wide-sense admissible. Let $\varepsilon > 0$. Argue as in the proof of Theorem 2.3.1 to obtain a precise control that is constant on intervals of the type $[nT, (n+1)T), n \geq 0$, with $T > 0$, so that the cost, finite horizon or discounted infinite horizon as the case may be, is within ε of the optimal. However, this situation can be mapped to a discrete time control problem as in Definition 7.4.4 for which the observation process is $\{\mathcal{Y}_n\}$, where $\mathcal{Y}_n \in C([0, T]; \mathbb{R}^m)$ is defined by $\mathcal{Y}_n := \{Y_t : t \in [nT, (n+1)T]\}$ for $n \geq 0$. This problem will have an optimal control that is strict-sense admissible [20]. This control in turn maps into a piecewise constant strict-sense admissible ε-optimal control for the original continuous time control problem.

Unfortunately, this argument which mimics that of [59] does not carry automatically over to ergodic control because the ergodic cost is not necessarily a lower semicontinuous function of the law of (π, U). We can, however, use the dynamic programming principle in Theorem 8.5.3 to achieve this objective. Let the expectation on the right-hand side of (8.5.5), i.e.,

$$\mathbb{E}^U \left[\int_0^T (\bar{\tilde{c}}(\pi_t, U_t) - \varrho)\, \mathrm{d}t + V(\pi_T) \,\middle|\, \pi_0 = \tilde{\pi} \right], \quad T \geq 0,$$

be viewed as a cost functional for the finite horizon control problem for π over $\mathfrak{U}_{\mathrm{ws}}$. Fix $\varepsilon > 0$. As in [59], there exists a strict-sense admissible ε-optimal $U^s \in \mathfrak{U}$ for this problem. Thus

$$\mathbb{E}\,[V(\pi_0)] \geq \mathbb{E}^{U^s}\left[\int_0^T (\bar{\bar{c}}(\pi_t, U_t) - \varrho)\,\mathrm{d}t + V(\pi_T)\right] - \varepsilon\,.$$

Extend U^s to $[0, \infty)$ by concatenating such ε-optimal strict-sense admissible segments on $[0, 1]$, $[1, 2]$, ... Then

$$\mathbb{E}\,[V(\pi_0)] \geq \mathbb{E}^{U^s}\left[\int_0^n (\bar{\bar{c}}(\pi_t, U_t) - \varrho)\,\mathrm{d}t + V(\pi_n)\right] - \varepsilon n\,, \qquad n \in \mathbb{N}\,.$$

It is easy to deduce from this and (8.5.4) that

$$\limsup_{T\to\infty} \frac{1}{T}\, \mathbb{E}^{U^s}\left[\int_0^T \bar{\bar{c}}(\pi_t, U_t)\,\mathrm{d}t\right] \leq \varrho + \varepsilon\,,$$

i.e., the strict-sense admissible control U^s is ε-optimal for the ergodic cost.

The situation is, however, not as unreasonable as it may seem. For the optimal control we obtained to be strict-sense admissible, what we need is that the corresponding controlled nonlinear filter should have a strong solution. This is not easy to come by, given that the situation is essentially "degenerate." Nevertheless, the optimal control is still legitimate in so far as it does not use any information that it should not, such as the state process. Implicitly, it uses additional randomization of its own to attain optimality.

8.6 Bibliographical note

Section 8.2. Our treatment of controlled nonlinear filters follows [28, section V.1]. See [124] for an extensive account of filtering theory.

Section 8.3. Here we follow [21].

Section 8.4. This is adapted from [91, chapter 5].

Section 8.5. The treatment of ergodic control with partial observations follows [33, 35] using weaker hypotheses.

Epilogue

We conclude by highlighting a string of issues that still remain open.

1. In the controlled martingale problem with ergodic cost, we obtained existence of an optimal ergodic process and optimal Markov process separately, but not of an optimal ergodic Markov process, as one would expect from one's experience with the nondegenerate case. This issue still remains open. In particular it is unclear whether the Krylov selection procedure of Section 6.7, which has been used to extract an optimal Markov family for the discounted cost problem under nondegeneracy, can be similarly employed for the ergodic problem. The work in Bhatt and Borkar [22] claims such a result under very restrictive conditions, but the proof has a serious flaw.
2. The HJB equation was analyzed in two special cases. The general case remains open. In particular, experience with discrete state space problems gives some pointers:

   (a) In the multichain case for Markov chains with finite state space S and finite action space A, a very general dynamic programming equation is available due to Howard [53], viz.,

   $$V(i) = \min_{u \in A} \left[c(i,u) - \varrho(i) + \sum_j p(j \mid i,u) V(j) \right],$$

   $$\varrho(i) = \min_{u \in A} \sum_j p(j \mid i,u) \varrho(j),$$

   for $i \in S$. Here the unknowns are the value function V and the state dependent optimal cost ϱ. An analog of this for the degenerate diffusion case could formally be written down as

   $$\min_{u \in \mathbb{U}} \left(\mathcal{L}^u V(x) + c(x,u) - \varrho(x) \right) = 0,$$

   $$\min_{u \in \mathbb{U}} \mathcal{L}^u \varrho(x) = 0.$$

   This has not been studied.

(b) Likewise, the linear programming formulation in terms of ergodic occupation measures has the following general form for the multichain case, due to Kallenberg [75]:

$$\text{Minimize} \qquad \sum_{i,u} \pi(i,u)c(i,u)$$

subject to:

$$\sum_{i} \pi(i,u)p(j \mid i,u) = \pi(j,u)\,,$$

$$\sum_{i,u} \pi(i,u) = 1\,, \qquad \pi(i,u) \geq 0\,, \quad \forall i,u$$

$$\sum_{i,u} \xi(i,u)p(j \mid i,u) - \sum_{u} \xi(j,u) = \sum_{u} \pi(j,u) - \nu(j)\,, \quad \forall j\,,$$

where ν is the initial law of the chain. This too has no counterpart for the degenerate diffusion case.

3. We have not considered controlled diffusions with the control also appearing in the diffusion matrix and have given our rationale for doing so. As we commented, while the associated HJB equation has been studied, at least for the simpler cost functions (see, e.g., [46]), it is the stochastic differential equation that poses well-posedness issues. In the nondegenerate case with Markov controls, one has existence of solutions [78], but not uniqueness [92]. This calls for a good selection principle.
4. Ergodic control of diffusions with jumps has been reported in [5, 89, 90]. However, the treatment is far from being complete.
5. Show that the value function for the partially observed problem is a viscosity solution of the appropriate infinite dimensional HJB equation.
6. The ergodic control problem in the presence of state constraints or singular controls presents a variety of open issues. See Kurtz and Stockbridge [82] for an important recent development in this direction.
7. In our study of singularly perturbed ergodic control, we did not allow the diffusion matrix of the slow variable to depend on the fast variable, in order to avoid difficulties akin to item 3 above. This remains open, as does the singularly perturbed problem in the degenerate case.
8. Controlled reflected diffusions on non-smooth domains. As pointed earlier, if the domain D is not smooth, the problem is quite challenging.
9. The relative value iteration scheme for controlled Markov chains with ergodic cost (see Puterman [97, p. 205]) suggests the following continuous time and space analog:

$$\frac{\partial V}{\partial t}(x,t) = \min_{u \in \mathbb{U}} \left[\mathcal{L}^u V(x,t) + c(x,u)\right] - V(x_0,t)\,,$$

where x_0 is a fixed point in the state space. Does $V(\cdot\,,t)$ converge to a solution of equation (3.6.4) as $t \to \infty$?

Appendix

Results from Second Order Elliptic Equations

In this appendix we summarize some essential results of strong solutions of second order elliptic equations. Standard references are Gilbarg and Trudinger [67] and Chen and Wu [43]. Good accounts of the interface between stochastic differential equations and partial differential equations appear in Bass [12], Friedman [63] and Stroock and Varadhan [115].

A.1 Nondegenerate elliptic operators

The model in (2.2.1) gives rise to a class of elliptic operators, with $v \in \mathfrak{U}_{\mathrm{sm}}$ appearing as a parameter. We adopt the following parameterization.

Definition A.1.1 Let $\gamma\colon (0,\infty) \to (0,\infty)$ be a positive function that plays the role of a parameter. Using the standard summation rule for repeated indices, we denote by $\mathfrak{L}(\gamma)$ the class of operators

$$\mathcal{L} = a^{ij}\partial_{ij} + b^i\partial_i - \lambda\,,$$

with $a^{ij} = a^{ji}$, $\lambda \geq 0$, and whose coefficients $\{a^{ij}\}$ are locally Lipschitz and $\{b^i, \lambda\}$ are locally bounded and measurable and satisfy, on each ball $B_R \subset \mathbb{R}^d$,

$$\sum_{i,j=1}^{d} a^{ij}(x)\xi_i\xi_j \geq \gamma^{-1}(R)|\xi|^2 \qquad \forall x \in B_R\,, \tag{A.1.1a}$$

for all $\xi = (\xi_1, \dots, \xi_d) \in \mathbb{R}^d$, and

$$\max_{i,j}\, |a^{ij}(x) - a^{ij}(y)| \leq \gamma(R)|x - y| \qquad \forall x, y \in B_R\,, \tag{A.1.1b}$$

$$\sum_{i,j=1}^{d} \left\|a^{ij}\right\|_{L^\infty(B_R)} + \sum_{i,j=1}^{d} \left\|b^i\right\|_{L^\infty(B_R)} + \left\|\lambda\right\|_{L^\infty(B_R)} \leq \gamma(R)\,. \tag{A.1.1c}$$

Also, we let $\mathfrak{L}_0(\gamma)$ denote the class of operators in $\mathfrak{L}(\gamma)$ satisfying $\lambda = 0$.

Remark A.1.2 Note that the linear growth condition is not imposed on the class $\mathcal{L} \in \mathfrak{L}(\gamma)$. The assumptions in (2.2.4) essentially guarantee that $\tau_R \uparrow \infty$ as $R \to \infty$, a.s., which we impose separately when needed.

A.2 Elliptic equations

Of fundamental importance in the study of elliptic equations is the following estimate for strong solutions due to Alexandroff, Bakelman and Pucci [67, theorem 9.1, p. 220].

Theorem A.2.1 *Let $D \subset \mathbb{R}^d$ be a bounded domain. There exists a constant C_a depending only on d, D, and γ, such that if $\psi \in \mathscr{W}^{2,d}_{\mathrm{loc}}(D) \cap C(\bar{D})$ satisfies $\mathcal{L}\psi \geq f$, with $\mathcal{L} \in \mathfrak{L}(\gamma)$, then*

$$\sup_D \psi \leq \sup_{\partial D} \psi^+ + C_a \|f\|_{L^d(D)} \,. \tag{A.2.1}$$

When $f = 0$, Theorem A.2.1 yields generalizations of the classical weak and strong maximum principles [67, theorems 9.5 and 9.6, p. 225]. These are stated separately below.

Theorem A.2.2 *Let $\mathcal{L} \in \mathfrak{L}(\gamma)$. If $\varphi, \psi \in \mathscr{W}^{2,d}_{\mathrm{loc}}(D) \cap C(\bar{D})$ satisfy $\mathcal{L}\varphi = \mathcal{L}\psi$ in a bounded domain D, and $\varphi = \psi$ on ∂D, then $\varphi = \psi$ in D.*

Theorem A.2.3 *If $\varphi \in \mathscr{W}^{2,d}_{\mathrm{loc}}(D)$ and $\mathcal{L} \in \mathfrak{L}(\gamma)$ satisfy $\mathcal{L}\varphi \geq 0$ in a bounded domain D, with $\lambda = 0$ ($\lambda > 0$), then φ cannot attain a maximum (nonnegative maximum) in D unless it is a constant.*

A function $\varphi \in \mathscr{W}^{2,d}_{\mathrm{loc}}(D)$ satisfying $\mathcal{L}\varphi = 0$ ($\mathcal{L}\varphi \leq 0$) in a domain D is called $\mathcal{L}$-harmonic ($\mathcal{L}$-superharmonic). We also say that φ is $\mathfrak{L}(\gamma)$-harmonic in D, if it $\mathcal{L}$-harmonic for some $\mathcal{L} \in \mathfrak{L}(\gamma)$. As a result of Theorem A.2.3, $\mathfrak{L}_0(\gamma)$-harmonic functions cannot attain a maximum nor a minimum in a domain unless they are constant.

Harnack's inequality plays a central role in the study of harmonic functions. For strong solutions this result is stated as follows [67, corollary 9.25, p. 250].

Theorem A.2.4 *Let D be a domain and $K \subset D$ a compact set. There exists a constant C_H depending only on d, D, K and γ, such that if $\varphi \in \mathscr{W}^{2,d}_{\mathrm{loc}}(D)$ is $\mathfrak{L}(\gamma)$-harmonic and nonnegative in D, then*

$$\varphi(x) \leq C_H \varphi(y) \qquad \forall x, y \in K \,.$$

If $\Phi = \{\varphi_i : i \in I\}$ is a family of nonnegative $\mathfrak{L}(\gamma)$-harmonic functions in a domain D, which are bounded above at some point $x_0 \in D$, i.e., $\sup_{i \in I} \varphi_i(x_0) < M_0$ for some $M_0 \in \mathbb{R}$, then they are necessarily equicontinuous on every compact subset of D. First, note that by Theorem A.2.4, Φ is dominated on compact subsets of D by the constant $C_H M_0$. To show equicontinuity we use the following well-known a priori estimate [43, lemma 5.3, p. 48].

Lemma A.2.5 *If $\varphi \in \mathscr{W}^{2,p}_{\mathrm{loc}}(D) \cap L^p(D)$, with $p \in (1,\infty)$, then for any bounded subdomain $D' \Subset D$ we have*

$$\big\|\varphi\big\|_{\mathscr{W}^{2,p}(D')} \le C_0\Big(\big\|\varphi\big\|_{L^p(D)} + \big\|\mathcal{L}\varphi\big\|_{L^p(D)}\Big) \qquad \forall \mathcal{L} \in \mathfrak{L}(\gamma)\,, \tag{A.2.2}$$

with the constant C_0 depending only on d, D, D', p, and γ.

Returning to the family Φ and applying Lemma A.2.5 with $p \equiv d$ and $D' \Subset D$, we obtain the estimate

$$\big\|\varphi\big\|_{\mathscr{W}^{2,d}(D')} \le C_0 C_H M_0 |D'|^{1/d} \qquad \forall \varphi \in \Phi\,.$$

This estimate and the compactness of the embedding $\mathscr{W}^{2,d}(D) \hookrightarrow C^{0,r}(\bar{D})$, $r < 1$, asserted in Theorem A.2.15, imply the equicontinuity of Φ. It is also the case that the limit of any convergent sequence of nonnegative $\mathcal{L}$-harmonic functions is also $\mathcal{L}$-harmonic. Indeed, if $\{\varphi_n\}$ is a sequence of nonnegative $\mathcal{L}$-harmonic functions in a domain D and $\varphi_n \to \varphi$, then since the convergence is uniform on any bounded subdomain $D' \Subset D$, it follows that $\{\varphi_n\}$ is a Cauchy sequence in $L^p(D)$ for any $p \in (1,\infty)$, and hence by (A.2.2) $\{\varphi_n\}$ is also Cauchy in $\mathscr{W}^{2,d}(D'')$ for any $D'' \Subset D'$. Therefore $\varphi \in \mathscr{W}^{2,d}_{\mathrm{loc}}(D)$ and $\mathcal{L}\varphi = 0$ in D.

We summarize these results in the following theorem.

Theorem A.2.6 *Let $D \subset \mathbb{R}^d$ be a domain. Any family of nonnegative $\mathfrak{L}(\gamma)$-harmonic functions in D, which is bounded at some point of D, is equicontinuous on compact subsets of D. The limit of any convergent sequence of nonnegative $\mathcal{L}$-harmonic functions in D, with $\mathcal{L} \in \mathfrak{L}(\gamma)$, is $\mathcal{L}$-harmonic. In particular, any monotone sequence of $\mathcal{L}$-harmonic functions in D which is bounded at some point of D converges uniformly on compacts to an $\mathcal{L}$-harmonic function.*

We turn now to the Dirichlet problem. Let D be a domain in $\mathbb{R}^d$. We seek solutions to the equation

$$\mathcal{L}\varphi = -f \quad \text{in } D\,, \qquad \varphi = g \quad \text{on } \partial D\,. \tag{A.2.3}$$

The special case $g \equiv 0$ is called *Poisson's equation* for $\mathcal{L}$.

We refer the reader to Gilbarg and Trudinger [67] and Chen and Wu [43] for results on the existence and uniqueness of solutions of the Dirichlet problem. We summarize some of these results below.

Theorem A.2.7 *Let D be a bounded C^2 domain in $\mathbb{R}^d$, $\mathcal{L} \in \mathfrak{L}(\gamma)$, $\lambda \ge 0$, and $f \in L^p(D)$.*

(i) *If $p \in (1,\infty)$ and $g \equiv 0$, then the Dirichlet problem* (A.2.3) *has a unique solution $\varphi \in \mathscr{W}^{2,p}(D) \cap \mathscr{W}^{1,p}_0(D)$. Moreover, we have the estimate*

$$\big\|\varphi\big\|_{\mathscr{W}^{2,p}(D)} \le C_0'\big\|f\big\|_{L^p(D)} \tag{A.2.4}$$

for some constant $C_0' = C_0'(d,p,D,\gamma)$.

(ii) *If $p \ge d$ and $g \in C(\partial D)$, then* (A.2.3) *has a unique solution $\varphi \in \mathscr{W}^{2,p}_{\mathrm{loc}}(D) \cap C(\bar{D})$.*

Another useful version of the Dirichlet problem is the following.

Theorem A.2.8 *Let D and $\mathcal{L}$ be as in Theorem A.2.7, and $p \in (1, \infty)$. For each $f \in L^p(D)$ and $g \in \mathscr{W}^{2,p}(D)$ there exists a unique $\varphi \in \mathscr{W}^{2,p}(D)$ satisfying*

$$\varphi - g \in \mathscr{W}_0^{1,p}(D) \quad \text{and} \qquad \mathcal{L}\varphi = -f \quad \text{in } D.$$

Moreover, the estimate

$$\|\varphi\|_{\mathscr{W}^{2,p}(D)} \le C_0'(\|f\|_{L^p(D)} + \|\mathcal{L}g\|_{L^p(D)} + \|g\|_{\mathscr{W}^{2,p}(D)}),$$

holds with the same constant C_0' as in Theorem A.2.7.

When the coefficients of the operator and the data are smooth enough, then classical solutions are obtained. This property is known as *elliptic regularity*. We quote a result from Gilbarg and Trudinger [67, theorem 9.19, p. 243].

Theorem A.2.9 *Suppose D is a $C^{2,1}$ domain, and that the coefficients of $\mathcal{L}$ and f belong to $C^{0,r}(\bar{D})$, with $r \in (0, 1)$. Then any $\varphi \in \mathscr{W}^{2,p}_{\mathrm{loc}}(D)$, with $p \in (1, \infty)$, satisfying $\mathcal{L}\varphi = -f$ in D, $\mathcal{L} \in \mathfrak{L}(\gamma)$, belongs to $C^{2,r}(\bar{D})$.*

Another useful property of solutions of elliptic PDEs that we use quite often is the following:

Lemma A.2.10 *Let D be a bounded $C^{0,1}$ domain and $\mathcal{L} \in \mathfrak{L}(\gamma)$. If $\varphi \in \mathscr{W}^{2,p}(D)$ for some $p \in (1, \infty)$, and $\mathcal{L}\varphi \in L^q(D)$, $q > p$, then φ belongs to $\mathscr{W}^{2,q}_{\mathrm{loc}}(D)$. It follows that if $\mathcal{L}\varphi \in L^\infty(D)$, then $\varphi \in \mathscr{W}^{2,q}(D)$ for all $q \in (1, \infty)$.*

In certain places, we employ some specialized results which apply to a class of $\mathcal{L}$-superharmonic functions, which is defined as follows:

Definition A.2.11 For $\delta > 0$ and D a bounded domain, let $\mathfrak{K}(\delta, D) \subset L^\infty(D)$ denote the positive convex cone

$$\mathfrak{K}(\delta, D) := \left\{ f \in L^\infty(D) : f \ge 0,\ \|f\|_{L^\infty(D)} \le \delta |D|^{-1} \|f\|_{L^1(D)} \right\}.$$

We quote the following theorem from Arapostathis *et al.* [7].

Theorem A.2.12 *There exists a positive constant $\tilde{C}_a = \tilde{C}_a(d, \gamma, R, \delta)$ such that if $\varphi \in \mathscr{W}^{2,p}_{\mathrm{loc}}(B_R) \cap \mathscr{W}_0^{1,p}(B_R)$ satisfies $\mathcal{L}\varphi = -f$ in B_R, $\varphi = 0$ on ∂B_R, with $f \in \mathfrak{K}(\delta, B_R)$ and $\mathcal{L} \in \mathfrak{L}(\gamma)$, then*

$$\inf_{B_{R/2}} \varphi \ge \tilde{C}_a \|f\|_{L^1(B_R)}.$$

The Harnack inequality has been extended in [7, corollary 2.2] to the class of superharmonic functions satisfying $-\mathcal{L}\varphi \in \mathfrak{K}(\delta, D)$.

Theorem A.2.13 *Let D be a domain and $K \subset D$ a compact set. There exists a constant $\tilde{C}_H = \tilde{C}_H(d, D, K, \gamma, \delta)$, such that if $\varphi \in \mathscr{W}^{2,d}_{\mathrm{loc}}(D)$ satisfies $\mathcal{L}\varphi = -f$ and $\varphi \ge 0$ in D, with $f \in \mathfrak{K}(\delta, D)$ and $\mathcal{L} \in \mathfrak{L}(\gamma)$, then*

$$\varphi(x) \le \tilde{C}_H \varphi(y) \qquad \forall x, y \in K.$$

We summarize some useful embedding results below [43, proposition 1.6, p. 211], [67, theorem 7.22, p. 167]. We start with a definition.

Definition A.2.14 Let X and Y be Banach spaces, with $X \subset Y$. If, for some constant C, we have $\|x\|_Y \le C\|x\|_X$ for all $x \in X$, then we say that X is *continuously embedded* in Y and refer to C as the *embedding constant*. In such a case we write $X \hookrightarrow Y$. We say that the embedding is compact if bounded sets in X are precompact in Y.

Theorem A.2.15 *For any bounded domain $D \subset \mathbb{R}^d$, the following embeddings are compact.*

(1a) *for $p < d$, $\mathscr{W}_0^{1,p}(D) \hookrightarrow L^q(D)$ for $p \le q < \frac{pd}{d-p}$;*
(1b) *for $p > d$, $\mathscr{W}_0^{1,p}(D) \hookrightarrow C(\bar{D})$.*

If D is a bounded $C^{0,1}$ domain and $k \in \mathbb{N}$, then

(2a) *if $kp < d$, then $\mathscr{W}^{k,p}(D) \hookrightarrow L^q(D)$ is compact for $p \le q < \frac{pd}{d-kp}$ and continuous for $p \le q \le \frac{pd}{d-kp}$;*
(2b) *if $\ell p > d$ and $\ell \le k$, then $\mathscr{W}^{k,p}(D) \hookrightarrow C^{k-\ell,r}(\bar{D})$ is compact for $r < \ell - \frac{d}{p}$ and continuous for $r \le \ell - \frac{d}{p}$ ($r \le 1$).*

In particular, $\mathscr{W}^{2,d}(D) \hookrightarrow C^{0,r}(\bar{D})$ is compact for $r < 1$, and $\mathscr{W}^{2,p}(D) \hookrightarrow C^{1,r}(\bar{D})$ is compact for $p > d$ and $r < 1 - \frac{d}{p}$.

A.3 The resolvent

Throughout the section we fix the class $\mathcal{L} \in \mathfrak{L}_0(\gamma)$ for some $\gamma\colon (0,\infty) \to (0,\infty)$, and for a generic $\mathcal{L} \in \mathfrak{L}_0(\gamma)$ we let $\mathbb{P}_x$ and $\mathbb{E}_x$ be the associated probability measure and expectation operator, respectively, for the diffusion process that starts at $x \in \mathbb{R}^d$.

Definition A.3.1 We define the λ-resolvent $\mathcal{R}_\lambda$, for $\lambda \in (0,\infty)$, by

$$\begin{aligned}\mathcal{R}_\lambda[f](x) &:= \mathbb{E}_x\left[\int_0^\infty \mathrm{e}^{-\lambda t} f(X_t)\,\mathrm{d}t\right],\\ &= \int_0^\infty \mathrm{e}^{-\lambda t} T_t f(x)\,\mathrm{d}t\,, \quad f \in L^\infty(\mathbb{R}^d)\,.\end{aligned}$$

Observe that $\mathcal{R}_\lambda[f]$ is also well defined if f is nonnegative and belongs to $L^\infty_{\mathrm{loc}}(\mathbb{R}^d)$. For a bounded domain D, and $f \in L^\infty(D)$, $p \in (1,\infty)$, we let

$$\mathcal{R}_\lambda^D[f](x) := \mathbb{E}_x\left[\int_0^{\tau(D)} \mathrm{e}^{-\lambda t} f(X_t)\,\mathrm{d}t\right], \quad x \in D\,.$$

Suppose $f \in L^\infty(\mathbb{R}^d)$. If $\psi \in \mathscr{W}^{2,p}_{\mathrm{loc}}(\mathbb{R}^d) \cap L^\infty(\mathbb{R}^d)$, $p \in (1,\infty)$, is a solution of *Poisson's equation* (in $\mathbb{R}^d$)

$$\mathcal{L}\psi - \lambda\psi = -f\,, \tag{A.3.1}$$

then it follows from the Itô–Krylov formula that $\psi = \mathcal{R}_\lambda[f]$. Indeed,

$$e^{-\lambda t}\,\mathbb{E}_x[\psi(X_t)] - \psi(x) = \mathbb{E}_x\left[\int_0^t e^{-\lambda s}(\mathcal{L}\psi(X_s) - \lambda\psi(X_s))\,ds\right]. \tag{A.3.2}$$

Letting $t \to \infty$ in (A.3.2), and using dominated convergence, we obtain

$$\psi(x) = -\,\mathbb{E}_x\left[\int_0^\infty e^{-\lambda s}(\mathcal{L}\psi(X_s) - \lambda\psi(X_s))\,ds\right],$$

and the assertion follows by (A.3.1). Conversely, as shown later in Lemma A.3.4, if $\mathcal{R}_\lambda[f] \in C(\mathbb{R}^d)$, then it is a solution of (A.3.1) in $\mathbb{R}^d$. As a result, if $f \in L^\infty(\mathbb{R}^d)$, then $\mathcal{R}_\lambda[f]$, for $\lambda \in (0,\infty)$, is the unique bounded solution in $\mathbb{R}^d$ of (A.3.1) (see Corollary A.3.6 below). Concerning resolvents in a bounded domain we have the following lemma.

Lemma A.3.2 *Let D be a bounded C^2 domain, $f \in L^\infty(D)$, and $\lambda \in [0,\infty)$. Then*

$$\varphi(x) = \mathbb{E}_x\left[\int_0^{\tau(D)} e^{-\lambda t} f(X_t)\,dt\right], \quad x \in D$$

is the unique solution of Poisson's equation (A.3.1) *in D, $\varphi = 0$ on ∂D, in the class $\mathscr{W}^{2,p}(D) \cap \mathscr{W}_0^{1,p}(D)$, $1 < p < \infty$.*

Proof By Theorem A.2.7, Poisson's equation on the domain D has a unique solution $\varphi \in \mathscr{W}^{2,p}(D) \cap \mathscr{W}_0^{1,p}(D)$. Let $\{D_n\}$ be an increasing sequence of C^2 domains with $\cup_{n\in\mathbb{N}} D_n = D$. Define $\mathfrak{s}_n = \tau(D_n)$. Recall that by Lemma 2.6.5, $\tau(D)$ is finite almost surely. Using the Itô–Krylov formula, we obtain

$$\begin{aligned}\mathbb{E}_x[e^{-\lambda(t\wedge\mathfrak{s}_n)}\varphi(X_{t\wedge\mathfrak{s}_n})] - \varphi(x) &= \mathbb{E}_x\left[\int_0^{t\wedge\mathfrak{s}_n} e^{-\lambda s}(\mathcal{L}\varphi(X_s) - \lambda\varphi(X_s))\,ds\right]\\ &= -\,\mathbb{E}_x\left[\int_0^{t\wedge\mathfrak{s}_n} e^{-\lambda s} f(X_s)\,ds\right].\end{aligned} \tag{A.3.3}$$

Let $n \to \infty$ and then $t \to \infty$ in (A.3.3) and use the fact that $\varphi = 0$ on ∂D to obtain the result. □

Next, we characterize the solution of the Dirichlet problem

$$\mathcal{L}\psi - \lambda\psi = 0 \quad \text{in } D\,, \qquad \psi = g \quad \text{on } \partial D\,. \tag{A.3.4}$$

Lemma A.3.3 *Let D be a bounded C^2 domain, $g \in C(\partial D)$, and $\lambda \in [0,\infty)$. Then*

$$\psi(x) = \mathbb{E}_x[e^{-\lambda\tau(D)} g(X_{\tau(D)})]\,, \quad x \in D$$

is the unique solution to (A.3.4) *in the class $\mathscr{W}^{2,p}_{\mathrm{loc}}(D) \cap C(\bar{D})$, $p \in (1,\infty)$.*

Proof By Lemma 2.6.5, $\tau(D) < \infty$ a.s. We use the argument in the proof of Lemma A.3.2, but let first $t \to \infty$ and then $n \to \infty$ in (A.3.3), to obtain by dominated convergence $\psi(x) = \mathbb{E}_x[e^{-\lambda\tau(D)}\varphi(X_{\tau(D)})]$. Then use the fact that $\mathbb{P}_x(\tau(D)) = 0$ for $x \in \partial D$. Uniqueness follows by Theorem A.2.7. □

Lemma A.3.4 *Let $f \in L^\infty_{\mathrm{loc}}(\mathbb{R}^d)$, and $\lambda \in (0,\infty)$. If $\mathcal{R}_\lambda[f] \in C(\mathbb{R}^d)$, then it satisfies*

$$\mathcal{L}(\mathcal{R}_\lambda[f])(x) + f(x) = \lambda \mathcal{R}_\lambda[f](x) \quad a.e. \tag{A.3.5}$$

Proof Let $R > 0$. Using the strong Markov property, decompose $\mathcal{R}_\lambda[f]$ as

$$\mathcal{R}_\lambda[f](x) = \mathbb{E}_x\left[\int_0^{\tau_R} \mathrm{e}^{-\lambda t} f(X_t)\,\mathrm{d}t\right] + \mathbb{E}_x[\mathrm{e}^{-\lambda\tau_R}\mathcal{R}_\lambda[f](X_{\tau_R})]\,. \tag{A.3.6}$$

Let $\varphi \in \mathscr{W}^{2,p}(B_R) \cap \mathscr{W}^{1,p}_0(B_R)$ be the unique solution to Poisson's equation (A.3.1), and $\psi \in \mathscr{W}^{2,p}_{\mathrm{loc}}(B_R) \cap C(\bar{B}_R)$ the unique solution to (A.3.4) with $g = \mathcal{R}_\lambda[f]$. Then, by Lemma A.3.2, the first term on the right-hand side of (A.3.6) equals φ, while by Lemma A.3.3, the second term on the right-hand side of (A.3.6) equals ψ. Adding (A.3.1) and (A.3.4) it follows that $\mathcal{R}_\lambda[f]$ satisfies (A.3.5) in every ball B_R. □

Let $\lambda \in (0,1)$ and $\varphi[f]$ denote the solution of the Dirichlet problem

$$\mathcal{L}\varphi - \lambda\varphi = -f \quad \text{on } B_R\,, \qquad \varphi = 0 \quad \text{on } \partial B_R\,,$$

for $R > 0$. By Theorem A.2.1,

$$\sup_{B_R} \bigl|\varphi[f]\bigr| \le C_a \bigl\|f\bigr\|_{L^d(B_R)}\,.$$

It follows that for each $x \in B_R$, the map $f \mapsto \varphi[f](x)$ defines a bounded linear functional on $L^d(B_R)$. By the Riesz representation theorem there exists a function $g_R(x,\cdot) \in L^q(B_R)$, $q = \frac{d}{d-1}$, such that

$$\varphi[f](x) = \int_{B_R} g_R(x,y) f(y)\,\mathrm{d}y\,.$$

The function $g_R(x,y)$ is called the Green's function for B_R relative to the operator $\mathcal{L} - \lambda$. Now let A be a Borel set in B_R and set $f = \mathbb{I}_A$. Let

$$\tilde{g}_R(x,A) := \int_A g_R(x,y)\,\mathrm{d}y\,.$$

By the maximum principle, $\tilde{g}_R(x,A)$ is a nondecreasing (nonnegative) function of R, and therefore $g_R(x,\cdot)$ is also increasing as a function of R, a.e. Set $g = \lim_{R\to\infty} g_R$. Using Lemma A.3.2 and (A.3.6), with $\lambda > 0$, we obtain, by monotone convergence

$$\begin{aligned}
\int_0^\infty \mathrm{e}^{-\lambda t}\,\mathbb{P}_x(X_t \in A)\,\mathrm{d}t &= \lim_{R\to\infty} \int_0^{\tau_R} \mathrm{e}^{-\lambda t}\,\mathbb{P}_x(X_t \in A)\,\mathrm{d}t \\
&= \lim_{R\to\infty} \int_A g_R(x,y)\,\mathrm{d}y \\
&= \int_A g(x,y)\,\mathrm{d}y\,.
\end{aligned}$$

Moreover, by Theorem A.2.12, it holds that

$$\inf_{x\in B_{R/2}} \int_A g_R(x,y)\,\mathrm{d}y \ge \tilde{C}_a |A|$$

for some constant $\tilde{C}_a$, which is independent of $A \in \mathscr{B}(B_R)$. We have the following theorem.

Theorem A.3.5 *For $\lambda \in (0,1)$, and $x \in \mathbb{R}^d$ define the probability measure $Q_\lambda(x,\cdot\,)$ by*

$$Q_\lambda(x,A) := \lambda \int_0^\infty \mathrm{e}^{-\lambda t}\,\mathbb{P}_x(X_t \in A)\,\mathrm{d}t\,, \quad A \in \mathscr{B}(\mathbb{R}^d)\,.$$

For each $x \in \mathbb{R}^d$, $Q_\lambda(x,\cdot\,)$ is equivalent to the Lebesgue measure, and moreover, there is a constant $C_{\mathfrak{q}} = C_{\mathfrak{q}}(d,\gamma,R)$, such that if $\mathfrak{q}_\lambda(x,\cdot\,)$ denotes its density, then

$$\mathfrak{q}_\lambda(x,y) \ge \lambda C_{\mathfrak{q}} \qquad \forall x \in B_R\,,\ a.e.\ y \in B_R\,.$$

If $f \in L^\infty(\mathbb{R}^d)$, then by the strong Feller property of the resolvent (see Theorem 5.2.9) $\mathcal{R}_\lambda[f](x) \in C_b(\mathbb{R}^d)$. It follows by Lemma A.3.4 that $\mathcal{R}_\lambda[f]$ is a solution for (A.3.1) in $\mathbb{R}^d$. This together with the discussion in the paragraph preceding Lemma A.3.2 shows the following.

Corollary A.3.6 *If $f \in L^\infty(\mathbb{R}^d)$, and $\lambda \in (0,\infty)$ then $\mathcal{R}_\lambda[f]$ is the unique solution of Poisson's equation in $\mathbb{R}^d$ in the class $\mathscr{W}^{2,p}_{\mathrm{loc}}(\mathbb{R}^d) \cap L^\infty(\mathbb{R}^d)$, $p \in (1,\infty)$.*

Now let $f \in L^\infty_{\mathrm{loc}}(\mathbb{R}^d)$, $f \ge 0$. The next theorem shows that if $\mathcal{R}_\lambda[f]$ is finite at some point in $\mathbb{R}^d$, then it satisfies Poisson's equation in $\mathbb{R}^d$.

Theorem A.3.7 *Suppose $f \in L^\infty_{\mathrm{loc}}(\mathbb{R}^d)$, $f \ge 0$, and $\mathcal{R}_\lambda[f](x_0) < \infty$ at some $x_0 \in \mathbb{R}^d$, with $\lambda \in (0,\infty)$. Then $\mathcal{R}_\lambda[f] \in \mathscr{W}^{2,p}_{\mathrm{loc}}(\mathbb{R}^d)$ for all $p \in (1,\infty)$ and satisfies* (A.3.1) *in $\mathbb{R}^d$.*

Proof With $\tau_n = \tau(B_n)$, $n \in \mathbb{N}$, define

$$\varphi_n := \mathbb{E}_x\left[\int_0^{\tau_n} \mathrm{e}^{-\lambda t} f(X_t)\,\mathrm{d}t\right], \quad n \in \mathbb{N}\,.$$

By Lemma A.3.2, φ_n satisfies (A.3.1) in B_n. Since $f \in L^\infty(B_n)$, then $\varphi_n \in \mathscr{W}^{2,p}(B_n)$ for all $p \in (1,\infty)$. Let $n' \in \mathbb{N}$ and $x_0 \in B_{n'}$. Observe that $\hat\varphi_k := \varphi_{k+n'} - \varphi_{n'}$, $k \in \mathbb{N}$, is an increasing sequence of $\mathcal{L}$-harmonic functions in $B_{n'}$. Since $\hat\varphi_k + \varphi_{n'} \le \mathcal{R}_\lambda[f]$ in $B_{n'}$, this sequence is bounded at x_0. Hence by Theorem A.2.6 it converges to an $\mathcal{L}$-harmonic function $\hat\varphi$. By monotone convergence, since $\tau_n \to \infty$ a.s. (see Remark A.1.2), we obtain

$$\begin{aligned}\mathcal{R}_\lambda[f](x) &= \lim_{n\to\infty} \varphi_n(x)\\ &= \lim_{k\to\infty} \hat\varphi_k(x) + \varphi_{n'}(x)\\ &= \hat\varphi(x) + \varphi_{n'}(x) \qquad \forall x \in B_{n'}\,.\end{aligned}$$

Thus $\mathcal{R}_\lambda[f]$ satisfies (A.3.1) in $B_{n'}$ for each $n' \in \mathbb{N}$. □

Remark A.3.8 It follows from Theorem A.3.7 and (A.3.6) that if $f \in L^\infty_{\mathrm{loc}}(\mathbb{R}^d)$, $f \ge 0$, and $\mathcal{R}_\lambda[f]$ is finite at some point in $\mathbb{R}^d$, then

$$\mathbb{E}_x[\mathrm{e}^{-\lambda\tau_R}\mathcal{R}_\lambda[f](X_{\tau_R})] \xrightarrow[R\to\infty]{} 0\,.$$

References

[1] Agmon, S., Douglis, A., and Nirenberg, L. 1959. Estimates near the boundary for solutions of elliptic partial differential equations satisfying general boundary conditions. I. *Comm. Pure Appl. Math.*, **12**, 623–727.

[2] Allinger, D. F. and Mitter, S. K. 1980. New results on the innovations problem for nonlinear filtering. *Stochastics*, **4**(4), 339–348.

[3] Anderson, E. J. and Nash, P. 1987. *Linear Programming in Infinite-Dimensional Spaces*. Wiley-Interscience Series in Discrete Mathematics and Optimization. Chichester: John Wiley & Sons.

[4] Arapostathis, A. and Borkar, V. S. 2010. Uniform recurrence properties of controlled diffusions and applications to optimal control. *SIAM J. Control Optim.*, **48**(7), 152–160.

[5] Arapostathis, A. and Ghosh, M. K. 2004 (Dec). Ergodic control of jump diffusions in $\mathbb{R}^d$ under a near-monotone cost assumption. Pages 4140–4145 of: *43rd IEEE Conference on Decision and Control*, vol. 4.

[6] Arapostathis, A., Borkar, V. S., Fernández-Gaucherand, E., Ghosh, M. K., and Marcus, S. I. 1993. Discrete-time controlled Markov processes with average cost criterion: a survey. *SIAM J. Control Optim.*, **31**(2), 282–344.

[7] Arapostathis, A., Ghosh, M. K., and Marcus, S. I. 1999. Harnack's inequality for cooperative, weakly coupled elliptic systems. *Comm. Partial Differential Equations*, **24**, 1555–1571.

[8] Arisawa, M., and Lions, P.-L. 1998. On ergodic stochastic control. *Comm. Partial Differential Equations*, **23**(11–12), 333–358.

[9] Arrow, K. J., Barankin, E. W., and Blackwell, D. 1953. Admissible points of convex sets. Pages 87–91 of: *Contributions to the Theory of Games*, vol. 2. Annals of Mathematics Studies, no. 28. Princeton, NJ: Princeton University Press.

[10] Bachelier, L. 2006. *Louis Bachelier's Theory of Speculation: The Origins of Modern Finance*. Princeton, NJ: Princeton University Press. Translated and with a commentary by Mark Davis and Alison Etheridge.

[11] Basak, G. K., Borkar, V. S., and Ghosh, M. K. 1997. Ergodic control of degenerate diffusions. *Stochastic Anal. Appl.*, **15**(1), 1–17.

[12] Bass, R. F. 1998. *Diffusions and Elliptic Operators*. Probability and its Applications. New York: Springer-Verlag.

[13] Beneš, V. E. 1970. Existence of optimal strategies based on specified information, for a class of stochastic decision problems. *SIAM J. Control*, **8**, 179–188.

[14] Bensoussan, A. 1982. *Stochastic Control by Functional Analysis Methods*. Studies in Mathematics and its Applications, vol. 11. Amsterdam: North-Holland Publishing Co.

[15] Bensoussan, A. and Borkar, V. 1984. Ergodic control problem for one-dimensional diffusions with near-monotone cost. *Systems Control Lett.*, **5**(2), 127–133.

[16] Bensoussan, A. and Borkar, V. 1986. Corrections to: "Ergodic control problem for one-dimensional diffusions with near-monotone cost" [Systems Control Lett. **5** (1984), no. 2, 127–133]. *Systems Control Lett.*, **7**(3), 233–235.

[17] Bensoussan, A. and Frehse, J. 1992. On Bellman equations of ergodic control in $\boldsymbol{R}^n$. *J. Reine Angew. Math.*, **429**, 125–160.

[18] Bensoussan, A. and Frehse, J. 2002. Ergodic control Bellman equation with Neumann boundary conditions. Pages 59–71 of: *Stochastic Theory and Control (Lawrence, KS, 2001)*. Lecture Notes in Control and Inform. Sci., vol. 280. Berlin: Springer.

[19] Bertoin, J. 1996. *Lévy Processes*. Cambridge Tracts in Mathematics, vol. 121. Cambridge: Cambridge University Press.

[20] Bertsekas, D. P. and Shreve, S. E. 1978. *Stochastic Optimal Control: The Discrete Time Case*. New York: Academic Press.

[21] Bhatt, A. G. and Borkar, V. S. 1996. Occupation measures for controlled Markov processes: characterization and optimality. *Ann. Probab.*, **24**(3), 1531–1562.

[22] Bhatt, A. G. and Borkar, V. S. 2005. Existence of optimal Markov solutions for ergodic control of Markov processes. *Sankhyā*, **67**(1), 1–18.

[23] Bhatt, A. G. and Karandikar, R. L. 1993. Invariant measures and evolution equations for Markov processes characterized via martingale problems. *Ann. Probab.*, **21**(4), 2246–2268.

[24] Billingsley, P. 1968. *Convergence of Probability Measures*. New York: John Wiley & Sons.

[25] Billingsley, P. 1995. *Probability and Measure*. Third edition. Wiley Series in Probability and Mathematical Statistics. New York: John Wiley & Sons.

[26] Bogachev, V. I., Krylov, N. V., and Röckner, M. 2001. On regularity of transition probabilities and invariant measures of singular diffusions under minimal conditions. *Comm. Partial Differential Equations*, **26**(11–12), 2037–2080.

[27] Bogachev, V. I., Rökner, M., and Stannat, V. 2002. Uniqueness of solutions of elliptic equations and uniqueness of invariant measures of diffusions. *Mat. Sb.*, **193**(7), 3–36.

[28] Borkar, V. S. 1989a. *Optimal Control of Diffusion Processes*. Pitman Research Notes in Mathematics Series, vol. 203. Harlow: Longman Scientific & Technical.

[29] Borkar, V. S. 1989b. A topology for Markov controls. *Appl. Math. Optim.*, **20**(1), 55–62.

[30] Borkar, V. S. 1991. On extremal solutions to stochastic control problems. *Appl. Math. Optim.*, **24**(3), 317–330.

[31] Borkar, V. S. 1993. Controlled diffusions with constraints. II. *J. Math. Anal. Appl.*, **176**(2), 310–321.

[32] Borkar, V. S. 1995. *Probability Theory: An Advanced Course*. New York: Springer-Verlag.

[33] Borkar, V. S. 2003. Dynamic programming for ergodic control with partial observations. *Stochastic Process. Appl.*, **103**(2), 293–310.

[34] Borkar, V. S. and Budhiraja, A. 2004a. Ergodic control for constrained diffusions: characterization using HJB equations. *SIAM J. Control Optim.*, **43**(4), 1467–1492.

[35] Borkar, V. S. and Budhiraja, A. 2004b. A further remark on dynamic programming for partially observed Markov processes. *Stochastic Process. Appl.*, **112**(1), 79–93.

[36] Borkar, V. S. and Gaitsgory, V. 2007. Singular perturbations in ergodic control of diffusions. *SIAM J. Control Optim.*, **46**(5), 1562–1577.

[37] Borkar, V. S. and Ghosh, M. K. 1988. Ergodic control of multidimensional diffusions. I. The existence results. *SIAM J. Control Optim.*, **26**(1), 112–126.

[38] Borkar, V. S. and Ghosh, M. K. 1990a. Controlled diffusions with constraints. *J. Math. Anal. Appl.*, **152**(1), 88–108.

[39] Borkar, V. S. and Ghosh, M. K. 1990b. Ergodic control of multidimensional diffusions. II. Adaptive control. *Appl. Math. Optim.*, **21**(2), 191–220.

[40] Borkar, V. S. and Ghosh, M. K. 2003. Ergodic control of partially degenerate diffusions in a compact domain. *Stochastics*, **75**(4), 221–231.

[41] Borkar, V. S. and Mitter, S. K. 2003. A note on stochastic dissipativeness. Pages 41–49 of: *Directions in mathematical systems theory and optimization*. Lecture Notes in Control and Inform. Sci., vol. 286. Berlin: Springer.

[42] Busca, J. and Sirakov, B. 2004. Harnack type estimates for nonlinear elliptic systems and applications. *Ann. Inst. H. Poincaré Anal. Non Linéaire*, **21**(5), 543–590.

[43] Chen, Y.-Z. and Wu, L.-C. 1998. *Second Order Elliptic Equations and Elliptic Systems*. Translations of Mathematical Monographs, vol. 174. Providence, RI: American Mathematical Society. Translated from the 1991 Chinese original by Bei Hu.

[44] Choquet, G. 1969. *Lectures on Analysis. Vol. II: Representation Theory*. Edited by J. Marsden, T. Lance and S. Gelbart. New York: W. A. Benjamin.

[45] Chung, K. L. 1982. *Lectures from Markov Processes to Brownian Motion*. Grundlehren der Mathematischen Wissenschaften, vol. 249. New York: Springer-Verlag.

[46] Crandall, M. G., Kocan, M., and Święch, A. 2000. L^p-theory for fully nonlinear uniformly parabolic equations. *Comm. Partial Differential Equations*, **25**(11–12), 1997–2053.

[47] Dellacherie, C. and Meyer, P. 1978. *Probabilities and Potential A*. North-Holland Mathematics Studies, vol. 29. Amsterdam: North-Holland.

[48] Dubins, L. 1962. On extreme points of convex sets. *J. Math. Anal. Appl.*, **5**, 237–244.

[49] Dubins, L. and Freedman, D. 1964. Measurable sets of measures. *Pacific J. Math.*, **14**, 1211–1222.

[50] Dudley, R. M. 2002. *Real Analysis and Probability*. Cambridge Studies in Advanced Mathematics, vol. 74. Cambridge: Cambridge University Press.

[51] Dunford, N. and Schwartz, J. T. 1988. *Linear Operators*. Part I. Wiley Classics Library. New York: John Wiley & Sons.

[52] Dupuis, P. and Ishii, H. 1991. On Lipschitz continuity of the solution mapping to the Skorohod problem with applications. *Stochastics*, **35**, 31–62.

[53] Dynkin, E. B. and Yushkevich, A. A. 1979. *Controlled Markov Processes*. Grundlehren der Mathematischen Wissenschaften, vol. 235. Berlin: Springer-Verlag.

[54] El Karoui, N., Nguyen, D. H., and Jeanblanc-Picqué, M. 1987. Compactification methods in the control of degenerate diffusions: existence of an optimal control. *Stochastics*, **20**(3), 169–219.

[55] Ethier, S. N. and Kurtz, T. G. 1986. *Markov Processes*. Wiley Series in Probability and Mathematical Statistics: Probability and Mathematical Statistics. New York: John Wiley & Sons.

[56] Fabes, E. B. and Kenig, C. E. 1981. Examples of singular parabolic measures and singular transition probability densities. *Duke Math. J.*, **48**(4), 845–856.

[57] Feller, W. 1959. Non-Markovian processes with the semigroup property. *Ann. Math. Statist.*, **30**, 1252–1253.

[58] Fleming, W. H. and Rishel, R. W. 1975. *Deterministic and Stochastic Optimal Control*. Berlin: Springer-Verlag.

[59] Fleming, W. H. 1980. Measure-valued processes in the control of partially-observable stochastic systems. *Appl. Math. Optim.*, **6**(3), 271–285.

[60] Fleming, W. H., and Pardoux, E. 1982. Optimal control for partially observed diffusions. *SIAM J. Control Optim.*, **20**(2), 261–285.

[61] Freidlin, M. I. 1963. Diffusion processes with reflection and a directional derivative problem on a manifold with boundary. *Theory Probab. Appl.*, **8**(1), 75–83.

[62] Friedman, A. 1964. *Partial Differential Equations of Parabolic Type*. Englewood Cliffs, NJ: Prentice-Hall.

[63] Friedman, A. 2006. *Stochastic Differential Equations and Applications*. Mineola, NY: Dover Publications.

[64] Ghosh, M. K., Arapostathis, A., and Marcus, S. I. 1993. Optimal control of switching diffusions with application to flexible manufacturing systems. *SIAM J. Control Optim.*, **31**(5), 1183–1204.

[65] Ghosh, M. K., Arapostathis, A., and Marcus, S. I. 1997. Ergodic control of switching diffusions. *SIAM J. Control Optim.*, **35**(6), 1952–1988.

[66] Gikhman, I. I. and Skorokhod, A. V. 1969. *Introduction to the Theory of Random Processes*. Translated from the Russian by Scripta Technica, Inc. Philadelphia, PA: W. B. Saunders.

[67] Gilbarg, D. and Trudinger, N. S. 1983. *Elliptic Partial Differential Equations of Second Order*. Second edition. Grundlehren der Mathematischen Wissenschaften, vol. 224. Berlin: Springer-Verlag.

[68] Gyöngy, I. and Krylov, N. 1996. Existence of strong solutions for Itô's stochastic equations via approximations. *Probab. Theory Related Fields*, **105**(2), 143–158.

[69] Has′minskiĭ, R. Z. 1960. Ergodic properties of recurrent diffusion processes and stabilization of the solution of the Cauchy problem for parabolic equations. *Theory Probab. Appl.*, **5**(2), 179–196.

[70] Has'minskiĭ, R. Z. 1980. *Stochastic Stability of Differential Equations*. The Netherlands: Sijthoff & Noordhoff.

[71] Haussmann, U. G. 1985. L'équation de Zakai et le problème séparé du contrôle optimal stochastique. Pages 37–62 of: *Séminaire de Probabilités, XIX, 1983/84*. Lecture Notes in Math., vol. 1123. Berlin: Springer.

[72] Haussmann, U. G. 1986. Existence of optimal Markovian controls for degenerate diffusions. Pages 171–186 of: *Stochastic Differential Systems (Bad Honnef, 1985)*. Lecture Notes in Control and Information Science, vol. 78. Berlin: Springer.

[73] Hernández-Lerma, O. 1989. *Adaptive Markov Control Processes*. Applied Mathematical Sciences, vol. 79. New York: Springer-Verlag.

[74] Ikeda, N. and Watanabe, S. 1989. *Stochastic Differential Equations and Diffusion Processes*. Second edition. North-Holland Mathematical Library, vol. 24. Amsterdam: North-Holland Publishing.

[75] Kallenberg, L. C. M. 1983. *Linear Programming and Finite Markovian Control Problems*. Mathematical Centre Tracts, vol. 148. Amsterdam: Mathematisch Centrum.

[76] Karatzas, I. and Shreve, S. E. 1991. *Brownian Motion and Stochastic Calculus*. Second edition. Graduate Texts in Mathematics, vol. 113. New York: Springer-Verlag.

[77] Kogan, Ya. A. 1969. The optimal control of a non-stopping diffusion process with reflection. *Theory Probab. Appl.*, **14**(3), 496–502.

[78] Krylov, N. V. 1980. *Controlled Diffusion Processes*. Applications of Mathematics, vol. 14. New York: Springer-Verlag. Translated from the Russian by A. B. Aries.

[79] Krylov, N. V. 1995. *Introduction to the Theory of Diffusion Processes*. Translations of Mathematical Monographs, vol. 142. Providence, RI: American Mathematical Society.

[80] Kunita, H. 1981. Some extensions of Itô's formula. Pages 118–141 of: *Seminar on Probability, XV (Univ. Strasbourg, Strasbourg, 1979/1980) (French)*. Lecture Notes in Math., vol. 850. Berlin: Springer.

[81] Kurtz, T. G. and Stockbridge, R. H. 1998. Existence of Markov controls and characterization of optimal Markov controls. *SIAM J. Control Optim.*, **36**(2), 609–653.

[82] Kurtz, T. G. and Stockbridge, R. H. 2001. Stationary solutions and forward equations for controlled and singular martingale problems. *Electron. J. Probab.*, **6**, no. 17, 52 pp. (electronic).

[83] Ladyženskaja, O. A., Solonnikov, V. A., and Ural′ceva, N. N. 1967. *Linear and Quasilinear Equations of Parabolic Type*. Translated from the Russian by S. Smith. Translations of Mathematical Monographs, Vol. 23. Providence, RI: American Mathematical Society.

[84] Lions, P.-L. 1983a. Optimal control of diffusion processes and Hamilton–Jacobi–Bellman equations. I. The dynamic programming principle and applications. *Comm. Partial Differential Equations*, **8**(10), 1101–1174.

[85] Lions, P.-L. 1983b. Optimal control of diffusion processes and Hamilton–Jacobi–Bellman equations. II. Viscosity solutions and uniqueness. *Comm. Partial Differential Equations*, **8**(11), 1229–1276.

[86] Lions, P. L. and Sznitman, A. S. 1984. Stochastic differential equations with reflecting boundary conditions. *Comm. Pure Appl. Math.*, **37**, 511–537.

[87] Liptser, R. S. and Shiryayev, A. N. 1977. *Statistics of Random Processes. I.* Applications of Mathematics, Vol. 5. New York: Springer-Verlag. Translated by A. B. Aries.

[88] Luenberger, D. G. 1967. *Optimization by Vector Space Methods*. New York: John Wiley & Sons.

[89] Menaldi, J.-L. and Robin, M. 1997. Ergodic control of reflected diffusions with jumps. *Appl. Math. Optim.*, **35**(2), 117–137.

[90] Menaldi, J.-L. and Robin, M. 1999. On optimal ergodic control of diffusions with jumps. Pages 439–456 of: *Stochastic Analysis, Control, Optimization and Applications*. Systems Control Found. Appl. Boston, MA: Birkhäuser Boston.

[91] Meyn, S. and Tweedie, R. L. 2009. *Markov Chains and Stochastic Stability*. Second edition. Cambridge: Cambridge University Press.

[92] Nadirashvili, N. 1997. Nonuniqueness in the martingale problem and the Dirichlet problem for uniformly elliptic operators. *Ann. Scuola Norm. Sup. Pisa Cl. Sci. (4)*, **24**(3), 537–549.

[93] Neveu, J. 1965. *Mathematical Foundations of the Calculus of Probability*. San Francisco, CA: Holden-Day.

[94] Parthasarathy, K. R. 1967. *Probability Measures on Metric Spaces*. Probability and Mathematical Statistics, No. 3. New York: Academic Press.

[95] Phelps, R. 1966. *Lectures on Choquet's Theorem*. New York: Van Nostrand.

[96] Portenko, N. I. 1990. *Generalized Diffusion Processes*. Translations of Mathematical Monographs, vol. 83. Providence, RI: American Mathematical Society. Translated from the Russian by H. H. McFaden.

[97] Puterman, M. I. 1994. *Markov Decision Processes: Discrete Stochastic Dynamic Programming*. Hoboken, NJ: John Wiley & Sons.

[98] Rachev, S. T. 1991. *Probability Metrics and the Stability of Stochastic Models*. Wiley Series in Probability and Mathematical Statistics: Applied Probability and Statistics. Chichester: John Wiley & Sons.

[99] Rishel, R. 1970. Necessary and sufficient dynamic programming conditions for continuous time stochastic control problem. *SIAM J. Control*, **8**, 559–571.

[100] Robin, M. 1983. Long-term average cost control problems for continuous time Markov processes: a survey. *Acta Appl. Math.*, **1**(3), 281–299.

[101] Rockafellar, R. T. 1946. *Convex Analysis*. Princeton Mathematical Series, vol. 28. Princeton, NJ: Princeton University Press.

[102] Rogers, L. C. G., and Williams, D. 2000a. *Diffusions, Markov Processes, and Martingales.* Vol. 1. Cambridge Mathematical Library. Cambridge: Cambridge University Press.

[103] Rogers, L. C. G. and Williams, D. 2000b. *Diffusions, Markov Processes, and Martingales.* Vol. 2. Cambridge Mathematical Library. Cambridge: Cambridge University Press.

[104] Ross, S. M. 1970. Average cost semi-Markov decision processes. *J. Appl. Probability,* **7**, 649–656.

[105] Rudin, W. 1973. *Functional Analysis.* New York: McGraw-Hill.

[106] Rugh, W. J. 1996. *Linear System Theory.* Second edition. Prentice Hall Information and System Sciences Series. Englewood Cliffs, NJ: Prentice Hall.

[107] Safonov, M. V. 1999. Nonuniqueness for second-order elliptic equations with measurable coefficients. *SIAM J. Math. Anal.*, **30**(4), 879–895 (electronic).

[108] Skorohod, A. V. 1989. *Asymptotic Methods in the Theory of Stochastic Differential Equations.* Translations of Mathematical Monographs, vol. 78. Providence, RI: American Mathematical Society.

[109] Stannat, W. 1999. (Nonsymmetric) Dirichlet operators on L^1: existence, uniqueness and associated Markov processes. *Ann. Scuola Norm. Sup. Pisa Cl. Sci. (4),* **28**(1), 99–140.

[110] Stockbridge, R. H. 1989. Time-average control of martingale problems: the Hamilton–Jacobi–Bellman equation. *Stochastics Stochastics Rep.*, **27**(4), 249–260.

[111] Stockbridge, R. H. 1990a. Time-average control of martingale problems: a linear programming formulation. *Ann. Probab.*, **18**(1), 206–217.

[112] Stockbridge, R. H. 1990b. Time-average control of martingale problems: existence of a stationary solution. *Ann. Probab.*, **18**(1), 190–205.

[113] Striebel, C. 1984. Martingale methods for the optimal control of continuous time stochastic systems. *Stoch. Process. Appl.*, **18**, 324–347.

[114] Stroock, D. W. and Varadhan, S. R. S. 1971. Diffusion processes with boundary conditions. *Comm. Pure Appl. Math.*, **24**, 147–225.

[115] Stroock, D. W. and Varadhan, S. R. S. 1979. *Multidimensional Diffusion Processes.* Grundlehren der Mathematischen Wissenschaften, vol. 233. Berlin: Springer-Verlag.

[116] Veretennikov, A. Yu. 1980. Strong solutions and explicit formulas for solutions of stochastic integral equations. *Mat. Sb. (N.S.),* **111(153)**(3), 434–452, 480.

[117] Veretennikov, A. Yu. 1987. On strong solutions of stochastic Itô equations with jumps. *Theory Probab. Appl.*, **32**(1), 148–152.

[118] Wagner, D. H. 1977. Survey of measurable selection theorems. *SIAM J. Control Optim.*, **15**, 859–903.

[119] Walters, P. 1982. *An Introduction to Ergodic Theory.* Graduate Texts in Mathematics, vol. 79. New York: Springer-Verlag.

[120] Willems, J. C. 1972. Dissipative dynamical systems. I. General theory. *Arch. Rational Mech. Anal.*, **45**, 321–351.

[121] Wong, E. 1971. Representation of martingales, quadratic variation and applications. *SIAM J. Control,* **9**, 621–633.

[122] Wong, E. and Hajek, B. 1985. *Stochastic Processes in Engineering Systems.* Springer Texts in Electrical Engineering. New York: Springer-Verlag.

[123] Wu, W., Arapostathis, A., and Shakkottai, S. 2006. Optimal power allocation for a time-varying wireless channel under heavy-traffic approximation. *IEEE Trans. Automat. Control,* **51**(4), 580–594.

[124] Xiong, J. 2008. *An Introduction to Stochastic Filtering Theory*. Oxford Graduate Texts in Mathematics, vol. 18. Oxford: Oxford University Press.

[125] Yosida, K. 1980. *Functional Analysis*. Sixth edition. Grundlehren der Mathematischen Wissenschaften, vol. 123. Berlin: Springer-Verlag.

[126] Young, L. C. 1969. *Lectures on the Calculus of Variations and Optimal Control Theory*. Foreword by Wendell H. Fleming. Philadelphia: W. B. Saunders.

[127] Zvonkin, A. K. 1974. A transformation of the phase space of a diffusion process that will remove the drift. *Mat. Sb. (N.S.)*, **93(135)**, 129–149, 152.

Index of symbols

Subject index

Printed in the United States
by Baker & Taylor Publisher Services